FARMING SYSTEMS OF THE WORLD

Farming Systems of the World

A. N. DUCKHAM

AND

G. B. MASEFIELD

ASSISTED BY
R. W. Willey and Kathleen Down

PRAEGER PUBLISHERS
New York · Washington

BOOKS THAT MATTER
Published in the United States of America
in 1970 by Praeger Publishers, Inc., 111
Fourth Avenue, New York, N.Y. 10003

Library of Congress Catalog Card Number: 73-100934

Printed in Great Britain

Contents

Part 1 Analysis of Location Factors

Part 2 Farming Systems

CONTENTS

Part 3 Conclusions

PLATES

The Plates appear between pages 238 and 239

1 (*a*) Dairying on a big modern intensive grassland farm in the Netherlands.

1 (*b*) Fat lamb production on intensive grassland in the North Island, New Zealand.

2 (*a*) Flood irrigation of a strawberry crop in California, U.S.A.

2 (*b*) Intensive horticulture under glass in the Westland district, Netherlands.

3 (*a*) Intensive tillage in the U.S.A. Corn Belt—maize and alfalfa (lucerne) on the contour for control of water erosion.

3 (*b*) Extensive tillage on the North American interior plains—strips of wheat and fallow on the contour for control of water and wind erosion.

4 (*a*) Extensive grazing in New Zealand—a 'high country' sheep station.

4 (*b*) Extensive grazing on sown land, South Australia.

5 (*a*) A typical piece of savanna, the most widespread vegetation type of the tropics.

5 (*b*) A peasant farm in Zambia under the system of shifting cultivation.

6 (*a*) An area denuded by overgrazing and eroded by gullying in Tanzania.

6 (*b*) A soil conservation measure used in Tanzania is to make tied ridges (also known as 'basin listing').

7 (*a*) A homestead in the dry area of northern Ghana.

7 (*b*) Ploughing in south India.

8 (*a*) A typical rice-growing landscape in Malaya.

8 (*b*) A Trinidad village.

viii

List of Sample Farms

LIST OF SAMPLE FARMS

Authors' Introduction

This book is intended for students and research workers in agriculture and its related disciplines, and for teachers of agriculture and geography in all parts of the world.

Its objects are:

(*a*) to help meet the need for a textbook on comparative agriculture, and, in doing so, to try and bridge the gaps between tropical agriculturalists, temperate agriculturalists and economists.

(*b*) to provide a reference book for field agriculturalists, research workers and agricultural administrators, who, in order to interpret or apply research or other data to their own problem, need to know the ecological and economic background of the country of origin of the data.

(*c*) to attempt, especially for temperate countries, which are mostly better documented than the tropics, to systematise the analysis and synthesis of the many variables influencing the location, input intensity and food output of farming systems, and to submit models thereof which are actually or potentially quantifiable.

We have provided, therefore, plenty of maps and diagrams, some photographs and a glossary (in the index, p. 529).

The text has grown out of a lecture course in 'Comparative Agriculture' which we have given for 13 years at Reading University, and is based on personal research, reading, and, in particular, on our first-hand personal experience in the field. One or other of us has lived and worked in, or made study visits to each of the countries or regions discussed in Part 2, except those in Chaps. 2.8, 2.9 and parts of 2.14. Partly for this reason we have only given references when quoting new, unusual or controversial authorities. Each chapter, however, concludes with a list for further reading.

We have not attempted to cover the whole world, nor all the countries we have visited, nor all the districts in each country we discuss. We have omitted communist countries because we have not visited any since 1939 and because reliable data on them are scarce. We have selected for study those areas which between them cover the major farming systems of the world, which seem to us of social, economic or ecological importance or interest, and which throw light on our main theme. This is that the nature, location and intensity of farming systems are the products of the interactions between and within three groups of factors, viz. (i) ecological, (ii) operational and (iii) socio-economic, which are listed in detail in Tables 1.4.1 and 1.7.1; that, as yet, in many cases neither the relative importance of, nor the size of the interactions between individual factors or groups of factors can be quantified, but that nevertheless one can usually identify the critical factors or interactions in any area and offer tentative models of the major interactions; and that the simplest and most convincing way of illustrating the influence of single factors or of a group of factors, is to hold as many of the others as possible constant.

The book has three parts. In the first we analyse the factors summarised in Tables 1.4.1 and 1.7.1; in the second we describe the socio-economic background and, in particular, the climate, soils, vegetation and farming systems of selected countries in the light of this analysis; in the third we suggest the principles which emerge from our study—principles which our successors may wish to use as the basis of a quantified synthesis—and we then examine present and future trends in temperate and tropical countries and their relation to the world food problem.

Measurements

As many of our readers will be English speaking and not necessarily scientists, we have throughout used the foot-pound-second (F.P.S.) system in preference to the centimetre-gramme-second (C.G.S.) or metre-kilogram-second (M.K.S.) systems. However, agricultural inter-system conversion factors are given on p. xiii and a comprehensive ready reckoner for quick conversions is at p. xiv, xv. Fuel energy is given as coal equivalent (Table 1.5.1), biological energy as kcal or Mcal, but solar radiation energy for reasons stated (p. 22), as moisture evaporation equivalent.

Acknowledgements

Much of our experience of other countries was obtained whilst one of us was employed by the (British) Colonial Agricultural Service, and the other by what is now the (British) Ministry of Agriculture, Fisheries and Food. We also have to thank, for financial assistance, research and personal facilities, or for hospitality, the universities of Oxford and Reading, the Carnegie and Nuffield Foundations, the F.A.O., the governments of Canada, France, India, Israel, Malaya, New Zealand, Nigeria, Uganda, the United Kingdom, the United States and many universities and individuals. We are indebted for information and ideas to several hundreds of professional men, farmers, merchants and officials in many lands, and, for criticism and comments on drafts, to those listed at the end of the relevant chapters. Thanks are due, for research and editorial assistance, to Susan Giles, H. Farazdaghi, Janet Z. Foot, S. Manrakhan, Sarah Rowe Jones, Audrey Duckham and Kathleen Down who also drew the diagrams and maps and compiled many of the tables, and, for secretarial help to Rena Manning, Betty Pitt, Dorothy Skidmore and Sheila Smith. R. W. Willey co-operated in the writing of Chaps. 2.4(b) and 2.5. J. C. Bowman, J. B. Dent, T. R. Morris and Peter Street of the Dept. of Agriculture, Reading University, helped one of us with the models and regressions in Chaps. 1.1, 3.1 and 3.2. Agricultural and other officers at U.K. diplomatic posts in Canberra, Copenhagen, Tel-Aviv, Washington and Wellington also helped.

Conversion Factors

METRIC		IMPERIAL AND U.S.	
1 millimetre	= 0·03937 inches	1 inch	= 25·4 millimetres
1 centimetre	= 0·3937 inches		= 2·54 centimetres
1 metre	= 3·2808 feet	1 foot	= 0·3048 metres
1 kilometre	= 3280·8 feet	1 mile	= 1609·34 metres
	= 0·62137 miles		= 1·60934 kilometres
1 hectare	= 2·47109 acres	1 acre	= 0·40468 hectares
1 kilogram	= 2·20462 lb	1 lb	= 0·45359 kilograms
		1 cwt (112 lb)	= 50·80 kilograms
1 metric ton (1,000 kg)	= 2204·62 lb	1 short ton (2,000 lb)	= 0·90718 metric tons
	= 1·10231 short tons	1 long ton (2,240 lb)	= 1·01605 metric tons
	= 0·98421 long tons		
1 litre (water)	= 2·1997 lb	1 lb (water)	= 0·4546 litres
	= 0·2642 U.S. gallons	1 U.S. gallon (8·33 lb)	= 3·78533 litres
			= 0·834 Imperial gallons
	= 0·2200 Imperial gallons	1 Imperial gallon (10 lb)	= 4·54596 litres
			= 1·2 U.S. gallons
1 cubic metre	= 0·0097 acre in	1 acre in	= 102·75 cubic metres

$$°C = 5\left(\frac{°F - 32}{9}\right) \qquad °F = 9\left(\frac{°C}{5}\right) + 32$$

METRIC		IMPERIAL AND U.S.	
1 Mcal	= 10^3 kcal (kilocalories)		
	= 10^6 gramme calories		
1 kcal	= 1,160 milli-watt hours	1 milli-watt hour	= 0·00086 kcal
	= 4,170 joules	1 joule	= 0·00024 kcal
1 kg starch equivalent (Kellner)	= 2,360 net kcal for fattening (N.K.F.)	1 lb starch equivalent (Kellner)	= 1,071 net kcal for fattening (N.K.F.)

xiii

WEIGHTS AND MEASURES CONVERSION TABLES

The centre column in bold figures represents either of the two columns beside it:
e.g. 7 km = 4.35 miles   7 miles = 11.27 km

1 mm = 0·03937 in / 1 in = 25·4 mm			1 m = 3·28084 ft / 1 ft = 0·3048 m			1 km = 0·62137 miles / 1 mile = 1·60934 km			1 ha = 2·47109 acres / 1 acre = 0·40468 ha		
millimetres		inches	metres		feet	kilometres		miles	hectares		acre
25·4	**1**	0·04	0·31	**1**	3·28	1·61	**1**	0·62	0·41	**1**	2·47
50·8	**2**	0·08	0·61	**2**	6·56	3·22	**2**	1·24	0·81	**2**	4·94
76·2	**3**	0·12	0·91	**3**	9·84	4·83	**3**	1·86	1·21	**3**	7·41
101·6	**4**	0·16	1·22	**4**	13·12	6·44	**4**	2·49	1·62	**4**	9·88
127·0	**5**	0·20	1·52	**5**	16·40	8·05	**5**	3·11	2·02	**5**	12·36
152·4	**6**	0·24	1·83	**6**	19·69	9·66	**6**	3·73	2·43	**6**	14·83
177·8	**7**	0·28	2·13	**7**	22·97	11·27	**7**	4·35	2·83	**7**	17·30
203·2	**8**	0·31	2·44	**8**	26·25	12·87	**8**	4·97	3·24	**8**	19·77
228·6	**9**	0·35	2·74	**9**	29·53	14·48	**9**	5·59	3·64	**9**	22·24
254·0	**10**	0·39	3·05	**10**	32·81	16·09	**10**	6·21	4·05	**10**	24·71
279·4	**11**	0·43	6·10	**20**	65·62	32·19	**20**	12·43	8·09	**20**	49·42
304·8	**12**	0·47	9·14	**30**	98·43	48·28	**30**	18·64	12·14	**30**	74·13
355·6	**14**	0·55	12·19	**40**	131·23	64·37	**40**	24·85	16·19	**40**	98·84
406·4	**16**	0·63	15·24	**50**	164·04	80·47	**50**	31·07	20·23	**50**	123·55
457·2	**18**	0·71	18·29	**60**	196·85	96·56	**60**	37·28	24·28	**60**	148·27
508·0	**20**	0·79	21·34	**70**	229·66	112·65	**70**	43·50	28·33	**70**	172·98
762·0	**30**	1·18	24·38	**80**	262·46	128·75	**80**	49·71	32·37	**80**	197·69
1,016·0	**40**	1·57	27·43	**90**	295·27	144·84	**90**	55·92	36·42	**90**	222·40
1,270·0	**50**	1·97	30·48	**100**	328·08	160·93	**100**	62·14	40·47	**100**	247·11
1,524·0	**60**	2·36	60·96	**200**	656·17	321·87	**200**	124·27	80·94	**200**	494·22
1,778·0	**70**	2·76	91·44	**300**	984·25	482·80	**300**	186·41	121·40	**300**	741·33
2,032·0	**80**	3·15	121·92	**400**	1,312·34	643·74	**400**	248·55	161·87	**400**	988·44
2,286·0	**90**	3·54	152·40	**500**	1,640·42	804·67	**500**	310·69	202·34	**500**	1,235·55
2,540·0	**100**	3·94	182·88	**600**	1,968·50	965·60	**600**	372·82	242·81	**600**	1,482·65
5,080·0	**200**	7·87	213·36	**700**	2,296·59	1,126·54	**700**	434·96	283·28	**700**	1,729·76
7,620·0	**300**	11·81	243·84	**800**	2,624·67	1,287·47	**800**	497·10	323·74	**800**	1,976·87
10,160·0	**400**	15·75	274·32	**900**	2,952·75	1,448·41	**900**	559·23	364·21	**900**	2,223·98
12,700·0	**500**	19·69	304·80	**1,000**	3,280·84	1,609·34	**1,000**	621·37	404·68	**1,000**	2,471·09
15,240·0	**600**	23·62	609·60	**2,000**	6,561·68	3,218·68	**2,000**	1,242·74	809·36	**2,000**	4,942·18
17,778·0	**700**	27·56	914·40	**3,000**	9,842·52	4,828·02	**3,000**	1,864·11	1,214·04	**3,000**	7,413·27
20,320·0	**800**	31·50	1,219·20	**4,000**	13,123·36	6,437·36	**4,000**	2,485·48	1,618·72	**4,000**	9,884·36
22,860·0	**900**	35·43	1,524·00	**5,000**	16,404·20	8,046·70	**5,000**	3,106·85	2,023·40	**5,000**	12,355·45
25,400·0	**1,000**	39·37	1,828·80	**6,000**	19,685·04	9,656·04	**6,000**	3,728·22	2,428·08	**6,000**	14,826·54
50,800·0	**2,000**	78·74	2,133·60	**7,000**	22,966·88	11,265·38	**7,000**	4,349·59	2,832·76	**7,000**	17,297·63
76,200·0	**3,000**	118·11	2,438·40	**8,000**	26,247·20	12,874·72	**8,000**	4,970·96	3,237·44	**8,000**	19,768·72
			2,743·20	**9,000**	29,527·56	14,484·06	**9,000**	5,532·33	3,642·12	**9,000**	22,239·81
			3,048·00	**10,000**	32,808·40	16,093·40	**10,000**	6,213·70	4,046·80	**10,000**	24,710·90
			6,096·00	**20,000**	65,616·80				8,093·60	**20,000**	49,421·80
									12,140·40	**30,000**	74,132·70
									16,187·20	**40,000**	98,843·60
									20,234·00	**50,000**	123,554·50

WEIGHTS AND MEASURES CONVERSION TABLES

1 litre milk = 2·2657 lb
1 lb milk = 0·4414 litres

litres		lb
0·44	1	2·27
0·88	2	4·53
1·32	3	6·80
1·77	4	9·06
2·21	5	11·33
2·65	6	13·59
3·09	7	15·86
3·53	8	18·13
3·97	9	20·39
4·41	10	22·66
8·83	20	45·31
13·24	30	67·97
17·66	40	90·63
22·07	50	113·28
26·48	60	135·94
30·90	70	158·60
35·31	80	181·26
39·73	90	203·91
44·14	100	226·57
88·28	200	453·14
132·42	300	697·10
176·66	400	906·28
220·70	500	1,132·85
264·84	600	1,359·42
308·98	700	1,585·99
353·12	800	1,812·56
397·26	900	2,039·13
441·40	1,000	2,265·70
882·80	2,000	4,531·40
1,324·20	3,000	6,797·10
1,765·60	4,000	9,062·80
2,207·00	5,000	11,328·50
2,648·40	6,000	13,594·20
3,089·80	7,000	15,859·90
3,531·20	8,000	18,125·60
3,972·60	9,000	20,391·30
4,414·00	10,000	22,657·00

Note:
Imp. Gallon Milk = 10·3 lb
Imp. Gallon Water = 10 lb

1 m³ = 0·0097 acre/in
1 acre/in = 102·75 m³

cubic metres		acre/in.
102·75	1	0·0097
205·50	2	0·0194
308·25	3	0·0291
411·00	4	0·0388
513·75	5	0·0485
616·50	6	0·0582
719·25	7	0·0679
822·00	8	0·0776
924·75	9	0·0873
1,027·50	10	0·0970
1,130·25	11	0·1067
1,233·00	12	0·1164
2,055·00	20	0·1940
2,466·00	24	0·2328
3,082·50	30	0·2910
3,699·00	36	0·3492
4,110·00	40	0·3880
4,932·00	48	0·4656
5,137·50	50	0·4850
6,165·00	60	0·5820
7,192·50	70	0·6790
7,398·00	72	0·6984
8,220·00	80	0·7760
8,631·00	84	0·8148
9,247·50	90	0·8730
9,864·00	96	0·9312
10,275·00	100	0·9700
11,097·00	108	1·0476
11,303·00	110	1·0670
12,330·00	120	1·1640

Note:
1 million m³ = 811 acre/feet
1 acre/foot = 1233 m³

1 kg = 2·20462 lb
1 lb = 0·45359 kg

kilogram		lb
0·45	1	2·20
0·90	2	4·41
1·36	3	6·61
1·81	4	8·82
2·27	5	11·02
2·72	6	13·23
3·18	7	15·43
3·63	8	17·64
4·08	9	19·84
4·54	10	22·05
9·07	20	44·09
13·61	30	66·14
18·14	40	88·18
22·68	50	110·23
27·22	60	132·28
31·75	70	154·32
36·29	80	176·37
40·82	90	198·42
45·36	100	220·46
50·80	112	
90·72	200	440·92
136·08	300	661·39
181·44	400	881·85
226·80	500	1,102·31
272·15	600	1,322·77
317·51	700	1,543·23
362·87	800	1,763·70
408·23	900	1,984·16
453·59	1,000	2,204·62
907·18	2,000	4,409·24
1,016·04	2,240	

1 kg/ha = 0·89214 lb/acre
1 lb/acre = 1·12088 kg/ha

kg/ha		lb/acre
1·12	1	0·89
2·24	2	1·78
3·36	3	2·68
4·48	4	3·57
5·60	5	4·46
6·72	6	5·35
7·84	7	6·24
8·96	8	7·14
10·08	9	8·03
11·21	10	8·92
22·42	20	17·84
33·63	30	26·76
44·84	40	35·69
56·04	50	44·61
67·25	60	53·53
78·46	70	62·45
89·67	80	71·37
100·88	90	80·29
112·09	100	89·21
125·54	112	
224·18	200	178·43
336·26	300	267·64
448·35	400	356·86
560·44	500	446·07
672·53	600	535·28
784·62	700	624·50
896·70	800	713·71
1,008·79	900	802·93
1,120·88	1,000	892·14
2,241·76	2,000	1,784·28
2,510·77	2,240	
3,362·64	3,000	2,676·42
4,483·52	4,000	3,568·56
5,604·40	5,000	4,460·70
6,725·28	6,000	5,352·84
7,846·16	7,000	6,244·98
8,967·04	8,000	7,137·12
10,087·92	9,000	8,029·26
11,208·80	10,000	8,921·40
22,417·60	20,000	17,842·80
33,626·40	30,000	26,764·20
44,835·20	40,000	35,685·60
56,044·00	50,000	44,607·00

$$1°C = 1·8°F$$
$$1°F = 0·5556°C$$

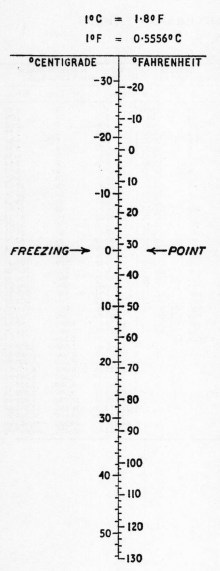

°CENTIGRADE | °FAHRENHEIT

FREEZING → 0 ← POINT

APPROXIMATE UNIT WEIGHTS OF CEREALS FOR EIGHT SELECTED COUNTRIES

Country	Volumetric unit	Wheat	Rye	Barley	Oats	Mixed Grain	Maize	Millet	Sorghum	Rice
U.S.A.	bushel	60 lb 27·215 kg	56 lb 25·401 kg	48 lb 21·772 kg	32 lb 14·515 kg		56 lb 25·401 kg	48·50 lb 22·23 kg	56 lb 25·401 kg	45 lb* 20·412 kg
Canada	bushel	60 lb 27·215 kg	56 lb 25·401 kg	48 lb 21·772 kg	34 lb 15·422 kg	40 lb 18·144 kg	56 lb 25·401 kg			
Australia	bushel	60 lb 27·215 kg	60 lb 27·215 kg	50 lb 22·680 kg	40 lb 18·144 kg		56 lb 25·401 kg	60 lb 27·215 kg	60 lb 27·215 kg	42 lb 19·051 kg
New Zealand	bushel	60 lb 27·215 kg	56 lb 25·401 kg	50 lb 22·680 kg	40 lb 18·144 kg		56 lb 25·401 kg			
Sweden	hectolitre		160·937 lb 73 kg							
Netherlands	hectolitre	165·346 lb 75 kg	154·323 lb 70 kg	143·300 lb 65 kg	110·231 lb 50 kg					
Uruguay	bolsa	143–151 lb 65–68 kg		99–110 lb 45–50 kg						
Japan	koku	301·759 lb 136·875 kg	311·294 lb 141·200 kg		173·615 lb 78·750 kg		289·358 lb 131·250 kg			

* Also in bags of 100 lb (45·359 kg)

Abbreviations and Symbols

A, A–E–T	=	mean Actual Evaporation in the T–G–S (p 22) unless context otherwise requires
B	=	Beckerman Index (p 472) of level of economic development
B^f	=	B adjusted for elasticity of demand for food (p 473)
C	=	Assumed conversion ratio of plant kcal into animal kcal (p 477)
c	=	circa, approximately
C.V.	=	Coefficient of Variation, i.e. Standard Deviation as % of the mean
Ca, CaO, $CaCO_3$	=	Calcium, Lime, Chalk, etc.
dec.	=	decreasing
D_s	=	Human population density per unit of farmed area
E	=	East or eastwards or easterly (i.e. winds from the E)
ed.	=	Edited by
ext.	=	Extensive (of farming)
°F	=	degrees Fahrenheit (temperature)
G	=	Photosynthetic capacity of cultivated plant genotype(s) (p 5, p 473)
h	=	Hydrologic ratio (p 475)
inc.	=	Increasing
int.	=	Intensive (of farming)
K, K_2O	=	potash salts (as fertiliser)
K_a	=	Kilocalories of edible animal food in human diets
kcal	=	Kilocalorie
K_e	=	Plant equivalent in kcal of human diets (p 10)
K_f	=	Kilocalories of total plant and animal food in human diets
K_p	=	Kilocalories of edible plant food in human diets
Lat°	=	Latitude, degrees of latitude
L.A.D	=	Leaf Area Duration
L.A.I.	=	Leaf Area Index (of plants)
Long°	=	Longitude, degrees of longitude
m	=	Million(s)
Mcal	=	1 Megacalorie = 1,000 Kilocalories = 1 million calories
N	=	North, see East
N	=	Nitrogen, nitrogenous manure
N.A.R.	=	Net Assimilation Rate (of plants)
N.P.K.	=	Fertiliser expressed as elemental N, P_2O_5 (phosphate) and K_2O (potash)
P, P_2O_5	=	Phosphate (as fertiliser)
P	=	Mean annual precipitation, unless otherwise indicated

xvii

ABBREVIATIONS AND SYMBOLS

P–E–T	=	Potential evapotranspiration (p 22)
pH	=	Hydrogen ion concentration. Neutrality = 7.0, acidity < 7.0, alkalinity > 7.0
R_1, R_2, R_3, R_m	=	Input Ratios (p 6, p 473)
S	=	South, see East
S_a	=	Farmed surface (actual or proportion of) in crops including grassland for livestock
S_d	=	Farmed surface per 1,000 human population
S_g	=	Actual or proportion of farmed surface in grassland or grazing
S_p	=	Farmed surface (actual or proportional) devoted to crops for direct human consumption
Spp	=	Species
S_t	=	Actual or proportion of farmed surface devoted to tillage crops
T	=	P–E–T
T–G–S	=	Thermal Growing Season (p 34)
T°F, T°C	=	Temperature in degrees Fahrenheit or Centigrade
W	=	West, see East
W_a	=	Wastages of energy into conversion of plants into edible animal food for humans
W_p	=	Wastages of energy in conversion of photosynthates (assimilates) into edible food for humans or into feed for livestock
$\simeq$	=	Approximately equal to
$>$	=	Greater than
$\geqslant$	=	Equal to or greater than
$\geq$	=	Much greater than
$<$	=	Less than
$\leqslant$	=	Equal to or less than
$\leq$	=	Much less than
Σ	=	Summation of equation

Part 1
ANALYSIS OF LOCATION FACTORS

CHAPTER 1.1

The Nature of Food and Agricultural Systems

Ecology, Socio-Economics and Food-producing Systems

Physical geographers and agricultural ecologists stress the importance of climate, altitude, soil and 'natural' flora and fauna as influents in the location of farming systems. Economic geographers often emphasise the locating roles of population density, distance from market, state of economic development and other social or economic influents. Agriculturalists, on the other hand, are basically concerned to combine (*a*) local physical or ecological environment with (*b*) local socio-economic conditions in order (*c*) to satisfy market demand with the maximum profit or domestic or social satisfaction. Ecological and socio-economic factors are, on a world scale, jointly important as influents of the location and intensity of farming systems. Ecological influents may limit the range of enterprises that are practicable at a given site, but socio-economic influents determine which of the feasible enterprises the farmer will choose and the input intensity with which he farms. Nature proposes, man disposes. Ecological influents, however, tend to be more important as locating and intensity factors in undeveloped areas than they are in advanced economies. In the latter, high levels of industrial and scientific inputs (fertilisers, pesticides, machinery, oil fuel, etc.) firstly, raise the level of input intensity which is practicable or profitable at a given ecological site; and, secondly, by extending man's control over his physical and ecological environment, increase the range of ecologically feasible enterprises on a farm and, in doing so, often further raise the practicable or profitable level of input intensity. The relation between these groups of influents and resultants is shown in Fig. 1.1.1 and, as models, at pp. 471ff.

Later in this chapter, the application of the above general thesis is illustrated by a sample village in Uganda and a sample farm in south-east England, but, before that, we need to consider briefly the object of food and agricultural systems. A biologically efficient food-producing system must (Duckham 1968):

(i) supply, in temperate areas, approximately 1,000 Mcal per person per year of dietary energy in food (K_f) of relatively concentrated form, i.e. within the limits of human dry matter and water intake. (In tropical climates, the annual need per head is less because heat loss is less, the people are usually smaller, and there is a higher proportion of children.) Probably not less than 10% of this energy should come from animal products (K_a) and the diet should also be appetising, adequate in minerals and vitamins, and non-toxic.

3

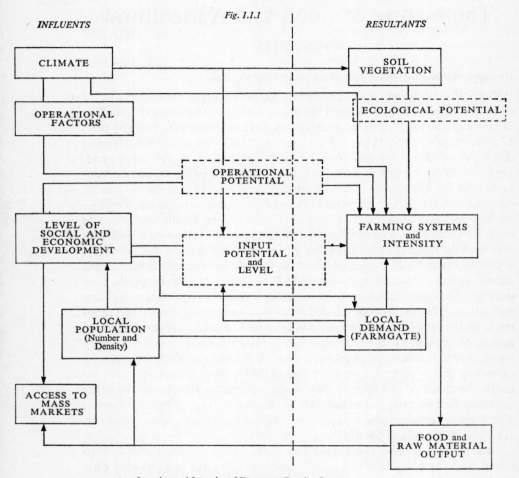

Fig. 1.1.1

INFLUENTS   *RESULTANTS*

CLIMATE

SOIL
VEGETATION

OPERATIONAL
FACTORS

ECOLOGICAL POTENTIAL

OPERATIONAL
POTENTIAL

LEVEL OF
SOCIAL AND
ECONOMIC
DEVELOPMENT

INPUT
POTENTIAL
and
LEVEL

FARMING SYSTEMS
and
INTENSITY

LOCAL
POPULATION
(Number and
Density)

LOCAL
DEMAND
(FARMGATE)

ACCESS TO
MASS
MARKETS

FOOD and
RAW MATERIAL
OUTPUT

Location and Intensity of Temperate Farming Systems
Summary of Influents and Resultants

(ii) because climate, and hence production, is highly seasonal, provide adequate storage and distribution facilities (with minimum wastages in each case) so that the average person can have about 3 Mcal of edible energy (K_f) every day (at least in cool temperature countries).

(iii) provide, with minimum wastages (W_p, W_a) 'off-the-farm', the processing methods and equipment and the cooking needed to render crops and animal products digestible by and attractive to man.

(iv) maximise plant growth ($A.G.$) and minimise plant and animal wastages (W_p and W_a) 'on-farm'.

(v) achieve (i) to (iv) above by applying the optimum input ratios (R_1, R_2 and R_3) of the energy in (or used in the production of) skill, man work, animal work, fossil fuels and scientific and industrial inputs (e.g. fertilisers) in order to maximise plant growth ($A.G.$) and minimise plant and animal wastages (W_p and W_a respectively). In addition, the system must also:

(vi) be reliable between *and* within years, months and weeks.

(vii) be persistent over decades (and centuries).

(viii) be capable of reduction, expansion or adjustment to meet changes, usually increases, in population (D_s) or other changes in demand (e.g. increased demand for animal products (K_a)), especially where land supply is static or decreasing by reason of increases in population or in living standards (e.g. new towns, aerodromes, industrial pollution, military devastation).

Of these eight objectives, the first three, being concerned with human nutrition and food technology, are mostly outside the scope of this book (but see Chaps. 3.1, 3.4 and 3.5). The last three are dealt with in Chaps. 1.4, 3.2, 3.3 and elsewhere in the text. The remaining two, viz. (iv) and (v) may be most simply expressed as an energy model.

A basic simplified model, in energy terms, for a food-production system* is:

$$R_1 \cdot A.G. - (R_2 \cdot W_p + R_3 \cdot W_a) = K_a + K_p = K_f$$

(Model 1.1.1.)

Where:

A is the limiting climatic parameter. This is assumed, unless otherwise stated, to be the mean effective transpiration A–E–T (p. 22) in the thermal growing season (T–G–S) (months at or above $42\frac{1}{2}°F$).

G is Mcal in the organic matter (i.e. photosynthate) formed per unit of A by the plant genotype(s) best adapted to maximise K_p and, through animals, K_a.

W_p is the sum of phytic wastages, i.e. energy losses due to respiration, pests and diseases, unharvested leaves, stubbles and roots, losses in processing, storage, distribution and in the home, etc.

W_a is the sum of animal wastages, i.e. energy losses due to basal metabolism, additional metabolism (e.g. in cold weather); pests and diseases; conversion of residual metabolisable energy into tissues or milk; loss of inedible

---

* In the hydro-neutral zone (p. 461) of temperate countries. For the effect of climate and other factors on this model, see pp. 462 ff.

Analysis portions at slaughter; other losses in processing, storage and distribution, and
of in the home, etc.

Location K_p, K_a are the dietary energy available to man in edible food of plant origin
Factors (K_p) and animal origin (K_a) respectively, and K_f is their sum.

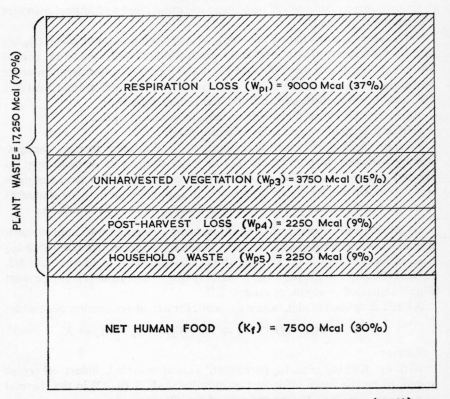

GROSS ENERGY DISPOSAL OF POTATO CROP (Mcal per acre)

Excluding pests and diseases

RESPIRATION LOSS (W_{p1}) = 9000 Mcal (37%)

UNHARVESTED VEGETATION (W_{p3}) = 3750 Mcal (15%)

POST-HARVEST LOSS (W_{p4}) = 2250 Mcal (9%)

HOUSEHOLD WASTE (W_{p5}) = 2250 Mcal (9%)

PLANT WASTE = 17,250 Mcal (70%)

NET HUMAN FOOD (K_f) = 7500 Mcal (30%)

TOTAL ORGANIC DRY MATTER FORMED = 24,750 Mcal (100%)

Fig. 1.1.2 Gross Energy Disposal of Potato Crop (Mcal per acre), excluding pest
and disease wastages

R_1 is the ratio of the actual to the optimal energy input in (and used in the
production of) the inputs of skills, man labour, animal work, scientific and
industrial inputs (but excluding inter-farm inputs, e.g. purchased feeding
stuffs) needed to maximise A.G.; and R_2 and R_3 are the ratios of the actual to
the optimal energy, etc. in the inputs needed to minimise plant wastages and
animal wastages respectively. R_1, R_2 or R_3 cannot exceed unity, which is the
optimal ratio for maximising the model.

Fig. 1.1.2 (Duckham 1968) shows how much of the total photosynthate

(organic dry matter) produced (i.e. ($R_1 . A.G.$) in the model) in a potato crop is
lost as plant wastages ($R_2 . W_p$), i.e. respiration, unharvested parts, storage loss,
household loss, etc., but excluding in this case pests and diseases. Only 30%
of the photosynthate formed or about 0·25% of the total solar energy received
is consumed as human food. Fig. 1.1.3 (Duckham 1968) shows how much of
the total photosynthate ($R_1 . A.G.$) produced by an intensive grass crop grazed

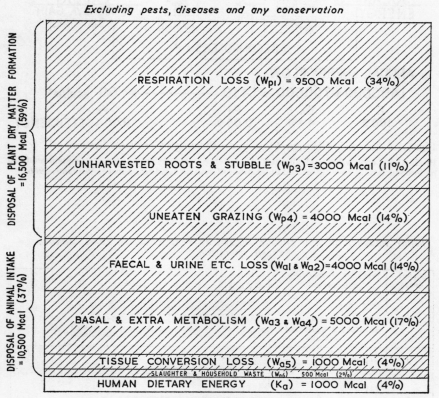

for beef production is lost as plant wastages ($R_2 . W_p$) and as animal wastages
($R_3 . W_a$); these include basal metabolism and extra metabolism for exercise,
wind-chill, cold-chill, etc., other metabolic losses, losses on slaughter (e.g. of
guts and inedible portions) and household waste. Only 4% of the photosynthate formed or about 0·02% of the total solar energy received becomes human
food. The energy flow in a comparable extensive grazing system is shown in

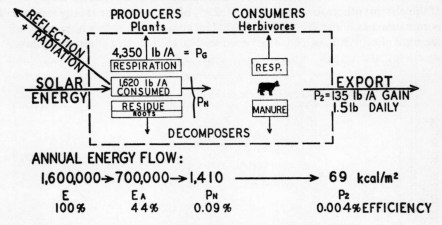

Fig. 1.1.4 Quantified Energy Flow in an Extensive Grassland System

Productivity, energy flow and efficiency of solar energy conversion in a range ecosystem modified by introduction of rose clover and application of sulphur fertiliser. *Source:* Williams, 1966

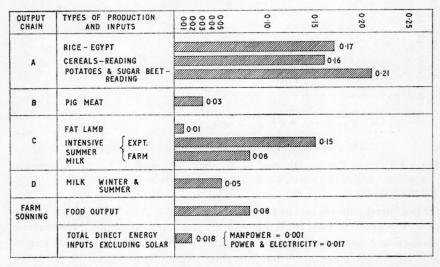

Fig. 1.1.5 Energy Production and Energy Inputs as Percentage of Total Annual Solar Radiation for various Food Chains and Enterprises

Fig. 1.1.4. The energy recovery for various temperate enterprises are shown in
Fig. 1.1.5 and Table 1.1.1; comparable data are not available for the tropics.

The four main food chains (Duckham 1963, and in Wilkins 1967) viz.
Chains A (tillage crops/man), B (tillage crops/livestock/man), C (grassland/
ruminants/man) and D (tillage crops and grassland/ruminants/man) are
shown in Fig. 1.1.6 (Duckham in Wilkins 1967). Their comparative efficiency
as energy and protein producers of human food is shown in Fig. 1.1.5 and
Table 1.1.1. The greater energetic efficiency of Chain A is obvious, but man

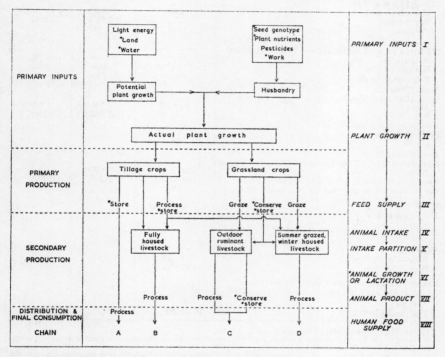

Fig. 1.1.6 The Four Food Chains

needs, or, at least when he can afford it, likes to eat animal products. Hence,
in advanced economies with high living standards, 30% or more of the human
energy intake may come from Chains B, C and D (see Fig. 1.1.8 and Table
1.5.2). Moreover (pp. 88 and 481) livestock may be the only effective
way of using areas with difficult climates or uneven relief to produce food or
other agricultural products (e.g. wool) for man. Further, working livestock
are essential sources of power in many developing countries where, therefore,
animal products (meat, milk) may be joint or by-products of the power
source used to produce Chain A tillage crops. The relation between these
chains and farming systems is shown in Row 17 of Fig. 3.1.1.

FARMING SYSTEMS OF THE WORLD

Table 1.1.1

Food Output (Energy and Protein) from the Four Food Chains

	Mcal of human food[1] (approx) per 100,000 Mcal of Solar Radiation	Mcal of human food[2] per acre	Edible protein in human food[2] lb per acre
UNITED KINGDOM			
Food Chain A			
(Tillage Crops/Man)			
Cereals	200	4,600	230
Sugar-beet ⎫	250	{ 5,900	n.a.
Potatoes ⎭		{ 6,500	
Food Chain B			
(Tillage Crops/Intensive indoor Livestock/Man)			
Bacon Pigs		⎧ 840	55
Barley-fed Beef ⎫	15 to 30	{ 440	50
Eggs and Fowls ⎭		⎩ 440	80
Food Chain C			
Grassland (Intensive)			
Meat	5 to 25	200†	20†
Milk	30 to 80	2,250‡	250‡
Ranching (California)			
(Williams 1966)			
Cattle	2 to 4		
Food Chain D			
(Grassland and Tillage Crops)			
Milk	30 to 50	800	90
Good mixed tillage, grass and livestock farms	50 to 80	2,200	130
Continuous wheat growing (if practicable) say	150		
EGYPT			
Food Chain A			
Rice	170 (210)*		
INDIA			
Food Chain A			
Rice—W. Bengal	70 (200)*		
Rice—Madras	190 (350)*		
DIRECT ENERGY INPUTS ON READING UNIVERSITY FARMS:			
Manpower	1		
Fuel and Electricity	10 to 20		

Sources: [1] Duckham 1968. [2] Duckham and Lloyd 1966.
* The figures in brackets are the results of high fertiliser experiments compared with state or national average unbracketed (see Holliday 1966).
† Ewes producing fat lamb.   ‡ University of Reading Farm, Sonning 1968–69, after including dry cows and estimated winter maintenance costs.

INPUT RATIOS AND ECONOMIC DEVELOPMENT

In poor, developing countries the greater nutritional dependence on Chain A foods is associated partly with the problems of tropical livestock (Chap. 1.2) but more with low input ratios (R_1, R_2 and R_3) in model (1.1.1). Input ratios are, in practice, positively correlated with social and economic development. Fig. 1.1.7 shows the relation between (*a*) several international indices of living standards and economic development, (*b*) cereal yields per unit area and (*c*) plant equivalent (K_e)* consumption per head of total population of the

* Plant equivalent has been calculated by assuming arbitrarily that each Mcal of animal product consumed as food has involved the intake by livestock of the equivalent, as grass, hay, cereal, oil-seed residues, etc., of 5·5 Mcal of cereals ready for human consumption.

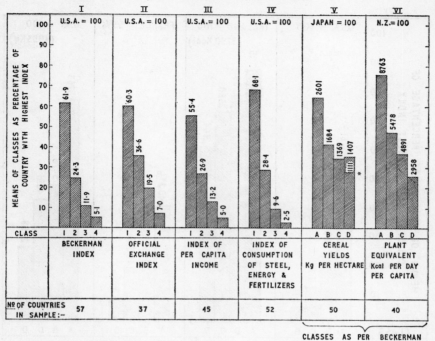

RELATION BETWEEN INDICES OF LIVING STANDARDS AND DEVELOPMENT, CEREAL YIELDS AND PLANT EQUIVALENT CONSUMPTION.

Fig. 1.1.7 Relation in 1960 between Indices of Living Standards and Development, Cereal Yields and Plant Equivalent Consumption

countries in each quartile (using the Beckerman Index as the parameter of economic development). Fig. 1.1.8 shows the relation between (*a*) the Beckerman Index (which has, for present purposes, certain technical advantages over the other indices of living standards and economic development (O.E.C.D. 1967)) and (*b*) food-consumption levels. Whilst Fig. 1.1.9 shows the relation between (*a*) the Beckerman Index, (*b*) the consumption per head of the *whole* population of some of the major industrial inputs of advanced farming, (*c*) total output of food per head in wheat equivalents* and (*d*) cereal yields per unit area.

Figs. 1.1.7, 1.1.8 and 1.1.9 relate to 1960 (or nearest year) data from Clark and Haswell 1964, Coppock 1966, F.A.O. 1962, U.N. 1962 and O.E.C.D. 1967; they reflect the well-recognised positive correlation between (*a*) levels

* Wheat equivalents roughly measure the total economic worth of a diet by expressing the price per unit of each food as a ratio of wheat prices, multiplying the *per capita* consumption by weight of each food by the relevant ratio and summing the resultants (see Clark and Haswell 1964).

11

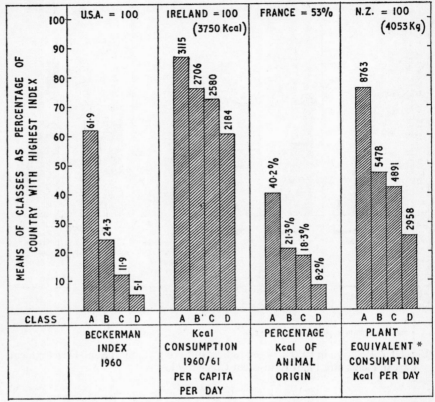

RELATION BETWEEN LIVING STANDARDS (BECKERMAN INDEX) AND FOOD CONSUMPTION

Fig. 1.1.8 Relation in 1960 between Living Standards (Beckerman Index) and Food Consumption

of economic development and overall living standards on the one hand, and (*b*) levels of food consumption and their nutritional adequacy on the other. But they also suggest that food output per head is positively correlated with level of economic development (Duckham 1968) and cereal yields per acre are, though less obviously, similarly correlated with development.† It may be that countries in the first quartile (i.e. with the highest living standards) also have, on the average, rather better farming climates than those in the other quartiles (i.e. those with lower living standards). But any such climatic differences

* See footnote p. 10.
† Linear, and in appropriate cases, quadratic regression equations and also coefficients of determination for the relations between some of the above parameters are given in Table 1.1.2.

Table 1.1.2

No. of observations	x	y	Linear equations ($y = a + bx$)	r^2	Quadratic equations ($y = a + bx + cx^2$)	R^2
(1) 37	Beckerman Index	Official exchange index of gross national product *per capita* per year	$y = 2{\cdot}09 + 0{\cdot}83x$	0·86	$y = 6{\cdot}25 + 0{\cdot}51x + 0{\cdot}003x^2$	0·87
(2) 56	Beckerman Index	Energy consumption kg *per capita* per year	$y = -199{\cdot}09 + 61{\cdot}80x$	0·81	$y = -159{\cdot}05 + 27{\cdot}50x + 0{\cdot}43x^2$	0·83
(3) 56	Beckerman Index	Steel consumption kg *per capita* per year	$y = -19{\cdot}11 + 5{\cdot}91x$	0·90	$y = -37{\cdot}26 + 7{\cdot}65x - 0{\cdot}22x^2{*}$	0·91
(4) 53	Beckerman Index	Food energy consumption kcal *per capita* per day	$y = 2337{\cdot}64 + 12{\cdot}37x$	0·50	$y = 2170{\cdot}12 + 28{\cdot}45x - 0{\cdot}20x^2$	0·59
(5) 40	Beckerman Index	Percentage of food energy of animal origin	$y = 10{\cdot}61 + 0{\cdot}46x$	0·66	$y = 4{\cdot}75 + 0{\cdot}93x - 0{\cdot}006x^2$	0·72
(6) 40	Beckerman Index	Total plant equivalent consumption kcal *per capita* per day	$y = 3515{\cdot}87 + 81{\cdot}32x$	0·63	$y = 2381{\cdot}50 + 172{\cdot}18x - 1{\cdot}09x^2$	0·70
(7) 30‡	Beckerman Index	Wheat equivalent consumption kg *per capita* per year	$y = 422{\cdot}25 + 12{\cdot}66x$	0·47	$y = 355{\cdot}32 + 25{\cdot}95x - 0{\cdot}45x^2{*}{*}$	0·51
(8) 52	Beckerman Index	Fertiliser consumption kg *per capita* per year	$y = 1{\cdot}13 + 0{\cdot}69x$	0·47	$y = -7{\cdot}94 + 1{\cdot}53x - 0{\cdot}01x^2$	0·54
(8) 52	Beckerman Index	Fertiliser consumption lb *per capita* per year	$y = 3{\cdot}23 + 1{\cdot}52x$	0·47	$y = -17{\cdot}20 + 3{\cdot}41x - 0{\cdot}02x^2$	0·54
(9) 51	Beckerman Index	Cereal yields in kg per hectare	$y = 1264{\cdot}66 + 19{\cdot}45x$	0·24	$y = 939{\cdot}03 + 48{\cdot}73x - 0{\cdot}36x^2$	0·30
(9) 51	Beckerman Index	Cereal yields in cwt (of 112 lb) per acre	$y = 10{\cdot}06 + 0{\cdot}15x$	0·24	$y = 7{\cdot}46 + 0{\cdot}39x - 0{\cdot}003x^2$	0·30
(10) 46	Beckerman Index	Wheat equivalent (food only) output kg *per capita* per year	$y = 438{\cdot}63 + 15{\cdot}76x$	0·42	$y = 363{\cdot}58 + 23{\cdot}25x - 0{\cdot}09x^2{\dagger}$	0·42
(11) 51	Fertiliser consumption lb *per capita* per year	Cereal yield in cwt (of 112 lb) per acre	$y = 10{\cdot}53 + 0{\cdot}83x$	0·33	$y = 8{\cdot}93 + 0{\cdot}17x - 0{\cdot}0004x^2$	0·39
(12) 44	Fertiliser consumption lb *per capita* per year	Wheat equivalent (food only) output kg *per capita* per year	$y = 410{\cdot}52 + 11{\cdot}69x$	0·71	$y = 498{\cdot}70 + 6{\cdot}06x + 0{\cdot}037x^2$	0·73
(13) 33	Wheat equivalent (food only) output lb. or kg *per capita* per year	Plant equivalent consumption kcal *per capita* per day	$y = 2849{\cdot}33 + 2{\cdot}94x$	0·64	$y = 545{\cdot}17 + 7{\cdot}33x - 0{\cdot}001x^2$	0·78
(14) 30	Wheat equivalent (food only) output kg *per capita* per year	Wheat equivalent consumption kg *per capita* per year	$y = 323{\cdot}85 + 0{\cdot}43x$	0·49	$y = 19{\cdot}92 + 1{\cdot}32x - 0{\cdot}0005x^2$	0·64

Sources: O.E.C.D., No. 26, 1967; F.A.O. *Production Yearbook 1962*; U.N. *Statistical Yearbook 1962*; Clark, C. and Haswell, M. R., 1964; Coppock, J. O., 1966.

T is above 1·5 in all except two asterisked cases.

* $T = 1{\cdot}4$.
† $T = 0{\cdot}8$.
‡ Figures for developed countries n.a.

13

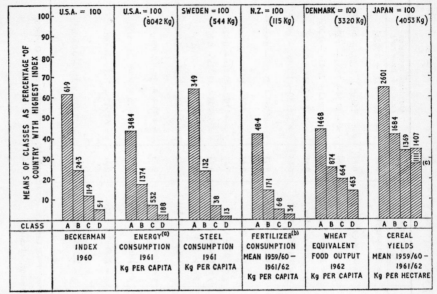

RELATION BETWEEN ECONOMIC DEVELOPMENT, FOOD OUTPUT AND CEREAL YIELDS

(a) – COAL EQUIVALENT
(b) – N + P₂O₅ + K₂O
(c) – EXCLUDING TAIWAN (FORMOSA) & EGYPT

Fig. 1.1.9 Relation in 1960 between Economic Development, Food Output and Cereal Yields

are, it is suggested, outweighed by the impact of social and economic development and hence of industrial and scientific inputs on farming, and especially on cereal yields per acre and food output *per capita*. Much of the difference in cereal yields per unit area and in food output *per capita* (as measured by wheat or plant equivalent) reflect, in fact, the level of social and economic development, social factors, and, in some cases, accessibility of markets (pp. 83 and 469). In brief, the input ratios (R_1, R_2, R_3) in model (1.1.1) represent the effective sum (excluding market access) of the influences of social and economic development on farming systems, intensities and outputs. Thus the model combines economic and social terms (R_1, R_2, R_3) with biological terms $(A.G., W_p, W_a)$.*

* The above, necessarily condensed arguments are firstly based on the inferential use, which has its dangers, of statistics, which may themselves be open to several objections, and secondly may imply causal relations which are, in our present state of knowledge, incapable of proof. Nevertheless, it is suggested that they provide, for present purposes, a working hypothesis that is worthy of a critical examination which is, however, outside the scope of this book.

To recapitulate, the *object* of any food-production system is (*a*) to optimise **Analysis** inputs (1) of human skill, (2) of man and animal work, (3) of abiotic materials **of** mainly of industrial origin (fertilisers, machines, pesticides, petroleum, etc.) **Location** (4) of suitable genotypes of plants and animals and (5) of inter-farm inputs **Factors** such as purchased feeding stuffs or breeding cattle with a view, in a given ecological situation, to (*b*) maximising crop plant growth, and (*c*) minimising plant and animal wastages so that (*d*) an adequate economic return (or food output on subsistence farms) is obtained, and (*e*) in such a way that this economic return or food output is reliable from year to year and persistent over decades or even longer.

The layout of this book is, firstly, to analyse the factors which determine the many different ways in which, in different ecological, social or economic environments, man achieves these objectives; secondly, to describe and illustrate, in Part 2, the resultant farming systems country by country, and thirdly, in Part 3, to suggest some of the principles by which the influents discussed in Part 1 lead to the resultant farming systems in Part 2, and to consider current agricultural, food and population trends.

FARMING SYSTEMS

The farming systems by which these objects are achieved fall into four main classes, viz. (i) plantation perennial, (ii) tillage, (iii) alternating (i.e. alternating between tillage and grassland phase (or the degradative and accumulative phases of soil fertility (Duckham 1958)) and finally (iv) grassland or grazing. Each of these systems may vary in input intensity between extensive and intensive. Their location and intensity is influenced by climate, soil, vegetation and other ecological factors (Chaps. 1.2, 1.3, 1.4), by socio-economic factors (Chap. 1.5) and by operational factors (Chap. 1.7). The synthesis of such systems is outlined in Chaps. 1.4, 1.7 and 3.1, and summarised in Tables 1.4.1, 1.7.1 and 3.1.1 and Fig. 3.1.11.

The results of such syntheses feature in Part 2 but it may help the reader to give examples of a simple system with low input ratios and a complex system with high input ratios.

A Village in Uganda

We may select as a sample village in the tropics, Kasilang in the Teso district of Uganda, $1\frac{1}{2}$ degrees north of the Equator, which was the subject of detailed agricultural surveys in 1937 and 1953 (Wilson and Watson 1956) and has been seen again more recently by one of us (G.B.M.). The term 'village' is here a misnomer, since, as is usual in East Africa, the people live in scattered homesteads and not in what a geographer would describe as 'nucleated settlements'. The total acreage of the area is 3,763, of which 2,529 acres are cultivable, the rest being taken up by swamps, hill-tops, house sites, roads and the necessary minimum of woodland to provide fuel and building poles. The mean annual rainfall is approximately 44 in, the area lying near the boundary between the 'dry' and 'wet' tropics as we shall define them later, but being too dry for perennial crops to be of any importance. The farming system is therefore definable as dry-land annual cropping.

The population of this area is recorded as 880 people in 173 families, a density of 150 people to the square mile; 311 of these were children. The people were all of the Teso or closely related tribes except for 12 herdsmen of the cattle-keeping Bahima tribe, who often look after cattle belonging to Teso owners in return for keeping the milk and an occasional calf. All the remaining population are engaged in full-time or part-time farming, but a number carry on other occupations (often part-time) as well. These include 6 headmen, 4 traders, 4 fish-sellers, 3 carpenters, 3 bricklayers, 1 bicycle repairer,

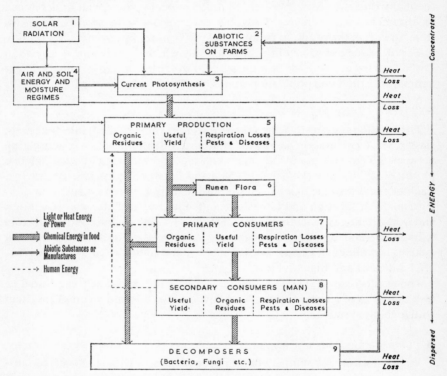

Fig. 1.1.10 Energy Flow in a Simple Farming System

1 messenger, 1 laundryman, 1 lorry driver, 1 tailor, 1 roadman, 1 schoolmaster and 1 chief. Land is held by the customary African tribal tenure, each cultivator enjoying the usufruct of a holding allotted to him by the tribal authority as long as he continues to need the use of it, without any payment of rent and equally without powers of sale or bequest.

The system is an alternating one with some extensive grazing. The method of cropping practised involves an alternation on any piece of land of approxi-

mately 3 years under cultivation and 3 years' fallow under regenerating
natural cover, mainly grass, some 45 % of the cultivable area being cultivated
annually. This more or less suffices to maintain soil fertility, although little or
no manure is applied to the land; cattle are kept, but with only one ox-cart
in the area transport of manure is too difficult a problem. Soil erosion, at any
rate in gross degree, is prevented by a system of contour strip cropping with
buffer strips of grass which has long been inculcated in this area by the agricultural extension service. The mean acreage under each crop per family in a
typical year was as follows:

Crop	Acreage
Cotton	2·21
Finger millet	3·01
Sesame	0·67
Sorghum	0·48
Cassava	0·42
Groundnuts	0·34
Sweet-potatoes	0·33
Cow-peas	0·29
Bananas	0·03
Maize	0·03
Total	7·81

Nearly all these crops are planted in the first and main rainy season each
year, but cow-peas (usually following finger millet) and sweet-potatoes are
planted in the second rains. Cotton is the main cash crop, but the people of
this district derive just about as much income from sales of cattle as they do
from sales of cotton. Finger millet is the chief food crop, and any surplus is
stored in wicker granaries as a famine reserve. Any surpluses of the other food
crops which accrue provide some subsidiary income by being sold in the chief
town of the district, 12 miles away. Crops are usually taken to market on the
back of a bicycle; the inhabitants between them owned 116 bicycles.

In this area the plough has now superseded the hoe almost entirely in opening new land for planting, though weeding is still done with the hoe. Since the
173 families only owned 105 ploughs, some have to borrow or hire ploughs
from their neighbours and cannot prepare the ground in the optimal period.
Both 2-ox and 4-ox ploughs are used, but with only 248 draught oxen in the
area, the tractive force is under strength. Three people are required to manage
one plough and its team, and succeeded in ploughing less than one-third of an
acre in a day. Goats, a few sheep, and hens are also kept by many families,
but their numbers have not been recorded. Cattle and goats live almost entirely on the grazing of the communally owned waste land, often losing weight

in the dry season which has to be slowly regained after the rains. Their burden of intestinal worms and of ticks is heavy; there has been some argument as to whether pastures in this district are overstocked or not, but the position is certainly marginal. The people would be reluctant to reduce numbers of stock, not only because of the income they provide but because of their use in paying bride-price.

This simple economy suffices to maintain the population without want or distress. The energy flow is illustrated by Fig. 1.1.10. The chief problems lie in the future, for it will not much longer be possible for the rapidly increasing population to be carried without resorting to practices such as the use of ferti-lisers which are at present uneconomic. There is little industrial employment in Uganda to absorb any surplus population, nor are empty lands available without moving to remote parts of the country, which will themselves only provide a temporary solution. By the end of this century or earlier, Uganda will be typical of many tropical countries in facing a critical position.

A Large Intensive Farm in South-east England

The Sonning Farm, one of the University of Reading's, has 440 acres with some outlying land, is divided by a road busy with automotive traffic, is within 4 miles of a large industrial town in a heavily populated area with a high living standard, and has not only easy access to market but also to industrial inputs and up-to-date technical and repair services. The annual precipitation aver-

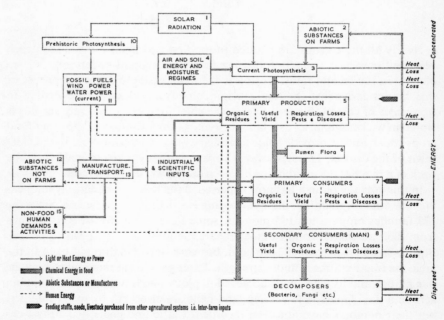

Fig. 1.1.11 Energy Flow in a Complex Farming System

18

ages 25 in but the mean growing season moisture deficit is 6 in. The mean A–E–T in the thermal growing season is about 22 in. The climate is cool humid. There are three farming systems, each intensive. The floodable alluvial soils, 90 acres, high in organic matter, adjoining the River Thames, and also 40 acres of outlying gravel soil, are in continuous *tillage* mainly spring barley. 60 acres of high-lying but irrigated gravel soils are in very *intensive non-leguminous grassland*. This is grazed by 270 dairy cows in milk, i.e. 4·5 per acre. Between these higher gravels and the river alluvium, 150 acres are in *intensive alternating* with 2 years in winter wheat, 1 in spring barley and 1 in early potatoes and then 3 or 4 years in mixed grass/clover swards for the 240 ewes for fat lamb, and for making silage and hay for the winter forage of the dairy cows and their replacements. The wheat is sold, the barley is

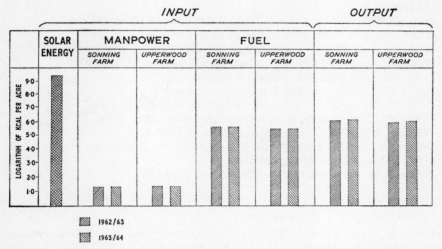

Fig. 1.1.12 Energy Inputs and Outputs per Farm Acre (Sonning and Upperwood Farms)

fed to cattle whilst inter-farm inputs of protein feeds and some cereals are purchased. All the land is heavily fertilised, the irrigated grass receiving up to 400 lb N/acre/year; pest-control chemicals are widely used; the field work is fully mechanised as is milking the cows. There is adequate storage and a small mill. There are 5 tractors, and modern cultivating, drilling and harvesting equipment; the commercial work force is 8 men. The crop yields are not outstanding, especially on the poorer soil, but intensive grassland yields up to 10,000 lb milk per acre. The food output per man is high. Thus, whereas one family on 8 acres in Uganda just about fed itself, at Sonning 8 men and their families provide enough to feed 220 families of 4 a diet with enough energy from animal products. That is one man on the farm and, being an advanced farm, another man or possibly 1½ men producing industrial, etc. inputs 'off-the-farm', produce food for 11 to 14 families including their own. If the whole farm were in cereals, potatoes and pulses (e.g. beans, peas) for human food, these figures would be trebled. The complex energy flow associated with such high output farming is shown in Fig. 1.1.11 and its energetic efficiency in Fig. 1.1.12.

Acknowledgement. To J. B. Dent and Peter Street for work on Table 1.1.2.

References

*Clark, C. and Haswell, M. R. 1964. *The Economics of Subsistence Agriculture.* London.

Coppock, J. O. 1966. *Atlantic Agricultural Unity.* New York. p. 135.

*Duckham, A. N. 1958. *The Fabric of Farming.* London.

Duckham, A. N. 1963. *Agricultural Synthesis: The Farming Year.* London.

Duckham, A. N. and Lloyd, D. L. 1966. *Farm Economist.* **11**. 2. 95–97.

Duckham, A. N. 1968. *Proc. Chem. and Ind.* 903–906.

F.A.O. 1962. *Production Yearbook 1962.* F.A.O. Rome.

O.E.C.D. 1967. *O.E.C.D. Observer.* **25**. Feb. 36–37. O.E.C.D. Paris.

U.N. 1962. *Statistical Yearbook 1962.* United Nations. New York.

Williams, W. A. 1966. *J. Range Management.* **19**. 1. 29–34.

Wilkins, R. J. (Ed.) 1967. *Fodder Conservation.* Occ. Symposium No. 3. British Grassl. Soc. Hurley, Berks., England.

Wilson, P. W. and Watson, J. M. 1956. *Uganda J.* **20**. 182.

Further Reading

(SEE ALSO ASTERISKED ITEMS)*

Baker, J. J. and Allen, G. E. 1965. *Matter, Energy and Life.* Addison-Wesley, Reading, Mass., U.S.A.

Phillipson, J. 1966. *Ecological Energetics.* London.

Energy and Moisture Regimes

Introduction

In any large geographical region (e.g. north-west Europe) the input of *solar energy* controls or greatly influences the intensity and daily duration of light energy receipts on the land, the heat regimes (temperature), the wind systems and the water regimes (precipitation and evapo-transpiration). These climatic factors are also regionally influenced by the effect of rotation of the earth on wind systems and by ocean currents, which are also affected by the distribution of the world's land masses. Nevertheless, *solar* energy and precipitation (P) are the main factors determining regional climates which, in their turn determine the 'natural' vegetation, the zonal soil group and, in part, the natural fauna. These last three also interact to some extent with each other.

Within a given climatic region there are local gradations and anomalies attributable to distance from seaboard, to altitude relief (land-form, including aspect) and to the influence of local geological and geomorphological factors which result in the formation of intra-zonal soils (e.g. a clay podsol or a gravelly podsol).

Hence, in any locality, the *ecologically possible systems* of farming are determined, in larger areas, mainly by climate, but local factors become of greater significance as the study area becomes smaller. The number of *actual systems* and farm enterprises found, however, are generally fewer than the ecologically possible systems or enterprises. This is by reason of operational factors (e.g. the unsuitability of the terrain for mechanisation) (Chap. 1.6) and of economic and social factors (Chap. 1.5), including the subjective decisions made by individual farmers whether or not to farm any given land and on what system.

In temperate areas there are, however, increasing numbers of exceptions where the range of ecologically possible systems is artificially increased by technology, i.e. industrial and scientific inputs, e.g. irrigation water in dry sunny areas, the housing of livestock, the drying of grain, the use of quicker and more powerful machinery, better weather forecasts, etc. Partial control of the environment is both locally reducing the farmer's dependence on natural climate and helping to 'weather-proof' some farming systems.

The effect of technology in abating climatic constraints in the tropics and sub-tropics is mainly seen in irrigation. This has been chiefly applied in high tropical latitudes and the sub-tropics, in such areas as northern India, West Pakistan and northern Africa, where it is in parts impossible to grow crops at all without irrigation. Since enough water is rarely available to irrigate all the year round, the crops grown are usually annuals such as cotton in the Sudan or rice in Mali which only occupy the ground for part of the year, and full advantage cannot be taken of the high solar radiation at these latitudes. Recent developments in the use of sprinkler irrigation in the wetter zones nearer the Equator, for example on bananas in Central America and sugar-cane in Uganda, show that here too inter-seasonal stability of production and total yield can be strikingly increased whilst exploiting the natural advantages of the tropical climate for growing perennials (cf. p. 44 below).

Nevertheless, the natural climate is still the major ecological influent on possible and actual temperate and tropical farming systems. Brichambaut and Wallen (1963) point out that there are currently three methods for studying and quantifying the relations between climate, vegetation and farming. These are (a) those based on the influence of single climatic factors, (b) those using simple formulae based on temperature and precipitation and (c) those based on estimated energy and water balances (Penman 1948, 1962; Prescott 1943; Thornthwaite et al. 1955, 1958; Smith 1964 and Turc 1961; see also Veih-meyer in Chow1964 and Gleizes 1965, who compares the various evapo-trans-piration formulae on which these balances are based). Though we have used methods (a) and (b) where appropriate, we have in general relied on (c) (energy and water balances), mainly because they can express climatic com-plexes simply and quantitatively in terms of moisture. We have used Thorn-thwaite's method (in which the energy receipts are expressed as millimetres or inches of potential evapo-transpiration (P–E–T) and are calculated from tem-perature and latitude) rather than, for instance, Penman's. This is not because we consider it the best—in many ways Penman's and Turc's are more accept-able—but because C. W. Thornthwaite Associates (1962–64) have now pub-lished world-wide estimates of average climatic water balances for some 13,000 stations* and because the formula demands the minimum of meteorological data.

All energy/moisture balance formulae are *substitutes* for the direct accurate

---

* Penman's method estimates moisture loss from combined heat balance and aero-dynamic equations, and is based on measurements of air temperature, humidity, run of wind and sunshine hours. Unfortunately, long-term records of these measurements (especially sunshine hours) are not available for a large and geographically well-distributed number of meteorological stations. Nevertheless, when it can be applied Penman's method has proved very useful and in simplified form, very suitable for day-to-day use by farmers.

Thornthwaite's formula for calculating potential moisture from a vegetation-covered surface is not so complex and depends on fewer measurements. It is based primarily on tem-peratures and day-length (latitude) for which long-term records are widely available or cal-culable. It tends to be inaccurate for coasts and islands, but is broadly valid over large parts of the world. Like Penman's, it expresses estimated radiation receipts in terms of milli-metres or inches of potential evaporation and transpiration, which is called Potential Evapo-transpiration (P–E–T). As long-term mean precipitation records are also widely published one can estimate (after allowing for any water available from the soil as a result of earlier precipitation), how much moisture is actually evaporated or transpired in a given period. This is the actual estimated evapo-transpiration (A–E–T). Thus, allowing for a fall of 1 in in *soil moisture*, if P–E–T for a month is say 4 in and precipitation (P) 2 in, then the A–E–T is 3 in. If the P–E–T is 4 in and the precipitation is 6 in, then the A–E–T for that month is 4 in.

measurement of solar radiation and of loss of moisture from (usually) grass-
covered ground. They are subject, in some cases, to appreciable errors in
annual or monthly or shorter-term estimates. Thus Stanhill (1963) has shown
that, whilst Penman's was a more accurate method of estimating actual moisture losses in central England and southern Israel than at Ibadan, Nigeria,
Thornthwaite's gave reasonable accuracy in England and Nigeria, but not in
southern Israel. But, nevertheless, both Penman's and Thornthwaite's formulae have proved very useful in farming and hydrological practice. In addition, for data on temperature and precipitation we have used the (British)
Meteorological Offices, M.O. 617 series (1958) and official publications of the
countries concerned.

In this chapter, therefore, energy receipts and balances, temperature, precipitation and energy/moisture balances are discussed in that order and briefly
related to climatic regions and farming systems. It concludes with a section
on the relation between climate and crop and livestock location. Weather
uncertainty as a locating factor is left until Chap. 1.4.

ENERGY REGIMES

The mean annual value in the Northern Hemisphere of solar radiation
energy, incident at the edge of the atmosphere on a horizontal surface, is
0·485 cal per cm² per minute. Of this only about one-fifth is visible light; the
remainder being ultra-violet or infra-red light.

The fate of this energy, on the average, is as follows (in percentage units)*:

Clouds allow through	17
Directly received through clear skies	24
Scattered by sky to earth (down scatter)	6
Reaching the earth	47
Clouds reflect back to space	25
Clouds absorb	10
Atmospheric absorption and scatter	18
Remainder	53

$$100\% = 0{\cdot}485 \text{ cal/cm}^2/\text{min}$$

* Gates 1962, p. 5.

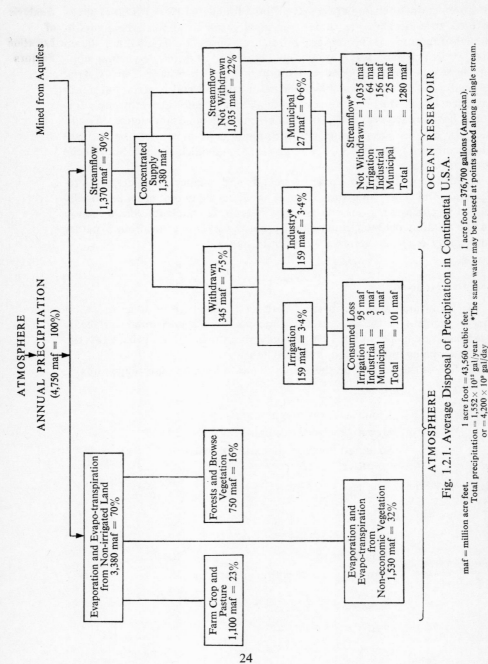

ATMOSPHERE

ANNUAL PRECIPITATION

(4,750 maf = 100%)

Mined from Aquifers

| Streamflow 1,370 maf = 30% |
| Concentrated Supply 1,380 maf |

| Streamflow Not Withdrawn 1,035 maf = 22% |
| Withdrawn 345 maf = 7·5% |

| Municipal 27 maf = 0·6% |
| Industry* 159 maf = 3·4% |
| Irrigation 159 maf = 3·4% |

| Streamflow* |
| Not Withdrawn = 1,035 maf |
| Irrigation = 64 maf |
| Industrial = 156 maf |
| Municipal = 25 maf |
| Total = 1280 maf |

| Consumed Loss |
| Irrigation = 95 maf |
| Industrial = 3 maf |
| Municipal = 3 maf |
| Total = 101 maf |

OCEAN RESERVOIR

| Evaporation and Evapo-transpiration from Non-irrigated Land 3,380 maf = 70% |
| Forests and Browse Vegetation 750 maf = 16% |
| Farm Crop and Pasture 1,100 maf = 23% |
| Evaporation and Evapo-transpiration from Non-economic Vegetation 1,530 maf = 32% |

ATMOSPHERE

Fig. 1.2.1. Average Disposal of Precipitation in Continental U.S.A.

maf = million acre feet.　　1 acre foot = 43,560 cubic feet　　1 acre foot = 376,700 gallons (American).
Total precipitation = 1,552 × 10¹² gal/year　　*The same water may be re-used at points spaced along a single stream.
or = 4,200 × 10⁹ gal/day

Source: Wolman Abel,1962.

24

Of the 47% reaching the earth's surface, the ultimate return to the atmosphere
is:

as long-wave radiation	14
as latent heat of evaporation	
(including transpired water)	23
as sensible heat	10
	——
	47%
	——

The latent heat of evaporation is not lost to the atmosphere because the re-condensation of water as rain or snow releases it. In fact evaporation from the oceans lifts 490,000 cubic km of water into the air per annum, 108,400 of which are precipitated on to the land and of this amount 37,300 flow to the sea as rivers or glaciers. (See Fig. 1.2.1 for the U.S.A.). Ultimately, however, all this heat energy escapes back to space from the earth's atmosphere but is replaced by further solar radiation. The fate of the energy reaching the earth's

Table 1.2.1

Provisional Balance Sheet: Day and Night
Expenditure of Energy in Summer (Soil Fully Moist)
(South-east England. Income = 100)

Reflection	20
Radiation	34
Evaporation and Transpiration	39
Heating Air	4
Heating Soil	2
Plant Growth	1 or less
	——
	100
	——

Source: Russell 1957.

surface in temperate regions is typified in Table 1.2.1. This shows that only about 1%* of the energy received in summer in the United Kingdom is converted by photosynthesis into living matter; that more than half is lost by reflection and radiation; that 6% goes in heating air, soil; and that two-fifths is used in the evaporation and transpiration of water. Outside the thermal growing season (T–G–S) or the hydrological growing season in arid areas, the percentage photosynthesised is almost negligible.

* The mean figure of 1% and less, for the efficiency of conversion of solar energy to dry matter, can be rather misleading because:
(1) only a limited range of wavelength (40–45% of total energy) is actually *utilisable* for photosynthesis, and,
(2) it greatly underestimates the mean efficiency which can occur during the height of the growing season (e.g. sugar-beet efficiency may be as high as 3–3½% during mid-season (see Monteith 1966, Duckham 1968)).

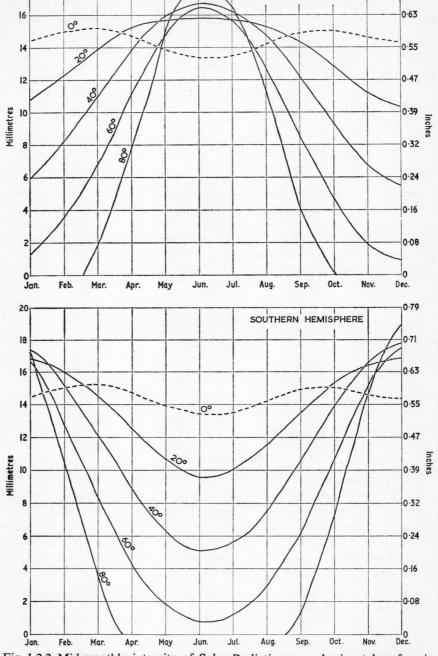

Fig. 1.2.2 Mid-monthly intensity of Solar Radiation on a horizontal surface in quantity of water evaporated per day. *Source:* as for Fig. 1.2.3.

There is, in most parts of the world, a high positive correlation between net energy receipts, moisture loss and potential photosynthesis. So the potential loss of moisture (P–E–T) from ground potentially completely covered with photosynthesising leaves is a useful measure of energy receipts (Figs. 1.2.2 and 1.2.3), and, if water and nutrients are non-limiting, of potential plant

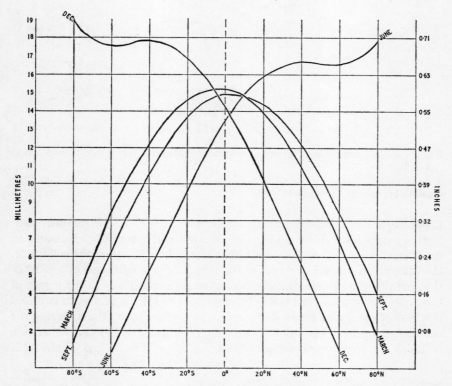

Fig. 1.2.3 Mid-monthly intensity of Solar Radiation on a horizontal surface (in quantity of water evaporated per day)

Values from a table by N. Shaw (multiplied by 0·86 and divided by 59 give the radiation in mm of water per day) in 'Comparative Meteorology', *Manual of Meteorology*, Vol. II, 2nd edn. (Cambridge) 1936.

production (i.e. net formation of dry matter excluding respiration losses) by crops.

Figs. 1.2.4 and 1.2.5 show the relation between latitude and longitude respectively on energy receipts as measured by the annual potential evapo-transpiration (P–E–T). The data is from Thornthwaite (1962–64). Fig. 1.2.4 shows that the mean annual P–E–T in mid Italy (Lat. 40–45°N) is 50% higher and within the tropics ($23\frac{1}{2}$°N to $23\frac{1}{2}$°S) it is about 3 times greater than in northern Sweden (> 65°N Lat.), which is inside the Arctic Circle. Fig. 1.2.5 shows that

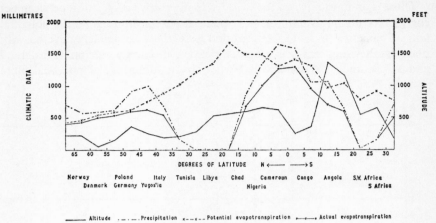

Fig. 1.2.4 Relation between precipitation, potential and actual evapo-transpiration and latitude

Average Altitude, Annual Precipitation, P.E. and A.E. of 650 Meteorological Stations in 27 countries between 10° 10′ E and 19° 59′ E

longitude has, as expected, much less influence on P–E–T than latitude. Latitude has important effects on the 18 % of solar energy receipts which are, on the average, scattered or absorbed by the atmosphere. The higher the latitude, the longer, on annual average, is the length of the light energy's passage through the atmosphere. But absorption rises geometrically as the length of this path increases arithmetically (Gates 1961, p. 35). At very high latitudes this loss of energy by atmospheric absorption more than offsets, during the summer, the greater length of summer day. Fortunately, however, at middle latitudes in midsummer the effect of this absorption is much less, Table 1.2.2.

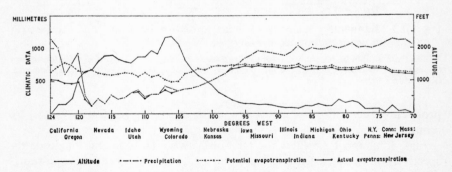

Fig. 1.2.5 Relation between precipitation, potential and actual evapo-transpiration and longitude (U.S.A.)

Average Altitude, Annual Precipitation, P.E. and A.E. of 390 Meteorological Stations across the U.S.A. between 37° 30′ N and 42° 30′ N

28

Table 1.2.2

Effect of Latitude on Annual and June Energy Receipts

	Latitude °N	Average Net Radiation Receipts at Earth's Surface in kcal/cm² and kcal/in²				Average Potential Evapo-transpiration (P–E–T) in mm and in			
		Annual total		June total		Annual total		June total	
		per cm²	per in²	per cm²	per in²	mm	in	mm	in
Lapland	65	20	130	8	50	500	20	89–102	3·5–4·0
N.W. Europe	50/60	20–40	130–260	8–10	50–65	610	24	102	4·0
Mediterranean Basin	35/45	60–80	390–520	8	50	710–1,015	28–40	122–132	4·8–5·2
Great Plains	40/50	30–50	190–320	8	50	560–660	22–26	112–122	4·4–4·8
S.E. of U.S.A.	30/40	60	390	8	50	840–1,270	33–50	140–160	5·5–6·3

Source: Net receipts estimated from Budyko (quoted by Gates 1962, p. 22); P–E–T from Thornthwaite 1962–64.

Solar Energy Receipts and Energy Balances in Temperate Zones

But annual P–E–T figures can be deceptive. For what matters (in those temperate climates where all the year production is not possible) is the energy receipts in the thermal growing season (T–G–S) and particularly at the height of the plant-production season, which is May to July for most temperate crops except in Mediterranean climates where it is about 2 months' earlier.

So at the potential height of the 'growing season' (if moisture is not a constraint), the energy receipts (P–E–T), and hence crop growth potential, are of the same order at the higher latitudes, despite marked differences in mean annual net radiation receipts. This is an important agricultural fact (see Sweden, p. 224) and location factor.* In winter, on the other hand, latitude makes a big difference. Thus, in the United Kingdom at Winchester (51°N) summer half-year energy receipts are three times those of the winter half-year; in the Hebrides (58°N) the summer/winter ratio is 4:1. (Duckham 1963, p. 161).

The influence of cloudiness. As noted, 35% of the energy reaching the atmosphere is, on average, reflected back or absorbed by clouds. Thus, the reflected or absorbed proportion is higher than this in cloudy maritime climates like that of the north-west of Europe where the annual mean of sunshine hours daily are only 4 to 5, and less than 35% in other temperate climates, Table 1.2.3.

Clouds are a disadvantage at higher latitudes where they may reduce light receipts and hence, also, temperatures below optima during the thermal growing season. At middle latitudes, especially in the interior of continents, some reduction of insolation due to cloud may be helpful in that in summer it lowers temperatures at the earth's surface and so decreases heat stress in animals, reduces moisture loss in crop plants and reduces radiation losses at

* The mean estimated actual evapo-transpiration (A–E–T) (i.e. P–E–T for which there is enough rain or soil-stored water) for the three months April–June when about two-thirds of the primary production takes place (Duckham 1963, p. 477) in these latitudes is estimated to be:

Place	Latitude	Estimated Actual Evapo-transpiration			
		April/June		Year	
Winchester, England	51°N	205 mm	8·0 in	580 mm	22·8 in
Edinburgh, Scotland	56°N	188 mm	7·4 in	560 mm	22·0 in
Copenhagen, Denmark	56°N	200 mm	7·9 in	528 mm	20·8 in
Outer Hebrides	58°N	189 mm	7·4 in	578 mm	22·8 in

Source: Thornthwaite Associates 1964.

Table 1.2.3

Sunshine Hours and Cloud Cover at Various Latitudes

Latitude		Month	Mean cloud cover (tenths covered)	Annual mean sunshine daily (hours)
50–60°N	N.W. Europe	July	6–7	4–5
40–50°N	Great Plains, U.S.A.	July	4–5	7
35–45°N	Mediterranean Basin	July	2–3	8
30–40°S	Australia, southern parts	Jan.	4–5	7–8
40°S	New Zealand	Jan.	5–6	5½

Source: Mainly Smith 1962.

night. Clouds may also be useful in winter. Thus, high sunshine values in winter in the United Kingdom lead to a lowering of the soil temperature and are often associated with cold weather.

Energy Balances in the Tropics

In considering the radiation received in the tropics and sub-tropics, the first point to be made is that the total number of daylight hours in a year is the same at all points on the earth's surface. Therefore, a perennial plant which is in the ground all the year round derives no advantage, as regards this factor considered alone, from being grown at any particular latitude. The case is very different with annual crops which can be grown in a period of 3 to 6 months. Such crops grown at the Equator never experience a day-length greater than 12 hours, whereas if grown in sub-tropical or warm temperate latitudes the greater summer day-length provides them often with greater radiation. Thus, any advantage over temperate climates provided by the constant high temperatures of the tropics is greater relatively for perennial than annual crops; we shall see later (p. 68) that there are other reasons, too, for using perennial crops, where possible, to exploit the tropical environment. In fact, the world's zones of highest annual energy receipts lie in the sub-tropics, and are associated with very arid or desert areas which are of little significance for agriculture (Fig. 1.2.4). In the Northern Hemisphere, these zones include the Sahara, Arabia and lower California; and in the Southern Hemisphere, the north Chilean desert, South West Africa and Botswana, and central Australia (Black 1956).

Within the tropics, the amount of solar radiation received is not always so much or so consistently greater than in the temperate zone as dwellers in the latter often believe. There are, for instance, at least two tropical areas, in central India and along part of the West African coast, where, because of a high degree of cloud cover during the rainy season, solar radiation received during the months from June to August is less than in the British Isles, these months being an important part of the growing season. The effects of such cloudiness on plant growth and farming practice in the tropics have never been properly evaluated. This factor might be expected to be particularly

30

important in its effect on the distribution of a crop such as cotton, where high
yields are believed to be particularly correlated with sunshine (Knight 1935).
In Brazil it has always been claimed that the reason for not growing coffee
without shade trees is because of the greater cloudiness of the climate there
than in for example, African coffee-growing countries.

An example of a latitudinal effect on plant physiology which has been reported by Hilditch (1951) is of particular interest because it concerns one of
the oil-seeds, a class of crops which are so important in the tropics. In this
case, the iodine number (or degree of unsaturation) of sunflower seed oil increased regularly with increase of the distance from the Equator at which the
plants were grown, the effect being apparently correlated with the rate at
which the seed matured.

TEMPERATURE REGIMES

Strictly speaking, temperature is not a production factor but a measurement of the average kinetic energy or average velocity of molecules of any
kind of matter. Heat is the sum of these kinetic energies and is thus a quantity
of energy as distinct from (light) radiation or chemical energy (in food). The
earth is heated by concentrated, short-wave radiation energy (photons) from
the sun which becomes less concentrated and longer waved on contact with
earth or living things. The air is heated by this long-wave energy from the
earth or by that which is given out or reflected by living things.

At the earth's surface, temperature derives firstly from the balance between
the incoming solar radiation receipts and the outgoing radiation discussed
above. Secondly, it derives from the history and force of current air (e.g. the
temperature of the area from which the incoming winds have arrived). This
component is largely related to the world's main wind belts and to local physical features (see below). Thirdly, altitude, aspect and soil factors (see below
and Chap. 1.3) are locally important.

Temperature in Temperate Regions

Relief: Altitude. Although solar radiation is more intense at high altitudes,
the mean temperature, nevertheless, drops, on average, 3·3°F per 1,000 ft rise.
This fact, together with greater cloudiness and the higher precipitation in
mountainous parts (which is due to orographical uplift, and some convectional uplift of moist air), largely explains the rapid changes in vegetation as
one climbs. Thus, in Kenya, there are coconut palms at Mombasa at sea-level,
S.100 white clover and English-type pasture swards at Molo at 10,000 ft and
the snowline at 15,000 ft. In the temperate zone, this same effect is well illustrated in the Alps and in the U.S. Rockies (Fig. 1.2.6 Gates 1961, p. 189).

For this reason, altitude, is, in many areas, a more important factor than
latitude in determining possible farming systems. Even in the United Kingdom, which has no real mountains, the altitudinal effect is very marked (about
5 days of growing season per 100 feet), especially at the start and the end of

the growing season, when small differences substantially effect growth potential (Duckham 1963, p. 199). The effect on altitude on grassland yields in
Britain is well shown in Table 1.2.4.

Table 1.2.4

Effects of Altitude on Grassland Yields

Altitude range (feet)	Estimated yield of starch equivalent (cwt/acre)	Range of yields in farms in sample (starch equivalent cwt/acre)
400–475	13·7	11·4/16·0
475–725	11·1	7·9/15·4
>725	10·0	8·0/12·0

Source: Wynne 1960.

Relief: Aspect and slope.   Whether land is level or slopes towards or away
from the sun has important agronomic effects. Hogg, W. H., in Johnson and
Smith 1965, quotes the following figures for a 30° slope in early May at 50°N.

Aspect	Maximum possible radiation receipts (cal/cm²/day)
Level ground	580
South	650
North	350
South-east	600
North-east	400

Azzi (1956, p. 153) suggests that compared with level ground temperatures on
north-facing slopes at Perugia, Italy, average 4·0°C less and those facing
south 1·0°C more than on level ground. East and west slopes averaged 0·7°C
and 0·6°C less. Olives grown on northern slopes had a lower percentage of oil
which had a higher iodine number, i.e. was less saturated than on southern
slopes. On the other hand, damage due to frost and cold was greater on south
slopes because frozen plant tissues thawed more quickly than on north slopes.
(The advantages of preventing a quick thaw, e.g. *by sprinkler irrigation* in the
early morning, are recognised by, for instance, early-potato growers in the
United Kingdom, whilst irrigation the previous evening (i.e. before frost)
produces a moist microclimate which prevents frosting of plants). The effects
of aspect are, of course, well recognised by farmers (one of the authors has a
farm on a slightly northern slope and he envies his south-facing neighbour on
the other side of the valley for his earlier harvests). Aspect can, in fact,
materially affect the location of enterprises.

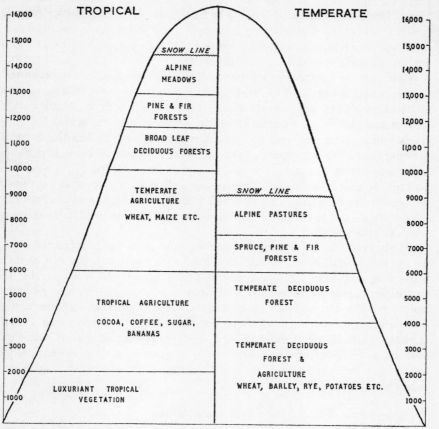

Fig. 1.2.6 Illustration to show how vegetation changes as elevation (in feet) increases on a mountain in a tropical climate (left) and temperate climate (right). From Gates, 1961

Relief: Barrier effects. The barrier effect of mountains and hills on winds, precipitation, temperature (through damming of stable air) and A–E–T can have, locally, as much influence on these four factors as latitude or intracontinental location (Table 1.4.1) (Critchfield 1967). It is, therefore, a locating and intensity influent of importance within climatic regions.

Soil factors. Within the great soil groups of the world (see Chap. 1.3 below), dark soils (e.g. fen or muck soils) tend to warm up more quickly after winter than light-coloured soils because they absorb more radiation. This may be an advantage in humid and semi-humid temperate climates. Soils with persistently high moisture (e.g. because they are low-lying or because they are moisture-retentive, heavy clays) may take a long time to become workable and to warm up in the spring because of the high specific heat and the latent

33

heat of evaporation of water. Soil colour and moisture content, therefore, locally influence systems of farming because this may delay drilling (as wet soils are not cultivable so early as drier soils), seed germination and the build-up of leaf area needed to make full use of radiation receipts. But these factors are not of great importance on a world scale.

Wind systems and ocean currents substantially influence temperatures. In the Pacific North West of North America and in north-west Europe, the combination of prevailing warm south-westerly winds with the North Pacific current and the North Atlantic Drift (the Gulf Stream) respectively, appreciably raise winter half-year temperatures and hence lengthen the thermal growing season. This 'temperature anomaly', combined with the unexpectedly high midsummer radiation receipts in higher latitudes (p. 29), is very important. It makes north-west Europe agriculturally viable, and combined with moderate but reliable rainfall and P–E–T and a high level of economic development, makes the North Sea Basin one of the most productive farming regions in the world.

Winds, however, are not only bringers but also removers of moisture. Thus, on the east side of New Zealand, the Westerlies (having shed much of their water orographically on the central mountain range), and in Israel, the hot desiccating winds from the adjoining deserts (p. 296) may remove moisture both from the soil surfaces by evaporation and from plants by increasing transpiration losses. Such winds can make local precipitation and P–E–T statistics unreliable indicators of local energy/moisture regimes. Thus, wind systems (as distinct from wind hazards (p. 70) such as hurricanes) have some locationary influence, especially when combined with warm or cold ocean currents.

The Thermal Growing Season in Temperate Regions

The geographic 'growing season' is, conventionally, the mean annual duration of air temperature at or above $42\frac{1}{2}°F$. This is the lower temperature threshold for many temperate crops, with some important exceptions such as, in the United Kingdom, winter wheat and Italian rye-grass (*Lolium multiflorum*). (This threshold should properly be applied to soil temperatures, which tend to move more slowly with the seasons than do air temperatures, but on a world scale the former are not so well-recorded and published, so the latter are more commonly used.) Moreover, the lower 'critical' temperature for adult-housed livestock is broadly speaking between 35°F and 45°F depending on species and rate of feeding. Below these temperatures the efficiency of feed conversion starts to fall because chemical energy which might be used for growth or lactation is converted to heat to maintain body temperatures. Further, in temperate countries, the months at or below $42\frac{1}{2}°F$ coincide with the period of expensive winter feeding. Thus, for both crops and livestock, $42\frac{1}{2}°F$ is a useful, if somewhat arbitrary, lower-constraining threshold and is used as such in Part 2.

No clear-cut upper threshold temperature can be suggested, especially for crops. But as, at high temperatures, plant respiration losses increase (Monteith 1966, Lemon, 1963, 1965) and as, at day temperatures above 75–80°F, evapo-transpiration losses are often high; as dairy cows may start to lose productivity efficiency (Findlay 1959 quoted by Duckham 1963, p. 27) (see also Israel, p. 296); as the rate of live-weight gain of full-fed pigs may fall, even if they are not actually distressed; as unshorn sheep may be very distressed, or liable to fly strike in hot humid weather, an upper constraint of $> 72\frac{1}{2}F°$ has been used in Part 2 (p. 108). This is not an upper critical temperature at which production by temperate crops and livestock falls abruptly, but an indicator of probable loss of productivity unless, at least with livestock, anti-heat stress precautions (e.g. shade from direct sun) are taken.

Therefore, in Part 2 the months at or above $42\frac{1}{2}°F$ have been taken as the thermal growing season, and with the number of months at or above $72\frac{1}{2}°F$ as possible heat-stress periods. Whilst, anticipating the next section, it may be suggested that, by analogy with Australia, p. 175, any months in which P–E–T less P exceeds 3 in (in temperate regions) or 5 in (on, say, tropical clays) are non-growing months, i.e. are outside the hydrologic growing season, by reason of moisture deficit. In addition, within the thermal growing season, biological seasonality may reduce the effective growing season. Thus, it has been estimated (Duckham 1963, p. 477) that in England and Wales, all cereal, 70% of pasture, 55% of potato and 20% of sugar-beet dry-matter formation takes place in the first half of the thermal growing season of 8–9 months. The same general seasonal pattern applies in all cool temperate areas and many warm temperate ones on eastern margins of continents. Production is also highly seasonal in warm dry climates with their marked summer droughts.

Thus, attempts to express the farming potential of an area (a) by the length of the thermal growing season, or (b) by the device of accumulated temperatures (e.g. the sum, on the days when temperature is $> 42\frac{1}{2}°F$, of the differences between the recorded mean for such days and $42\frac{1}{2}°F$), or (c) by the P–E–T during the thermal growing season are of limited value. However, as rough indicators of differences on a world scale, both (a) and (c) are used in Part 2, together with A–E–T (see below p. 40).

Temperature Factors in the Tropics

The world's highest temperatures occur in the sub-tropics and in high tropical latitudes during their summer season. If moisture is non-limiting, all the annual crops of the tropics can also be grown in the sub-tropics and even in the warm temperate zone (e.g. south-east of the U.S.A.) wherever summers are sufficiently long and hot. There is no class of annual 'tropical' crops which is strictly limited to the tropics. With some of these crops, such as rice and cotton, higher yields are generally obtained outside the tropics (e.g. rice in Australia, cotton in California) than within them; with others, such as tobacco, groundnuts and other pulses, yields seem to be about the same in both environments, when husbandry is equally good. Caution must, however, be exercised in comparing the yields of a single crop within and outside the tropics. Outside the tropics that crop will usually be the only one to be grown during the year. Within the tropics another crop may often be taken off the same ground during the year, and it may be more desirable to compare the total production from an acre of land over a year than the yield of a single crop.

With perennials, the case is quite different. Most tropical perennial crops are strictly limited to the tropics, and, with fairly high temperatures enabling them to maintain growth throughout the year, some of them are capable of very high yields indeed. Thus the oil palm yields more oil per acre than any other crop, tropical or temperate. The maximum rate of dry-matter production per unit area by plants anywhere in the world which Blackman and Black (1959) found recorded was by sugar-cane in Hawaii, near the outer limit of the tropics. The fact that sugar-cane, by a wide margin, yields more calories of human food per acre than any other crop, and gives its highest yields in the tropics, indicates one probable direction of agricultural development as the human population of the globe grows inexorably greater (see Chaps. 3.3 and 3.4). Perennial grasses in the tropics, which have been recorded to yield up to 24 tons of dry matter per acre per year, far outyield grasses in temperate regions where 8 tons is about the most that can be obtained. But a few tropical perennials are particularly restricted to the equatorial zone, presumably by the temperature factor; thus coconuts will not give a commercially profitable yield at more than about 20° from the Equator, nor much above 1,000 ft altitude,

PRECIPITATION

It was noted above (p. 23) that the equivalent of almost half the radiation reaching the earth's surface in the Northern Hemisphere returns to the atmosphere as the latent heat of evaporation of water from plants (transpiration water), land and open waters. Following the adiabatic cooling of ascending air "masses, occurring without loss or gain of heat to the whole atmosphere, this" moisture is precipitated. Regions of ascent of air masses are mainly either (a) zones of converging horizontal air flows (i.e. towards the Equator and the poles where low-pressure areas yield convectional and frontal rains (Fig. 1.2.7), or (b) areas along the windward sides of mountains, which yield orographic precipitation.

The regions of precipitation descent are determined by:

(i) the belts of ascending air at the Equator and at polar fronts (Fig. 1.2.7);
(ii) the direction and kind of moisture-bearing winds (e.g. over a long ocean passage as in New Zealand) and their relation to ocean currents;
(iii) the distance from the main source of water;

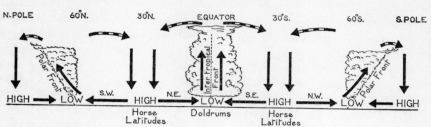

Fig. 1.2.7 Pressure Systems and Latitude. From Gates, 1961

(iv) the height of the land over which wet winds blow, and
(v) air temperatures which govern the potential moisture content of the air masses.

The results are shown in Figs. 1.2.8 and 1.2.9.* Fig. 1.2.7 shows that many *temperate farming areas* are influenced by the N.W. and S.W. wind belts for at least part of the year. In such areas precipitation is often brought in by these westerly belts, the site and amount of its fall being determined by distance from seaboard, by water mass, by altitude and by air temperatures. Thus orographical precipitation tends to be more important than convectional precipitation, particularly on western seaboards with high relief, e.g. western England, western New Zealand and British Columbia. Finally, in cool and cold temperate climates much of the winter precipitation may fall as snow which, especially if the ground is frozen, may lie until the spring thaw. This (see below, p. 112) precludes winter cultivations (e.g. in Canada) and may delay them in the spring until the thaw is complete and the land dried out somewhat.

Tropical rainfall. Rainfall conditions at low latitudes are rarely ideal for agriculture. A general disadvantage is the tendency of rain in these regions to fall in storms of high intensity; quantities of water, greater than the soils can absorb during the time in which it falls, have to pass away as surface run-off which is unavailable for crop growth and causes the high degree of risk of soil erosion which is characteristic of the tropics and sub-tropics. Roughly speaking, any rainfall at an intensity greater than about 0·4 inches per hour is liable to be dangerously erosive, and such intensities are a commonplace at these latitudes. The sub-tropics and outer tropics contain vast areas with very low rainfall where the growing of crops is impossible or precarious. The farmers' difficulties are increased by the fact that rainfall tends to have the highest coefficients of variation where it is lowest, and that rainfall figures below the mean occur in more years than those above the mean; this is the part of the world where crop failures from drought are most frequent and livestock may die from lack of water and grazing. In these 'dry' tropics advantage cannot

* Figs. 1.2.8 and 1.2.9 are based on Finch *et al.* 1957, Gates 1961 and Smith 1962.

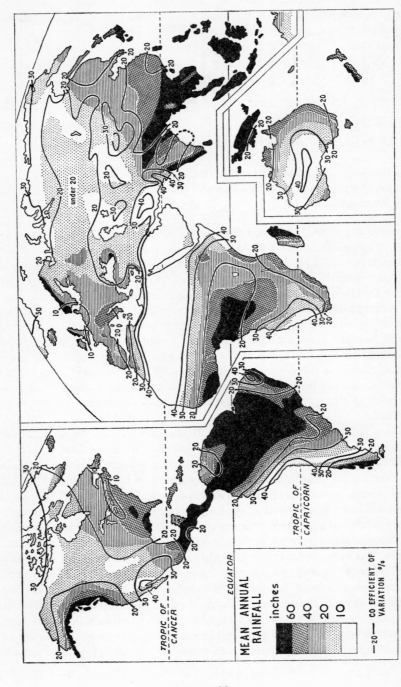

Fig. 1.2.8 World Map of Annual Precipitation and Coefficient of Variation. *Source:* Finch *et al.* 1957. (See References on p. 46)

be taken of the radiation and thermal suitability of the climate for perennial crop production, because the limited rainy season only enables the growing of annual crops during one part of the year. The 'wet' tropics, although not lacking moisture for crop growth, suffer from another set of disadvantages. The high intensity of even larger quantities of rainfall can lead to extreme erosion in the ultimate form of landslides. The combination of continuously wet and hot conditions produces excessive leaching of plant nutrients from the soil, massive weed growth, and a plethora of crop and livestock diseases and pests. There is also a direct effect on livestock. Bergmann's rule that hot humid conditions will produce miniature animals is well exemplified in south-east Asia, Ceylon and coastal West Africa where dwarf cattle, goats and

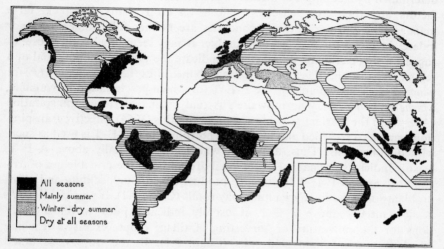

All seasons
Mainly summer
Winter - dry summer
Dry at all seasons

Fig. 1.2.9 World Map of Seasonal Distribution of Precipitation. *Source:* Gates, 1961

poultry are a contrast to the much larger animals of dry tropical or of temperate climates.

In contrast to cool and some warm temperate areas where seasonal operations hinge on insolation and temperature levels, the whole annual rhythm of agriculture in the tropics is determined by the incidence of rain seasons, when crops may be planted, and of dry ones when harvesting takes place. Along the line of the Equator, most land surfaces receive a fair quantity of rain each month, though two peaks of rainfall following the equinoxes are sometimes perceptible. Some of these areas are so seasonless that planting and harvesting may take place in any month of the year. A few degrees further north or south, two distinct post-equinoctial rainy seasons of 2–3 months are punctuated by two dry seasons of mild severity. These two zones may be said to constitute the 'wet' tropics. Perennial crops can be grown, and two crops of annuals a year, or even more of short-term vegetables, can often be taken. Towards the outer limits of the tropics, the two rainy seasons tend to coalesce into one long one of 4–6 months, whilst the remainder of the year is dry and sometimes practically rainless. The monsoon system of southern Asia gives India (p. 390) a special climate with most of its rain in the months of the northern summer.

ENERGY AND MOISTURE BALANCES

When soil and operational factors (see Chaps. 1.3 and 1.6) are non-limiting, the plant production potential and, unless feeding stuffs are imported or exported, also the livestock production potential of a given locality are largely determined by a complex of meteorological factors. The interactions of these factors and their 'weather-chain' effects on crop plants, livestock, farm management and agricultural economics are neither fully understood nor indeed, in many cases, as yet measurable. It may seem unwise, therefore, to suggest a single climatic parameter to indicate the production potential of a region. What we need is a parameter which measures the precipitation (P) for which P–E–T is available *or* the P–E–T for which precipitation is available, *whichever is the less*. Thornthwaite's Actual Estimated Evapo-transpiration (A–E–T) in the thermal growing season or Smith's (1964) 'effective transpiration' can be so regarded and in Part 2 Thornthwaite's A–E–T is used as such, with qualifications. Thus, a P–E–T, which is substantially above A–E–T, usually indicates permanent or seasonal drought and the possible use of irrigation to exploit P–E–T to enhance production—though salinity and allied problems of irrigation agriculture may result (Chap. 3.1). If precipitation (P) is substantially above P–E–T, it usually indicates poor working conditions for the cultivation or harvesting of tillage crops, excessive leaching of plant nutrients and acid soils (Chap. 3.1). In the tropics, as noted above, neither P–E–T nor temperature are major constraints and annual P or rather the timeliness, mean amount and reliability of P is a better guide.

CLIMATIC REGIONS

The accepted climatic classifications are not much used by agriculturalists. Every farm or local district seems, to those who farm it, to have a climate of its own. At first sight, therefore, to be agriculturally useful, any climatic classification would need a very detailed breakdown. But there are relatively few climates in which one finds farming that is not a variant of one of the four main farming systems. Thus, the farming systems in the North East, in the Corn Belt and in the Pacific North West of the U.S.A. are essentially variants of systems which evolved in north-western Europe in the 18th and 19th centuries. We only need recognise, therefore, four classes of temperate agro-climate and two classes of tropical agro-climate (Fig. 1.2,10).

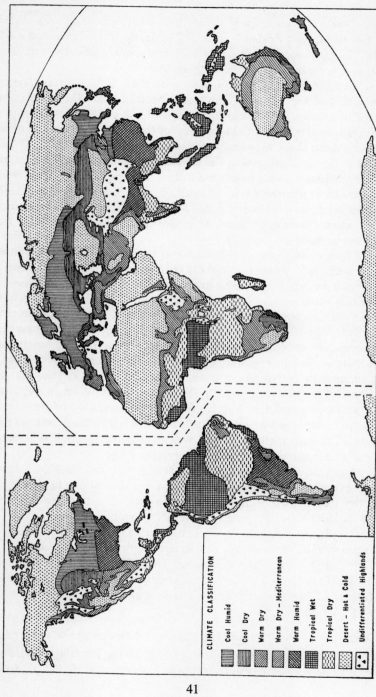

Fig. 1.2.10 World Map of Agricultural Climate Classification. From Finch *et al.* 1957

CLIMATE CLASSIFICATION

Cool Humid

Cool Dry

Warm Dry

Warm Dry – Mediterranean

Warm Humid

Tropical Wet

Tropical Dry

Desert – Hot & Cold

Undifferentiated Highlands

Temperate agro-climates. The classes are (Fig. 1.2.10):

(i) *Cool Humid* is a good agro-climate. It occurs in Europe, Canada, the U.S.A., Tasmania, New Zealand, parts of Siberia, northern Japan and southern Chile. The main constraints are P–E–T, winter cold, short-term moisture deficits and, in some cases, poor working weather.

(ii) *Cool Dry* (continental) is found in Canada, the U.S.A., Australia, Argentine, Uruguay and the U.S.S.R. The main constraints are the low summer rainfall and its unreliability.

(iii) *Warm Humid* occurs in the south-east of the U.S.A., in South America, South Africa, Australia, China and Japan.

(iv) *Warm Dry* (including warm dry Mediterranean) is semi-arid to arid with hot dry summers and is found in southern Europe, North Africa, South Africa, central and western Asia, southern parts of South America, California and the south-west of the U.S.A., Mexico, and the southern parts of Australia. Mediterranean climates (e.g. northern Israel), however, have markedly wet winters.

Of these four classes, the cool humid and warm humid have reasonably reliable agro-climates, whilst the cool dry temperate and warm dry temperate ones become increasingly unreliable as the average precipitation declines (see p. 38). In this classification, climates in areas of low or nil farming productivity, e.g. polar, mountain, cold temperate and most of the hot deserts are omitted. They may be technically interesting, but socially and economically they are, today, unimportant, even though hot deserts may one day be irrigated.

Tropical agro-climatic regions. Agriculturally there are only two major climatic regimes in the tropics, these colloquially known as the 'wet' and 'seasonally dry' tropics (Fig. 1.2.10). These are vague terms, though they indicate fundamentally different agricultural systems and ways of life.

(v) The 'wet' tropics may be empirically and usefully defined as those parts where rainfall is sufficiently high and well distributed through the year to make it possible to grow perennial crops throughout the year.

(vi) The 'seasonally dry' tropics are where this is not possible. Note that it is not possible to grow any crop on a rainfall of less than about 12 in annually and the 'dry tropics', therefore, include some areas where (unless irrigation water is available) extensive grazing is the only possible land use.

The wet tropics are, in general, concentrated in a zone within a few degrees of the Equator, and the seasonally dry areas occupy the outer latitudes of the tropics, but there are important exceptions. Thus, the Equator passes through a very dry area in Somalia and Kenya, whilst in south-east Asia wet conditions extend from the Equator northward to Thailand and Burma and even beyond the tropic to Assam.

CROP AND LIVESTOCK DISTRIBUTION IN TEMPERATE CLIMATES

Apart from some nomadic pastoralism (e.g. in tundra fringe and desert fringe) and some very extensive ranching (e.g. in Arizona, U.S.A.), farming of any type is not at present practicable where (*a*) the thermal growing season (T–G–S) is less than 3 months, *or* (*b*) where, without irrigation, the precipitation or A–E–T per annum is less than 10 in, *or* (*c*) where the annual P–E–T is less than 15 in *or* (*d*) where annual precipitation exceeds 120 in. Any one or more of these four constraints appears to be effectively absolute. If operational and social constraints are non-limiting within these four constraints, the farmer usually has an increasing choice of economic species, and hence of enterprises, as he moves away from any of these constraints towards the more favourable climate of the 'hydro-neutral zone' (Table 3.1.1 and Fig. 3.1.11) with its high A–E–T. What matters to him is not so much which particular species are ecologically possible, but whether he has a choice of a reasonable number of species which will thrive and yield economically under his local conditions. A wide choice enables him to meet local climatic abnormalities; disease, insect and weed problems; soil nutrient balance problems; personal or market preferences, etc. Even if, for profit motives, he specialises, the fact that he can switch to other enterprises often justifies permanent investment in buildings, irrigation schemes, etc.

No attempt is made, therefore, to outline the climatic limits of particular economic species. It is enough that, in cool humid and warm humid climates and in the wet tropics, there is a wide choice of genotypes, adapted to local conditions, of cereals, pulses, forage grasses, forage legumes, high-yielding starch or sugar crops, sheep, goats, cattle, pigs and fowls. In cool humid and warm humid climates, amongst the cereals, wheat, barley and oats can be grown almost anywhere, except where relief is high or precipitation is excessive, as winter or spring crops, or both. But maize (for grain) is mostly found in warm humid climates, although it is also grown in warm dry climates and in cool humid areas such as the Corn Belt of the U.S.A. Here May temperatures are high enough for its rapid germination; late frosts are unusual; both radiation receipts and mean temperatures (70°F) are fairly high in the weeks preceding flowering, and moisture deficits are unusual in July and August. Sorghums and millets are also practicable in maize climates, but are more competitive in lower rainfall areas.

Warm humid oil-seeds include tung, groundnut and soya-bean (the latter is also grown in maize climates). In cool humid climates of higher latitude than maize areas, linseed and rape-seed are usually cheaper ways of producing vegetable oils (Canada, p.130 ; Sweden, p. 222) than the above crops whilst animal fats (butter, lard) can usually be more competitively produced than in lower latitudes. Irish potatoes (*Solanum tuberosum*) and sugar-beet (*Beta vulgaris*) are characteristic carbohydrate crops in cool humid, in warm humid or in irrigated warm dry areas, though in frost-free warm humid areas (e.g. Louisiana, U.S.A.) sugar-cane is competitive. Other 'tropical' crops grown in warm humid or irrigated warm dry regions are tobacco, cotton and rice, which yields very well in irrigated Mediterranean climates. Amongst tree crops, apples, pears and plums are grown all over cool and warm humid areas

FARMING SYSTEMS OF THE WORLD

with citrus, peaches, apricots and vines added in warm temperate humid or irrigated warm dry districts. In all temperate humid areas the choice of grasses legumes and forage crops is wide. In the cool dry and warm dry climates, the choice of forage grasses and legumes is good, but that of cereals is limited to wheat, barley, oats in cool areas supplemented by sorghums, millets and sometimes maize in warm dry areas. Starchy or sugar crops, and oil-seeds are only produced under irrigation in dry zones though in the Canadian Prairies linseed and rape-seed are found as far west as Portage la Prairie.

As to *livestock*, cattle are less well adapted to extreme climates than locally adapted genotypes of sheep or goats. One can find a breed of sheep (from the Soay of Shetland to the Navajo of Arizona or the Awassi of Jordan) to fit almost any temperate climate but with cattle one is perhaps more restricted. But European breeds of dairy cow (especially the Holstein/Friesian) and of beef cows are found all over the cool temperate zone and, where environment can be controlled as (e.g. in California, Israel), in warm dry climates. Channel Islands breeds (which *may* be of partly tropical origin) seem better suited to warm humid temperate areas, so Jersey cows were popular in the southeast of the U.S.A., whilst in Australia and the U.S.A. tropical breeds of Zebu type (or crosses with tropical breeds) are increasingly popular for beef production in both humid and warm dry temperate climates. Pigs, and fowls are, with modest housing, well adapted to cool humid conditions, but need more protection against cold in cool dry and against heat in warm humid or warm dry climates.

CROP AND LIVESTOCK DISTRIBUTION IN THE TROPICS

In the tropics, rainfall is the main determinant in the choice of crops. A very clear ecological series of cereals is discernible with decreasing rainfall. Whenever rainfall is high rice is usually the staple cereal; in slightly less wet regions, maize takes this position; where temporary droughts are liable to impose a check on plant growth, the farmer's response is to grow sorghum; arid conditions limit the choice mainly to finger millet (*Eleusine coracana*) and finally, on the very fringes of the desert, to bulrush millet (*Pennisetum typhoideum*). Root crops can only provide an important part of the food supply where rainfall is sufficient for them to give the high yields of which they are capable; since some of the most important tropical root crops, such as yams and cassava, require a crop period of more than half a year, this means in effect the perennially wet climates. These are also necessary for the perennial crops and are exemplified in coastal West Africa, the West Indies, south-east Asia and the Pacific islands. Shorter-period annuals, including many pulses, oil-seeds and industrial crops, such as cotton and tobacco, can as well be grown in countries, like Rhodesia, with one reliable rainy season, although there may be six almost rainless months in the year, as in much wetter ones.

A climatic factor which should not be overlooked is the prevalence, frequent in the wet equatorial zone, of heavy morning and evening mists which maintain atmospheric humidity and reduce transpiration, making it often possible to grow particular crops with a lower rainfall than is stated to be necessary in the textbooks. Altitude imposes the other main climatic restraint in tropical agriculture, but it is difficult to be precise about its effect on particular crops as this varies so greatly with latitude. Thus coconuts are commercially grown in Ceylon at 7°N latitude up to 1,700 ft, although yields may already be less than at sea-level, but in Jamaica at 18°N, 400 ft is about the limit. Most crops of the tropical lowlands disappear progressively from the agricultural scene at altitudes between about 4,000 and 7,000 ft at the Equator, and by the time 10,000 ft is reached, foodstuffs are being mainly provided by such temperate crops as wheat and Irish potatoes (Fig. 1.2.7).

With livestock, the position is not quite so plain as their distribution is controlled not only by climate but by disease, whose incidence, however, is itself often dependent on the climatic factor. But a very clear distinction can be made between the reactions of two different classes of livestock to the tropical environment. In the first class, which includes pigs and poultry, temperate genotypes can be transferred to most tropical environments with complete success. In the second class (cattle, sheep, goats) this is not usually possible until the environment has been modified, at least by the control of disease. It should be noted that two classes of livestock—goats and donkeys—assume their greatest importance in the tropics and sub-tropics, partly because of their ability to subsist on the poorest grazing in arid areas, and partly because of their relative immunity to disease.*

This chapter has necessarily generalised a very complex subject. For instance, as Elston 1967 points out, most climatic parameters vary gradually with distance and any climatic classification is based on arbitrary limits (Figs. 3.1.1 and 3.1.11). The local influence of one or more climatic factors, as will be seen in Part 2, may, like the effect of climatic uncertainty (Chaps. 1.4 and 1.7), have major effects on intensity and location of farming systems and on the farmer's choice of economic species (i.e. enterprises). Finally, the use of 'derived' parameters such as Thornthwaites' P–E–T and A–E–T can, like climatic means, be misleading; Thornthwaite, as noted, is used here for ease rather than universal accuracy.

Acknowledgements. For comments and criticims to J. Elston, L. P. Smith, R. Willey and K. Down.

* For maps and notes on the climatic factors influencing world distribution of crops and livestock, see Van Royen 1954 and Wilsie 1962; Whyte 1960 discusses the interactions between climatic factors, genetic factors and plant physiology. U.S.D.A. 1964 includes world maps of land-forms, annual average precipitation, primary groups of soils, natural vegetation, approximate cropland area, population, levels of food consumption, acreage and numbers of principal crops and livestock and world trade therein.

References

Azzi, G. 1956. *Agricultural Ecology.* London.

Black, J. N. 1956. *Archiv für Meteorologie, Geophysik u. Bioklimatologie*, Ser. B. **1**. 165.

Blackman, G. E. and Black, J. N. 1959. *Ann. Bot.* **23**. 131.

* Brichambaut, C. P. de and Wallén, C. C. 1963. *A Study of Agro-Climatology in Semi-Arid and Arid Zones of the Near East.* Technical Note No. 56. No. 141 T.P. 66. World Meteorological Organisation, Geneva.

Critchfield, C. J. 1967. Personal Communication. Western Washington State College, U.S.A.

Duckham, A. N. 1963. *Agricultural Synthesis: The Farming Year.* London.

Duckham, A. N. 1968. *Chem. and Ind.* 903–906

Elston, J. 1967. Personal Communication. Univ. Reading.

*Finch, V. C., Trewartha, G. T., Robinson, A. H. and Hammond, E. H. 1957. *Elements of Geography: Physical and Cultural.* New York and London. 4th edn.

*Gates, D. M. 1962. *Energy Exchange in the Biosphere.* New York.

*Gates, E. S. 1961. *Meteorology and Climatology for Sixth Forms and Beyond.* London.

Gleizes, C. 1965. *Bull. Technique d'Information des Ingenieurs des Services d'Agriculture.* Paris. **201**. 569–78.

Hilditch, T. P. 1951. 'Rational Grading of Seed Oils', *Chemistry and Industry*, p. 846.

Johnson, C. G. and Smith, L. P. (Eds.) 1965. *The Biological Significance of Climatic Changes in Britain.* London and New York.

*Kendrew, W. G. 1954. *The Climates of the Continents.* Oxford.

Knight, R. L. 1935. *Emp. J. exp. Agr.* **3**. 31.

Linton, D. L. 1965. *Geography.* **50**. 3. 197–227.

Meteorological Office 1958. *Tables of Temperature, Relative Humidity and Precipitation for the World.* Parts I to VI. (M.O. 617*a* to M.O. 617*f*). H.M.S.O. London.

Monteith, J. L. 1966. *Agric. Progress.* **41**. 9–23.

Penman, H. L. 1948. *Proc. Roy. Soc.* (A). **193**. 120.

Penman, H. L. 1962 (Woburn Irrigation 1951–59). *J. Agric. Sci.* **58**. 343, 349, 365.

Prescott, J. A. 1943. *Trans. Roy. Soc. South Australia.* **67**. 1–6.

Russell, E. J. 1957. *The World of the Soil.* London.

Smith, L. P. 1962. *Weather and Food.* Freedom from Hunger Campaign. Basic Study No. 1. World Meteorological Organisation. Geneva.

Smith, L. P. 1964. Meteorological Office, Agricultural Memoranda, Nos. 101, 102, 104.

Stanhill, G. 1963. *J. Inst. Water Engineers.* **17**. 1. 36–44.

*Taylor, J. A. 1967. *Weather and Agriculture.* London.

Thornthwaite, C. W. and Mather, J .R. 1955. 'The Water Balance', *Publications in Climatology.* **8**. 1. Laboratory of Climatology, Centerton, N. J., U.S.A.

Thornthwaite, C. W., Mather, J. R. and Carter, D. B. 1958. *Publications in Climatology.* **11**. 1. Laboratory of Climatology, Centerton, N. J., U.S.A.

Thornthwaite, C. W. and Associates, 1962, 1963 and 1964., 'Average Climatic Water Balance Data of the Continents', *Publications in Climatology.* Laboratory of Climatology, Centerton, N. J., U.S.A. **15**. 2 (Africa), **16**. 1 (Asia, excluding the U.S.S.R.), **16**. 2 (U.S.S.R.), **16**. 3 (Australia, New Zealand and Oceania), **17**. 1 (Europe), **17**. 2 (North America, excluding the U.S.A.), **17**. 3 (U.S.A.).

Turc, L. 1961. *Ann. Agronomiques INRA.* **12**. 1.

*U.S.D.A. 1964. *A graphic summary of World Agriculture.* Misc. Pub. No. 705. U.S. Dept. Agric., Washington, D.C.

*Van Royen, W. 1954. *The Agricultural Resources of the World.* New York.

Whyte, R. C. 1960. *Crop Production and Environment.* London.

*Wilsie, C. P. 1962. *Crop Adaptation and Distribution.* San Francisco and London.

Wolman, A. 1962. *Water Resources. A report to the Committee on Natural Resources of the National Academy of Science.* National Research Council. Pub. 10008, Washington.

Wynne, J. 1960. *J. Brit. Grassl. Soc.* **15**. 3. 216–19.

General Reading

(SEE ALSO ASTERISKED ITEMS)*

Beckinsale, R. P. 1957. 'The Nature of Tropical Rainfall, *Trop. Agr., Trinidad.* **34**. 76.

Gates, D. M. 1963. 'The Energy Environment in Which We Live', *American Scientist.* **51**. 3. 327–48.

Hills, E. S. (Ed.) 1966. *Arid Lands: A Geographical Appraisal.* London and Paris.

Lemon, E. 1963 in Evans, L. T. (Ed.) *Environmental Control of Plant Growth.* New York.

Lemon, E. 1965 in Steward, F. C. (Ed.) *Plant Physiology.* New York.

Rose, C. W. 1966. *Agricultural Physics.* London.

Russell, E. W. 1967. 'Climate and Crop Yields in the Tropics, *Cotton Growing Review.* **44**. 87.

CHAPTER 1.3

Vegetation, Soils and Air

Vegetation

Vegetation* is broadly the product of the climate, though the underlying parent rock may have some influence (see below). Hence, as Fig. 1.3.1 shows, there is a fairly close correlation between climate and vegetation and between these and the major soil groups (zones) of the world. Descriptions and excellent illustrations of types of natural vegetation are given by Riley and Young 1966.

Vegetation (Fig. 1.3.2) is important as a locating and intensity factor in four ways. Firstly, it may, especially in dry or seasonally dry grasslands or savanna areas, provide grazing, and even fodder for conservation, on which extensive livestock systems can be supported. Secondly, it may limit or has limited farming or its intensity because of the problem or cost of clearing the forest completely (e.g. eastern Canada, south-east Australia, the groundnut scheme in Tanzania in the late forties and the oak forests of England before iron axes became available), or because, for instance, grassland could not be used for tillage in the Interior Plains of the U.S.A. until the digger plough was evolved, or cleared Australian bush could not be cultivated until the advent of the stump-jump plough. Thirdly, when, especially in the tropics it is near cropped or grazed areas, local vegetation may provide habitats or hosts for crop and livestock pests and diseases (e.g. tsetse-fly) or predatory animals. Fourthly observation and analysis of the vegetation may give one some idea of the potential productivity, climate (Budowski 1966) and sometimes soils. Although this method has many pitfalls, for centuries farmers have used vegetation as useful guide lines; it is often a useful indicator of the plant production potential of an area (see Table 1.4.2 and p. 65). Replacing the natural plant cover by cropping may, however, have serious effects on hydrological factors (e.g. encourage floods) and promote soil erosion (Russell 1966).

Soil Zones and Soil Types

The concept of zonal, intra-zonal and azonal soils was developed by Russian pedologists in support of a hypothesis that each major climatic region of the world (Fig. 1.3.1) has a characteristic group or zonal type (Fig. 1.3.3) and that climate (and to a lesser extent vegetation) were more important than the parent rock in determining the soil zones. But this zonal concept is not very precise (Russell 1961, p. 565) and, in practice, the location and intensity of

* Vegetation here excludes farm crops, etc.; it means 'natural' vegetation as influenced by man.

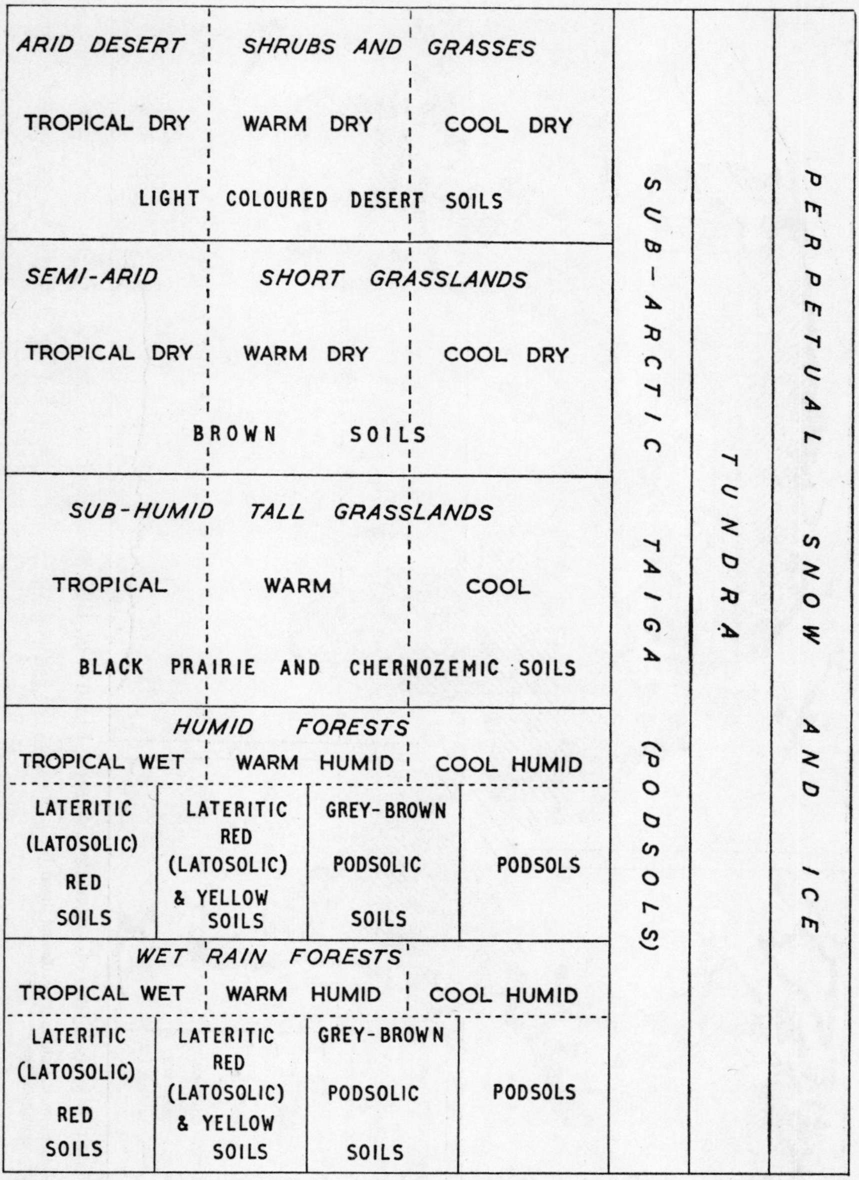

			SUB-ARCTIC TAIGA (PODSOLS)	TUNDRA	PERPETUAL SNOW AND ICE
ARID DESERT	*SHRUBS AND GRASSES*				
TROPICAL DRY	WARM DRY	COOL DRY			
	LIGHT COLOURED DESERT SOILS				
SEMI-ARID	*SHORT GRASSLANDS*				
TROPICAL DRY	WARM DRY	COOL DRY			
	BROWN SOILS				
SUB-HUMID TALL GRASSLANDS					
TROPICAL	WARM	COOL			
BLACK PRAIRIE AND CHERNOZEMIC SOILS					
HUMID FORESTS					
TROPICAL WET	WARM HUMID	COOL HUMID			
LATERITIC (LATOSOLIC) RED SOILS	LATERITIC RED (LATOSOLIC) & YELLOW SOILS	GREY-BROWN PODSOLIC SOILS	PODSOLS		
WET RAIN FORESTS					
TROPICAL WET	WARM HUMID	COOL HUMID			
LATERITIC (LATOSOLIC) RED SOILS	LATERITIC RED (LATOSOLIC) & YELLOW SOILS	GREY-BROWN PODSOLIC SOILS	PODSOLS		

Fig. 1.3.1 Correlation between climate, vegetation and major soil groups

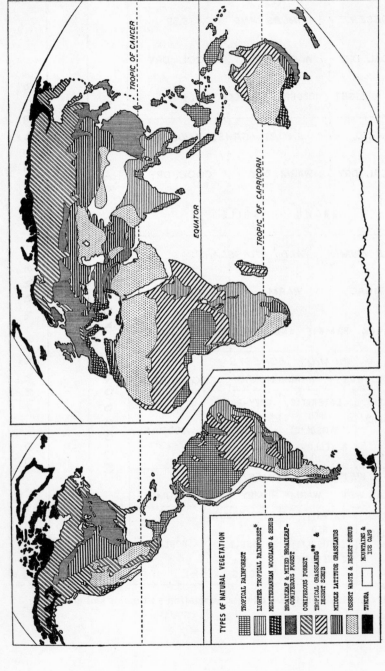

Fig. 1.3.2 World Map of Types of Vegetation. *Source:* Finch *et al.* 1957.

TYPES OF NATURAL VEGETATION

TROPICAL RAINFOREST
LIGHTER TROPICAL RAINFOREST*
MEDITERRANEAN WOODLAND & SHRUB
BROADLEAF & MIXED BROADLEAF–
CONIFEROUS FOREST
CONIFEROUS FOREST
TROPICAL GRASSLANDS** &
DESERT SCRUB
MIDDLE LATITUDE GRASSLANDS
DESERT WASTE & DESERT SHRUB
TUNDRA
MOUNTAINS &
ICE CAPS

TROPIC OF CANCER

EQUATOR

TROPIC OF CAPRICORN

* Includes semi-deciduous, deciduous, scrub and thorn
** Wooded savanna and savanna

farming systems is more influenced by climate, weather and local soil variations (soil types) than by the great soil groups.

Within each *zonal* group (Fig. 1.3.1), e.g. Chernozems, Podsols, there are
soil *series* with broadly similar and characteristic *soil profiles* (see illustrations
at p. 48 in Riley and Young 1966). Within each soil series, these *soil types*
differ primarily (*a*) in *soil depth*; (*b*) in *soil texture* (i.e. in the proportion of
sand, silt and clay particles) and *structure*, i.e. in the proportion of soil aggregates and in permeability to water and gases (structure is a less permanent
feature than soil texture proper; the proportion of larger soil aggregates of
good size is usually higher in soils that are high in organic matter); (*c*) in the
bio-geochemical status, i.e. in pH; in the content and availability of major
(N, P, K, Ca) and minor plant nutrients; in the fauna (e.g. worms) and flora
(bacteria, fungi) which break down organic matter; and in the infection with
soil-borne pests and diseases (e.g. potato root eel-worm); (*d*) in their *moisture
status*, i.e. their content of, or their ability to hold moisture and release it to
crops; (*e*) in their *stability*, i.e. their capacity to withstand wind or water erosion or to remain productive under saline conditions.

LOCATION EFFECTS IN TEMPERATE AREAS

As locating and intensity factors, (*a*) (soil depth), (*b*) (soil texture and structure) and (*c*) (bio-geochemical status) are in general more important than (*d*)
(moisture status) and (*e*) (stability) partly because the two latter are largely
resultants of (*a*), (*b*) and (*c*). Nevertheless, in some cases (*d*) and (*e*) can be
locationally critical. The locational influence of these five soil factors are
summarised in Fig. 1.3.4 and in Nos. 13 to 17 in Table 1.4.1.

Soil Depth

Shallow soils occur on slopes that have been badly eroded by natural forces
or by the action of man, and in heavily glaciated areas such as the Canadian
Shield (p. 124), and parts of Sweden. Such soils limit the depth of mechanical cultivation and crop rooting. Outcrops of rock, boulders or uncertainty
about the depth of soil, make mould-board ploughing or even disc ploughing
difficult and sometimes dangerous to the tractor driver (p. 88). Further,
these soils have less total bulk per acre, and hence tend to contain, at field
capacity, less moisture so that tillage or grassland yields are lower (or have a
higher coefficient of variation) than deeper rooting or even similar crops
grown on deeper soils. In other words, good soil depth can be a 'buffer'
in a crop-production system.

Soil Texture and Structure

Clay soils are hard to cultivate, primarily because of their high moisture
retention, their low permeability to water and the way they bake hard in hot,
dry weather. The heavy clay soils of England (Jackson *et al.* 1963) and northern Holland are partly, for this reason, safest in grassland where the moisture

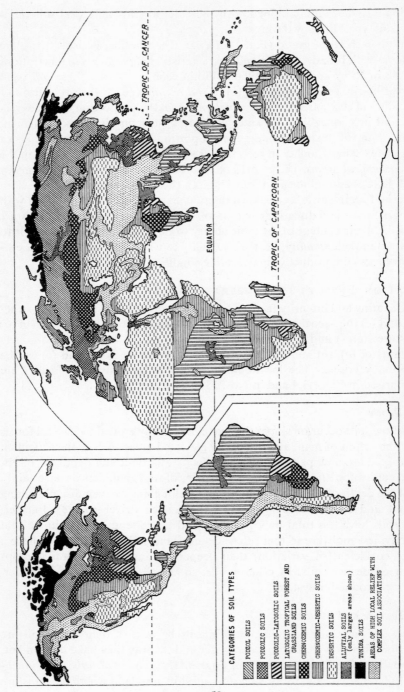

CATEGORIES OF SOIL TYPES

PODZOL SOILS

PODZOLIC SOILS

PODZOLIC-LATOSOLIC SOILS

LATOSOLIC TROPICAL FOREST AND
GRASSLAND SOILS

CHERNOZEMIC SOILS

CHERNOZEMIC-DESERTIC SOILS

DESERTIC SOILS

ALLUVIAL SOILS
(only larger areas shown)

TUNDRA SOILS

AREAS OF HIGH LOCAL RELIEF WITH
COMPLEX SOIL ASSOCIATIONS

TROPIC OF CANCER

EQUATOR

TROPIC OF CAPRICORN

Fig. 1.3.3 World Map of Categories of Soil Types. *Source:* Finch *et al.* 1957.

retention in late summer may be some protection against drought. But clay soils are often high in chemical status and yield well under cereals, especially wheat. Moreover, good management, land drainage and highspeed powerful modern machinery have extended the range of crops on clay soils to include e.g. potatoes, sugar-beet, etc. However, in countries where land is plentiful and manpower expensive heavy clay soils are sometimes just not used at all, e.g. some of the 'gumbo clays' in the U.S.A. Low permeability tends to make clays less receptive to rain than, for example, sands; so there is usually more loss of water by run-off on clay soils on slight slopes whilst if the slope is severe and the rainfall intense, they are more liable to water erosion than coarser-textured soils (Wilsie 1962, p. 271).

Sandy soils are more receptive of rain but more liable to wind erosion than clays. Free-draining sandy or gravel soils do not readily retain water, so may dry out early in the summer and tend to have a lower chemical status than clays. They may not be suitable for intensive grassland except in areas with high A–E–T which are well-watered or are irrigated. They can, especially if well fertilised, give good yields of a wide range of horticultural and tillage crops; their tendency to high variability of crop yield can often be controlled by irrigation. They are easier to work (i.e. are more cultivable and more trafficable than clays). Jutland in Denmark, and the 1,500-mile eastern seaboard of the U.S.A., are examples of what well-managed sandy soils can do. Loam soils, with a balanced mixture of sand, silt and clay particles, are very popular with farmers. Fen (or 'muck') soils (e.g. in the English Fens), with their high organic matter, are also favoured, but are sometimes liable to wind erosion, whilst the risk of winter and spring floods may limit choice and intensity of use.

Bio-geochemical Status; Major Nutrients

Under natural forest, part of the leaf and other litter that rots on the forest floor becomes incorporated (e.g. by the action of earth-worms) in the surface soil and is ultimately mineralised by the action of the soil flora. As many forests have high precipitation, which is in excess of P–E–T, the nitrogen and other nutrients are leached down. When the forest is cleared for cultivation, much of this *nitrogen* (N) may be 'out of the reach' of the roots of tillage and grass plants and N may be deficient in the surface soil unless it receives urine and faeces from livestock grazed or fed locally or is dressed with fertiliser N. Arid soils are also naturally low in N, partly because the climate cannot support much vegetation unless irrigated. Until the advent of the fertiliser industry, nitrogen was, therefore, almost certainly the major nutrient most limiting to crop production in the world.

The main exceptions to this constraint were the regions with temperate continental grassland climates. Here the shortage of moisture in summer delayed the rate of mineralisation of organic matter by the soil micro-flora, whilst the frozen ground in winter prevented leaching. So, although their

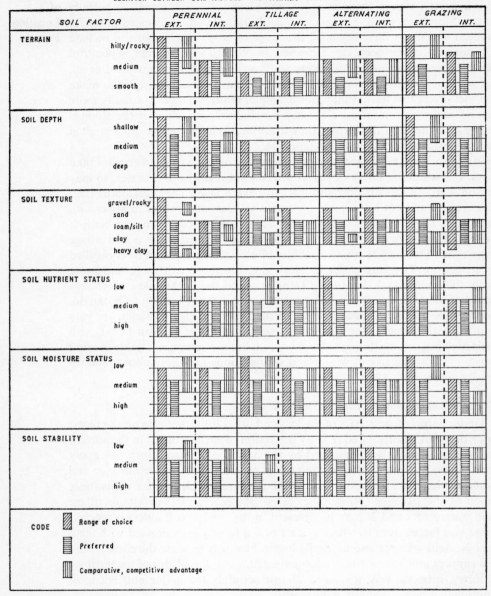

RELATION BETWEEN SOIL FACTORS AND FARMING SYSTEMS

Fig. 1.3.4 Relation between Soil Factors and Farming Systems

54

mean A–E–T is not high, the N content of the Interior Plains of the U.S.A. and Canada, the Steppes of Russia, the Pampas of the Argentine and the natural grasslands on the east of the South Island of New Zealand, is or was relatively high. But continuous tillage cropping in such areas reduces N; firstly, because pre-drilling cultivations stimulate mineralisation at a time of the year when some leaching is likely and secondly, by the sale of the resultant cereal crops. Thus, at an experimental farm in the Canadian Prairies, the N content of the soil fell in 22 years from 6,940 lb/N/acre to 4,750 lb/N/acre—a loss of 2,200 lb/N/acre, of which only 700 lb/N/acre were recovered in crops. This loss of one-third of the N was paralleled by a similar loss in total organic matter, though the soils still remain eminently cultivable and well structured. In temperate areas, except where fertiliser N is expensive (e.g. Australia), or where legume N is an unreliable source (e.g. on heavily leached, acid soils), nitrogen is not now a major factor in system or enterprise location though it may influence intensity.

Under natural conditions phosphate (P_2O_5) in plants and animals comes (unlike nitrogen) from parent rock, and not from the air. P_2O_5 is deficient or unavailable, particularly for intensive cropping and stocking, in many parts of the world. It affected the location and intensity of grassland farming in New Zealand, Australia and South Africa, where 40 years ago, Theiler's classic work showed that a serious livestock disease was attributable to the low P content of the sward on which they grazed—a sward which also had low N. In Australia, the correction of this P_2O_5 deficiency encouraged subterranean clover (*Trifolium subterraneum*) which built-up legume N, led to good swards, an effective fat lamb industry and a viable alternating system, viz. wheat and pasture (for fat lamb). This system revitalised the falling yields on the ex- hausted single-crop wheat lands and established a 'wheat–sheep' system which now dominates Australian wheat production (p. 183). Much the same occurred in New Zealand (p. 199).

Potash (K_2O), though deficient in many soils, rarely seems to act as a locat- ing factor of either farming systems or enterprises. *Calcium* (Ca) is an im- portant plant and animal nutrient associated especially (in combination with P_2O_5) with bone formation. As a base, it tends to reduce soil acidity and raise pH. In humid climates, calcium and other bases get leached through the soil and, unless these deficiencies are remedied, they may limit stocking rates on grassland or the range and yields of tillage crops. Thus oats, potatoes and grasses are more tolerant of lime deficiencies than are sugar-beet, wheat and legumes (and especially alfalfa/lucerne, which demands a higher soil pH), whilst the temperate legumes also have a high physiological need for calcium. Fortunately in most parts of the temperate world, lime, where needed, is available near by, and though Ca and pH may influence the choice and in- tensity of enterprises *within* systems it has not had such dramatic effects on farming patterns as N and P_2O_5. In some arid areas calcium may be in excess in the soil.

Minor elements. In addition to the carbon, oxygen and hydrogen, and to the above macro-nutrients (N.P.K.),* sulphur, magnesium, iron, manganese, copper, zinc, sodium, molybdenum, chlorine, vanadium and cobalt are needed by plants. All these,† plus iodine, have an essential function in the animal body, whilst fluorine, selenium and chromium are increasingly regarded as necessary in livestock nutrition (Maynard and Loosli 1962, p. 121). Geographically, the distribution of these minor elements in the soil is very variable. Deficiencies or excesses affecting plants, and through them livestock, have substantial influences on enterprise choice and intensity. Examples of such imbalances are the responses of sugar-beet and wheat to manganese on fen (high organic) soils in the United Kingdom, of wheat to sulphur in the Canadian Prairies, and of many fruits to various elements in the major fruit-growing areas of the world. In ruminants, especially grazing dairy cows, hypomagnesaemia is a frequent and sudden, but now largely avoidable, trouble in north-west Europe, the Pacific North West of the U.S.A. and elsewhere; this deficiency may be associated in some way with N and K intakes. In Australia, New Zealand, the U.S.A. and the United Kingdom, research has eliminated 'ill thrift' and other diseases, especially in sheep, due to lack of cobalt or copper, or both. In parts of New Zealand, the administration of selenium commercially increases the lambing rate. In Australia, lack of some minor elements may be absolute constraints on farming systems. Thus, there are areas which without small dressings of zinc will not produce either tillage crops or pasture swards.

Some trace elements occur in excess; for instance, selenium in the Interior Plains of the U.S.A. and molybdenum in the 'teart' pastures of Somerset in the United Kingdom and Florida in the U.S.A. There are, also, often complex interactions between elements. Thus, molybdenum poisoning can be stopped by the administration of copper (which is dangerous in excess), whilst copper poisoning (which can occur in sheep fed largely on concentrates) can be corrected by molybdenum. Sometimes an excess of calcium may reduce the needed uptake of manganese by crops. As a United Kingdom jingle has it:

> *The Government, in times of ease,*
> *Pays farmers large lime subsidies,*
> *Which starve our crops of manganese*

Thus minor elements can be significant locating factors.

Saline soils. Finally, there is, in arid-irrigated areas, the grave problem of salinity which affects tens of millions of acres and is, for instance, creating grave social and economic difficulties in Pakistan and elsewhere. Under these conditions, salts may rise from the lower profiles to the soil surface. If drainage is good, salts can be kept low by flushing 'accumulations of soluble salts

---

* In agricultural literature, P_2O_5 and K_2O are often shortened to P and K.
† Except vanadium.

from the surface soil into the ground water' (Russell 1961, p. 608) in the non-
growing season, when P–E–T is low, and subsequently by using irrigation
water with a low salt content, and also by maintaining ground-water
level at well below the soil surface. In bad cases, acidifying agents, such as
powdered sulphur or gypsum (Ca So_4, $H_2 O$) can be used along with special
water and cropping regimes. High salt, especially sodium, content destroys
the structure and lowers the permeability of the soil and reduces the power of
crop plants to absorb water; the plants are dwarfed and stunted. Crop plants
vary widely in their tolerance of salts. Allison 1964, summarised the tolerance
of various crops to salinity, sodium and boron. Thus, barley and rice are
more tolerant than maize or wheat, lucerne more tolerant than white clover.
Boyko 1967, discusses salt-tolerant species that might be grown with salty or
sea-water on sandy soils. Salinity is thus an important locating and intensity
factor. It influences what can be grown and the intensity of inputs, especially
where enough water (and hence fertilisers) can safely be applied to make full
use of the P–E–T. In some parts of south-west Asia, it has probably led to the
abandonment of land for farming.

AIR, SOIL AND WATER POLLUTION IN TEMPERATE AREAS

Air pollution. Although, of course, there are some saline soils that are not
man-made, in many, if not most, cropped areas they are. Somewhat analogous
man-made toxicity occurs in industrial countries from car exhaust fumes and
through smoke from coal fires and furnaces (and from industrial plants, e.g.
brick-works), which may not only reduce the light energy reaching the plant,
but sulphur oxides in the effluent may lead to crop damage and soil acidity;
in some parts of industrial western Europe rainfall has a pH as low as 4·0
(*The Times*, 27 November 1967). Other elements occurring in atmospheric
industrial effluents, e.g. nickel, zinc, chromium, manganese, may have adverse effects on crop plants though tolerances vary, e.g. sugar-beet is more
susceptible than grasses to these. In the U.S.A., air pollution, mainly due to
fuel combustion and car exhaust fumes, is estimated to have cost agriculture
$500 million in 1966. The main pollutants, especially in photo-chemical
'smog', appear to be ozone (which damages certain types of tobacco), sulphur
dioxide, nitrogen oxide and pheno-oxyacetyl-nitrate (PAN). Near urban
areas the problem has become so great that, for example, in parts of New
Jersey spinach can no longer be grown. Accumulations in crops and grass
may lead to livestock disease, e.g. fluorosis in cattle (Blood 1963, Webster
1967), but once identified, the correction of these toxicities and excesses is
usually practicable, though it may involve the avoidance of particular enterprises (e.g. no grazing cattle near brick-works.) To this extent, though costly,
these toxicities *now* have only minor effects on the location and conduct of
farming systems, but their influence may increase. On the other side of the
coin, farm pollutants (e.g. of pesticides, animal wastes) may create problems
for the rest of the community (Brady 1967).

Soil and potential land use. Soils are, with climate, relief and other physical factors, important in determining the relative actual or potential value to the community of a given site for farming, urban housing and industry, recreation, water storage, airfields and roadways, etc. In densely populated areas such as the Netherlands, the United Kingdom or the north-east of the U.S.A., these often competing demands may influence the present and future location or intensity of farming systems or enterprises (p. 500). As a basis for land-use policy, several methods of assessing land-use capability are in use; these are based on physical and/or economic data and in some cases operational ease (Chap. 1.6). The Ministry of Agriculture for England and Wales recently reviewed six systems used in various countries and recommended a seventh for agricultural land. (M.A.F.F. 1966—this reference has a useful short bibliography). Bartelli *et al.* (1966) reviewed the problem in the U.S.A. But all the assessment methods involve subjective (value) judgements and further research is needed to put them on a quantified scientific or an econometric basis.

THE CHANGING LOCATIONAL INFLUENCE OF SOIL FACTORS

In advanced temperate countries, modern science and technology, and their input industries, have reduced and will further reduce the influence of soil factors in determining the location and intensity of farming systems. Jacks (1956, 1962) argues that energy-rich industrial civilisations raise soil fertility through fertilisers, genetic improvement, herbicides, mechanisation, irrigation, etc. He does not see soil fertility as limiting world population. In fact, apart from irrigated areas with warm dry (Mediterranean) climates, the highest tillage crop yields are found on initially unfavourable podsolised soils in forest climates with moderate to high A–E–T's, i.e. on soils on which it is easier, given adequate industrial inputs, to raise crop yields than in the 'naturally' fertile grassland chernozemic soils where A–E–T is low. Thus, in the podsolised, but well-watered south-east of the U.S.A., cereal production (Wilsie 1962, p. 288) increased by 25% from 1935 to 1950 compared with 14% in the short grass, better soiled, Great Plains of the U.S.A. The biggest rises in crop yields in the U.S.A. have come from naturally infertile soils (Wilsie 1962, p. 287). Technology and industrial inputs are, in advanced economies, diminishing the yield difference due to soil type. In advanced countries the importance of the land and of other traditional inputs are falling; those of technological inputs of off-the-farm origin are rising (Schultz 1964, Duckham 1966). The story, as shown below, may be very different in less-developed countries; here the nature of the soil was, and still is, more important to the production of food and fibre than in modern industrial societies.

SOIL AND VEGETATION IN THE TROPICS

The first generalisation that can be made about the soils of the tropics is that they are mainly poor. This is partly because most of the land masses in

the tropics have geologically very old surfaces, which have been subject to
long ages of weathering and leaching. High temperatures and often high
humidity have also promoted these processes and the result is a general poverty in plant nutrients, and particularly in nitrogen and phosphorus. The organic matter content of cultivated tropical soils is usually low, owing to the
rapid degradation of such material under tropical conditions. Corbet (1935)
has restated Jenny's law to the effect that the nitrogen and organic matter
content of soil varies inversely with temperature and the amount of solar
radiation, so that above 78°F the decomposition of humus proceeds faster
than its formation. There are, however, many exceptions to this rule of low
fertility in tropical soils. Some volcanic soils in the tropics are very fertile, as
in Indonesia, parts of Africa, and some Pacific and West Indian islands.
Underlying limestone rock, which is rare in the tropics, provides some good
soils in south-east Asia and the West Indies; some excellent soils are derived
from coral, as in Barbados, but raw coral soils can be most intractable for
farming, as in parts of Zanzibar or the Pacific atolls.

In the great world groups of soils (Fig. 1.3.1), the tropics chiefly figure as
providing the locus for latosols (Fig. 1.3.3). These are a range of soils whose
extreme form, viz. true laterite characterised by a very low ratio of silica to
iron and alumina, is very infertile. Partially laterised soils of red colour and
easy to work, although often having a high clay content, are of much more
extensive occurrence in the tropics. Some of them are of excellent fertility, as
in parts of Kenya and Uganda, whilst West African examples are generally
less fertile. The kaolinite clay fraction of these soils has less ability to store and
release plant food than the montmorillonite clays of the temperate regions
(Swanson 1955).

But there are many other soil types in the tropics. Soils derived from
ancient granites and gneisses are common and not very fertile as is exemplified
in Malaya. In Africa vast stretches of rather poor sandy soils stretch from
Rhodesia to the Congo basin, and in South America occupy much of the
Amazon and Orinoco basins, rendering these regions less attractive for agricultural development. Tropical black clays, sometimes known as 'black
cotton soils', occur locally in poorly drained situations; they are of fair fertility but extremely difficult to work because they become so sticky when wet
and so hard when dry. Alluvial soils are important in some countries and may
be of good fertility, as in India where they support immense masses of population along the valleys of the Ganges and other Indian rivers. Peat soils are
found in wet tropical climates in Indonesia, Malaya, Guyana and elsewhere
and are useful for agriculture but unsuitable for perennial tree crops, which,
as the peat shrinks with cultivation and drainage, often lose their roothold.

The soils of the sub-tropics differ from those of the tropics proper, chiefly
in that so many of them are formed under conditions of very low rainfall. In
a desert environment, with no plant growth and no downward movement of
water in the soil, soils have a featureless profile and little structure. The pre-

dominance of evaporation over drainage leads to a concentration of salts in the upper layers of the soil. This sometimes produces alkali soils (p. 56), which are useless for agriculture; but many desert soils, if irrigation can be provided, are found to be remarkably fertile.

The tropical farmer has to work against a background of three main vegetation types. The first of these is the tropical rain forest, which is the natural vegetational climax in three great regions: equatorial South America, western equatorial Africa and south-east Asia from Indonesia northwards to Burma. The clearing of these forests is excessively laborious for the sparse populations who inhabit them, especially with the crude hand-axes which are still usually the only tool available. These areas, therefore, constitute to a large degree the last great reserves of cultivable land which have not yet been developed for agriculture, especially in the basins of the Amazon and Orinoco, the Congo basin, and a region straddling the frontiers of Burma, Thailand and Laos. Conditions are particularly unfavourable for animal husbandry, since there is no natural pasturage, and even if clearings are made for grassland, a constant struggle (for which the usually low level of animal production will hardly pay) has to be waged to prevent their re-invasion by woody plants.

The second vegetation type is the savanna, a community of mixed trees and grasses which overlaps the boundary of what we have defined as the 'wet' and 'seasonally dry' tropics. Under the high rainfall conditions, trees predominate to the extent of forming a rather open woodland; where rainfall is lower, the vegetation assumes the aspect of grassland with scattered trees. Savanna is the vegetation type which covers most of the tropics, but is not always a natural climax, since it is thought to be in many areas a sub-climax which is prevented from developing into forest by man's activity in grazing and burning, grass in many regions being deliberately fired annually by graziers partly with the object of preventing tree growth. The savanna is easily enough cleared for shallow cultivation, and these areas contain most of the cattle population of the tropics.

The third vegetation type is the desert scrub which occupies vast areas, particularly in Africa and the Middle East, often around the fringes of the true deserts. In this plant community, small shrubs are more important than grasses. Crop production is precarious or impossible owing to the low rainfall. The grazing potentialities are best used by animals which are better browsers than cattle, such as goats or camels which are, however, destructive of their own environment if overstocked. Some have thought that greater production can be obtained in these areas, and in some of the drier savannas, from wild game animals suitably managed than from domestic livestock, but this remains to be proved experimentally.

Technology in the tropics. The use of fertilisers to remedy the poverty in plant nutrients of tropical soils is an obvious step, although they do not always show their full potentiality in this environment owing to the rate at which they are leached in the wet tropics and the frequent absence of moisture

to enable crops to give their full response in the seasonally dry tropics. Nevertheless, many experiments have shown their high value, and their use is slowly
expanding from the plantation industries to small farmers. The chief deter-
rent is the high cost of fertilisers, especially where much inland transport is
involved, compared with the value of the crops which can be produced.

The luxuriant growth of weeds, particularly in areas which are perennially
hot and wet, is one of the major problems of tropical agriculture, necessitating
much labour in their control. Hence the use of selective herbicides appears in
the long run to have an even brighter future in tropical than in temperate
countries, though at present only a bare beginning has been made. The use of
modern mechanical equipment for the clearance of vegetation in forest and
savanna areas would appear, in theory, to make the development of these
areas much easier. However, experience in the East African groundnuts
scheme and elsewhere has shown that these methods are still very expensive
compared with the output gained, and hand clearance still remains the predominant method. An additional technical burden is placed upon tropical
agriculture by the risk of water erosion wherever the land is not completely
flat. The necessity may be for extra skill, as in the case of contour ploughing,
or for extra labour, as in the work of terracing. Science has done its work in
elucidating methods of soil conservation, but this has merely made the additional expenses more obligatory.

References

Allison, L. E. 1964. *Advances in Agronomy.* **16.** 139.

Bartelli, G. J., Klirigebiel, A. A., Baird, J. V. and Heddleson, M. R. (Eds.) 1966.
Soil Surveys and Land Use Planning. Soil Science Society of America and American
Society of Agronomy. Madison, Wis.

Blood, J. W. 1963. *N.A.A.S. Quarterly Review.* **14.** 50. H.M.S.O. London, p. 97.

Boyko, H. 1967. *Scientific American,* **216.** 3. 89–96.

*Brady, H. C. (Ed.) 1967. *Agriculture and the Quality of our Environment.* (American
Ass. Adv. Sci.) Washington, D.C. and London.

Budowski, G. 1968 in *UNESCO Agro-Climatological Methods.* Natural Resources
Research, No. 7. UNESCO, Paris.

Corbet, A. S. 1935. *Biological Processes in Tropical Soils with Special Reference to
Malaysia.* Cambridge.

Duckham, A. N. 1966. *J. Royal Agric. Soc.* **127.** 7–16.

Jacks, G. V. 1956. *Adv. Sci.* **13.** 50. 137–45.

Jacks, G. V. 1962. *J. Soil and Water Conservation.* **17.** 4.

Jackson, B. G., Barnard, C. S. and Sturrock, F. G. 1963. *The Pattern of Farming in
the Eastern Counties,* Occ. Papers No. 8. Farm Econ. Branch, School of Agric.,
Univ. of Cambridge.

M.A.F.F. (Ministry of Agriculture, Fisheries and Food). 1966. Agricultural Land
Service. Technical Report No. 11. *Agricultural Land Classification.* London.

Maynard, L. A. and Loosli, J. K. 1962. *Animal Nutrition*. New York, 5th edn.
*Riley, D. and Young, A. 1966. *World Vegetation*. Cambridge (England).
*Russell, E. W. 1961. *Soil Conditions and Plant Growth*. London. 9th edn.
Russell, E. W. 1968 in *UNESCO Agro-Climatological Methods*. Natural Resources Research, No. 7. UNESCO, Paris.
*Schultz, T. W. 1964. 'Transforming Traditional Agriculture', *Studies in Comparative Economies* No. 3. New Haven and London.
Swanson, C. L. W. 1955. 'An Inside Look at the Soil', *World Crops*. 7. 247.
U.S.D.A. 1967. *Agric. Research*. 15. 7. U.S. Dept. of Agric. Washington, D.C.
*Wilsie, C. P. 1962. *Crop Adaptation and Distribution*. San Francisco and London.
*Webster, C. C. 1967. Agric. Res. Council. *The Effects of Air Pollution on Plants and Soil*. H.M.S.O. London.

* Asterisked items are for further reading.

CHAPTER 1.4

Ecological Interactions, Productivity and Stability

Ecological Locating Factors

Chaps. 1.2 and 1.3 showed how climatic factors interacted with each other and how they influenced soil and vegetation which also, to some extent, interacted. We now need to consider how these actions and interactions influence the location and intensity of farming systems. Rows and columns 1 to 17 of Table 1.4.1 list the major ecological factors and show, in the relevant cells, which ones act upon each other under natural, non-farming conditions. The influents are listed on the left and the factors influenced across the head of the table. Thus cell 4·7 shows that relief aspect (e.g. southern face) influences solar energy receipts. Columns 18 and 19 show which of the factors listed on the left influence or determine local vegetation and local fauna (including predators, pests and diseases of crops and livestock), whilst column 20 lists those which influence the net primary production of vegetation (Table 1.4.2). Resultant items 18, 19 and 20 are, in effect, the integration or synthesis which evolution has made over the millennia for any given locality. Rows and columns 21 to 24 show which of factors 1 to 20 influence the choice of farming systems and also the effect of man as a farmer on these natural conditions. Thus cell 22·16 shows that choice of crop influences soil moisture. Numbers 21 to 24 are the resultants of influent numbers 1 to 20. But these resultants are, as shown in Chaps. 1.1, 1.5, 1.6, 1.7 and Parts 2 and 3, also very substantially influenced by non-ecological factors. Table 1.4.1 is necessarily a subjective and rather arbitrary summary. It will repay careful scrutiny because a farming system may be regarded, ecologically, as a function of these twenty-three summary factors and their interactions, and because it shows, better than many words, the number and complexity of the relevant interactions, and is more meaningful than a *complex* model which cannot, as yet, be quantified (but see Chap. 3.1, for some simple models).

SUMMARY

The interactions shown in this table and the content of Chaps. 1.2 and 1.3 may be summarised thus:

(1) The correlation between the major climatic regions, soil groups (zones) and vegetation regions of the world is fairly close (Chaps. 1.2 and 1.3). Climate emerges as the dominant and, in most cases, the causal partner of the triad.

Part 1
Analysis
of
Location
Factors

(2) *Climate.* The major climatic variables influencing a farming system are numbers 7 to 12. At a given site, the major natural buffers to climatic stresses are the survival capacity of the crops and livestock, aided by the soil factors (13 to 17). In temperate farming, these buffers are enhanced by industrial and scientific or by inter-farm inputs.

(3) *Soil.* But on the local scale, soil (Nos. 13 to 17) may determine what crops are ecologically practicable, though science and industrial inputs are making soil factors less important as constraints (p. 58).

(4) Nevertheless, the interactions between climate and soils remain important as locating factors. Thus, in an area of high or moderate precipitation, heavy-textured moisture-retaining soils (14), even if cultivable, may be so slow

Table 1.4.1

Summary of Ecological Factors and their Effects on Farming Systems

Code: Small action — Roman figures   Medium action — Italic figures
Large action — Bold figures

Ecological factor	No.	6	7	8	9	10	11	12	13	14	15	16	17	18	19	20	21	22	23	24
Influents																				
Latitude	1	**6**	7	8	9	10	*11*	12				16		18	19	20	21	22		24
Intra-continental Location	2		**7**	**8**	**9**	10	*11*	**12**			15	16		18	19	20	21	22	23	24
Relief–Altitude	3		7	8	**9**	10	*11*	12			15			18	19	20	21	*22*	23	24
Relief–Aspect	4		7	8	?9	?10	*11*	12			15			18	19	20	21	*22*	23	24
Relief–Slope and Surface	5a		7	8	9	*10*	*?11*	*?12*	*13*		15	16	17	18	19	20	21	*22*	23	24
Relief–Barrier Effect	5b			8	9	10		12						18	19	20	21	*22*	23	24
Day Length	6		**7**	8				12						18	19					
Solar–Energy Receipts	7			8	9		*11*	12			15	16		18	19	20	21	*22*		24
Temperature	8				9	*10*	**11**	**12**			15	16	17	*18*	19	20	21	**22**	**23**	24
Precipitation	9							**12**			15	*16*	17	**18**	19	20	21	*22*	23	24
Wind	10						11	**12**				16	17	18	19	20	**21**	22		
P–E–T	11							**12**				16		18	19	20	*21*	22	23	24
A–E–T	12										15	16		*18*	19	20	*21*	22	23	24
Soil–Depth	13										15	16*		*18*		20	*21*	22	23	24
Soil–Texture	14										15	16*	17	18		20	*21*	22	23	24
Soil–Bio-geochemical	15													18	19	20	*21*	**22**	23	24
Soil–Moisture Status	16							12*		14	15		?17	18	19	20	21	**22**	23	24
Soil–Stability	17													?18			21	22	23	24
Resultants																				
Vegetation	18									14	*15*	16	17		19	20	21		23	24
Fauna	19										15		17	**18**			21		23	24
Net Primary Production of Vegetation	20																	22	23	24
Potential Farming Productivity	21																	22	23	24
Actual Crop Enterprises	22										*15*	*16*	*17*						23	24
Actual Livestock Enterprises	23										*15*		*17*					22		**24**
Actual Farming System	24										*15*	*16*	*17*							

* Through soil moisture reserves and their availability and in the case of the action of 10 on 12 through its influence on transpiration.

This table may be regarded as a tabular expression of a catena or series of catenae. Each cell shows whether the relevant factor named at the left of the row acts on the factor identified at the top of the columns and if so, how strongly. Only direct actions, i.e. apparently causal relationships (or closely correlated even if not causally related events) are shown in the cells. Indirect effects for example of latitude on choice of farming systems are, if only for simplicity, omitted. Farming is such a varying, complex web of interacting forces that if one included all the secondary, tertiary and more remote effects of an action, there would be few empty cells in, for instance, Table 1.7.1. Thus, it could justifiably be argued that economic infra-structure acted on water-control facilities and that this should be shown in cell (10·4) of Table 1.7.1. But the more direct influences on water control (column 4) are climate, land, soil regime, operational facilities, agrarian structure (e.g. size and layout of farms), social structure (e.g. attitude to the social discipline involved in irrigation or drainage) and the price of water (rows 1, 2, 3, 5, 7, 9, 11). Of these, all except climate, land and possibly soil regime, are influenced by economic socio-economic factors.

to dry out in the spring that only light sandy soils which dry out quickly are cropped. In a drier area, the low moisture-retaining capacity of sandy soils (16) may preclude tillage, whereas the moisture-holding capacity of heavy soils is an advantage. Or again, on slopes (5a), on clay soils (14), or on loess soils (that are liable to cap) much of the rain may run-off or the soils may themselves be eroded.

(5) *Vegetation* may be of some use to the farmer as an indicator of productivity. If the climate is suitable, forest can, and in fact largely has been in tillage or cultivable grassland where it is or has been economic to do so. If the relief, or the local price situation, is unfavourable for tillage, then grazing livestock are usually found on 'natural' grasslands. Overgrazing may, however, expose the soil surface and thus permit wind or water erosion (17). So may the clearing of forests, if the trees are not replaced by crops or other controls which check erosion. To this extent, there is, in many parts of the world, considerable interaction between vegetation and soil—an interaction generated by climatic and aggravated by human factors.

(6) *Relief.* Local climates may, of course, differ greatly *within* climatic regions due to altitude (3), aspect (4), slope (5a) and barrier effect (5b). Thus, though soils, aspect and altitude may be the same, slope effects (5a), e.g. on frost drainage,* may make citrus, for instance, practicable on one slope but not on another only a few hundred yards away, which has, therefore, to be used to grow olives or some other 'second choice' crop (see Israel, p. 311).

(7) Thus, (*a*) climate × soil, and (*b*) climate × soil × vegetation interactions are important, especially where erosion is a problem (17) or where the moisture receptivity and retention capacity of a soil (16) is marginal for agriculture.

Temperate Productivity Parameters

Can the causal ecological relations and interactions summarised above be quantified? The answer, broadly, is 'No'. However, amongst the 17 influents in Table 1.4.1 the most statistically available and perhaps ecologically useful seem to be P–E–T (11), Precipitation (9) (and its variability), A–E–T (12) and the length of the growing season. Each of these (except the last) can be expressed as millimetres or inches of water and are, therefore, used in Part 2. Soil texture (14) is also important and can sometimes be quantified.

Thornthwaite's P–E–T (or other authors' formulae for the estimation of mean potential evaporation and transpiration) in the thermal growing season (T–G–S) in effect integrates, for a given site, factors 1 to 8. Precipitation data are needed to estimate A–E–T and may, where P–E–T is always non-limiting, e.g. in much of Australia, be more useful than A–E–T. Annual mean precipitation (as distinct from that in the T–G–S) is used here because some of it

---

* i.e. the movement of heavier cold air downhill often to valleys which are 'frost pockets'.

helps to bring the soil moisture up to field capacity in the non-T–G–S. The variability of precipitation is discussed below (p. 70). Thornthwaite's A–E–T (or other models for estimating mean effective transpiration) (p. 40) in the thermal growing season is, with qualifications, a useful single climate parameter. It indicates either (a) the amount of moisture for which light energy is available for plant production (e.g. in deserts) or (b) the amount of light energy which is available to make use of precipitation for plant production (e.g. in Japan or the uplands of Britain) where precipitation substantially exceeds energy receipts expressed as P–E–T in the T–G–S. But, as shown in Chap. 3.1 an excess of P–E–T *or* of precipitation over A–E–T may be a key locating factor. Soil texture is a useful indicator because of its role in soil moisture, storage and release, because light soils are often poor in bio-geochemical status (15) and because (see Chapter 1.6) it affects cultivability, drainage problems, etc.

Table 1.4.1 suggests that nearly all the first 19 influents act on net primary production (20) which, as it were, synthesises the ecological factors and as such should be a useful parameter. Unfortunately, such measurements are not generally available, but Odum 1959 has collated the available data, some of which are given in Table 1.4.2. This is an attempt to show the relation between Annual Precipitation (9), P–E–T (11) and A–E–T (12), on the one hand, and net primary production of vegetation (20) and wheat yields, on the other. It indicates, in broad terms, the correlation between natural net primary production of vegetation and mean farm yields of wheat grain. The relation between natural and augmented (by irrigation) A–E–T and crop yields is shown in Fig. 1.4.1 for a U.S. transect. Note the falls in yields (except in the Pacific North West, p. 147, or where irrigated) west of 95°W as A–E–T falls.

Where net primary production or the A–E–T is high, or can be made high by irrigation, then the temperate farmer usually has a wide range of farming systems and intensities of input to choose from, unless precipitation is excessive. Where net primary productivity and A–E–T are low, his choice is normally limited to extensive grazing or extensive tillage systems, with low inputs, and he compensates by farming a large enough acreage to make a family living (see also Chap. 3.1).

ECOLOGICAL PROBLEMS IN THE TROPICS

When we put together the factors conditioning tropical agriculture which we have considered in Chaps. 1.2 and 1.3, they go far to justify the view that the tropics provide a relatively unfavourable environment for agriculture. It may be well at this point to summarise the reasons which enable such a view to be held:

(1) *Rainfall.* Over large areas of the tropics this is so low and unreliable (p. 37) as to make crop production a risky business. Where it is sufficient, it falls

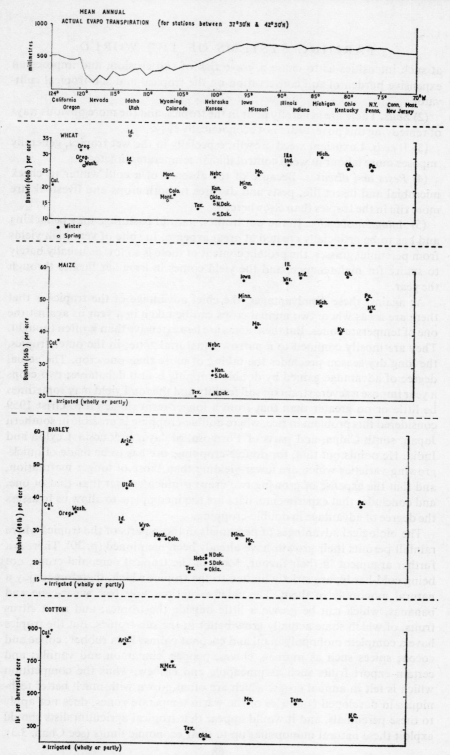

Fig. 1.4.1 Relationship between Actual Evapo-transpiration (Thornthwaite) and Average Crop Yields (1963/64), U.S.A., from U.S.D.A. Agric. Stats. 1965

at such intensities as to cause a grave risk of soil erosion, and imposes an expensive burden of soil conservation on the tropical and sub-tropical culti-vator.

(2) *Soils*. These are generally poor in the tropics, and the more obvious ways of remedying this poverty are not economically easy.

(3) *Weeds*. Luxuriant weed growth, especially in the wet tropics, generally mposes more labour in weed control than in temperate climates.

(4) *Pests and diseases*. Because of the absence of a cold winter to check microbial and insect life, pests and diseases of both crops and livestock are more rife in the tropics than elsewhere.

(5) *Animal husbandry*. In the wet tropics, natural pasturage may be lacking and has to be artificially created at great expense. In spite of very high yields from perennial grasses, the protein content of these is so low as usually barely to suffice for maintenance, and the yield comes in irregular flushes through the year.

As against these disadvantages, the chief advantage of the tropics is that there are areas where two annual crops can be taken in a year as against the one of temperate zones. But these areas are less extensive than is often thought. They are mostly confined to a narrow equatorial zone; in the outer tropics, the long dry season precludes the taking of more than one crop. The actual degree of advantage gained by double-cropping is also debatable; two crops a year impose a severe strain on soil fertility and the total yield may sometimes be little or no greater than that from a longer-term single crop. Grist 1959 considered this problem in rice, where double-cropping is possible in southern Japan, south China, and parts of Formosa, Malaya, Indonesia, Ceylon and India. He points out that, for double-cropping, use has to be made of quick-growing varieties which are lower-yielding than those of longer maturation, and that the expense of growing two crops is much greater than that of one, and concludes that experimental data are too incomplete to allow us to assess the degree of advantage in double-cropping.

The biological advantages of perennials in those parts of the tropics where rainfall permits their growth have already been mentioned (p. 30). There is a further argument in their favour. Most of the tropical perennial crops, not being cold-hardy, are strictly limited to the tropics which, therefore, enjoy a natural monopoly in them. The chief exceptions are: tea; sugar-cane and bananas, which can be grown a little outside the tropics; and most citrus fruits, of which some actually grow better in the sub-tropics. But the tropics have a complete monopoly in oil and coconut palms; Para rubber; coffee and cocoa; spices such as nutmeg, cloves, pepper, cinnamon and vanilla; and certain export fruits such as pineapple and cashew. Thus the competition which is felt in annual crops, which are often grown with much better tech-niques in developed countries of the warm temperate zones, does not apply to these perennials, and it would appear that tropical agriculturalists should exploit these natural monopolies up to their economic limits (see Chap. 3.3).

Table 1.4.2

Relation in Temperate Zones Between Climate, Vegetation and Net Primary Production (Biological Yield)

Vegetation and Climate	Mean Annual Precipitation (inches)	Mean P–E–T in Thermal Growing Season (inches)	Mean A–E–T in Thermal Growing Season (inches)	Net Primary Production* (Biological Yield) of dry organic matter Grammes/sq metre per year	per day over year	per day in growing season	Economic Yield Range of Mean Wheat (Grain) (Winter or Spring) Yields (lb/acre) and Coefficient of Variation (C.V.)
Desert (Hot dry)	<10	>40	<10	<50		<0·3	
Short grasslands							
Cold or cool dry	10–20	>25	<20	50–70		0·3 to 0·5	900 (C.V. 35%)
Warm dry (including dry warm forest)	<20	>30	<20				
Tall grasslands							
Transitional to cool humid	25–35	>25	<25 }	70–500		0·5 to 3·0	900 to 1,500 (C.V. 35% to 15%)
Transitional to warm humid	30–45	35	>30 }				
Forest							
Cool humid	40–80	>25	<25 }	500 to 3,000		3·0 to 9·0	1500 to 4,500, (C.V. <15%)
Warm humid	40–70	35	>30 }				
Warm dry (see above)							
Wheat — world average				344	0·94	2·3	
— Netherlands				1,250	3·43	8·3	
Rice — world average				497	1·36	2·7	
— Italy and Japan				1,440	3·95	8·0	
Sugar-beet — world average				765	2·10	4·3	
— Netherlands				1,470	4·03	8·2	
Sugar-cane — world average				1,725	4·73	4·7	
— Hawaii (average)				3,430	9·40	9·4	
— Hawaii (maximum)				6,700	18·35	18·4	

Sources: Various, including Odum 1959; Wilsie 1962, pp. 424, 425; Chap. 1.2 and Part 2.
* Gross formation of dry organic matter less respiration losses.

FARMING SYSTEMS OF THE WORLD

Uncertainty and Instability as Locating Factors

Short-term uncertainty* has important effects on the farmer's choice of a farming system, especially on its input intensity and on the proportion of land or capital in 'risk sensitive' enterprises. Longer-term instabilities (e.g. due to soil erosion or salinity) also affect, often cumulatively, the ecological and commercial viability of particular systems, and are therefore also locating factors.†

Classes and Sources of Short-term Uncertainties in Temperate Areas

Ecologically, uncertainties include genetic variability; pests, diseases and weeds; weather and weather chains. Economic and political uncertainties may be of ecological or non-ecological origin. The farmer likes his genotypes to be as dependable as possible, though he depends, of course, on genetic variability for breed and cultivar improvements, and for random resistance to unfavourable conditions. (Sometimes the advent of a new genotype, e.g. hybrid maize in the U.S. Corn Belt, or leucosis-resistant poultry, have great effects on farming systems. So do new and unexpected (genetic) variations in pests, diseases or weeds, e.g. new physiological races of the cereal rust *Puccinia glumarum*.) The risks of infection or invasion by pests, diseases and weeds condition the detail of many farming systems and have locational influences, of which examples are given in Part 2.

The problem of weather risks has been examined by Duckham 1963, in Johnson and Smith 1965, 1966, and in Taylor 1967, as it affects the location and intensity of particular enterprises and systems, including indirect effects through capital investment. Thus, in areas with predictably good cereal harvest weather such as Western Australia, one combine can harvest several times the acreage the same combine could achieve in the uncertain harvest weather of north-west Europe. The low and uncertain precipitation and yields of the former area are partially offset by the absence of the need to have excess machine capacity in case the harvest weather is adverse. If the harvest weather were as uncertain in Western Australia as it is in north-west Europe, then wheat growing in the former would probably be uneconomic. Or again, in warm dry climates of the south-west of the U.S.A., cattle can be kept outdoors in dry lots, whereas in southern England cattle are usually housed in winter often expensively, because of the risk that snow or rain or cold (cold-chill), or cold wind with rain (wind-chill), will excessively raise the energy requirement for maintenance of the cattle. This comparative weather

---

* As ecological uncertainties and instabilities are, in many countries, as much concern to the farmer as political risks or as economic uncertainties of non-ecological origin, uncertainty is dealt with here and not in Chap. 1.7.

† Contrary to the practice of some economists who define risk as quantifiable probabilities and uncertainties as possibilities which cannot be quantified, no distinction is made here between risk and uncertainty.

advantage is one of the reasons why feed-lot fattening of beef is growing so rapidly in the drier parts of North America.

But perhaps the main impact of weather uncertainty on systems and their intensity can be seen as one moves from humid to dry areas in the U.S.A., Australia and Israel. The lower the mean annual precipitation and the mean A–E–T in the T–G–S, then the more variable and less reliable both these parameters usually are (Figs. 1.2.8, 2.3.4 and 2.4.2 and Table 3.1.1. It is probably their unreliability as much as their low averages in, for example, the Western Great Plains of the U.S.A. that accounts for these extensive tillage or extensive grazing systems. In a season of good moisture supply, wheat yields *could* reach those of a humid climate, but the moisture probabilities do not commercially justify the higher seed rates and the higher use of fertilisers to exploit the higher A–E–T of a moist year. The same thing happens in temperate areas where the thermal growing season is short and unreliable, e.g. the northern edge of the Canadian Prairies, the north of Sweden and the 'mountains' of Britain. Thus, it is often the unreliability rather than the mean of the critical parameters that make intensive systems too risky for farmers in such localities, unless they have enough capital (or credit) and courage to hazard a run of bad years.

But most farmers are unwilling or unable to take such risks, partly because they or their bank managers probably tend (Curry 1962, 1963) to overestimate the hazards. In other words, they farm at a lesser intensity than the probabilities, if calculated, would justify. This is why meteorological feasibility studies of the type done by Manning (1958) in Uganda for cotton and maize, by Kates (1962) for flood-liable areas of the U.S.A., by Slatyer (1960) for the Yass Valley in Australia, by Curry (1962, 1963) for dairy farming in the North Island of New Zealand, by Duckham (in Taylor 1967) for irrigated intensive milk production from grass in south-east England, are urgently needed, especially for areas that are normally regarded as 'marginal' for intensification (e.g. Western Nebraska, U.S.A., the uplands of Wales, the north of Finland). But the problem must not be exaggerated. For marginal and extensive farming areas tend to have enterprises which have 'built-in' weather and other buffers. Farmers with extensive tillage systems (Chain A) can often adjust cropped acreage to year-to-year variations in soil-moisture status (see U.S.A., p. 157, and Israel, p. 316) and, of course, given good credit and silo facilities, cereal grains can be stored for long periods to stabilise market conditions. Intensive systems, whether tillage, alternating or grazing, have similar or more sophisticated buffers. Thus, intensive grazing systems (Chain C) can and do use such buffers as variation in the intensity of grazing, the compensating growth capacity of non-lactating ruminants, the mobility of cattle and sheep, the purchase of feed, the variation of selling or slaughter weights and degree of fatness (finish) and the use of processing and cold-storage plants (Duckham in Wilkins 1967).

Investment in 'weather-proofing', including sprinkler irrigation, higher

FARMING SYSTEMS OF THE WORLD

fertiliser use, grain drying and fodder conservation, and faster, more powerful machines, is growing rapidly in most intensive advanced agricultures, where there has also been a shift to less weather-sensitive enterprises such as indoor pig-keeping (Chain B, p. 9) and winter milk production (Chain D). All these cut down the effects of weather-induced uncertainties. The direct yield on the capital put into 'weather-proofing' may not be attractive but, indirectly, the greater certainty gives one confidence to intensify inputs and the courage to experiment; these psychological dividends may turn out to be very profitable in cash terms and lead to drastic change of farming system.

Scientific and industrial input thus mitigate short-term disease, pest and weather uncertainties and help to stabilise both yields and net farm income (see Table 2.2.4). When supplemented by governmental and other institutional price supports, they have had remarkable effects, in, for instance, the U.S.A., in reducing short-term economic instability (Table 1.4.3). Though no figures seem to be available for other countries,* the Canadian and the U.S.A. data (Ware 1963, Hathaway 1963) suggest that, in humid areas in the last 30 or 40 years, year-to-year yield variability has fallen by a third, and income uncertainty has been practically halved. Dry areas, however, with low mean annual precipitation, a high coefficient of variation and low mean annual A–E–T, still have annual incomes with high coefficients of variation.

Long-term Instabilities in Temperate Areas

In the past, long-term climatic changes have affected the location and intensity of farming systems, but with modern technology it should be practicable to cope with any likely climatic changes and any new pests and diseases of natural origin. Most sources of long-term ecological instability are, in fact, man-made. Soil erosion and salinity occur naturally but are (p. 56) aggravated by man and are at least partly avoidable or remediable. The build up of pests and diseases by persistent overstocking or by overcropping with a single species can be avoided, like the loss of soil structure, by 'good husbandry' or remedied by temporary abandonment of the causal enterprise or the land. The macro-nutrients and micro-nutrients removed when crops and livestock are sold off the land to the city, and the loss in heavy rainfall areas by leaching, can, in advanced economies, be made good by applying N, P, K, CaO and minor elements. Thus, most types of long-term ecological instability (except soil erosion and perhaps salinity) can be avoided or remedied without serious loss of food production, or alternatively, they are self-correcting; as, when yields fall too low, the land is abandoned and allowed to recuperate or switched to a restorative land use such as grassland. The persistence of farming for five or six thousand years, e.g. the Plains of Megiddo in Israel,†

---

* However, Langley 1967 suggests that the variability of U.K. aggregate farm incomes in real terms may be decreasing.

† The long agricultural history of the Nile Valley is a special case; the annual flooding of the river brings in new soil and further nutrients.

in India, China, etc. supports these apparently complacent claims. Even in cool humid climates, some land at least has been farmed for several thousand years. One of the authors lives on a farm where recent archaeological 'digs' suggest some form of agricultural activity, 4,000 years ago. (Slade 1962.) Thus (with erosion, its consequences and salinity excepted), although long-term ecological instability may have temporary effects, lasting many decades, on

Table 1.4.3

Instability of Annual Net Income by Selected Types of Farms
By decades—U.S.A.—1930–60

Climate and natural vegetation	Farming systems	Type of farm*	Coefficient of Variation %†		
			1930–39	1940–49	1950–60
Cool humid. Forest	Int. Grassland	Dairy Farms —Central North East —Eastern Wisconsin	33·3 45·5	38·8 34·9	10·3 17·5
Cool humid. Tall grassland	Int. Tillage and Int. Alternating	Corn Belt —Cash Grain —Hog and beef fattening	58·6 64·6	35·5 46·2	14·6 17·5
Cool dry. Short grassland	Ext. Tillage and Ext. Grassland	Northern Plains —Wheat and other small grains; livestock —Sheep ranches	513·6 52·0	47·2 29·7	32·4 49·8
Warm humid. Forest	Int. Tillage and Int. Alternating	Southern Piedmont —Cotton, Mississippi Delta —Small cotton farms	39·5 n.a.	36·0 23·3‡	22·0 14·8
Warm dry. Short grassland	Ext. Tillage and Ext. Grassland	Southern Plains —Wheat and sorghum Texas High Plains —Non-irrigated cotton Southern Plains —Cattle ranches	201·5 n.a. 443·2	58·7 55·1 43·5	67·5 56·8 39·4
Warm. Hot dry desert	Ext. Grazing	South West —Cattle ranches —Sheep ranches	n.a. n.a.	54·1 22·5	85·8 72·8
Warm. Dry Mediterranean	Int. Tillage	San Joaquin Valley, California—Irrigated large cotton farms	n.a.	n.a.	15·2

* Reclassified from 'Crops and Returns on Commercial Farms', *Statistical Bull*. 297. Economic Research Service, U.S.D.A., Washington, 1961.
† Hathaway 1963, pp. 45 and 46.
‡ 1944–49 only.

the location and intensity of farming systems; in the very long run they are not important, especially with modern technology and industrial inputs (see Chap. 3.2).

On the other hand, the ecological effect of urban man is of growing importance. The farming problems attributed to industrial and domestic smoke, to motor vehicle fumes and to water pollution are important (p. 504), but are under control or potentially controllable. More serious is the loss, in populous industrial areas, of farming land for factories, roads, airports and to house and provide recreation for the growing populations. Such land (unlike e.g. overcropped acreage) is permanently lost to farming and is usually located in flat areas well suited to intensive farming (see also Chap. 3.2).

Finally comes long-term economic and political instability. Slow changes in consumer taste with rising incomes (e.g. the growing demand for beef in western Europe), and well-marked trends in international trade (e.g. the increasing imports of wool and meat by Japan and the growing use of dried milk from New Zealand in south-east Asia), and the competition of man-made fibres with wool, cotton, jute, etc., are having or will have effects on location and intensity of systems. More dramatic are the changes which follow major regional political-economic decisions such as the setting up of the European Economic Community. But such changes, together with periodic economic recessions or booms, are inevitable in a developing world and do not seriously retard progress. More insidious and unwelcome is the chronic political instability and civil or military disturbances, with their consequential lack of economic stability, found in many parts of Asia, Africa and Latin America. If the farmer is uncertain whether he will be able to reap what he sows, neither he nor his bankers will have confidence to increase investment and he will have little incentive to intensify or improve his farming even if he has, or could have, as in Argentine, adequate access to industrial inputs. Nor will local investment in the input industries (e.g. fertilisers) be encouraged.

Uncertainty and Instability in the Tropics

Farming in the tropics is, on the whole, subject in a greater degree to natural risk than is temperate farming, owing to the greater incidence of droughts, floods, hurricanes, locust visitations and epizootics of animal diseases, so that a tropical farmer needs (but rarely has) an especially large capital reserve to enable him to survive these economically. The degree of risk is in general being steadily lessened by the progress of technology. Irrigation, the provision of water supplies for livestock by dams and boreholes, and increasing veterinary control of animal diseases have all helped in this regard; there is, for example, less need than there used to be for tropical stock-owners to maintain over-large herds which encourage overgrazing and erosion, simply as an insurance against occasional catastrophic losses. The red locust and the migratory locust are under a high degree of control by internationally supported organisations, and it is to be hoped that serious invasion by them is a thing of the past; but the desert locust is still a rampant threat to many countries in Asia and Africa. On the other hand, there are cases where technology has increased instability. This has happened, for instance, where large irrigation schemes have tempted populations to settle and multiply on them only to find, a generation or two later, that the land had become incapable of supporting them owing to increased salinity, water-logging or silting of the reservoirs. India and Pakistan have in this way lost, from cultivation, millions of acres of irrigated land in the last twenty years. Methods of preventing such disasters are now better understood, but make irrigation much more expensive.

Acknowledgements. For comment and criticism: H. J. Critchfield, S. Fox, L. P. Smith, J. A. Taylor and E. W. Russell.

References

Curry, L. 1962. *Geog. Rev.* **52**. 2. 174–94.

Curry, L. 1963. *Econ. Geog.* **39**. 2. 96–118.

*Duckham, A. N. 1963. *Agricultural Synthesis: The Farming Year*. London.

Duckham, A. N. 1966. *The role of agricultural meteorology in capital investment decisions in farming*. Study No. 2. Farm Mgmt. Section, Dept. of Agric., Univ. Reading.

Grist, D. H. 1959. *Rice*. London. 3rd edn.

Hathaway, D. E. 1963. *Government and Agriculture: Public Policy in a Democratic Society*. Macmillan. New York.

Johnson, C. G. and Smith, L. P. (Eds.) 1965. *The Biological Significance of Climatic Changes in Britain*. London and New York.

Kates, R. W. 1962. *Hazard and Choice Perception in Flood Plain Management*. Univ. Chicago. Dept. Geog. Res. Paper No. 78. Chicago.

Langley, J. A. 1967. *Risk, Uncertainty and the Instability of Incomes in Agriculture*. Agric. Econ. Dept., Univ. Exeter.

Manning, H. L. 1958. Conf. Directors, etc. of Overseas Dept. Agric. Col. Office, Misc. No. 531. H.M.S.O. London, p. 47.

*Odum, E. P. 1959. *Fundamentals of Ecology*. Philadelphia and London. 2nd edn.

Slade, C. F. 1962. Personal Communication. History Dept. Univ. Reading.

Slatyer, R. O. 1960. *Agricultural Climatology of the Yass Valley*. Divsn. Land Res. and Regional Survey. Tech. Paper No. 6. C.S.I.R.O. Melbourne.

Taylor, J. A. (Ed.) 1967. *Weather and Agriculture*. London.

Ware, D. W. 1963. *The Variability of and Sources of Farm Cash Income. Canada and Provinces, 1926–1960*. Canada. Dept. Agric. Economics Div. Ottawa.

Wilkins, R. J. (Ed.) 1967. *Fodder Conservation*. Occ. Symposium No. 3. British Grassl. Soc. Hurley, Berks., England.

Wilsie, C. P. 1962. *Crop Adaptation and Distribution*. San Francisco and London.

Further Reading
(SEE ALSO ASTERISKED ITEMS)*

Boyko, Hugo (Ed.) 1966. *Salinity and Aridity: New Approaches to Old Problems*. Junk. The Hague.

Bracey, H. E. 1963. *Industry and the Countryside*. London.

Brady, N. C. (Ed.) 1967. *Agriculture and the Quality of our Environment*. (American Ass. Adv. Sci.) Washington, D.C. and London.

Gourou, P. 1961. *The Tropical World*. London. 3rd edn.

Webster, C. C. and Wilson, P. N. 1966. *Agriculture in the Tropics*. London.

Social and Economic Factors

Each farm and the genotypes and enterprises on it, and each farming system, is the product of human reactions to three particular sets of ecological, operational, economic and social circumstances. Chaps. 1.2, 1.3, 1.4 and 1.6 are about the first two of this triad. This one is about the third, i.e. the impact of economic and social factors. These, as might be expected from p. 12, have profound effects on the infra-structure, input supply and market demand and hence on the location, intensity and efficiency of farming systems.

Infra-structure and Economic Development

Infra-structure literally means the 'underneath structure' on which manufacturing industry, agriculture, military strategy or a national economy is built. Thus, an air-line needs not only aircraft, but also aerodromes, pilots, technical ground-staff, engineering workshops, telecommunications, meteorological services, catering facilities, motor roads, public safety and educated local staff. Agricultural infra-structure can be broken down into social, economic and agrarian sectors.* These three sectors are positively correlated and each must be good for advanced farming systems to be practicable.

SOCIAL INFRA-STRUCTURE

In less-developed countries in the tropics (see below) and in parts (e.g. southern Italy, Iran, Egypt) of the temperate zone, farming constraints which flow from the social infra-structure may be as important, or more so, than economic infra-structure. Religion, language, caste, tribe, nationality, the inertia of tradition and customs, political instability (p. 74), liability to civil disturbance, educational levels, the frequency of literacy, inadequacy of diet and the locally accepted attitudes to risk and uncertainty and to the concept of progress may limit production or resist change (Bunting 1967). Examples of social constraints in the temperate zones are seen in Spain, Chile (p. 324) and some of the poverty pockets of the U.S.A. (p. 109); tropical instances are given below. But in advanced farming areas less obvious, subtler constraints are numerous. Thus, in the south-west of the United Kingdom many small farmers who would benefit financially from capital injection, will not, for moral principles, seek, or accept bank credit. Again, in the Netherlands advisory services have to be duplicated in some areas to cater for the Protestants and the Catholics, whilst one of the present writers had to administer a comparable situation in Northern Ireland 30 years ago.

The level of housing, nutrition, hygiene and endemic nutritional or trans-

* For a detailed survey, particularly of the agrarian sector, see E.E.C. 1960.

missible diseases, influence the physical, and indirectly the mental, productivity of farm and urban populations, especially around the Mediterranean and
in the tropics, (because) 'a poor diet . . . reduces working efficiency by (a) decreasing the workers' resistance to disease, (b) increasing the rate of absentee-
ism, (c) causing lethargy, lack of initiative and drive, and (d) increasing accidents'. (F.A.O. 1962, see also F.A.O. 1963 and Clark and Haswell 1964).
People so affected cannot easily adapt traditional farming systems to counter
sources of instability or take advantage of new technologies or markets
(Schultz 1964).

Social infra-structure cannot be expressed numerically by a single parameter
but Tables 1.5.1 and 1.5.2 Figs. 1.1.8 and 1.1.9, give some useful indications.
In broad terms, a mean daily (energy) intake per head of less than 2,000 kcal,
of which less than 10% are of animal origin; an infant mortality over 30 per
1,000 live births; a housing density of more than 1·5 persons per room; and a
newspaper circulation of less than 100 per 1,000 population, suggest a social
infra-structure which is incapable of supporting, at present, advanced systems
of farming.

Table 1.5.1
Food Consumption and Economic Infra-structure of Selected Countries [1]

| Country | Indicative statistics, 1965 | | | | | Indices of 'Real' private consumption per head in 1960 (U.S.A. in 1960 = 100) [2] | |
	kcal/head/ day*	% kcal Animal origin*	kg Steel consumed/ head/ annum	Energy consumed/ head/ annum (kg coal equiv.)	Newspaper circulation/ 1,000 pop	Index based on Beckerman method	Index at official exchange rates
Afghanistan	2,050	13	—	25	6	—	—
Algeria	—	—	23	300	15	13·8	—
Argentine	3,040	30	114	1,341	148	23·8	18·5
Australia	3,160	43	514	4,795	373	65·4	57·2
Canada	3,090	43	531	7,653	227	77·0	73·9
Chile	2,370	22	70	1,089	118	16·9	20·9
Denmark	3,330	44	361	4,172	347	59·2†	46·6†
France	3,050	—	331	2,951	245	54·3†	47·4†
Greece	2,960	23	85	784	125	12·7	16·4
Guyana	—	—	—	811	93	—	—
India	1,980	6	16	172	12	3·1	—
Iran	2,050	12	30	391	15	7·3	—
Israel	2,830	20	187	2,239	143	27·8	45·8
Japan	2,320	11	294	1,783	451	28·7	12·6
Jordan	2,280	4	—	291	8	—	—
Netherlands	2,890	29‡	313	3,271	293	45·0	31·3†
New Zealand	3,410	52	239	2,530	399	58·6	56·0
Nigeria	—	—	6	44	7	2·6	—
Sweden	2,950	42	682	4,506	505	77·4	54·5
U.A.R.	2,930	6	26	301	15	6·4	—
U.K.	3,260	42	424	5,151	479	61·7†	49·9†
Uruguay	2,970	44	24	916	314	16·2	—
U.S.A.	3,140	38	656	9,201	310	100.0	100.0
U.S.S.R.	—	—	—	3,611	264	—	—

Sources:
[1] United Nations 1966.
[2] O.E.C.D. 1967. It is apparent that considerable differences exist between the series of figures in the last two columns, e.g. the Japanese index, following the Beckerman method, which is based on modified non-monetary technique, is more than double that of the official rate of exchange, whereas the position in other countries is the reverse (see also Chap. 1.1).
 * Latest available figures, varying from 1960 to 1965.
 † Countries for which estimates shown are obtained by extrapolation from those established by Gilbert and Associates for 1955 and not from equations of the proposed method. (See O.E.C.D. 1967.)
 ‡ 1954–56.

ECONOMIC DEVELOPMENT AND INFRA-STRUCTURE

Roads, railways and shipping, electricity and other public utilities, press and telecommunications, credit, insurance, banking and other financial services, engineering and technical services, are essential parts of the infra-structure of all advanced farming systems, i.e. systems which have high inputs and output per acre, per man or per unit of capital invested, or of all three. Further, the high level of industrial inputs (i.e. excluding inter-farm inputs) (Fig. 2.3.1.)

Table 1.5.2

Social Infra-structure of Selected Countries
(Indicative statistics, 1965)

Country	Population		Infant mortality (Under 1 year) per 1,000 live births	Persons per room	Rural housing	
	Density/ sq km (approx)	Increase* (annual rate %)			% dwellings with:	
					Electricity	Piped water
Afghanistan	23	1·8‖	—	—	—	—
Algeria	5	1·9‖	36·4 (1960)	—	83·9†	99·7†
Argentine	8	1·6	—	1·7	29·2	14·3
Australia	1	2·1	18·5	0·7	—	—
Canada	2	2·0	23·6	0·7	65·9	65·5
Chile	11	2·3	114·2	2·0	23·9	11·4
Denmark	110	0·8	18·7	0·7	96·6	100·0‡
France	89	1·3	22·0	1·0	96·8	58·7
Greece	65	0·6	34·3	1·6	13·5	49·9
Guyana	3	2·8	42·3	2·1	—	—
India	159	2·3‖	—	2·6	—	—
Iran	14	1·9‖	—	—	—	—
Israel	124	3·6	22·7	1·8‡	70·5	85·1
Japan	265	1·0	18·5	1·1	—	46·8
Jordan	22	3·2‖	—	—	1·4	13·5
Netherlands	366	1·4	14·4	0·8	96·3	78·3
New Zealand	10	2·1	19·5	0·7‡	—	85·5‡
Nigeria	62	2·0‖	—	3·0†	81·3†	—
Sweden	17	0·6	13·3	0·8	—	79·4
U.A.R.	30	2·6‖	—	1·6†	37·8†	39·5†
U.K.	224	0·7	19·6	0·6	96·7§	95·0¶
Uruguay	15	1·4‖	?	?	?	?
U.S.A.	21	1·5	24·7	0·7	95·0	81·2
U.S.S.R.	10	1·6	27	1·5†	—	—

Source: United Nations 1966 (unless otherwise indicated).
* Includes net migration change.
† Urban houses.
‡ Combined urban and rural.
§ Farm consumers—England and Wales.
‖ Estimates of questionable reliability.
¶ Indoor piped water.

characteristic of such systems (Duckham 1966) depends on urban industrialisation. Local manufacturing industries are, theoretically, not essential if inputs can be imported. But it is difficult to find examples of advanced farming systems which have no supporting local manufacturing industries. Thus, although New Zealand may import industrial farming inputs, its local factories meet special local farming needs such as superphosphate 'fortified' with cobalt or other trace elements or with pesticides for the control of the grass-grub of rye-grass (p. 214).

No single parameter can measure the level of economic development or infra-structure. But the gross national product (G.N.P.) per annum per head of population, Beckerman-type indices (p. 11), energy and steel consumption per head per annum, and motor vehicles are useful indicators. Perusal of Table 1.5.1 and Figs. 1.1.8 and 1.1.9 suggests that countries with a Beckerman Index of less than 25, or a steel consumption of less than 100 kg/head/annum or an energy consumption of less than 1,400 kg coal equivalent/head/annum have economic infra-structures and development levels which could not, at present, support advanced farming systems.

AGRARIAN INFRA-STRUCTURE

This term covers farming structure, rural communications and public utilities, ancillary (input and marketing) industries and technical services. *Farming structure* includes land, tenure system and the size and cohesion of individual farms. Land tenure systems, especially share-cropping or métayage, can have a paralysing effect (e.g. in Chile until recently, p. 331) on production methods and on the uptake of innovations, and hence on the nature, intensity and location of farming systems. But cash tenure, too, may affect choice and intensity of farming system, for example, because the landlord may be unwilling or financially unable to install or to improve fixed equipment such as irrigation facilities or livestock housing. Thus, in the United Kingdom, some cash tenant farmers, who would like to enlarge their dairy herds, cannot readily acquire the buildings needed to do so. Even the owner-occupier is often limited, for example, by resistance to farm amalgamation schemes or by farmers' unwillingness (Firey 1960, p. 138) to accept the communal discipline which soil or water conservation demands. *Size of farm*, as influenced by custom, law or governments, may also constrain choice. Thus, in the United Kingdom many small farmers stay in milk production because their holdings are not large enough to make cereal growing a profitable alternative and because capital is 'trapped' (Hathaway 1963) in dairy buildings and cannot be transferred to tractors, drills and combine harvesters. In Denmark, where for social reasons the upper size of farm is restricted, the small mean area of farms forces many farmers into intensive tillage with livestock when they might be better off and more efficient on alternating systems on large farms. The *fragmentation* or dispersal of individual farms into several physically separated fractions is prevalent in much of Europe and Asia, and often limits both choice and efficiency of farming systems. Thus, in the Netherlands, where the mean size of farm is about 20 acres, the mean distance of the individual fields from the farm dwelling is 0·6 miles, whilst in West Pakistan in a sample survey, three-fifths of the fields were more than 0·6 miles 'from the main plots' (Chisholm 1962). In such cases, the more distant fields are usually less well farmed or in less labour-consuming enterprises than the near-by ones.

Government action to remedy defects of farming structure is particularly active where the pressure of rural population on land is high and where the

D   79

national policy is for farm people to move to industrial areas (e.g. Italy), or where the farm population can be partly re-deployed (e.g. in the Netherlands or in France). Too often, however, the radical land reforms of today become the agrarian anachronisms of tomorrow. Fortunately, in advanced temperate countries, the decreasing importance of land (Schultz 1964) as a resource, may mitigate the adverse effect of poor farming structure.

Communications, Public Utilities, Ancillary Industries and Technical Services

Transport facilities have great impact on the location and intensity of farming systems. In North America, the railroads turned extensive grazing (ranching) land into vast extensive tillage areas. Refrigerated shipping altered the farming face of New Zealand and of the Argentine, where meat production largely replaced extensive production of wool, hides and skins. In many parts of Africa, the first step to farming progress is a good road to bring in education, experts, new ideas and simple inputs and to take out produce and, in due course, underemployed farm people. Telecommunications are almost as important as transport, for example, broadcasting for education and for farming and market news, telephones for summoning the artificial inseminator or the spraying contractor or medical services. Both transport and telecommunication facilities are usually good in industrialised areas. In areas where they are poor (e.g. the U.K. uplands, much of Canada, Australia and, of course, most less-developed countries), the farmer has, irrespective of ecological constraints, fewer choices of farming systems. Well-developed public utilities also widen choice; thus, efficient market milk production is, in these days, almost unthinkable without piped water and mains electricity. Finally, good communications and public utilities permit the economic growth of local ancillary industries.

Ancillary processing industries are usually well developed where economic development is marked. It is, however, often difficult to say whether the existence of ancillary processing industries was or is attributable to the local presence of suitable product outputs and to good communications and public utilities or whether the ancillary industry (e.g. a sugar-beet factory) promotes the needed volume of product output and better communications, etc. Often, of course, it is a mixture of both. Thus, in 1933, the pig industry in Northern Ireland was based on the Ulster pig that could not travel to market, was killed on the farm and sold in carcase form to small local bacon curers. One of the present writers simultaneously promoted a switch from the Ulster pig to the Large White breed which could travel and made good Wiltshire-cut bacon and also the introduction of large bacon factories which handled live pigs. These changes revolutionised the local pig industry and substantially affected farming systems, because Large White bacon pigs largely replaced the production of milk for manufacture. Denmark, New Zealand and the Argentine provide bigger and more classic examples of the same type of interaction between ancillary output industries and farming systems.

Agricultural education, advisory, research and diagnostic services, pest and disease control, genotype improvement, machinery repair, spare-part supply and similar services tend, like ancillary industries, to be good where economic and social infra-structure, including general education, are good. They often reflect the attitude of governments to agriculture and that of the local farm population to technical progress. They can have great influence on the intensity and efficiency of farming systems, but their availability is more often the result rather than the cause of the location or intensity of particular systems (e.g. artificial insemination schemes are generated by the location of milk production). One of the great challenges of today is, in fact, to make education, technical and health services *causal* factors in the nature, location and intensity of farming systems. This involves, as Schultz 1964 and others have shown, planned investment in farm people. The man 'who is bound by traditional agriculture cannot produce much food no matter how rich the land. . . . To produce substantially more requires that the farmer has access to, and the skill and knowledge to use, what science knows.' This knowledge and that involved in producing the industrial inputs needed for greater food output 'is a form of capital, which entails investment—investment not only in material inputs in which a part of this knowledge is embedded, but importantly also investment in farm people'. (Schultz 1964, p. 206.) (See also Chaps, 3.1, 3.2, 3.3 and 3.4.).

INPUT ECONOMY AND ACCESS TO INPUTS IN TEMPERATE ZONES

The change over the last half-century in the nature of farming inputs has been dramatic (Table 1.5.3). In most advanced agricultural economies, expenditure on inputs of industrial origin is as great as, or more than, that on land (which, of course, includes climate) and often heavier than land costs and inter-farm inputs (cereals, seed, livestock) combined (see Table 2.3.4). In temperate countries, therefore, access to, and the prices of, industrial inputs and often inter-farm inputs are almost as important as manpower, land supplies, access to market, market prices, and the decrease in the influence of ecological influents (p. 70). They are also (p. 71 and 72) important stabilising 'buffers' and, as such, influence location and the confidence to intensify. Fortunately, access to industrial and inter-farm inputs usually presents fewer problems than market access (p. 83) and their influence on location and intensity is correspondingly less.

Water has a low value/ton but can be relatively cheap to transport in canals or pipes and, whilst not strictly an industrial input, its use increasingly involves large public investment in capital works and the purchase of sprinklers, etc. by the farmer. Irrigation water can, if its quantity, quality (p. 57) and price are favourable, dramatically widen enterprise choice and thus the location and intensity of farming systems, especially in areas where P–E–T is high but precipitation, and hence the A–E–T, is low. Thus, in warm dry (Mediterranean) climates, surface or overhead irrigation makes possible intensive

Analysis systems (e.g. California). In short-grass areas such as Lethbridge, in Alberta,
of Canada, which would only be fit for extensive grazing or cereals, irrigation
Location makes intensive tillage and a sugar-beet factory practicable. Sub-irrigation
Factors influences farming systems in the Netherlands (p. 238) and the Fenlands of
England, whilst even in hydro-neutral zones (p. 461) (e.g. western Europe,
the eastern half of the U.S.A.) sprinkler irrigation, which is economical of

Table 1.5.3

Index Numbers of Values of Inputs in U.S. Farming
(Base 1957–59 = 100)

	1 Farm labour	2 Land (including climate)	3 Power and machinery†	4 Fertilisers and lime†	5 Miscel- laneous	6 Total inter-farm inputs*	7 Total inputs
1910	212	88	20	12	56	16	82
1966‡	70	99	103	185	128	130	105

Source: U.S.D.A. 1967.
* Purchased feed, seed and livestock.
† Industrial inputs.
‡ Preliminary data.

water, is increasing yields and the range of farmers' choices. Water for livestock can limit intensity in arid areas, e.g. outback Australia, but is rarely a
constraint in intensive areas.

Fertilisers. Chalk and lime have a relatively low value/ton; N, P, and K
fertilisers a medium value (of the same order as cereals), and trace elements a
high value/ton. Transport costs rarely hinder the use of the latter, but they do
influence inputs of lime and sometimes the use of fertilisers. Nevertheless, in
fairly remote extensive farming areas where A–E–T and yields do not justify
high fertiliser use, small doses of P_2O_5 in particular are applied (e.g. ammonium phosphate to wheat on the Canadian Prairies (p. 132), superphosphate
to the alternating wheat and sheep system in Australia, and from the air to
upland pastures in New Zealand). In the two latter cases, P_2O_5, together with
the resultant legume N, are strong locating factors. In more intensive systems,
fertiliser prices may have locational influences. Donald 1960 showed that
fertiliser N was about twice as expensive to the farmer in the Netherlands or
Australia and about $2\frac{1}{2}$ times more in New Zealand than it was in England,
California (U.S.A.) or Hawaii (U.S.A.). Thus, if fertiliser N was as cheap in
New Zealand as it is in England, farmers in the former country (p. 216) might
depend less on legume N in their pastures, and some land now in intensive
grazing might be in intensive alternating or even intensive tillage farming,
especially as cereal markets are growing in the Far East. Generally, however,
in temperate areas with *good* infra-structures, access to high value/ton industrial and scientific inputs (e.g. genotypes, machinery, oil fuel, herbicides) does
not greatly influence the location or intensity of farming systems.

Inter-farm inputs. Access to inter-farm inputs may be locationally important. Thus, in Colorado, and elsewhere in the U.S.A., feeder (store) cattle from the ranches in the mountains and the drier plains, and feed grains from for example, the Corn Belt, are brought together on large open air 'feed-lots' for beef production, particularly where some bulky feed (e.g. hay) can be grown or purchased locally. Comparable 'feed-lot'-type liquid milk enterprises with cut and carted bulky feeds and heavy feed grain purchases are found in the Mediterranean climates of California and Israel. In general, however, subject to price and security of cereal supplies, intensive pig and poultry (Chain B) and market (liquid) milk (Chain D) systems tend to develop as part of the farming system in intensive tillage farming areas (e.g. Denmark, the Corn Belt of the U.S.A., eastern England) especially with good market access. Near good consumer markets (e.g. western England, the north-east of the U.S.A., New South Wales in Australia) seasonal summer grazing (Chain C) is supplemented by inter-farm inputs (feed grains and even purchased cows, etc.) to supply year-round market (liquid) milk (i.e. to become Chain D).

Demand and markets in temperate countries. Useful indicators of the level of general economic demand per head for food, by a given domestic or foreign population, are G.N.P. per head, Beckerman Index, the consumption of steel, fuel energy, plant or wheat equivalent consumed per head, mean dietary energy per head, and the percentage of the latter which is of animal origin (Tables 1.5.1 and 1.5.2 and Fig. 1.1.8). Demand per head of population is thus high in advanced economies; in most of these, if not all, it is also in effect, augmented by production or commodity subsidies to the farmer or consumer and by other protective devices. Such support is rare in tropical and other less-developed countries (e.g. Iran) where governments rarely have enough 'cash-flow' under their control to subsidise specific farm products or shape the agricultural output 'product mix'. The absence of such assistance is one of the major economic differences between advanced and less-developed agricultures.

Second is market access. Market access* (viz, L_m in Model 3.1.1 on p. 473) often has critical influence on the location and intensity of farming systems. It can sometimes, but not always, be quantified by comparing local and 'distant' prices or by creating marketing cost models. The classical tendency (Chisholm 1962) for intensive systems to be located round large urban centres with high living standards and for extensive systems to be in less populated and sometimes poorer areas, is being modified by advances in storage and transport. Thus, the environs of big conurbations are now often partly semi-intensive (e.g. much of the north-east of the U.S.A.) because they

* Market access comprises (*a*) customs tariffs and import/export quotas; (*b*) market structure (governmental and other regulating or trading monopolies, access to capital for marketing improvements); (*c*) processing and preserving costs and facilities; (*d*) transport rates (as influenced by governments, shipping conferences and competition); (*e*) market promotion schemes; (*f*) distance from local processing depots, selling depots, rail head, sea-port, together with the nature of the terrain and the number and all-weather reliability of the roads or transport system. Visher 1955 (*Econ. Geog.*, **31**, 82–86) (quoted by Wilsie 1962) subjectively ranks market access. His ranking is used in the transect tables in Part 2.

cannot compete with more remote, better favoured, less polluted (p. 504), lower cost and less populated but intensively farmed areas (e.g. the U.S. Corn Belt), or with even more distant areas (e.g. Argentina, New Zealand) despite customs tariffs, etc.

SOCIAL AND ECONOMIC FACTORS IN THE TROPICS

In a few remote districts of tropical countries, the people are still subsistence farmers, growing enough food and fibre for their own needs, collecting fuel and building materials from natural sources, and hardly entering into the cash economy at all. Historically, there have been two methods of breaking this routine. The first is that recommended by Lord Lugard in his classic primer of British colonial policy 'The Dual Mandate'. It is to impose taxation on such people, thereby forcing them to obtain cash either by growing cash crops or going out to work. The second is to tempt them into wanting cash by the availability of suitable consumer goods; not least important nowadays is the desire for money to pay school fees for children. An adequate supply of consumer goods in all country places is still a very important incentive in tropical agriculture.

However, the transition from a purely subsistence economy to some production for market has now been made by most tropical farmers. What remains important is that the subsistence outlook still survives so strongly amongst a large proportion of tropical farmers who produce some commodities for sale. As long as a farm is still thought of mainly as a source of food for the family, with a small output for sale as quite subsidiary, there is little real incentive for improvement because a family cannot eat, and does not require, more than a certain quantity of food. Not until farms are regarded primarily as a source of income, for which man's desire is infinitely elastic, does it really become attractive to use fertilisers and practise soil conservation, and abhorrent to allow fragmentation. It is this transition to a commercial farming outlook which is still so rare in the tropics, though it is beginning to be found in a few emergent farmers in some relatively advanced countries such as Rhodesia.

The vast majority of tropical farms are still run entirely by family labour. A difference from more advanced countries is the amount of field work put in by women, which in some African countries exceeds that by men. The great exception to this is in those Moslem countries where women are in more or less strict purdah and hence do not take part in farm work. Children also contribute to the tropical labour force, but their contribution is declining because of increased attendance at school, and this is often particularly felt in the herding of animals. The number of hours worked per year is often low, particularly in Africa where surveys have revealed that farmers often work as little as 4 hours a day and the total work for the year commonly ranges from about 650 to 1,450 hours (Clark and Haswell 1964), whilst labourers on plantations may only work 25 hours a week. (In advanced temperate areas the weekly hours range between 45 and 65.) There are many reasons to explain this low input of work: climatic stress, malnutrition (p. 517), chronic illness, lack of tasks in the dry season and sometimes the limited size of holdings. But in Jamaica the median amount of work put in by small farmers was found to be 2,500 hours per year, and in tropical Asia, Chinese farmers work for very long hours in a most enervating climate.

Social factors have a particular effect on animal husbandry where livestock are valued or disvalued for other than commercial reasons. In India where the slaughter of cattle is prohibited for religious reasons, much of the available pasture land has to be devoted to the maintenance of senile animals whose output for human consumption is nil. In Africa, much overstocking of pastures is due to the prestige conferred by ownership of mere numbers of cattle, and to their use in paying bride-price. Amongst Moslem populations pigs are not kept; strict Buddhists may make no use of animal food; some African tribes taboo the use of certain animal foods, especially to women.

The lack of industrial development in tropical countries (Chap. 1.1) is a disadvantage to agriculture in two ways. It deprives farmers of any large local market for their commodities and it means that agricultural requisites, i.e. industrial inputs, are not manufactured locally. Thus, even such simple tools as hoes and ploughs usually have to be imported, and mechanical equipment and fertilisers are enormously increased in price by having to be imported from distant industrial countries. Fertiliser manufacture has only begun in a very few tropical countries; and even India and Pakistan, countries lying partly in the tropics which have recently built up a sizeable production, still have a need for imported fertilisers as well to meet their requirements. Tractor manufacture is at the same early stage of development; tractors are assembled in Brazil, Mexico, India and the Philippines, but of these only Brazil and India use a majority of domestically produced components.

Export of produce usually involves its transport to one of the two great consuming areas, western Europe and North America. To reach these areas from many parts of the tropics, canal dues payable in the Suez or Panama canals add an extra item to landed costs and give some advantage to countries from which this is not necessary. Some tropical countries are land-locked, and in such cases as Uganda and Zambia produce may have to travel up to 1,000 miles by road and rail before it even reaches the seaboard. In such circumstances exports often have to be limited, not by what can be grown, but by the need for produce with a relatively high value per unit of weight such as cotton, coffee or tobacco, whilst the opium poppy provides an answer in some Asiatic hill areas. Animal products may be the only possible solution to this problem in some remote areas: hides and skins in many places, cream from some remote Kenya farms, ghee or bacon in other cases. With a few products, location of their production is almost determined by transport factors. Bananas, even in specially equipped ships, will not stand a very long voyage and so each temperate area is supplied from the tropical zone nearest to it: Europe from West Africa and the Caribbean, the east coast of North America from the Caribbean, the west coast of North America from western South American countries of which Ecuador is the chief, Argentina from Brazil, South Africa from Natal, Australia from its own tropical and sub-tropical areas, New Zealand from Fiji.

Over-production of tropical crops for export, with a resultant depressing effect on prices, has been historically a constantly recurring feature of tropical agriculture. This is no less so today, when almost every application of agricultural science or attempt to lower costs of production by increasing yield, tends to produce a greater quantity of the product concerned. The position is aggravated because so many tropical products come into the category of semi-luxury commodities rather than staples; this is particularly true of the beverage products (coffee, tea, cocoa), the fruits and the spices. A comparatively slight economic recession in the consuming countries can cut the consumption of such products sharply, and they are consequently liable to particularly violent price fluctuations. Tropical agriculture has also suffered severely from the competition of synthetic substances; thus the demand for indigo and other natural dyes has been reduced by synthetic dye-stuffs, for vegetable oils by detergents; for cotton by man-made fibres, and for natural rubber by synthetic rubber. Finally, some tropical products which are mainly limited to one use such as rubber for tyres or sisal for twine are extremely vulnerable to price fluctuations dictated by sudden changes in demand.

Internal marketing suffers from as severe transport difficulties as do exports. There are farmers in the tropics who live two or three days' walk from a road, and human porterage is the only available means of evacuating produce; wheelbarrows or pack-animals would provide a slight advantage, but not apparently a sufficient one to be an incentive to their use. Many farmers have within reach of them no traders willing to buy their produce; the volume is too small, and the profit margins of export too low to attract them to operate. Better roads, railways or waterways are still the top priority in many parts of the tropics for agricultural development.

Imperfect systems of land tenure hamper agricultural improvement in many parts of the tropics. Share-cropping has some merits in spreading the loss from crop failure between tenant and landlord. But in Ceylon it has been shown that weeding will increase rice yields to an extent that would be profitable if the whole of the increase remained with the tenant, but ceases to be so when he has to give half the increase to the landlord. Even in such a well-run organisation as the Gezira irrigation scheme in the Sudan, it has only proved possible to pay the tenant 40% of the proceeds from his cotton production, which must blunt his incentive to improvement. Insecurity of tenure by land occupiers must also discourage improvement and the planting of perennial crops, and has compelled the Kenya Government to undertake a huge programme of land registration in recent years. Fragmentation of holdings is another evil which may reach extremes under Moslem laws of inheritance. A situation as in parts of Kenya, where one farm may consist of 15–20 plots

scattered up to a radius of 5–10 miles from the farmer's house, must waste his
time, makes alternate husbandry impossible and encourages predial larceny.
It is, however, possible to overstress the importance of reform of land tenure.
Extreme examples of soil exhaustion and soil erosion can be found in the
tropics under almost every conceivable form of land tenure. A good system of
land tenure makes good farming possible, but it does not itself guarantee it.

References

Bunting, A. H. 1967. Personal Communication. Univ. Reading.

Chisholm, M. 1962. *Rural Settlement and Land Use: An essay in location.* London.

*Clark, C. and Haswell, M. 1964. *The Economics of Subsistence Agriculture.* London.

Donald, C. M. 1960. *J. Aust. Inst. Agric. Sci.* December. **26**. 4. 319–38.

Duckham, A. N. 1966. *J. Royal Agric. Soc.* **127**. 7–16.

E.E.C. (European Economic Community). 1960. Agricultural Commission. *Principal Conditions of Agricultural Production.* Brussels. (In French.)

F.A.O. 1962. Freedom from Hunger Campaign. Basic Study No. 5. *Nutrition and Working Efficiency.* Rome.

F.A.O. 1963. Freedom from Hunger Campaign. Basic Study No. 12. *Malnutrition and Disease.* Rome.

Firey, W. 1960. *Man, Mind and Land: a Theory of Resource Use.* Free Press of Glencoe, Illinois, U.S.A.

Hathaway, D. E. 1963. *Government and Agriculture: Public Policy in a Democratic Society.* Macmillan. New York.

O.E.C.D. 1967. *O.E.C.D. Observer.* **26**. Feb. 36–37. O.E.C.D. Paris.

*Schultz, T. W. 1964. *Transforming Traditional Agriculture.* New Haven and London.

U.S.D.A. 1967. *Changes in Farm Production and Efficiency: A Summary Report.* Stat. Bull. 233. Washington, D.C.

United Nations. 1966. *Statistical Yearbook 1966.* New York.

Wilsie, C. P. 1962. *Crop Adaptation and Distribution.* San Francisco and London.

Further Reading

(SEE ALSO ASTERISKED ITEMS)*

Allen, W. 1965. *The African Husbandman.* London.

Bauer, P. T. and Yamey, B. S. 1957. *The Economics of Underdeveloped Countries.* London.

Bunting, A. H. 1969. (Ed.) *Proc. International Symposium on Change in Agriculture.* (London.)

Hutchinson, J. 1966 (Land and Human Populations). *Adv. Sci.* **23**. 111, 241–54.

CHAPTER 1.6

Farm Work

This chapter is about the impact of relief, soil, vegetation, mechanisation and manpower, and infra-structure on farm work and hence on the location and intensity of farming systems.

Relief

ROUGH TERRAIN

The land-forms of the world are shown in Fig. 1.6.1.

Slope (or lack of it) may create operational problems in erosion control (see above, p. 65) and hence an enterprise choice; whilst at high altitudes (p. 32) high precipitation may, especially on slopes away from the sun, adversely affect working weather at seed time and harvest (Duckham 1963, p. 177) and so rule out tillage crops. In almost all climates, steep slopes, outcrops of bare rock or shallow soil overlying rock make tillage and especially mechanised tillage difficult or impossible, and thus restrict land use to tree crops or grass land systems (p. 54). Thus, in Hawkes Bay (New Zealand) a good tillage climate is used for grazing, partly because the terrain is too rough for mechanisation or too liable to erosion. Nevertheless, tillage is found on steep or rough terrain where manpower is plentiful (e.g. terrace farming in Indonesia) or where special mountain-type walking tractors are available (e.g. Switzerland).

Thus, rough terrain, unless in forest or tree crops, is often in 'natural' or cultivated grassland which checks erosion, and which, in really rough country, is grazed by sheep or goats; these are more agile, often hardier and less liable to accidents than cattle. The aeroplane has advanced the use of such terrain; for example, in rapidly sowing grass seed from the air after forest and grassland fires in the U.S.A., and in New Zealand in applying P_2O_5 and trace elements in deficiency areas, in dropping fencing posts and wire and even in pest and disease control. But whilst industrial inputs (e.g. Land Rover motor vehicles) can overcome many relief difficulties, rough terrain tends to preclude the good local infra-structure (e.g. in the Highlands of Scotland) on which intensive farming depends.

Although, to avoid the problems of flat country (below), many tillage farmers prefer gently rolling land; slope, if too great, can add to costs and hence reduce comparative advantage. Table 1.6.1 shows its effect on the efficiency of combine harvesting of cereals.

88

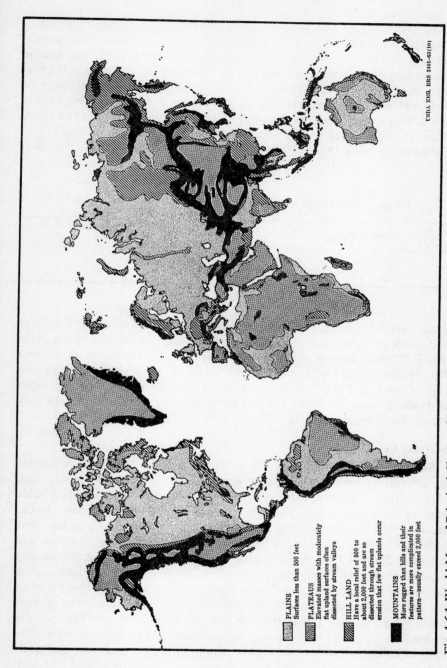

Fig. 1.6.1 World Map of Principal Classes of Land-forms. *Source:* U.S. Dept. Agric.

PLAINS
Surfaces less than 500 feet

PLATEAUS
Elevated masses with moderately
flat upland surfaces often
dissected by stream valleys

HILL LAND
Have a local relief of 500 to
about 2,000 feet and are so
dissected through stream
erosion that few flat uplands occur

MOUNTAINS
More rugged than hills and their
features are more complicated in
pattern—usually exceed 2,000 feet

USDA NEG. ERS 2401-63(10)

FLAT COUNTRY

Hidore 1963 found, in mid-western U.S.A., a high positive correlation between acreage in cash grain and the percentage of flat land; statistically, the amount of flat land accounted for 'over 50%' of the variation in the percentage of cash grain on farms. However, in flat country poor drainage or slight water-filled depressions may, as in the Canadian Prairies, limit the extent to which mechanisation can open up potential tillage land. In the flood-plains of rivers, flooding may be a hazard but the statistical analysis of frequency of flooding (Kates 1962) may help farmers to decide whether such

Table 1.6.1

Effect of Field Slope on Efficiency of Combine Harvesting of Barley

Average slope	Net throughput, ton/hour		Grain losses, lb/acre
	Straw	Grain	
Level	3·0	4·0	19·2
Across 1 : 5·1	2·6	3·6	55·7
Up 1 : 4·4	1·3	N.R.	46·6

Source: Modified from N.I.A.E. 1963

land is worth cropping. Poor drainage land reduces soil workability for tillage and raises its poachability and thus limits grazing, whilst in irrigated areas (e.g. California) it may aggravate salinity (p. 56). Land drainage is expensive but the Mississippi Delta, much of the humid tall grass country of the rich soils of the U.S. Corn Belt, the Fenlands of the United Kingdom and much of Holland, were originally too wet or flood-liable for cropping until they were drained or protected from flood.

Soil and Vegetation

Soil texture and structure (p. 51) may limit workability and be a constraint on enterprise choice and farming systems, especially when they have a high clay fraction. However, industrial inputs have opened heavy or flood-liable land to more enterprises. Thus, 15 years ago flood-liable meadows at the Reading University's Sonning Farm could only be extensively grazed, but now carry annual tillage crops 'between each winter flooding'. As shown in Chaps. 1.2 and 1.3 much favourable farming country is 'naturally' heavily wooded or has tall grass. These have to be removed before the land can be farmed, or at least cultivated. The tall grass prairies of the Interior Plains of the U.S.A. could only be extensively grazed until the prairie-buster plough made tillage farming practicable. One of the reasons for the failure of the groundnut scheme in East Africa after the Second World War was the difficulty of killing and then removing the stumps of the heavy, hard wood trees. Although short summers and hard winters are added constraints, the high cost of clearing woodland limits the northern expansion of extensive tillage

into the partly wooded belt to the north of the Canadian Prairies. Even
where trees have been killed by ringing or felling, their stumps may preclude
tillage and intensive grassland management; under-used pastures dotted with
stumps or dead trees can be seen in the north-east of the U.S.A. and eastern
Canada, in parts of Australia and New Zealand where 'second-growth' reversion to forest precludes mechanised pasture improvement. (The 'stump-jump'
plough originated in Australia.) In brief, the costs of removal and control of
'heavy' vegetation limits the choice of enterprise and farming system, though
modern machinery and chemicals are reducing this constraint.

Weather and Seasonal Climate

The same applies to weather. Cold stress (e.g. freezing of crops or seeds,
cold-chill, rain-chill and wind-chill of outdoor livestock), heat stress, especially in livestock (p. 35), moisture loss due to wind (e.g. through added transpiration or drying out the soil) are examples of weather hazards, against
which protection is usually possible (Blaxter and Duckham in Johnson and
Smith 1965, and Webster and Park 1967). But such evasive action or 'weather-
proofing' investment (p. 71), whilst reducing the influence of weather on
farming systems, adds to the manpower and/or capital needs, and hence
affects the intensity and location of farming systems. Thus a combine that
can harvest 160 acres of cereals in south-east England, can only do 60–90 in
the wet Welsh uplands, but 750 acres in Western Australia where harvest
weather is highly predictable and lasts for months instead of weeks as in the
United Kingdom (Duckham 1963, p. 344). Herschfield 1962 shows, for the
U.S.A., a high positive correlation between total precipitation and the number of precipitation days. In medium or high rainfall areas, with much cloud
and frequent showers, reduced soil and crop workability adversely affects
tillage. It is one of the reasons why cereals are not grown in the Welsh uplands and why the main tillage area of the United Kingdom is the south and
east where the number of effective working days is higher than in the north
and west of the country (Duckham 1963).

As to regular seasonal features, in areas where the winter is severe and the
land covered with snow or ice and the soil frozen (e.g. the northern half of
North America, the northern part of Scandinavia), both winter field work and
grazing of, for example, winter forage crops, are impracticable. This has some
effect on choice of tillage enterprises but more effect on the location of livestock, particularly ruminants, Conversely, in seasonally drier climates, the
soil, for example, round the Mediterranean, may, after cereal harvest in May/
June, be too hard to plough until the autumn rains soften the soil. This may
preclude a possible summer crop, e.g. sorghum.

Finally, human work and mental energy tend to be limited when day temperatures are over 80°F, especially when humidity is also high. Such weather
obtains not only in equatorial conditions but for operationally important
parts of the year in some of the tropics, in some warm dry Mediterranean

climates near the sea, in the east of Australia and parts of the U.S.A. Mechanisation and air-conditioning are, however, easing this constraint in advanced
countries. But in less-developed warm or hot areas, heat stress and/or humidity can affect not only man, but also his work animals, particularly if
much of the feed of work cattle consists of fibrous straw or hay which has a
high heat increment (heat of digestion), so that their appetite and work output are low.

Work Conditions in the Tropics

This brings us to the tropics. Whilst the plough is the main instrument of
cultivation in the lowlands of tropical Asia, the hand hoe is still predominant
in tropical Africa, parts of the American tropics and many hill regions of
Asia. Failure to use the plough is not necessarily due to a lack of technology
and inventiveness on the part of the peoples concerned. Many factors may
militate against it: steep relief, a high incidence of animal disease precluding
the keeping of draught animals, the inability of farmers to feed such animals
to the standards required for work, smallness of farms and fields (especially
where land tenure has led to extreme fragmentation of holdings), and, in the
Old World, the activities of mound-building termites are amongst these. Some
of these factors pose even greater obstacles to the mechanisation of cultivation. A long dry season, during which many tropical soils bake so hard as to
be uncultivable except with the heaviest mechanical equipment, goes far to
explain both the small number of days worked annually by some tropical
farmers and the high cost of tractor cultivation when work can only be found
for the tractor during limited seasons of the year. Nevertheless, mechanisation
becomes more feasible and more imperative in tropical agriculture as the
wages of manual workers rise. This is well exemplified in the sugar-cane industry. Replacement of manual by mechanical harvesting of cane began first
in the cane-growing countries where labour is most expensive, Hawaii and
Australia. It is now spreading to those West Indian islands which have the
highest wage levels, such as Jamaica and Trinidad.

Many of the agricultural operations of the tropics are carried out within a
framework of shifting cultivation, which still occupies over 200 million people
on more than 30% of the world's exploitable soils (Nye and Greenland 1960).
When forest or savanna are cleared for only temporary cropping, cultivation
by the plough may not be feasible because of the presence of unsuspected
boulders in the soil, tree-stumps or standing trees, an often dense network of
plant roots, and termite mounds which it is not worthwhile to eradicate. In
such systems, fire becomes an important tool of agriculture. It is used to help
in clearing land of vegetation, the extreme example being the 'chitamene'
system of Zambia in which tree branches lopped over a large area are burned
on one small plot to fertilise the following crop with their ash. In the 'hariq'
system of cultivation in the Sudan, grass is burned and the seeds sown in the
resultant ash without the use of any tools at all. Fire also has to be used in

grassland management. In the absence of any means of conserving the flush of grass growth in the wet season, which is usually too much for the animals who have survived the dry season to eat, the resultant mat of dead grass is burned off to remove obstruction to new growth and to reduce the danger of uncontrolled conflagrations.

Wasteful in labour and in the use of land though shifting cultivation may be, it does maintain soil fertility and the problem of replacing its function in this respect is not easily solved in the environments where it is practised. Nor is it easy to wean the population from its practise, though this has to be attempted where it causes erosion and degradation of the soil, as amongst the Iban of Sarawak who are being encouraged to change to the cultivation of swamp rice in the river valleys (Freeman 1955). But such changes are not always the most urgent priority in tropical agriculture, since shifting cultivation does preserve a population, which is sufficiently sparse, in ecological balance with its environment.

References

Duckham, A. N. 1963. *Agricultural Synthesis: The Farming Year*. London.

Freeman, J. D. 1955. *Iban Agriculture*. H.M.S.O. London.

Herschfield, D. M. 1962. *J. App. Meteorology*. **1**. 4. 575–78.

Hidore, J. J. 1963. *Econ. Geog*. **39**. 1. 84–89.

Johnson, C. G. and Smith, L. P. (Eds.) 1965. *The Biological Significance of Climatic Changes in Britain*. London and New York.

Kates, R. W. 1962. *Hazard and Choice Perception in Flood Plain Management*. Univ. Chicago. Dept. Geog. Res. Paper No. 78. Chicago.

N.I.A.E. (National Institute of Agricultural Engineering). 1963. Test Report No. 371, Silsoe, Beds., England.

Nye, P. H. and Greenland, D. J. 1960. *The Soil Under Shifting Cultivation*. Tech. Comm. 51, Commonwealth Bureau of Soils.

Webster, A. J. F. and Park, C. 1967. *Animal Prod*. **9**. 4. 483–90.

Further Reading

Bates, W. N. 1957. *Mechanisation of Tropical Crops*. London.

The Synthesis and Comparative Analysis of Farming Systems

We now turn from the analysis of locating factors to their interactions and to their synthesis into farming systems.

Synthesis

In any locality, the farming systems are the results of past or present decisions by individuals, communities or governments and their agencies. These decisions are usually based on experience, tradition, expected profit, personal preferences and resources, social and political pressures and so on; they reflect the combined effects, on men, of the location and intensity factors discussed in previous chapters. They are, in fact, the *de facto* answers to the questions:

Ecologically feasible enterprises	(a)	What crop or livestock enterprises are, with present pest- and disease-control methods (i) ecologically practicable here, and (ii) at what level of input intensity? (Chaps. 1.2 and 1.3.)
Constraining ecological interactions	(b)	What ecological interactions occur between these enterprises? Are they such that one or more otherwise feasible enterprises must be eliminated, or such that the chosen enterprises must be combined in a special way (e.g. rotations) in the farming system (Chap. 1.4.)?
Constraining infra-structure factors	(c)	Are any of the ecologically feasible enterprises (as modified by (b) ruled out by infra-structure factors? (Chap. 1.5.)
Economically preferable enterprise combination and intensities	(d)	Which, of the enterprises now remaining on the list, are most profitable (or yield most food, etc.)? In what combinations and at what levels of input-ratios (e.g. R_1, R_2 and R_3 on p. 473) would they make best use of local land, climate and input resources in (i) the short term, and (ii) the long term, bearing in mind (iii) the degree of food and income security required by the individual farmers and the community? (Chaps. 1.4 and 1.5.)

Operational (*e*) What operational factors rule out or amend the size
constraints on   and method of any of the economically preferable
economic   enterprises and combinations thereof? (Chap. 1.6.)
preferences

Personal (*f*) Finally, are the enterprise combination, farming system
preference   and input ratios suggested by this process of the indi-
                  vidual farmer compatible with his own skills, enterprise
                  preferences health, age, capital, etc.?

This selection process is shown schematically in Fig. 1.7.1.

In its simpler arithmetical forms or in more sophisticated mathematical
forms (linear programming, dynamic programming, etc.) the above approach
is implicit in most contemporary planning techniques for individual farms or
districts. Under the term 'spatial programming', these techniques have been
applied to regional farm programmes by Egbert and Heady 1961 in the U.S.A.
by Randhawa and Heady 1962 in India, and by Birowo 1965 and Folkesson
and Renborg 1965 in Sweden. But in practice the questions about ecological
and operational factors and infra-structure can rarely be answered quantitatively even if they are asked, which is unusual. Locally acceptable assumptions on constraints, and on input/output ratios are, perforce, accepted in
lieu of local experimental or survey data, or of generalised scientific and economic data. Farm management planning is still, in effect, mostly the application of quantitative logic to systematised local experience, i.e. to accumulated
and collective subjective judgements of local conditions.

But, systematised experience cannot be fully used to measure the potential
of an area, or to quantify its current farming systems in scientific (e.g. energetic) terms. If enough were quantitatively known about (i) the constraints,
(ii) the input/output ratios in (*a*) to (*e*) above, and (iii) the interactions between
them, then, in theory, one ought to be able to synthesise models of farming
systems for any part of the world and compare such predictions with actuality.
But the best that we can do today is to accept systematised experience as valid
and enhance and evaluate it where possible by quantitative scientific and
economic data. We can then analyse and describe, in part quantitatively,
different localities and the many variables (and their interactions) that are
synthesised into farming systems. We can even suggest (Chap. 3.1) some possible predictive models.

Table 1.7.1 summarises (in the form of a tabulated catena) the groups of
factors, discussed in Chaps. 1.2 to 1.6 (rows 1 to 12), which have to be synthesised. It shows which factor group acts on other factor groups (columns 1
to 12) and its relative influence on ecologically feasible species on actual enterprises, and actual farming systems (rows and columns 13, 14, 15). In each

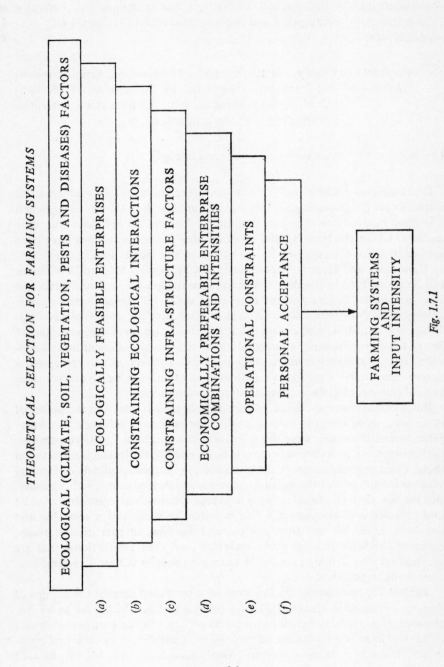

THEORETICAL SELECTION FOR FARMING SYSTEMS

(a) ECOLOGICAL (CLIMATE, SOIL, VEGETATION, PESTS AND DISEASES) FACTORS

(b) ECOLOGICALLY FEASIBLE ENTERPRISES

(c) CONSTRAINING ECOLOGICAL INTERACTIONS

(d) CONSTRAINING INFRA-STRUCTURE FACTORS

(e) ECONOMICALLY PREFERABLE ENTERPRISE COMBINATIONS AND INTENSITIES

(f) OPERATIONAL CONSTRAINTS

PERSONAL ACCEPTANCE

FARMING SYSTEMS AND INPUT INTENSITY

Fig. 1.7.1

relevant cell there is a subjective indication of the importance of the influence on the row factors upon the column factors. The table suggests that climate (1), land-form (2), operational facilities (7), agrarian (8), social (9) and economic (10) infra-structures, access to and prices of inputs (11), access to and prices at markets (12), and feasible species (13) are the more important in determining actual enterprises and systems.* The validity of this conclusion

Table 1.7.1

Summary of Interactions of Locating Factors

Code: Small action — Roman figures   Medium action — Italic figures
Large action — Bold figures

Locating factor	No.	Factor in left-hand column acts on factors:														
		1	2	3	4	5	6	7	8	9	10	11	12	13	14	15
Effective Climate	1			**3**	*4*	*5*	*6*	?7				11	12	**13**	**14**	**15**
Land (Physical Form)	2	1		*3*	*4*	*5*		7			*10*	11	12	**13**	*14*	*15*
Land (Biotic or Bio-geochemical Status)	3				*4*	*5*	6							13	*14*	*15*
Moisture Control Facilities (Irrigation, Drainage, Floods)	4	**1**				*5*	6	7	8					13	*14*	*15*
Soil Stability (Erosion, Salinity, etc.) and Texture	5				*4*		6	7						13	*14*	*15*
Unwanted Species (Weeds, Predators, Pests, Diseases)	6							7						*13*	*14*	*15*
Operational Facilities	7				*4*	*5*	*6*							13	*14*	*15*
Agrarian Infra-structure	8				*4*		6	7							*14*	*15*
Social Infra-structure	9				*4*	*5*	*6*	7	8		10	11	12		*14*	*15*
Economic Infra-structure and Development	10							7	8	9		*11*	*12*		*14*	*15*
Inputs—Prices and Access	11				*4*	*5*	*6*	7	8					13	*14*	*15*
Markets—Prices and Access	12								*8*						*14*	*15*
Ecologically Feasible Species of Crops and Livestock	13			3											*14*	*15*
Actual Enterprises	14			3		*5*	*6*					11	12			**15**
Actual Farming Systems	15			3		*5*	*6*					11	12		14	

Note: Factor groups 1 to 6 are ecological (Chaps. 1.2, 1.3); 7 is operational (Chap. 1.6); 8 to 12 are socio-economic (Chap. 1.5).
See explanatory note on Table 1.4.1.

must be judged from evidence in Part 2.† The relative importance of these factor groups will obviously vary between localities. Thus, on irrigated land in Pakistan where salinity is critical, soil regime may be a dominant constraint on enterprise and system choice. In subsistence economies, markets (row 12) may be less important and so on. The table does not, of course, show *how*, in different areas, actual farming systems are synthesised (see Chapter 1.1 and Duckham 1958, 1963, p. 424 ff), nor *why* the synthesis differs between localities, or, as noted above, varies between individuals, nations and localities‡.

We can, however, throw some light on the how and why of synthesis by

---

* The need to distinguish between enterprises and systems arises because farmers (individually or as groups) with say 3 enterprises may organise their production on several different systems. Thus, at the Sonning Farm of Reading University there are 3 systems, viz. intensive tillage, intensive alternating, intensive grassland (p. 18) for 5 enterprises, viz, wheat and barley; cows' milk; 'baby beef'; fat lamb; potatoes. But it could work with one system, viz. intensive alternating. The same enterprise, for example, cereal growing, may, of course, comprise different species in different places (wheat, sorghum, maize, rice, etc.).

† The classification and order of factors in Table 1.7.1 does not strictly follow that used in the analysis in Chaps. 1.2 to 1.6. This is because for summary purposes a more practical grouping seems advisable.

‡ Firey 1960, offers a logical, mainly sociological, approach to this process.

comparative analysis of the response of farmers to differing local conditions, i.e. to different incidences of factor groups. Comparative studies should show, firstly, what correlations there are in fact between locating and intensity factors and their interactions and, secondly, how far reality bears out the theoretical expectations discussed in this and earlier chapters. To this end, one needs to reduce the number of variables investigated simultaneously. Thus, one can, firstly, take ecologically similar areas (e.g. with warm dry Mediterranean climates) and study the apparent effects of operational and socio-economic factors. Or secondly, one can take areas where the socio-economic factors (other than input and market access) are fairly homogeneous (e.g. the U.S.A. or Australia or Western Pakistan) and study the apparent effects of varying ecological and operational factors.*

Comparative Analysis

Effect of Socio-economic Factors in Ecologically Similar Areas

The first course is not easy as it is sometimes difficult to find ecologically similar areas with widely different socio-economic backgrounds. But there are some places where such comparisons are possible and valid. Thus, the differences between the Israel and Jordan sides of their frontier or crossing the Rio Grande from the U.S.A. at El Paso into Mexico, are both dramatic examples of the influence of good infra-structure, economic development and favourable input and market regimes on the efficiency and intensity of farming systems. An equally dramatic example is that of the teams of Japanese farmers working in India (who are not associated with the Ford Foundation's Programme) 'farming on the same land, under the same climatic situation as the Indian farmers (but) with much greater capital intensity'. Some have put in pumps to improve water supply. 'All of them have small tractors, improved tillage tools, power duster and sprayers, and mechanical harvesters and threshers and modern pesticides. They do many things that Indian farmers can duplicate in the area of better husbandry. There can be no doubt about their being good farmers. Technically they are better educated and trained than the average Indian cultivator. . . .' (Finfrock 1965).

The tropics, of course, abound with examples of diversity of land use for socio-economic reasons in a uniform ecological environment because of the frequent juxtaposition of different races with totally different cultures. Extreme examples are the contrast between the pygmy hunters and the ordinary African farmers of the Congo, or between the primitive Sakai and the relatively sophisticated Malay and Chinese farmers of Malaya. Another contrast is between nomadic pastoralists and settled cultivators, a contrast which is not

* The defects of these two types of analysis are appreciated; for example, causal factors may, in any example, be confounded. But this would also be true of an historical approach. Until sufficient data are available to justify sophisticated statistical analysis, these two comparative approaches seem the most useful.

always related to differences in the quality of soil or grazing, for the Masai in East Africa in the 19th century chased the Kikuyu out of areas where the latter had practised successful arable cultivation and converted these to pastoral use. Equally striking is the contrast between the farming systems of European settlers or planters and those of the indigenous inhabitants amongst whom they settled, as in parts of Africa or in New Guinea and Melanesia. These do indeed sometimes reflect slight differences of ecological environment. In South Africa and Rhodesia, it may be true that the lands occupied by European farmers contain an undue proportion of the more fertile soils, though the impression received is often exaggerated by the fact that the African farmers have allowed much more soil exhaustion and erosion to occur than the Europeans. In Zambia, European farmers benefited economically from their lands being situated along the 'line of rail', whilst the Africans had poorer access to markets. In Kenya, on the other hand, much of the best arable land was already in African occupation when European settlers arrived, and the latter had often to be content with areas of lower rainfall which were sometimes only suited to pastoral use. But the contrast between the modern agronomic methods employed by European farmers and the simpler practices of the Africans can be attributed almost entirely to socio-economic factors, and above all to the access to capital and the education and technical knowledge enjoyed by the Europeans.

In temperate areas with broadly similar climates, for example, along the north of the Mediterranean basin, Spain and Greece are relatively backward, whereas France, northern Italy and Yugoslavia are moderately to well advanced. Spain, southern France and Italy (south of the Po Valley) have broadly the same ecological and operational features, though Spain is perhaps the least ecologically favoured of the three. Fifty years ago they were agriculturally relatively homologous. Today, however, southern France and southern Italy are changing rapidly, either through local industrialisation and development as in the Bas Rhône scheme (p. 282), or through planned movement of population to industrial areas and farm amalgamation schemes as in Italy. Most countries north of the Mediterranean contrast favourably with those south of it (where the climate, however, is less reliable and the economic development much poorer) but unfavourably with the advanced 'Mediterranean' agriculture of Australia which is, in turn, in some respects, less advanced in enterprise selection and system organisation than California and the south-west of the U.S.A. (see Chaps. 2.2, 2.3, 2.4 and 2.7). Whilst in Egypt, the annual inundation from the Nile effectively raises the A–E–T and crop yields, and the Nile Valley supports a dense population which has a low standard of living (p. 77); vigorous attempts to control population growth, to modernise agriculture and to industrialise are, however, in hand (Mansfield 1965).

The countries surrounding the North Sea basin (Chaps. 2.5 and 2.6) are in a relatively homogeneous ecological area, especially eastern U.K., Denmark, southern Sweden, the Netherlands, coastal West Germany, Belgium, the plains of northern France and Brittany. Over all this region technology is highly interchangeable. Thus, many genotypes are used in common (especially cereals, grasses, Friesian cattle, Landrace pigs) and there is considerable inter-country trade in agricultural machinery, etc. All have, by world standards, reasonable access to inputs, so the differences are presumably attributable to agrarian structure (especially farm size, rural population density and land tenure systems) and to prices at, and access to, markets.

The cool humid climate of the North Sea basin is broadly reproduced in the South Island of New Zealand (p. 213), in coastal British Columbia (p. 133) and the Pacific North West of the U.S.A. (p. 147). All these areas have good infra-structures and developed economies. The drier eastern counties of the United Kingdom (where intensive tillage and alternating systems with considerable livestock predominate) are ecologically comparable to the Canterbury Plains of New Zealand and to the Willamette Valley of the Pacific North West. Yet in the Canterbury Plains one finds an intensive grazing system with only a little tillage; this is largely because market access is poor (p. 473). In the Pacific North West manpower shortage and the need to specialise to exploit capital result in specialised one- or two-enterprise farms under perennial fruit, tillage, alternating or grazing systems (p. 147). Finally, there are some areas which are ecologically, economically and socially similar, and where the inter-country differences in farming are almost negligible, e.g. between Scania in southern Sweden and the isles of Denmark and across the frontier between Canada and the U.S.A.

EFFECT OF ECOLOGICAL AND ACCESS FACTORS IN AREAS WITH HOMO-
GENEOUS INFRA-STRUCTURES AND ECONOMIC DEVELOPMENT

A more rewarding approach is the second, i.e. to examine national areas where the socio-economic factors can be regarded as a constant and the main variables are ecological and access factors. Such a study can usefully take the form of transects along selected longitudes or latitudes through countries which have a wide or informative range of ecological conditions. This is done in Part 2.

Acknowledgements. For comment and criticism: H. J. Critchfield, S. Fox, L. P. Smith, J. A. Taylor, E. W. Russell.

References

Birowo, A. T. 1965. *Lantbrukshögskolans Annaler. 31.*
Duckham, A. N. 1963. *Agricultural Synthesis: The Farming Year.* London.
Egbert, A. C. and Heady, E. O. 1962. *Regional Analysis of Production Adjustments in the Major Field Crops.* U.S.D.A. Tech. Bull. 1,294, Washington, D.C., see also

Chap. 12 in E. O. Heady *et al.* (Ed.) 1961. *Agricultural Supply Functions.* Iowa Univ. Press, Ames, Iowa.

Finfrock, D. W. 1965. Personal Communication. Ford Foundation: Intensive Agriculture Districts Programme in India.

Firey, W. 1960. *Man, Mind and Land: a Theory of Resource Use.* Free Press of Glencoe, Illinois, U.S.A.

Folkesson, L. and Renborg, U. 1965. *Optimal Inter-Regional Allocation of Agricultural Production and Resources in Sweden.* (Paper presented at Symposium on Operational Research. Dublin, 1965.) Mimeo.

Mansfield, P. 1965. *Nasser's Egypt.* Penguin African Library, Harmondsworth, England.

Randhawa, N. S. and Heady, E. O. 1962. *J. Farm Economics.* **46.** 1. 137–49.

Renborg, U. 1962. *Studies on the Planning Environment of the Agricultural Farm.* Agric. Econ. Dept., Agri. Coll. of Sweden, Uppsala.

Part 2

FARMING SYSTEMS

CHAPTER 2.1

The Classification of Farming Systems

As indicated on p. 100, this part of the book studies the influence of eco-logical factors (Chaps. 1.2, 1.3 and 1.4) when socio-economic factors (Chap. 1.5) can be regarded as held constant (e.g. within a national or regional bound-ary). Each study of a country or region begins with a summary of its economic development, infra-structure and chief farm products; it then examines the relation between (i) ecological parameters or classes, and (ii) farming systems by following, when possible, suitable transects. The former (i) have been out-lined in Chaps. 1.2 and 1.3; the classification used for the latter (ii) is given below (p. 106).

There is no recognised international farm classification, although the Com-mission for Agricultural Typology of the International Geographical Union may produce one. Each branch (and sometimes each individual) of the agricultural or geographical profession designs one to meet its own needs. Most classify by intensity or land use or both. Some classify *intensity* by climate or by land capability; others by type or rate of production (e.g. grain yields, gross or net primary production, proportion of cash sales from differ-ent enterprises, value of sales per unit areas, etc.); others again by particular inputs (e.g. manpower or capital per unit area, etc.), or by size of farm (e.g. small farms are (doubtfully) assumed to be more intensive than large ones), or, in respect of grassland, by stocking rate (which is, biologically, more a measure of conversion capacity of livestock than of production or input, though economically it is obviously a capital input). Each of these classifica-tions can be, in part at least, quantified, but they are of limited value, at least in temperate countries where productivity and farming systems are increas-ingly influenced by the sum and the balance of the inputs (see Tables 2.3.4 and 2.3.5). In this book, intensity means either the actual sum of inputs (other than 'natural' ecological factors) or the input ratios as defined on p. 6. Neither actual total inputs or input ratios can usually be quantified on inter-nationally comparable terms, so our classification into VERY-EXTENSIVE, EXTENSIVE, SEMI-INTENSIVE and INTENSIVE (see Table 2.1.1) is subjective and arbitrary.

Land-use classifications show an equal diversity, but apart from those which classify solely on species populations (e.g. pigs per unit area) most distinguish between arable (though this term sometimes includes temporary or short-term leys) and grassland or pasturage. In this book we use four classes of farming land use: PERENNIAL TREE OR SHRUB CROPS, TILLAGE (annual crops), GRAZING OR GRASSLAND (pasture and ruminant livestock) and ALTERNAT-ING between tillage and either grassland, fallow or bush which is sometimes

Table 2.1.1

Classification of Farming Systems

	Tree crops		Tillage with or without livestock		Alternating tillage with grass, bush or forest		Grassland or Grazing of land consistently in 'indigenous' or man-made pasture	
	Temperate	Tropical	Temperate	Tropical	Temperate	Tropical	Temperate	Tropical
Very Extensive	Cork collection from Maquis in southern France	Collection from wild trees, e.g. shea butter	—	—	Shifting cultivation in Negev Desert, Israel	Shifting cultivation in Zambia	Reindeer herding in Lapland. Nomadic pastoralism in Afghanistan	Camel-herding in Arabia and Somalia
Extensive Examples	Self-sown or planted blueberries in the north-east of the U.S.A.	Self-sown oil palms in West Africa	Cereal growing in Interior Plains of North America, pampas of South America, in unirrigated areas, e.g. Syria	Unirrigated cereals in central Sudan		Shifting cultivation in the more arid parts of Africa	Wool-growing in Australia. Hill sheep in the U.K. (Sheep in Iceland.) Cattle ranching in the U.S.A.	Nomadic cattle-herding in East and West Africa. Llamas in South America
Semi-Intensive Examples	Cider apple orchards in the U.K. Some vineyards in France	Cocoa in West Africa. Coffee in Brazil	Dry cereal farming in Israel or Texas, U.S.A.	Continuous cropping in congested areas of Africa. Rice in S.E. Asia	Cotton or tobacco with livestock in the south-east of the U.S.A. Wheat with leys and sheep in Australia	Shifting cultivation in much of tropical Africa	Upland sheep country in North Island, New Zealand	Cattle and buffaloes in mixed farming in India and Africa
Intensive Examples	Citrus in California or Israel	Rubber in S.E. Asia. Tea in India and Ceylon	Corn Belt of the U.S.A. Continuous barley growing in the U.K.	Rice and vegetable growing in south China. Sugar-cane plantations throughout tropics	Irrigated rice and grass beef farms in Australia. Much of the east and south of the U.K., the Netherlands, northern France, Denmark, southern Sweden	Experiment stations and scattered settlement schemes	Parts of the Netherlands, New Zealand and England	Dairying in Kenya and Rhodesia highlands
Typical Food Chains	A	A	A, B	A	A, B, C, D	A (C)	C (D)	C

called alternate or rotational husbandry. *Tree and shrub-crop systems* imply
perennial crops of tree fruit, beverage crops, sisal, etc., which may or may not
be intertilled or intergrazed. *Tillage* systems, broadly defined, are where 75 %
or more of the ploughable land is in tillage crops or one year fallow; *grassland*
systems, where 75 % or more of the ploughable land is in temporary leys or
permanent pasture or where non-ploughable land is in 'cultivated' grassland
or grazable shrub, scrub or 'natural' grasses; *alternating* systems where less
than 75 % and more than 25 % of the ploughable land is in tillage which is
alternated with grassland (mostly temporary leys) or with long-term fallow
or forest regeneration. *Livestock* can be found on all systems but land
use on grassland systems is usually confined to ruminants (cattle and
sheep). A fifth class, *factory*, viz. practically landless factory-type production
of milk, pigs, poultry, feed-lot beef, mushrooms, etc. is often additive but not
interactive with any of the four major land-use classes. The above land-use
classification, though less subjective than intensity classing, is also arbitrary
and in practice difficult to define. Thus, a farm in the Interior Plains of the
U.S.A. may appear statistically to be Extensive Alternating but in practice the
cereals are grown on the flat cleared land and the livestock are grazed on
'native' pasture land which is not suitable for, or not accessible enough for
tillage. Again, as noted on p. 97, Reading University's Sonning Farm would,
from acreage and livestock statistics and from annual accounts, appear to be
Intensive Alternating, but in fact one-quarter of the area is in Intensive
Tillage and a fifth in very Intensive Grassland.

But despite these drawbacks, the simple classification in Table 2.1.1 into
(*a*) Intensive or Extensive and (*b*) Tree crops, Tillage, Alternating and Grassland, with a further subdivision into (*c*) Temperate or Tropical, seems most
practicable for this book.

NOTES ON TABLES IN PART 2

(A)

Tables of Crop Acreages, Livestock Numbers and Yields

The tables, unless otherwise indicated, are from F.A.O. 1966, *Production Yearbook*
1966, Vol. 20, F.A.O. Rome, and from United Nations 1966, *Statistical Yearbook*
1966, U.N., New York. The values in these tables are for 1965 or the mean of 1963,
1964 and 1965, but values in the text are, in some cases, for later dates or from
other sources.

(B)

Climatic Tables

Sources are primarily the U.K. Meteorological Office 1958 (M.O. 617 Series).
(*Tables of Temperature, Relative Humidity, Precipitation for the World.* (which gives
values for selected stations in each continent), H.M.S.O. London), Thornthwaite
(below and p. 122), or local sources. Brooks, C. E. P. 1960, *Climate in Everyday Life*,
London, has been used for mean sunshine hours. The Moisture Indices (I_m) are
calculated by the method of Thornthwaite and Mather 1955 and Thornthwaite
1962–64 (see below).

Transects Tables (Definitions and Sources)

Notes on Rows

3 and 4   Approximate start and end of each sector.

5   *Miles*. Distances 'as the crow flies'.

6   *Climatic-types*. See p. 40.

7a   *Thermal Growing Season*

     —*Months at or above* $42\frac{1}{2}°F$ *(means)*. See p. 34 and Duckham, A. N. 1963 (*Agricultural Synthesis: The Farming Year*, London, p. 165). Data in rows 7a, 7b, 8, 9, 10 are from Meteorological Office, London, M.O. 617 Series.

7b   —*Months at or above* $72\frac{1}{2}°F$ *(means)*. See p. 35.

8a   —*Potential Evapo-transpiration (P–E–T)* in the thermal growing season. See p. 27.

8b   —*Actual Estimated Evapo-transpiration (A–E–T)* in the thermal growing season. See p. 40. Rows 8a and 8b are abstracted from C. W. Thornthwaite Associates, 1963–64.

     ('Average Climatic Water Balance Data of the Continents', *Publications in Climatology*, **15**. 2; **16**. 1; **16**. 2; **17**. 1; **17**. 2. Laboratory of Climatology Centerton, N.J., U.S.A.)

9   —*Growth (in the thermal growing season) limited by*: Growth includes animal production.

10   *Precipitation — annual mean inches*. Given as range for the districts in the sectors. Sources: Finch, V. L., Trewartha, G. T., Robinson, A. H. and Hammond, E. H., 1957, *Elements of Geography: Physical and Cultural* (New York and London. 4th edn.) or M.O. 617 Series; or as indicated in text.

11   *Precipitation—C.V.%*. Coefficient of variability of annual rainfall; mostly from Finch *et al*. 1957, p. 143.

12   *Weather Reliability*. Adapted by Wilsie, C. P. 1962, *Crop Adaptation and Distribution*. San Francisco and London, p. 411, from Visher, *Econ. Geog*. **31**. 82–86, 1955.

     Visher's Rating: 0 = lacking; 1 = poor, little or low; 2 = medium, average or fair; 3 = above average, good; 4 = superior, excellent, large.

13   *Soil Zones/Groups*. From Finch *et al*. 1957 and sources in countries concerned.

14   *Soil — pH*. Various sources.

15   *Vegetation*. From Finch *et al*. 1957, but also local sources, e.g. C.S.I.R.O. 1960 (*The Australian Environment*, Melbourne) and McLintock, A. H. (Ed.) 1959 (*Descriptive Atlas of New Zealand*, Wellington, N.Z.).

16   *Relief*. Finch *et al*. 1957, and other sources.

17   *Market Access*. Visher's classification. See note on row 12. Assumed to apply also to input access.

18   *Farming Systems*. See Table 2.1.1. From sources indicated in references in chapters in Part 2, correspondence and other sources, including personal observations.

19   *Major Enterprises*. As for row 18 and various economic reports.

CHAPTER 2.2

Canada and the U.S.A.

Economic and Social Background

Canada and the U.S.A. have a population of some 220 million. The original Amer-Indians and Eskimos now form less than 1% of this total but are increasing rapidly. Since the advent, first of Spanish and later of northern European intruders and settlers, the population now mainly consists of white Europeans with, in the U.S.A., a strong minority of descendants of the Negro slaves which the early settlers bought from Africa. Since 1492, a sparsely populated area then inhabited by hunting or primitive agricultural tribes has become the most advanced and powerful industrial bloc in the world. By grafting the efficient production of industrial inputs onto the inventions of private industry and on to the research and advisory services established in the U.S.A. at the Land Grant Colleges (founded in 1862) and their Canadian equivalents, North America has given modern farming many of its more familiar tools (e.g. the tractor, the combine harvester). It has a technologically advanced labour-saving agriculture that is second to none. The farmers who affected this transformation had four great assets; ingenuity, fortitude, plenty of virgin land (which they had to find out how to farm), and lastly, a partial break with European social constraints of feudal or even earlier origins.

Both the U.S.A. and Canada have high outputs per man in factory and in farming and vast natural resources of land, water, mineral and power supplies, which have, perhaps, been too profligately exploited, but many of which (e.g. in northern Canada) are only just being tapped. The result is high material living standards (Table 1.5.1). On the other hand in the U.S.A. at least, the concept of State-aided social welfare (e.g. national health schemes) as part of the standard of living is not so highly developed as it is in western Europe or Australasia. Paradoxically, however, American farmers are as dependent for their annual incomes on the activities of government as are the farmers in any other non-communist country, whilst it is only the technical and regulatory services of the U.S.D.A. (United States Department of Agriculture) that make their high use of industrial inputs practicable and profitable. Although the Canadian economy, and especially its farming, is spread in a thin strip 4,000 miles across the continent, and although some Canadian provinces (especially French-speaking Roman Catholic Quebec) are, or were, conservative, Canadian agriculture rivals that of the U.S.A. in technical efficiency and services. The social, economic and administrative differences between Canada and the U.S.A. are, on a world scale, small. For present purposes we can regard these two countries as one large area where, despite some major pockets (e.g. the south-east of the U.S.A., Quebec) with special social, political or economic problems, socio-economic conditions can be held constant whilst we examine

the effects of ecological, input access and market access factors on a continental scale.

Climate, Vegetation, Soils

Figs. 2.2.1, 2.2.2 and 2.2.3 and Tables 2.2.3, 2.3.2 and 2.3.3, demonstrate a
very wide range of ecological factors which are reflected in an equally wide
range of farming systems (Fig. 2.2.4). The uplands and mountains of the east
and west are separated by the Interior Plains,* which are drained to the south

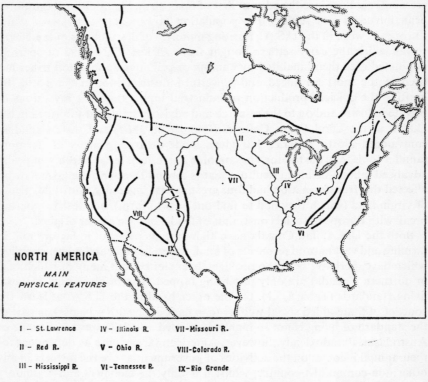

I — St. Lawrence	IV – Illinois R.	VII–Missouri R.
II – Red R.	V – Ohio R.	VIII–Colorado R.
III – Mississippi R.	VI – Tennessee R.	IX–Rio Grande

Fig. 2.2.1 North America. Indication of some of the Main Physical Features

by the Mississippi River system and to the north by rivers which run to the
Arctic Ocean. These plains are the agricultural heart-land of North America;
they are complemented by the farming in the Appalachian uplands and the
coastal belt on the east, and by the ranching and the irrigated farming in the
Rocky and allied mountain systems and valley floors of the west. The climate
and natural vegetation ranges from alpine over 12,000 ft in the Rockies, and
tundra and sub-boreal forest in the north through cool humid forests and

* The Interior Plains comprise the Great Plains and Prairies west of 100°W *and* the Mid-
West between the Great Plains and the Appalachian Mountains.

110

cool dry grasslands to warm humid forests in the south-east and hot dry desert
or Mediterranean climates and vegetation in the south-west. The soil groups,
soil pH, and the farming systems reflect this great climate range. A transect
(Fig. 1.2.5) from east to west at latitude 40°W shows well how the mean
P–E–T* stays fairly constant, except in the mountains, and how, as mean
rainfall decreases as one moves away from the coast(s) so mean A–E–T† in the

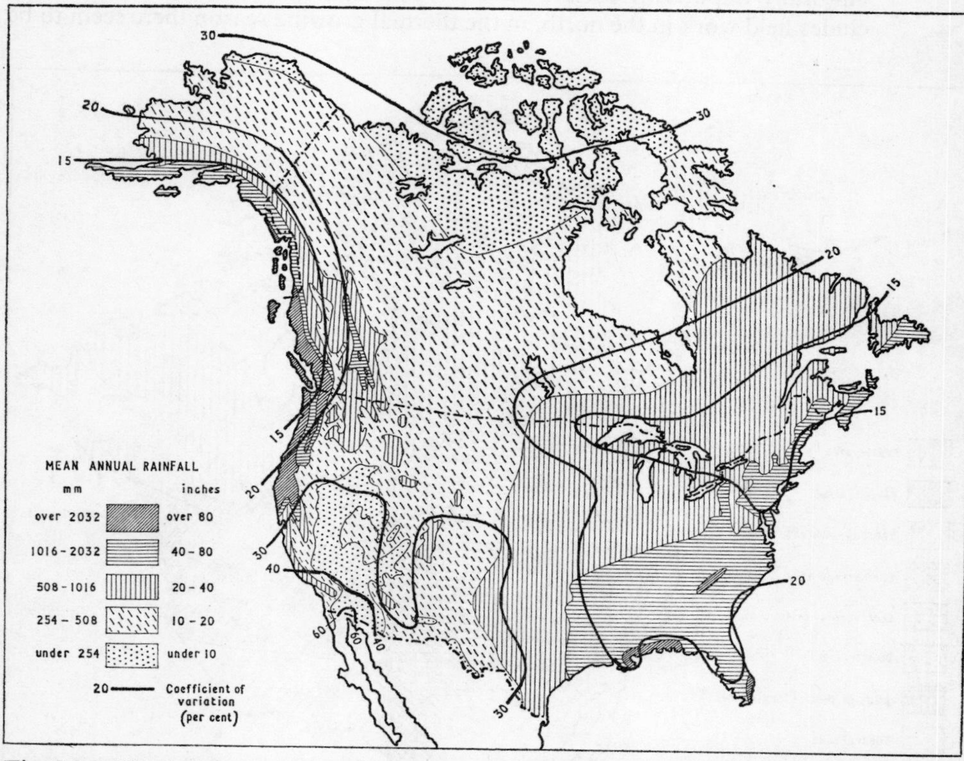

Fig. 2.2.2 North America. Mean Annual Rainfall (and its variability). From Finch *et al.* 1957.

thermal growing season falls because the mean annual precipitation is lower.
Moreover, as the mean annual precipitation falls, its year-to-year reliability
decreases (Fig. 2.2.2 and Herschfield 1962) (compare Australia), so except
where irrigation is used, the farming systems become either extensive tillage
or extensive grazing (Fig. 2.2.3, Tables 2.2.3, 2.3.2 and 2.3.3) in contrast with
the intensive systems of the better-watered areas. Drought is, therefore, a
major hazard over most of the land west of 100°W, except in the humid north-
west and except where the snow-fed rivers of the western mountain systems
can be used for flood or sprinkler irrigation in valleys or on favoured plains,

* P–E–T = Potential Evapo-transpiration (Thornthwaite), (see Chap. 1.2, p. 27).
† A–E–T = Actual Estimated Evapo-transpiration (Thornthwaite).

E

FARMING SYSTEMS OF THE WORLD

e.g. in Alberta. Even in humid areas, to counteract short-term droughts, sprinkler irrigation, often using ingenious labour-saving equipment, is rapidly growing in importance.

The climate as a whole is more extreme, and serious weather hazards are greater than in New Zealand or western Europe. Crop insurance, reserve stocks of grain, for example, and disaster precautions are, therefore, agriculturally important. On the other hand, though the long hard winter precludes field work in the north, in the thermal growing season there seem to be

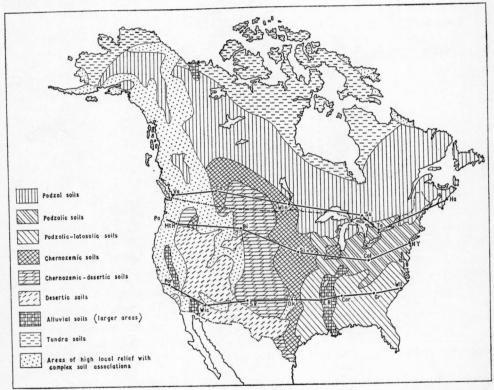

Fig. 2.2.3 North America. Principal Soil Zones with lines of transects. *Source:* Finch *et al.* 1957

more periods of good working weather than in western Europe. This is partly because rainfall intensity tends to be high. This means that there are more hours of sunshine or rainless periods per inch of rain (see also New Zealand), but, in sloping and hilly country, this advantage is offset by major water erosion problems which make soil conservation projects necessary. In the plains, conservation against wind erosion is important.*

* For climatic detail, see the climatic maps published by the U.S. Census Bureau and Canada Dept. of Forestry 1966.

The soil groups (Fig. 2.2.3) follow the expected pattern from podsolic in the east and Pacific North West, to desertic in the dry areas. pH is, as expected, high in low- and low in high-rainfall areas. Within the main soil zones there are, of course, variations in texture but heavy, hard-to-work clays are infrequent and, as a whole, North American soils are probably freer working than those of western Europe, though many soils in the north-east are very stony. With land plentiful and manpower scarce, patches of heavy or low-lying land are often simply not farmed at all, except where there are large-scale land-drainage schemes as in south-east Missouri or where flood-protection works play key roles, for example in the lower Mississippi valley. Since the advent, last century, of the prairie-buster plough, the tall grass prairies of the Interior Plains have been cultivable, whilst human sweat and latterly tree-clearing machinery have disposed of the forests on most good, workable, well-watered soils in the east and Pacific North West. But where the terrain is difficult to work, the soilpoor or the A–E–T is not high or reliable enough for cropping, natural, cut-over and man-planted forests occupy over 1,500 million acres, either as untilled patches on farms or in commercially or nationally owned forests. Similarly, in the natural grassland zones, much of the land is not tilled or even grazed because it is too uneven for mechanisation, lies too wet, or is badly shaped. Only 39% of the farmland in the U.S.A. is cropped or cultivated, and in Canada 60% of occupied farmland is listed as improved.

Agrarian Structures

In both Canada and the U.S.A. the owner-occupied family farm is not only the commonest but also ideologically the most acceptable form of tenure, which is supported by comprehensive government-sponsored co-operative and private mortgage facilities. Share-cropping, once typical of the cotton-growing areas of the south-east, is now history, but crop-share farming is of growing importance in the Corn Belt (Fig. 2.2.4), in the irrigated valleys of the west and other areas where high initial capital investment makes such renting helpful to the young go-ahead farmer. Further, to achieve the economies of scale and specialisation, there is a marked tendency for large, labour-employing farms to grow in number, especially in ranching and in some irrigated areas. Vertically integrated organisations where, for example, a large feedstuff firm capitalises and controls the activities of small men and pays them, for example, on an egg or broiler output basis, are also developing rapidly despite the resistance of the farmers' organisations. But in general, the family-farmer who produces and sells his produce to large industrial firms or farmer co-operatives, has shown a great capacity to secure the benefits of a highly industrialised society without excessive loss of his economic freedom.

Farm size varies greatly from a few acres in specialist poultry farms to 40 or 50 acres in one-man irrigated fruit farms up to thousands of acres on live-stock ranches (see Tables 2.3.4 and 2.3.5). In the eastern half of the continent a farm size of 200–300 acres is modal, whilst in the Great Plains 1,000–2,000-

Part 2
Farming
Systems

acre family farms are more typical. In 1966, the average size of 3,176,000 farms in the U.S.A. was 359 acres; but half these farms produced 90% of the output. Farm size is negatively correlated with favourable climate. Where A–E–T and P–E–T are roughly equal, or where A–E–T can be raised to P–E–T by irrigation, farms are usually 300 acres or less. Where A–E–T is substantially lower than P–E–T (e.g. on the plains west of 100°W) few farms are under 1,000 acres. Farm sizes are, however, increasing rapidly as mechanisation speeds ahead and economic pressures simultaneously permit and enforce amalgamations. In this, the relatively unconstrained land tenure system provides a flexibility to meet technological and economic changes which custom, law, population pressure or land shortage precludes in many, if not most, other parts of the world.

A comparable flexibility has spread recently to manpower which is now probably occupationally and geographically more mobile in North America than anywhere else in the world. Some farmers work seasonally or occasionally for other farmers or at other jobs. In the Great Plains a 'side-walk' farmer may work in the town for most of the year but comes to his land just long enough to drill and later to harvest a spring wheat crop. A 'suit-case' farmer moves seasonally between his several farms, e.g. main crop potatoes in Maine in the north-east and early potatoes in Florida in the south, or winter wheat in Texas and spring wheat in Montana. Specialised contractors play a big role in some areas (e.g. the strawberry and lemon farms in California mentioned in pp. 162 and 166). Finally in the U.S.A. comes a seasonal labour force (in 1963 = 617,900 of which 115,450 were migratory). This is particularly important where vegetables, tree fruit or sugar-beet are grown, but is declining rapidly since the advent of cotton-picking machines, of plant varieties suited to mechanical harvesters and of stricter control of Mexican immigrants.

This vast and varied structure of farms and farm manpower is well served by efficient, progressive and usually highly competitive input industries, marketing and processing firms, banks, motor and rail transport, etc. These service industries have, along with the farms, the use of efficient air, telecommunications, water and power networks. Farmers benefit from a sound general educational structure, high literacy, and in the U.S.A., a unique system whereby Federal, State and County governments co-operate in providing technical education, and increasingly high quality research and advisory (extension) services to agriculture through the Land Grant Universities in each state. These are backed by the extensive research, market information and regulation, disease control and commodity price support (based originally on the concept of parity of prices between what the farmer sells and what he buys but now too complex for description here), commodity storage, soil and water conservation, and the numerous other services of the vast U.S.D.A. Skilled private advisory, managerial and market information services such as Doanes and the Washington Farm Letter, help to raise both productivity and profits on many U.S. farms. In Canada despite some defects in the advisory services, a lower mobility of manpower and the handicaps of a harsher climate and scattered farm lands (p. 134), the infra-structure and technical services are good.

In brief, ample land and raw material resources, good access to industrial inputs, flexibility of land tenure and manpower, good general and agrarian infra-structure give both Canadian and U.S. farmers opportunities to achieve high output per man and, in well-watered areas, high yields per acre.

Mechanisation and Manpower

Supported by this infra-structure, both crop and, increasingly, animal production in these two countries are highly mechanised to save man hours and to reduce muscular work. Fig. 2.3.3 and the sample farms outlined below (pp. 128, 130, 143, 145 and 160) illustrate vividly the lengths to which field and farmstead mechanisation has gone. But the economic viability of the resultant heavy investment in buildings, tractors and machinery depends, in large part, on increasing firstly, the mean size of farm business (in order to reduce overheads per unit output) and secondly, greater specialisation. The former tends to generate chronic surpluses (Hathaway 1963, whilst, without *ad hoc* governmental price support, the latter raises economic vulnerability, although of course the resultant division of labour enhances both productive skills and output per man.

The added value per man on the farm in Canada and the U.S.A. is about 50% higher than it is in the more advanced countries of western Europe (p. 289, Van den Noort 1968, G. Sharp and C. W. Capstick 1969) where climates, husbandry practices, yields and annual rates of yield increase are as good as those of comparable areas of Canada and the U.S.A. This greater output per man is attributable to high capital intensity, mechanisation, and other industrial inputs, specialisation, flexibility of farm size, mobility of labour, rapid responses to cost/price factors and to less mental rigidity than is found in north-west Europe. In brief, the differences in man productivity must be sought in social and economic and not in ecological factors (see Chap. 3.1). But though output per man hour has more than doubled in 20 years, there are socially and politically important pockets of excessive farm population, viz. the southern Appalachians and southern Atlantic states in the U.S.A., Quebec and the Maritime provinces in Canada.

One result of the overall fall in farm population has, in both countries, been a loss of the key political power which farmers' organisations and the farming vote exercised a generation ago. Further, whereas in Canada the monolithic Canadian Federation of Agriculture persists, in the U.S.A. the three rival farmers' organisations have been supplemented by splinter 'commodity' interests, whilst the redistribution of political constituencies is also reducing the power of the farmers' votes. In neither country is hired labour effectively organised; American farmers, in particular, have no time for trade unions.

In both countries farm living standards are relatively high (see Table
1.5.2). Apart from the poverty pockets, nearly every farmer or regular farm
worker has a house with electricity and indoor water supply, telephone, refri-
gerator and one or more motor-cars. Most of the farm population, apart
from those of Negro origin who are mainly in the south-east, are of northern
European origin (especially British, Dutch, German and Scandinavian), although on the U.S. West Coast there are Japanese, Filipino, Portuguese and
Italo-Swiss farmers. Italian or Jewish farmers are rare.

Land Utilisation

Of the United States land surface of 2,271 million acres, 434m are in non-
agriculture use, 746m are in forest and woodland, 359m in tillage crops and
fallow, 66m in grassland and 34m in erosion control, etc., crops. Of the cultivated area of 460m acres, about one-third is drained and about 35m are
irrigated, mainly in the arid western half of the country. By contrast with the
Netherlands where nearly all and the U.S.A. where 20% of the land surface
is cultivated, in Canada this figure is about 5% (or 100m acres) though the
area in farms including 'natural' grassland is 170m acres, the rest (not in
farms) is tundra, mountains and forests.

CROPS AND GRASSLAND

Table 2.2.1 summarises acreages, livestock numbers and yields. Although
vast areas are in ranches and the 'cultivated' grassland acreage is significant,
it is cereals, oil-seeds and specialised cash crops (such as cotton, tobacco,
fruits and vegetables) that dominate Canadian and American cropping even
if sales of livestock products dominate gross income.

East of 100°W, the main centres of production lie in the 'dairy' belt (Fig.
2.2.4) round the Great Lakes, along the St Lawrence estuary and in the northern Appalachians—each of which has large urban populations; in the deep-
soiled Corn Belt lying to the south of the Canadian Shield; on the eastern edge
of the Great Plains; on the Atlantic seaboard and Gulf Coast; and along the
Missouri and Mississippi Valleys. West of 100°W, the irrigated valleys of the
western mountain systems are of great and growing importance. The vast
areas of extensive grazing, and the greater part of the spring and winter wheat
belts are less productive (see also Table 3.1.1).

The dominant cereal is *maize* (for grain or fodder) which reaches yields of
6,500 lb or more grain per acre on better farms in the Corn Belt, but is grown
wherever moisture is adequate south of about latitude 45°N, and is still an
important item of human diet in the south-east of the U.S.A. Elsewhere the
major outlet for grain maize is livestock. North of 45°N, where moisture permits, maize may be grown for silage, but the mean frost-free period and the
thermal growing season are too short, and July/August temperatures may not
be high enough for the grain to ripen, or at least to dry out enough. In warm
humid areas of the U.S.A., *small grains* (wheat, barley, oats) yield less than

116

maize but they are widely grown, often as a break crop and to spread seasonal work, in the maize-growing districts. In the semi-arid regions, such as the spring and winter wheat belts, in the Palouse area of Washington State and irrigated valley bottoms in California, wheat and barley are the dominant small grains; but, in more arid regions with hotter summers (e.g. Texas High Plains), grain sorghum, especially now that better adapted varieties are available, is increasingly popular as a summer crop, often in rotation with winter wheat if there is enough moisture in the soil (see Israel, p. 316). Wheat is the major crop in western Canada and is likely to remain so. Hard red wheats for bread and breakfast cereals are grown in the spring and winter wheat regions (Fig. 2.2.4), durum wheats in the spring wheat region, soft white or red wheats in the Palouse (Washington), in irrigated western valleys, in parts of the Dairy Belt and in Ontario.

Thailand, Burma and the U.S.A. are the world's leading *rice* exporters. Rice is grown in the Mississippi Delta area of Arkansas, Mississippi and Louisiana and in Texas, but the highest yielding and most important rice area is in California where the crop is highly mechanised and the seed is often dropped into the water from aeroplanes. Here, as in Australia (p. 186), the Camargue (France) (where American machinery is used) (p. 283) and the Po Valley (Italy), irrigated yields are much higher than in the rice-growing areas in the tropics. This is because P–E–T is greater at the summer peak in Mediterranean climates, and because technology, inputs and infrastructure are superior in U.S.A., Jennings 1966 states that breeding and selection have largely erased the differences between the temperate varieties of the *japonica* and *indica* types; much of the American rice originated from the latter.

In the U.S.A. *oil-seeds* (soya-beans in the Corn Belt, groundnuts in the south-east, as well as cotton-seed, linseed and sunflower) are important sources of protein supplements for livestock whilst their oil is used for human consumption and margarine is highly competitive with butter. Linseed and rape-seed are grown in western Canada.

Sugar-beet is grown under irrigation (usually surface irrigated but sometimes sprinkled) in all of the western states plus Kansas, Oklahoma and Texas. It is grown without irrigation in the Red River Valley of the north, and in Michigan, Ohio and New York. Beet is also grown in Quebec, south-west Ontario and the Canadian Prairies. *Sugar-cane* is grown in Louisiana and Florida on the Mainland and in Hawaii.* In the latter, very high yields are obtained from high solar energy receipts, adequate rain and very sophisticated science-based methods. Very early Irish (white) *potatoes* are found in the south-east (especially Florida) and in the irrigated valleys of the west (e.g. Kern County, California.) Thence, moving progressively through later harvests as one moves north, one comes to such important main-crop areas as New York, Presque Isle in Maine and adjoining New Brunswick and Prince Edward Island in the north-east, the Red River Valley (North Dakota and

* Alaska, Hawaii and Newfoundland are not included in this survey.

Minnesota) and the irrigated acreage in Idaho, Oregon in the north-west, as
well as Alberta.

Vegetables and *soft fruits* (i.e. market garden or truck crops) are grown, in
areas with good market access, all along the eastern seaboard, round the
Great Lakes, on the Gulf Coast and elsewhere in the east, sometimes with
spray irrigation. In the west, their distribution is mainly confined to the
Willamette (Oregon), and Okanagan (B.C.) valleys and to the irrigated valleys
of California and the south-west generally. *Tree fruits*, starting with apples in
the north and moving through peaches, cherries, etc., down to citrus in the
south, are grown on suitable Appalachian slopes or on coastal flats from
Nova Scotia down to Florida in the east, on the eastern shores of Lake
Michigan and, usually with spray (in the north) or surface (in the south)
irrigation, from British Columbia down to southern California in the
west.

Unirrigated *cotton* is the traditional crop of the south-east of the U.S.A.
and particularly of the Mississippi Delta (p. 152) where the climate is hot and
warm and where, a generation ago, Negro and white labour was plentiful.
The advent of the cotton-picking and other machines, of modern chemical
pesticides and of genotypes suited to Mediterranean climates have shifted the
court, but not the whole kingdom, of King Cotton to California and the irrigated valleys of the other south-west states. Here yields of lint are higher and
cotton farms more specialised and sophisticated than in the south-east.
Tobacco is the main cash crop in the eastern half of the general farming belt
to the north of the Cotton Belt (Fig. 2.2.4). Although the 'shade' tobacco for
cigar wrappers is grown in the Connecticut River Valley of Connecticut and
Massachusetts and the burley tobacco is produced in the southern mid-Atlan-
tic states, nearly half the North American tobacco is produced in North
Carolina, but some is grown in southern Ontario, Canada. This highly
specialised crop, with its specially needed skills, usually, as on French (p. 288)
and Australian tobacco farms, occupies only a small part of the holding. To
control disease it is rotated with cereals (usually maize) and grassland.

Grassland. On ranches the herbage consists mainly of indigenous and invader species which have become adapted to extensive grazing systems (Cos-
tello in Crisp 1964). Some sown species (e.g. lucerne or lucerne and brome or
orchard grass (*Dactylis glomerata*)) may, however, be grown in valley bottoms
to make hay or silage for winter feed. Despite the poor look of much of the
grazing, many of these ranches are, ecologically, very well managed. Careful
balances are maintained between vegetation and grazing pressure; the need
to have reserve feed for an extra dry summer or severe winters; and the control of erosion and fire hazards.

On the more intensive farms, however, grassland was, until quite recently,
not well managed. Too often, apart from fresh air and exercise, it only provided grazing in summer and hay in winter for maintenance; the livestock
production came, in summer as well as winter, from purchased or home-grown

grain and protein supplements. One can still see plenty of low producing or poorly utilised pasture. But the grassland revolution which has swept New Zealand, Australia and north-west Europe, is taking a firm grip in Canada and the U.S.A. Better genotypes, higher fertilisation and, above all, appreciation of the high animal production potential of well-managed pastures, are evident; whilst manpower shortage and the urge to mechanise have resulted in new concepts in grass handling and conserving equipment that are now also to be seen on many progressive farms in the homelands of the grassland revolution.

In the humid north-east, the long cold winter and the hot summer preclude the use of the rye-grasses (*Lolium* spp) and white clovers on which intensive grassland is based in New Zealand and much of north-west Europe. Cock's-foot (orchard grass, *Dactylis glomerata*), rough-stalked meadow grass (Kentucky blue grass, *Poa pratensis*), various fescues (*Festuca* spp, including *F. elatior*) and, in drier parts, brome grasses (*Bromus* spp) together with lucerne, Ladino white clover and bird's-foot trefoil (*Lotus uliginosus* and *L. corniculatus*) are popular. Fertilisation rates are often high at up to 400 lb elemental N per acre plus phosphate, lime and potash. In the Pacific North West, however, clovers and rye-grasses (*Lolium* spp) are important as the climate is usually milder.

In the south-east, cool humid species such as rye-grasses and red and white clovers and other legumes are grown in winter for feed and erosion control and followed by summer cash crops. For summer grazing and conservation, coastal Bermuda grass (*Cynodon dactylon*) is popular, as well-watered, well-fertilised pastures can yield up to 20,000 lb of dry matter per acre.* The warm humid climate of the south-east is well suited to intensive pasture production and, now that many of the external and internal pests of this climate are controllable, livestock production, especially for beef, is making great strides. In the irrigated areas of the Southern Plains and the south-west and, on higher pH soils of the west, lucerne with or without a companion grass is the dominant forage crop, whilst forage sorghums and grazed winter cereals are important in unirrigated parts.

Crop husbandry standards. Though there are, of course, many indifferent areas and bad farmers, the overall standard of crop production is high. The emphases are on genetically good disease and pest-resistant varieties; high fertiliser inputs (where P–E–T and moisture are adequate); the use of large, highly powered labour-saving precision machines which cover large acreages per man per day; and heavy reliance on herbicides and pesticides. Neither the Canadian nor the American farmer likes to use his muscles when he can use a machine or to employ 'good husbandry' for the control of weeds, plant

---

* Southern grasses include (Kennedy 1967) Bermuda grass (*Cynodon dactylon*), Dallis grass (*Paspalum dilatatum*), Bahia grass (*Paspalum notatum*). Among the legumes *Lespedeza* (*L. striata*, *L. stipulacea* and *L. cuneata*), Kudzu (*Pueraria thunbergiana* and *P. phaseoloides*) and clovers (crimson *T. incarnatum* and berseem *T. alexandrinum*) are important.

diseases or insect pests when he can rely on good genotypes or chemical sprays (from the ground or from the air) to do the job as well, or better .

Crop areas. In the U.S.A., broadly half the cultivated area (tillage and improved pastures) is in cereals and row crops. Of the cereals, one-quarter of the acreage is in food grains (mainly wheat), whilst maize* (with yields averaging twice those of other grains) is the dominant feed grain and accounts for half the cereal acreage. Other feed grains are oats, barley and grain sorghum. Oilseeds* rank next to cereals in acreage, followed by cotton,* vegetables,* Irish potatoes* and sugar crops.* The small tobacco* acreage is important as it has a high cash output per acre. The asterisked crops (thus *) occupy one-third of the cultivated acreage; they are intertilled row crops, i.e. crops which can be kept free of weeds by inter-row cultivation or, more usually now, by herbicides. But, as these crops leave more soil surface exposed than small grains (like wheat) or grass and forage crops, they are more subject to water erosion. This fact and the high intensity per hour of rainfall and the slope of much cropland outside the Interior Plains make water erosion a major problem. On the plains themselves, wind erosion is a difficulty. Erosion control measures, sometimes combined as in the Tennessee Valley with hydro-electric, flood control and navigation schemes, are an important part of national land policy. In Canada, of the cultivated areas, about one-quarter is in spring wheat, one-quarter in other cereals (oats, barley, rye and some maize), one-quarter in 'cultivated' grassland for grazing or conservation and one-quarter (mainly in the Prairies) in fallow. Minor acreages include oil-seeds (soya-beans and, in the Prairies, linseed and rape-seed (cf. Sweden)), Irish potatoes, tree fruits, especially in the Ontario peninsula with its water-moderated climate, and market gardening (soft fruits and vegetables).

Livestock. In New Zealand and Australia, and to a lesser extent in Ireland, Britain and the Netherlands, the function of ruminant livestock is to make use of land that is ecologically unsuitable for or is uneconomic for tillage and to convert grazing and hay or silage into cashable products. This is also true of the range lands of North America. But the main function of pigs, poultry and most ruminants in both Canada and the U.S.A. is to convert grains (and oilseed products) into cash as the animal products demanded by the high living standards of the populace. In the intensive areas, grassland, provides, as noted, little more than *maintenance*; the production portion of the ration is more dependent on concentrates than it is in western Europe.

Dairy cattle are concentrated mainly in the Dairy and Corn Belts (Fig. 2.2.4) and on some of the irrigated land in the west. The dominant breed is the Holstein (Friesian) of a milkier type than the dual-purpose strains found in the Netherlands and Britain. Jerseys, however, are popular in the south-east and seem better able to withstand heat stress than Friesians. Milk production is technically advanced with scientific feeding, artificial insemination, milk recording, genotype improvement, disease control and mechanisation all at high levels. Liquid milk consumption is high in both countries whilst butter consumption is declining, especially in the U.S.A. where *per capita* consumption is less than half that for most western European countries. Cheese, however, is no longer just something that you eat with apple pie. With rising living standards, the demand for high-quality *beef* is growing rapidly. Like many Europeans, neither American nor Canadian consumers are content with beef from dairy cows and dairy-type steers. Hence there is a great accent on the beef breeds, especially Herefords and Aberdeen Angus. The former dominate the cattle ranches which produce 1½-year-old stores (feeders) and weaned calves for yard or indoor fattening in the Corn Belt (p. 141) or in large outdoor feed-lots. These latter have sprung up near population or good transport centres, especially in the low rainfall areas west of 100°W.

Sheep. Lamb is not in great demand by Canadian or American consumers. This, combined with the need for good fencing, the fact that the sheep is hard to keep alive or on its feet when fattened largely on concentrates and, in the north, the need for winter housing with its health problems, reduces the attraction of fat lamb production. Sheep are, therefore, relatively unimportant in the east though small farm flocks of 50–100 ewes are found. From the western U.S.A. there is a major movement of store (feeder) lambs, born on mountain ranches, across the Southern Plains where they may spend a few weeks grazing winter wheat before being fattened in the Corn Belt or further east. Rambouillet (Merino type) sheep and various crosses are kept on these ranges for wool and store (feeder) production. In western Canada, however, predators (coyotes) and late spring snowstorms discourage sheep keeping. Sheep are, except on very poor or inaccessible land, less competitive than cattle as utilisers of land of low-carrying capacity (usually of low A–E–T), especially as the latter are easier to handle and need less labour.

Pigs and *poultry* are, with cattle, the main vehicles for cashing the vast output of feed grains. In the U.S.A., maize and oil-seed residues figure largely in the pigs' diet and the fat tends to be soft (unsaturated) and 'reasty'. Formerly, large, short, fatty 'lard'-type genotypes were favoured. But lard is feeling the competition of vegetable fats and public taste is changing, so strains of, for example, Hampshire, Duroc-Jersey, and new breeds with more lean meat and less fat are now popular. The Canadian pig industry, however, was built on the Large White breed to export lean Wiltshire-cut sides to the United Kingdom and was based largely on barley, less maize and more local milk by-products. The long, lean pig also meets Canadian taste and has persisted even though bacon exports have fallen away drastically since 1945.

Eggs and broiler meat are popular with consumers in both countries. Ample supplies of cereals and protein-feeds and of timber for housing; the scientific control of housing, nutrition and disease; mechanised feeding, manure removal and egg handling; the suitability of the fowl for quick genetic improvement (breeds as formerly known, e.g. Wyandottes, have disappeared in favour of numbered genetic strains often of secret ancestry) have been combined to make a large and very efficient poultry industry.

Table 2.2.1

Crop	CANADA		U.S.A.	
	Area (000's acres)	Yield (average for 1963–65) (lb per acre)	Area (000's acres)	Yield (average for 1963–65) (lb per acre)
Wheat	28,281	1,400	61,058	1,559
Oats	8,656	1,568	18,478	1,490
Barley	6,037	1,635	9,143	1,859
Rye	746	1,129	1,468	1,126
Maize	751	4,222	55,330	3,830
Sorghum	—	—	13,030	2,576
Rice	—	—	1,794	4,122
All cereals	46,027	1,504	149,026	2,520
Sugar-beet	84	26,611	1,248	35,123
Sugar-cane	—	—	474	53,850
Potatoes	301	16,218	1,404	19,891
Vegetables	131		1,579	
Dry peas and beans	141	1,340	1,730	1,480
Soya-beans	264	1,658	34,448	1,443
Other oil-seeds*	3,825	699	16,393	912
Tobacco	99	1,765	976	1,995
Cotton (lint)	—		13,615	520
		Production 000's lb		Production 000's lb
Tree crops (fruit)†		1,379,865		36,636,070
Livestock	Numbers 000's	Yield (average for 1963–65) (lb per animal)	Numbers 000's	Yield (average for 1963–65) (lb per animal)
Total cattle	11,908	1,017‡	107,184	1,010‡
Dairy cows	2,885	6,353§	17,592	8,081§
Pigs	5,577	209‡	53,132	238‡
Sheep	852	95‡	26,590	97‡
Wool		6·7		8·6
Goats			4,060	
Poultry (chickens)	67,440		375,424	

Source: F.A.O. *Production Yearbook 1966.*
* Cotton-seed, linseed, rape-seed and sunflower seed.
† Apples, pears, plums, prunes, cherries, peaches, apricots, oranges, tangerines, grapefruit, lemons, limes and other citrus and grapes (grape yields—Canada, 3,158 lb per acre; U.S.A., 14,568 lb per acre).
‡ Live-weight of animals slaughtered.
§ Milk per milking cow.

Canadian Transect (*Fig. 2.2.4 and Table 2.2.3*)

EASTERN CANADA (*Appalachian and Great Lakes, Sectors (a) and (b)*)

In eastern Canada the climate is cool humid. The relief is low mountains with river valleys in the Appalachians and rolling plains in the Great Lakes Sector (see Table 2.2.3). The 'natural' vegetation is coniferous or mixed forest. This had to be cleared, a difficult job, but in compensation there was wood for fuel, building and the zig-zag nail-less wooden rail fences once so characteristic of eastern Canada and adjoining parts of the U.S.A. The soils are mainly podsolic and, like those of coastal B.C., and the northern wooded fringe of

Table 2.2.2

Climatic Data — Canada

Station	Bright Sunshine hours/day	Thermal Growing Season length in months	Thermal Growing Season Mean Temp °F	Mean Months ≥72½ °F	Thermal Growing Season mean P–E–T (inches)	Thermal Growing Season mean A–E–T (inches)	Mean Annual Precipitation (inches)	Moisture Index Im	Annual Precipitation Coeff. Var. (C.V.)	Max. recorded rain/day inches (month of occurrence)
Halifax	—	6	57·4	0	21·6	21·5	55·6	137 per-humid	< 10%	5·4 (August)
Ottawa	5·5	6	59·8	0	22·0	20·8	34·5	50 humid	< 10%	3·1 (September)
London	5·7 (Toronto)	7	58·6	0	23·4	22·3	38·2	61 humid	< 10%	4·9 (July)
Sioux Lookout	—	6	54·7	0	20·0	19·1	24·5	22 humid	15–20%	2·4 (October)
Regina	5·8 (Winnipeg)	5	58·0	0	19·4	13·1	14·8	−30 semi-humid	20–25%	3·7 (June)
Cranbrook	—	6	54·7	0	20·2	13·5	14·4	−32 semi-humid	15–20%	1·7 (November)
Vancouver	5·3 (Victoria)	9	54·0	0	24·1	22·1	57·4	126 per humid	10–15%	4·4 (November)

Canadian Transect
Halifax N.S., Lat. 45°N, to Vancouver B.C., Lat. 49°N

	(a) *Appalachians*	(b) *Great Lakes*	(c) *Canadian Shield*	(d) *Prairies*	(e) *Rocky Mountains*	(f) *Coastal Valleys and Islands*
1. Sector	Appalachians	Great Lakes	Canadian Shield	Prairies	Rocky Mountains	Coastal Valleys and Islands
2. Name						Lower Fraser Valley
3. From	Halifax	Toronto, Ont.	Sudbury, Ont.	Winnipeg	Calgary	Vancouver Island
4. To	Toronto, Ont.	Detroit, Mich. Sudbury, Ont.	Winnipeg	Calgary	Lower Fraser Valley	
5. Miles	800	200	900	750	400	150
6. Climate type(s)	Cool Humid	Cool Humid	Cold Humid	Cool Dry	Mountain	Cool Humid
Thermal Growing Season:						
7a.—months at or > 42½°F	6	7	—	7	—	9
7b.—months at or > 72½°F	0	0	—	0	—	0
8a.—P–E–T (in) in T–G–S	21–22	22–24	—	18–20	—	20–24
8b.—A–E–T (in) in T–G–S	21–22	20–22	—	12–14	—	16–22
9. Growth limited by	Winter cold	Winter cold	Winter cold	Winter cold. Rainfall	—	Winter cold. Dry mid summer
Precipitation:						
10.—mean in (annual)	30–40	25–40	15–35	10–20	15–35	40–60
11.—C.V.% of Precipitation (annual)	10–15	10–15	15–20	15–25	15–20	10–15
12. Special Climate Features	—	Moderated by lakes	—	Eroding winds	—	Some summer drought
13. Weather Reliability	3	3	(1 ?)	2	2	4
Soil:						
14.—Zone/Group	Podsol	Podsol	Podsol	Chernozemic and Chernozem Desertic	Complex	Complex
15.—pH	Low	Low	Low	High	—	Low/Medium
16.—Special Features	Leached	—	Very thin soil over glaciated rock	Chernozems have high organic matter	—	Leached
17. Vegetation	Coniferous forest	Mixed forest	Coniferous forest	Short grassland	Coniferous forest	Coniferous forest
18. Relief	Low mountains	Rolling plains	Very rugged	Flat plains	High mountains	Valleys; coastal plains
19. Market Access	3	4	(1 ?)	2	1	3
20. Farming Systems	Intensive grazing and intensive alternating	Intensive alternating and intensive tillage		Extensive tillage and extensive grazing	Extensive grazing, some intensive tree crops	Intensive alternating intensive tillage and intensive tree crops
21. Major Enterprises	Milk. Horticulture (incl. potatoes) in some coastal valleys. Some pigs	Milk, pigs, poultry and semi-intensive sheep and beef. Horticulture in S. Ontario	Mining and forestry	Spring cereals and mixed arable where moisture enough, elsewhere cattle ranching	Some cattle ranches. Some forage crops. Some irrigated orchards	Level milk. Horticulture

the Prairies, respond to phosphates and lime. This area of clays and loams was heavily glaciated and the resultant soils are largely mixed and pulverised limestones, sandstones and shales.

The P–E–T in the thermal growing season (T–G–S) averages 21–24 in and is of the same order as in western Europe. But the thermal growing season is shorter with only 6 or 7 months at or above $42\frac{1}{2}°F$. The winters (Table 2.2.2) are much colder, but the summers are sunny and warmer. The mean temperature during the growing season, for example, at London, Ontario (Lat. 43°N) is, at $58\frac{1}{2}°F$, five or six degrees higher than the same parameter in the British Isles. This helps to offset the shorter growing season, whilst the higher solar altitude and lower cloud cover counter the shorter day. The mean annual precipitation decreases as one moves from the coast (40–60 in) to western Ontario (25–30 in) and is spread well over the year. The coefficient of variation of annual precipitation is 10–15 % i.e. much as in north-west Europe. But rain tends to be more intensive, i.e. precipitation per hour of rain is often high and there are more flooding and erosion problems than for a similar annual rainfall in the United Kingdom. Much of the winter precipitation comes as snow which lies until snow melt when there is some run-off erosion. The winter freeze up reduces leaching of nutrients—in contrast, for instance, to the south-east of the U.S.A. or western parts of the United Kingdom. The only parts of the country that are not encased in frozen snow for four or more months in winter are south-western Ontario, the coastal areas of British Columbia and the south-western part of Alberta where the Chinook winds occasionally melt the snow and facilitate winter grazing on range land. Everywhere melting snow is an important, sometimes embarrassing, source of spring moisture.

In the east, as in the west, the farmer's problem is to get on to the land as soon as possible after snow melts, soil temperatures start to rise and the land dries enough to drill (May). He can thus build up leaf area within 5–6 weeks so that he has a high L.A.I. at midsummer when N.A.R. is highest and ear emergence starts. Grain crops must generally be harvested before the end of September, and preferably in August, so the farmer has only 4 months, mid-May to mid-September, for spring cereals, compared with six or seven months in, for instance, south-east England. As no preparatory work can be done on the land in winter, working weather is all important when one week is 6 % of the effective drilling to harvest (i.e. crop duration) period, as against 4 % or less in south-east England, but the Canadian climate appears to provide more cultivating days in spring and harvest days in late summer and autumn. In midsummer, potential transpiration losses per week are, especially in the southern half of Ontario, high, and though annual precipitation exceeds P–E–T, short-term moisture deficits often develop. There is growing interest in sprinkler irrigation for high-value crops, for example, in the Ontario peninsula.

The combination of low winter temperature, snow-cover, wind and snow-storms means that all livestock, including sheep, have to be well housed and

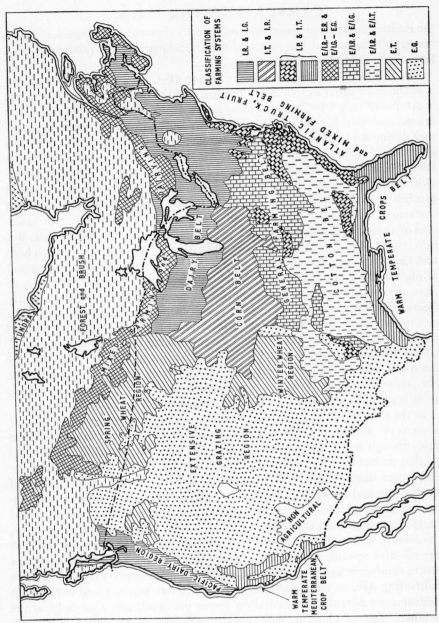

Fig. 2.2.4 North America. Generalised Types of Farming Systems

Classification of Farming Systems:
- I.R. & I.G.
- I.T. & I.R.
- I.P. & I.T.
- E/I.R. – E.R. & E/I.G. – E.G.
- E/I.R. & E/I.G.
- E/I.R. & E/I.T.
- E.T.
- E.G.

Map labels: TUNDRA; FOREST and BRUSH; MIXED FARMING; DAIRY BELT; SPRING WHEAT REGION; MIXED FARMING REGION; CORN BELT; WINTER WHEAT REGION; GENERAL FARMING BELT; COTTON BELT; ATLANTIC TRUCK, FRUIT and MIXED FARMING BELT; WARM TEMPERATE CROPS BELT; EXTENSIVE GRAZING REGION; NON AGRICULTURAL; PACIFIC DAIRY REGION; WARM TEMPERATE MEDITERRANEAN CROP BELT

	Tree Crops (P)	Tillage (T)	Alternating (R)	Grazing (G)
Extensive (E)	E.P.	E.T.	E.R.	E.G.
Semi-intensive (E/I)	E/I.P.	E/I.T.	E/I.R.	E/I.G.
Intensive (I)	I.P.	I.T.	I.R.	I.G.

From Van Royen, 1954

hand-fed (or increasingly, self-fed) during the winter (from October to May).
As the grazing season is rarely longer than 5 months, there are seven or more
months concentrate feeding and at least six months full winter feeding. In the
summer high day temperatures and humidities may sometimes distress grazing
stock.

EASTERN CANADA—*Farming Systems*

There are four main systems: Semi-intensive grassland, Intensive Alternating, Intensive Tillage and Intensive Tree crops.

Semi-intensive grassland milk production. This is found on the rougher,
boulder-strewn, thin-soiled uplands or in districts which have not easy access
to liquid milk markets. The main outputs are cream, or manufacturing milk,
combined with sales of heifers or beef production. Subsidiary enterprises are
often pigs or poultry. Some oats or barley may be grown for feeding on the
farm. Most of the land is in more or less permanent grass of indifferent quality.
Often the stumps of trees, left when the forest was cleared, are still standing.
The most common sward, often established by invasion rather than by proper
seeding, is Kentucky blue grass (*Poa pratensis*) red top (*Agrostis gigantea*)
and some white clover. There is little real grassland farming and not much
interest in leys. With land plentiful, manpower short and relatively cheap feed
grain available from the Prairies, there is little incentive for grassland improvement. The main winter bulky feed is hay. Recently there has been an increase
in the use of grass silage. But especially in southern Ontario and through
Quebec as far as Montreal production of maize for grain and fodder is increasingly widespread. Fertiliser use is limited, but lime and phosphates are
sometimes applied and the residual fertilising value of the grains bought in
from the Prairies and imported oil-seed residues must help.

Intensive alternating systems occur in the liquid milk-sheds of big cities and
on the lower, more cultivable lands of, for example the St Lawrence lowlands
and southern Ontario. Three or four years in grass are followed by 2 or 3
years in barley, oats and maize for silage or for cash, e.g. sweet-corn for
canning. The use of fertilisers and other technological inputs is higher than
on the semi-intensive grass farms. Milk or sometimes beef is the main enterprise with perhaps subsidiary pigs or poultry. The yields per animal and the
level of animal husbandry particularly on alternating farms, are relatively
high, but even on good farms it is unusual to obtain more than 200 'cow-
grazing days' per acre/annum, whilst average tillage crop yields are not outstanding. But one must remember that high output per man rather than high
outputs or inputs per acre or even per animal, is often the main objective,
especially where farm acreage is not limiting. In any case the short effective
growing and cropping season does not (for reasons stated above) favour
high yields or justify high fertilisation.

A Dairy Farm near Montreal, P.Q. (225 acres)

Climate, soils, etc. The local climate is, broadly, as described above. The low-lying, deep, heavy clay, boulder-free soil of marine alluvium, formed from glacial drift, is podsolic and liable to flood and has to be drained. This is done, not by tiles underground, but by open ditches and the 'ridge and furrow' layout of 'lands'. The fields are (as is common in Quebec) elongated rectangles.

Cropping, rotations and livestock. The main object, apart from a small acreage of cash row crops and some seed corn, is to provide summer pasture and winter bulky feed plus some grain for the dairy herd. The usual rotation is: *1st year*, row crops (including maize for ensilage) and oats; *2nd year*, oats and barley under-sown, typically with 5 lb lucerne (alfalfa), 6 lb timothy (*Phleum pratense*), 2 lb brome grass (*Bromus inermis*), 3 lb (red) clover per arpent;* *3rd and 4th years*, ley for hay and silage and aftermath grazing, and *5th year*, ley for grazing. The red clover generally dies out in the second year but is partly replaced by 'volunteer' wild white clover. Dung from the cattle is mechanically cleared from the cowhouse gutter and carted and spread when opportunity permits. Oats and barley receive 200/300 lb of superphosphate per arpent. Lime is occasionally applied. About 30 Friesians, mainly pedigree, are served by the bull or artificially inseminated. All heifers are kept as replacements and served to calve at 2–2½ years; bull calves are sold for veal at 3 weeks old. Summer grazing (some of which is rotational) is supplemented by concentrates. In winter, hay, grass silage and maize silage are fed for maintenance and purchased or own-grown concentrates for production. A few poultry are also kept.

Manpower, equipment and the farming year. The owner, wife and son, with occasional help from their three younger children (especially for hoeing row crops), work the farm. It has a wooden two-storeyed barn and tower silos. Two tractors and the usual cultivation equipment are supplemented by a maize-seed planter, a forage harvester, a silage blower and special ditch-cleaning devices. No field work is possible till early May when oats, barley and direct seed leys are drilled. Maize is drilled in late May. Grass silage and hay making are in June and July. Oats are harvested in early August, barley somewhat earlier, maize (for canning or silage) in September. The grass comes in late April but, owing to the risk of poaching, is not grazed until mid-May or later. The cattle are then on grass until late October when the cessation of growth, poaching damage and bad weather make housing necessary. (*Courtesy:* Association of Agriculture, London.)

Intensive tillage and tree crops. The lower latitude, adequate rainfall and the moderating influence of the mass of water in the Great Lakes makes the peninsula of southern Ontario (particularly the coastal belts) well suited to

---

* An areal arpent is 0·844 acre; a linear arpent is 191·9 ft.

intensive tillage and to tree and soft fruits. Vegetables, tobacco, grapes, potatoes and sugar-beet are also produced. This horticultural industry is technically advanced, strictly commercial and helped by a protective tariff. (The ecological factors immediately south of the Canadian/U.S.A. frontier are much the same but the farming on the U.S. side is more mixed and less horticultural. Though there are (or were a few years ago) some grape-producing farms and wine factories on the American side of Lake Erie, the lakeside farms of Ohio and Pennsylvania could hardly compete with the highly efficient but distant horticulturalists of Florida and California.)

CANADIAN SHIELD (*Summarised in Table 2.2.3, Sector (c)*)

THE PRAIRIES (*Table 2.2.3, Sector (d)*)

The climate is cool dry continental and precipitation has a high C.V. Temperatures show great extremes; minus 48°F and below have been recorded at Winnipeg. Though 4 to 5 degrees of latitude nearer the pole, July temperatures are much the same as in the eastern sectors. But January temperatures are much lower and more variable; Winnipeg averages 3°F. The standard deviation (S.D.) of July temperatures is about 2·3°F but that of January temperatures is 7·0 to 10·5°F, depending on location. As there are no crops in the land or field work in winter, these low and variable temperatures are not critical. But the S.D. of the months at the start and end of the mean thermal growing season (T–G–S) is high at 3·0 to 4·0°F,* and this has important consequences on drilling and harvesting dates. The T–G–S averages up to 7 months in the south but only 4 months in parts of the Park Belt of the north, where there may be only 12–16 frost-free weeks. Both cloud (sunshine hours/day in the growing season average over 8) and the number of rainy days are low, so the N.A.R. is high in midsummer if soil moisture is adequate. Often, however, in the south it is not. At Regina (Sask.) the mean thermal growing season P–E–T is 19·4 in, whilst the A–E–T is only 13·1 in—a mean deficit of 6 in. Annual precipitation averages about 20 in in the east decreasing to 10 in in the west, with coefficients of variation rising from 15% in the wetter areas to 25% in the drier parts of the south and west. Moreover, the western Prairies tend to be dehydrated by westerly winds that have lost their moisture coming over the Rockies.

The rain is favourably distributed. On average, about two-thirds of the annual precipitation occurs in the five months May–September. This factor, together with snow-melt moisture at seeding time enables farmers to obtain better cereal yields than the annual precipitation means would suggest.

The combination of unreliable starts and ends to the growing season, and of low and unreliable precipitation during it, has substantial effects on field operations in the drier areas. If the spring thaw is late and coincides with heavy rains, drilling may be delayed until early June. Late summer hail-storms are a hazard to unharvested crops, whilst, if winter frosts start earlier than

* Based on extrapolation from Atlas of the U.S.A. (U.S. Census Bureau).

average (mid-September) and a late sown crop is not then fully matured, it
may be frost damaged. If one needs 14 days for pre-drilling operations and 21
days for harvesting, one still needs 4½ months between spring thaw and winter
close-down. Hence the emphasis by plant-breeders on wheat varieties which
can be drilled, grow and ripen in 100 days.

There are few natural wind-breaks in the flat landscape where the only
features on the skyline are often the scattered grain elevators (silos), and,
apart from any effect wind has in raising transpiration and moisture loss, wind
erosion is a problem (see below, p. 131). Most of the farmhouses have shelter
belts, but wind-breaks in the fields encourage snowdrifts which may be slow
to thaw and then water-log the land when they finally do so. Soils tend to be
phosphate-deficient and in some areas crops also respond to sulphur, borax
and other trace elements. Nitrogen comes from the soil organic matter.

The Prairies—Farming Systems

Semi-intensive tillage. In the Red River Valley (Manitoba) and as far west
as the Portage Plains, the *system* is not very dissimilar from that found in the
drier parts of north-west Europe or the Corn Belt of the U.S.A. though there
are fewer livestock. The cash crops include wheat, barley, oats, linseed and
rape-seed (for oil and protein feeds), potatoes, sugar-beet and some vegetables.
With the growth in the size and living standards of the population, cows for
liquid milk, beef cattle and to a lesser extent, pigs and poultry are becoming
important. (There is also intensive tillage in the dry-surface-irrigated area
round Lethbridge (Alberta) with sugar-beet, vegetables, etc.; other large irrigation schemes, e.g. the St Mary's–Milk River project, are in progress. Whilst
in the north (55°N upwards) where market access is poor and the growing
season short, a good forage seed industry (high value per pound crops) has
developed.

A Grain and Cash Crop Seed-growing Farm in the Eastern Prairies

This farm of 1,500 acres lies 66 miles west of Winnipeg on the tall grass
prairies. The climate is as described for the eastern Prairies and in mid-winter
the ground is frozen down to 5 to 6 ft. There are 2 ft of lacustrine (black earth)
soil high in organic matter (8–10 %) laying over free-draining sandy loam. But
some of the land has high clay content (gumbo soil), and is less easy to work.
There are no serious soil erosion problems.

Cropping and rotations, etc. In 1963 the farm carried 900 acres of crops, viz.
wheat (280 acres), barley (200 acres), linseed (flax-seed) (140 acres), hay for
sale (120 acres), sugar-beet (80 acres), peas harvested dry (50 acres) and
fallow of 430 acres. As the farmer is a specialist seed grower there are more
enterprises than on many Prairie farms. The rotation is broadly 2 years of
cash crops, then 1 of summer fallow. The latter consists of a moisture-con-
serving stubble-trash cover and is cultivated for weed control (often six times
in summer). The soil is high in lime, potash and organic nitrogen, so these are

not applied, but ammonium phosphate is applied each year to about two-thirds of cropped acreage. Now that broad-leaved dicotyledons are controllable by the usual range of herbicides, the main weed problem is wild oats (*Avena fatua*). This is more or less controlled by weed-germinating autumn and spring cultivations plus extensive use of a special but costly herbicide applied before drilling. Stink-weed (*Thlaspi arvense*) is also a problem.

Manpower equipment and the farming year. The farm is worked by the owner and his son plus casual labour (local Indians), for example, for cereal harvest or for thinning and weeding sugar-beet. The equipment includes 2 five-furrow mould-board ploughs, 2 one-way disc ploughs ($7\frac{1}{2}$ and 9 ft wide), 1 twenty-one ft discer-seeder (which broadcasts seeds), 2 fourteen ft drills, 1 thirty-five ft harrow, 1 fifty-six ft. wide weed sprayer, 2 sixteen ft swathers, 2 self-propelled combine harvesters, 1 six-row sugar-beet drill, but the beet harvester is hired. Of the two wells for water, one is salinised and unfit for humans. The ground is normally thawed out by mid-April. Wheat is drilled in May, barley in May or June, peas by the end of May. But all drilling may be delayed to early June (latest safe date for useful yield) if the thaw is late or if heavy May rains make soil uncultivable. In June and July there are cultivations of fallows, hay making, thinning and hoeing of sugar-beet and other row crops. The grain and other seed crop harvest normally starts in August, or sometimes early September. It is completed by early October which is not too late if the grain is fully mature before mid-September frosts. The grain is first cut with a swather, which lays the crop on the stubble to dry for a week or two and is then combined at upwards of 40 acre/day per combine. Other seed crops (including peas, rape-seed) are harvested direct by combine. The winter is spent cleaning (dressing) seeds, repairing machinery, etc., and taking a rest and a holiday. (*Courtesy:* Assoc. of Agric., London.)

Extensive Tillage

On cereal farms in the drier areas the usual rotation is *1st year:* spring wheat; *2nd year*, fallow, but where moisture permits, it is often lengthened to wheat, wheat or barley, fallow. This rotation is also found in the U.S. Spring Wheat Belt (p. 144) and, with earlier harvest, in the U.S. Winter Wheat Belt (p. 156). Because of the problems of loss of organic matter, of low moisture, of wind erosion, and of high labour costs and because of the need to complete the whole crop cycle between mid-May and mid-September, the land is not ploughed. Immediately after harvest, or in the spring of the fallow year, the grain stubble is broken by a heavy disc plough. This does not really turn the sod under but leaves large lumps of soil which help to break the force of the wind. If erosion control is not an overriding factor, this fallow is cultivated (*a*) to reduce water-transpiring weeds and hence their competition for *next* year's moisture, and (*b*) to make a stubble-mulch (from the straw left by the combine) which appears to reduce wind erosion and moisture loss by evaporation.

Low-volume herbicides are used, even in the fallow year, as moisture conser-
vers. The most troublesome species of broad-leaved weeds were probably
Canada thistle (*Cirsium arvense*) and poverty weed (*Iva axillaris*), but are now
largely chemically controlled. The real problem now is wild oats (*Avena fatua*)
(see above).

In the spring of the next, i.e. the cropping year, wheat is drilled into the
fallow, usually in May, almost before the snow has finished melting. A discer-
seeder makes a rough seed-bed, drills the grain and applies ammonium phosphate. It is followed by a harrow, and whole 'gang' is drawn by a large tractor.
In this way the whole of the spring work is done in one operation, and, using
'gangs' up to 30 ft wide, 75–100 acres can be seeded by one man in one day.
To control wind erosion the land is worked in alternate strips, up to several
'gangs' wide, of wheat and fallow which are laid, when possible, at right-angles
to the prevailing, mainly westerly, winds so that any soil picked up by the
wind is trapped in the adjoining strip of wheat or stubble. Harvesting is by a
swath cutter; this is followed a few days later by a combine harvester which
leaves the chopped straw as a mulch and delivers the grain direct into an
accompanying lorry or trailer. Yields of wheat average 1,200 lb to the acre,
but (as noted) vary greatly from year to year, mainly with the weather. For the
plant-breeders have done marvels introducing disease-resistant cultivars, and
the never-ending war on the new physiological variants of stem or red rust
(*Puccinia* spp) has had some success, although the fight against leaf rust (*P.
rubigo-vera tritici*) has not done so well. Solid-stemmed wheats have been bred
for use in areas susceptible to the wheat stem saw-fly (*Cephus cinctus*) which
used to be a menace. Grasshoppers are still a burden, though not so troublesome as in the U.S.A.

Family farms, operated by the occupier with occasional help from neighbours or from a hired man, are the rule. On a two-year (wheat then fallow)
rotation, one man can manage 1,200–1,700 acres in a few weeks of really intensive work in the spring and late summer.

Extensive Grazing of Livestock—Western Prairies

But this system of mono-cultural extensive tillage has both biological and
economic dangers and the search for a mixed extensive system is active. Unfortunately in 5 months it is not easy to establish and harvest most clovers or
cultivated grasses, although some cultivars of lucerne, sweet clover and Agropyron crosses succeed and enable some farmers in the drier areas to 'retire'
some of their wheat land in favour of drought-resistant leys for raising feeder
(store) cattle and even fattening them.

Extensive grazing of cattle is found on the drier, lighter brown soils of the
south and west and in the foothills of the Rockies. The system is very similar
to that found south of the border in the U.S.A. (see p. 158). There are, in total.
some 6 million cattle on the Prairies where most animals are not finished at
home on the ranch but shipped to the east, or the U.S.A., or finished in the

local open-air 'feed-lots' which, however, provide shelter from wind and drifting snow and are increasingly popular. Though winter air temperatures are well below the critical temperature of fasting cattle, the air is dry, and they seem to thrive in highly mechanised outdoor 'feed-lots' if fed on high concentrate rations supplemented with hay or possibly grass silage.

ROCKY MOUNTAINS (*summarised in Table 2.2.3 Sector (e)*)

BRITISH COLUMBIA—COASTAL SECTOR (*Table 2.2.3, Sector (f)*)

Finally, between the Rockies and the Pacific, lie the well-watered coastal lowlands and valleys and islands of British Columbia with their cool humid climate, reliable rainfall (C.V. 10–15%), rather acid soils and coniferous forests. Mean sunshine hours (about 2,000 hour/year) and July temperatures are mostly higher than round the North Sea basin of western Europe. Further, though the annual rainfall (at 40 to 60 in) is good, it averages low in July and August. This combination means that drought is sometimes a pasture problem for the dairy farmers of, for instance, the fertile and beautiful lower Fraser River valley. In compensation, maize can be grown for silage. Intensive alternating farming for level milk production (based on grass, maize silage and purchased concentrates) for liquid consumption is, in fact, highly developed and so are vegetables, soft fruits and tree fruits. The growing and well-paid urban population of Vancouver and its environs provides good markets.

Stability and Efficiency—Canada

Ecologically, the most notable Canadian problems are two: the short growing season and the long, cold winter. The universal uncertainty of weather has added importance; for a few days' delay in spring seeding or an early fall of frost or snow can cause a significant reduction in the length of the growing and work season. There are bio-energetic limitations on the full exploitation of radiation receipts during the growing season; these limitations affect both the development of high leaf area by the summer solstice and good leaf area duration. Under these conditions, crop yields (including grass) are restricted, and high inputs per acre, particularly of fertilisers, are not economically justified, especially where moisture may be limiting. There would appear, however, to be ample room to raise efficiency and obtain a better bio-energetic fit (see Duckham 1963, p. 305) when economics permit higher inputs.

The long, hard winters prevent production of winter cereals except in southwestern Ontario, and to a minor extent in southern Alberta. Cold resistance is of extreme importance in all perennial crops; lack of this factor restricts the varieties that can be grown. Winter temperatures in many areas are below the 'critical level', even for grown ruminants. This tends to lower the conversion efficiency of crops into animal products and requires added investment in protective housing.

In the south-central region of the Prairies, the major wheat-producing area,
rainfall is low and variable. The wheat-summer fallow rotation was developed
to conserve moisture, control weeds and build up nitrogen. Wind erosion has
been controlled by avoiding the mould-board plough and developing trash-
cover cultivation. Plant-breeders have mastered the twin hazards of stem and
leaf rust by developing varieties resistant to these and other diseases. Increasingly sophisticated and widespread use of herbicides, pesticides and fungicides,
and steady increase in fertiliser inputs, have had stabilising effects. Though
wheat yields have not increased as rapidly as elsewhere, the trend is upward,
and it seems probable that coefficient of variation of wheat yields will decrease.
During the first half of the century, wheat farms drew heavily on the capital
represented by a highly fertile virgin soil. There was loss of organic matter,
and soil structure deteriorated. Improved cultivation methods and fertilisers
are slowing down this trend, but the original fertility is far from depleted, and
the Prairies produced by far their largest wheat crop in 1966. Despite the cost-
price squeeze and the distance to markets, this highly mechanised and efficient
wheat-producing area seems destined to make increasing contributions to
world grain needs (Anderson, J. A. 1967).

In the vast peripheral area that surrounds the wheat-growing heart of the
Prairies, farming becomes more mixed. Grain production is still significant,
but gives place to greater emphasis on forage crops, feed grains, beef, pigs.
dairy cattle (in the milk-sheds of major cities), poultry, and a range of other
crops, e.g. flax-seed, rape-seed, sunflower seed, sugar-beets, potatoes and
vegetables. Rainfalls are substantially higher, and the hazard of rust disap-
pears as one moves north and west. A broadly based and stable agriculture
defies the climatic hazards. To the west, in British Columbia, the agriculture
is small and scattered by Canadian standards. In southern Ontario where the
climate permits a very wide range of crops, and farms are close to markets, a
relatively prosperous and mixed agriculture continues to thrive. Tobacco,
grapes, peaches, maize and soya-beans can readily be grown, to name a few
crops not available on the Prairies. As one moves further east in Ontario and
into Quebec, the small size of original farms creates economic problems. The
dairy industry dominates and is now substantially subsidised. Still further
east, the typical ribbon development along the shore-line and river valleys
presents problems. Farms are often too small to provide the high Canadian
living standard. Grain for livestock is imported from western Canada. The
Federal and Provincial Governments' policy is to return to forestry or develop for recreation those areas least suited to farming, and to consolidate
farms into larger units in the better areas (Anderson, J. A. 1967).

In brief, despite the difficult climate, the ribbon of settlement and the
resultant problems of access to inputs and markets, Canadian farming is
an efficient response to great challenges. The combined effect of technological
progress and government actions has been substantially to reduce both ecological and economic uncertainty in Canada (Table 2.2.4) as in the U.S.A.

Table 2.2.4

Canada. Coefficient of Variation (%) of Farm Cash Income Received from Sale of Selected Farm Products

Period	Cattle and calves	Wheat	Dairy products	Hogs (pigs)	Eggs and poultry
1926–41	27	40	11	33	16
1942–47	15	35	6	16	11
1948–60	13	27	6	9	11

Source: Ware 1963.

and elsewhere (p. 72). Moreover, this increased reliability has been associated with rising productivity, particularly per man.

Conclusions and Prospects

Canada is an advanced industrial country, with a population of 20 million, enormous resources and a high and rising standard of living. It can support the range of industrial and scientific inputs required by modern farming, whilst proximity to the U.S.A., with its similar ecological problems in areas contiguous with Canada, and with its vast research organisation and large industrial production of agricultural inputs, is of enormous aid to Canada. But Canadian farming ribbons across the country, and scattered farmsteads cannot easily profit from industrial resources and urban markets that lack the geographic concentration of such areas as western Europe or much of the U.S.A. This problem is pressing in the Prairies, where distances to export outlets (across the Rocky Mountains to Vancouver, or east to the Great Lakes, the Welland Canal and the St Lawrence River), and even to eastern Canadian markets, are very great. Steps are being taken to deal with these and related problems; the Agricultural Rehabilitation and Development Act, Prairie Farm Rehabilitation Act, price stabilisation legislation, insurance against hail damage and crop failure and some promising irrigation schemes. But the basic problems remain, a short growing season, long, cold winters and low rainfall in the south-central Prairies. Nevertheless, as a whole, Canada, which has maintained its agriculture on a substantial net export basis, against world competition, without the heavy subsidisation so prevalent in many other countries, must be classed as having every prospect of continued agricultural development (Anderson, J. A. 1967).

Acknowledgements, p. 160.
References and Further Reading, p. 160.

Canada and the U.S.A

CONTINUED

U.S.A.—Northern Transect (*Fig. 2.2.4 and Table 2.3.2*).

The Northern Transect, Table 2.3.2, starts with the specialised farms on Long Island, crosses the great New York Milk Shed, the highly productive Corn Belt, which together with the North East produces 70–75% of milk and 85% of pigs in the U.S.A., and, passing further westward, moves over the extensive cereal growing of the Northern Great Plains and the ranching and specialised areas of the Rocky Mountains to the Pacific North West coast. Climatic data are in Table 2.3.1.

COASTAL LOWLANDS (*Table 2.3.2, Sector (a)*)

The narrow coastal lowland is flat or gently rolling. The soils are acid (pH 4 to 5) podsols, mainly glacial deposits of morainic gravels and sands. The original vegetation was mixed deciduous and coniferous forest. The cool humid climate has, on average, at least 180 frost-free days per annum. The mean January temperature is just below freezing-point, the July mean 70–75 °F. The thermal growing season is about 8 months, during which the P–E–T averages 28 in and A–E–T 27 in. The annual precipitation is adequate (40–45 in) and well distributed through the year with an annual C.V. of 10–15%.

Farming systems. This useful climate, the light soils and the very large adjoining New York market encouraged intensive tillage systems with early and main-crop potatoes, vegetables (such as cabbages, early autumn cauliflowers, brussels sprouts, lettuces, asparagus and others) and the raising of nursery plants as the main enterprises. The advent of modern processing, storing and transport techniques means, however, that this area competes with better seasonal climates (e.g. Florida) a thousand or more miles away. Large-scale egg production, however, persists in New Jersey (see Fig. 2.3.1 and Col. II, Tables 2.3.4 and 2.3.5 for details) and broiler production has grown in the Delaware–Maryland–Virginia peninsula to the south of the transect. Other livestock enterprises on Long Island are the duckling industry, dependent on feed shipped in. Very little of the liquid milk now consumed in New York City travels less than 50 miles.

THE APPALACHIANS (*Table 2.3.2, Sector (b)*)

The Appalachian Mountains are here 400 miles wide. After the Blue Ridge and a series of ridges and valleys running N.E.–S.W. comes the rolling Allegheny Plateau (500–2,000 ft) dissected by valleys, of which the larger ones have workable bottom lands, often, however, with drainage problems. The soils are mainly grey brown podsols and, on the Plateau, are thin, stony and rather acid, and often high in iron oxides, which makes P_2O_5 relatively unavailable. The original cover was mixed deciduous and coniferous forest. Much land is, in fact, still forested, some of it regrown after competition from the western states forced the steeper and poorer soiled land out of farming from the 1880s onwards. Temperatures average between 25–30°F in January and 70–75°F in July, with a frost-free period of 140–160 days and a mean thermal growing season of 7–8 months, during which the mean P–E–T is 28 in and A–E–T 26 in. However, more than half of the 35–45 in of annual precipitation (C.V. 10–15 %) is in the first half of the year, and drought often occurs in late summer.

Farming systems. The eastern part, especially the limestone area of the Great Valley, together with the Blue Grass area of Kentucky and the Nashville basin in Tennessee to the south, are the best farming districts in the Appalachians, where the percentage of land surface in farms varies from 70 in the limestone to 30 in the mountains of south-west Pennsylvania. But only half the farm area is actually cultivated, mostly on fairly *intensive grassland* or *alternating* systems. The rest is used for rough grazing or not at all. As most of the terrain is too rough for mechanised crop growing and cannot compete with cereals grown further west, as the climate favours grass and as the large north-eastern liquid milk market is near by, more than half the cultivated area is pasture or hay. Supplemented by purchased concentrates, this grassland is the basis of the milk production from the Friesian (Holstein) cows which dominate this sector. In the Great Valley, maize (1st year), mainly for grain, then wheat (or oats on the ridge land) (2nd year), are followed by 1- to 3-year leys based on lucerne or clover and timothy or Kentucky blue grass (*Poa pratensis*). In places, potatoes, vegetables and apples are grown, whilst cigar tobacco is locally important in the beautiful Susquehanna river valley. On the Allegheny Plateau less land is cropped and cereals are less important and the dairy farming resembles that of the cut-over areas of the Lake states. Clover/timothy swards prevail, although there is some lucerne, whilst the better land has permanent pastures of Kentucky blue grass (*Poa pratensis*) and white clover. On less fertile land, sweet vernal (*Anthoxanthum odoratum*) and Canada blue grass (*Poa compressa*) are the major species, whilst large areas of poor hilly land, stretching from west Virginia to Vermont, only support swards dominated by poverty grass (*Danthonia spicata*) which indicates a need for phosphorus, lime and sometimes potassium.

This is a sector of small (200 acre average) family dairy farms with some subsidiary enterprises, frequently poultry (see Col. I, Tables 2.3.4 and 2.3.5). The farms have up to 60 cows per man. Pasture, maize, silage and hay supply maintenance; the production part of the ration comes mainly from purchased concentrates. This intensive, mainly grassland dairy belt covers (Fig. 2.2.4) much of the north-east of the U.S.A. and eastern Canada (sample farm, p. 128).

Table 2.3.1

Climatic Data—U.S.A. Northern Transect

Altitude (ft)	Station	Bright Sunshine hours/day (mean annual)	Thermal Growing Season (>42½°F) length in months	Thermal Growing Season Mean Temp °F	Mean Months >72½°F	Mean Annual Precipitation (inches)	Annual Precipitation Coeff. Var. (C.V.)	Evapo-transpiration In Thermal Growing Season Potential (mean inches)	Evapo-transpiration In Thermal Growing Season Actual (mean inches)	Moisture Index Im	Max. Recorded rain/day inches (month of occurrence)
314	New York (New York)	7·3	8	62	2	43·0	10-15%	27·7	27·0	52 humid	9·4 (October)
724	Columbus (Ohio)		8	62	2	36·3	15-20%	28·0	26·3	29 humid	3·9 (July and September)
823	Chicago (Illinois)	7·2	7	62·5	1	32·9	15-20%	25·9	24·5	24 humid	6·2 (August)
1,093	Sioux City (Iowa)		6	63	2	25·4	20-25%	26·9	25·0	−7 dry sub-humid	5·1 (June)
1,718	Pierre (South Dakota)		7	62	2	16·6	20-25%	26·1	16·5	−37 semi-arid	3·7 (August)
4,021	Sheridan (Wyoming)		7	37	—	16·0	25%	22·9	15·9	−30 dry sub-humid	4·4 (July)
3,893	Helena (Montana)		7	55	—	12·4	20-25%	22·3	12·3	−45 semi-arid	3·7 (June)
949	Walla Walla (Washington)		9	59	2	16·1	20-25%	29·0	15·7	−45 semi-arid	2·7 (May)
154	Portland (Oregon)	5·9	9	57	—	42·2	20%	26·3	21·2	52 humid	7·7 (December)

Sources: M.O. World Tables (p. 2.1.00); Hershfield 1962; Thornthwaite (p. 2.1.00).

Table 2.3.1. (continued)

Climatic Data — U.S.A. Southern Transect

Altitude (ft)	Station	Bright Sunshine hours/day (mean annual)	Thermal Growing Season (>42¼°F) length in months	Thermal Growing Season Mean Temp °F	Mean Months >72½°F	Mean Annual Precipitation (inches)	Annual Precipitation Coeff. Var. (C.V.)	Evapo-transpiration in Thermal Growing Season		Moisture Index Im	Max. recorded rain/day inches (month of occurrence)
								Potential (mean inches)	Actual (mean inches)		
72	Wilmington (North Carolina)		12	63·5	4	50·0	20%	35·7	35·6	40 humid	9·5 (September)
1,054	Atlanta (Georgia)		12	61·5	4	48·1	15-20%	34·6	33·0	39 humid	7·4 (March)
1,254	Oklahoma City (Oklahoma)		9	67	4	31·9	25-30%	34·4	30·4	−9 dry sub-humid	7·9 (October)
3,590	Amarillo (Texas)		9	63	3	20·7	35%	30·7	20·2	−34 semi-arid	4·4 (May/June)
1,083	Phoenix (Arizona)		12	70	5	7·5	30-35%	45·6	7·5	−83 arid	5·0 (July)
658	Las Vegas (Nevada)		12	64·5	4	4·0	30-35%	44·4	4·0	−91 arid	2·0 (July)
178 below sea-level	Greenland Ranch (California)		12	76	7	Cow Creek 2·1	30-35%	Cow Creek 54·4	Cow Creek 2·1	Cow Creek −96 arid	1·4 (November)
331	Fresno (California)		12	63	4	9·5	35%	37·1	9·5	−74 arid	2·9 (November)
312	Los Angeles (California)		12	62·5	0	14·9	40-45%	33·1	14·9	−55 semi-arid	7·4 (December)

Sources: M. O. World Tables (p. 107); Hershfield 1962; Thornthwaite (p. 108).

139

Table 2.3.2

U.S.A. Northern Transect

40°43′N ← → 45°N
New York City — Oregon Coast

	N (a) *Atlantic Coastal Lowland*	N (b) *Appalachians*	N (c) *Corn Belt*	N (d) *Northern Great Plains*	N (e) *Rocky and adjoining Mountains*	N (f) *Pacific Coastal*
1. Sector						
2. Name						
3. From	East Long Island	New York City	Columbus	Sioux City	Billings	Mt. Hood
4. To	New York City (N.Y.)	Columbus (Ohio)	Sioux City (Iowa)	Billings (Mont.)	Mt Hood (Ore.)	Oregon Coast
5. Miles	50	500	800	750	600	150
6. Climate-type(s)	Cool Humid	Cool Humid	Cool Humid	Cool Dry	Cool Dry	Cool Humid
Thermal Growing Season:						
7a. —months at or > 42½°F	8	7–8	7	7	5–7	9–11
7b. —months at or > 72½°F	2	1	2	2	—	—
8a. —P–E–T (in) in T–G–S	28	28	26–27	23–27	22–29	26
8b. —A–E–T (in) in T–G–S	27	26	25	16–25	11–16	21
9. Growth limited by	Winter cold	Winter cold	Winter cold, some hot dry summers	Winter cold, some hot dry summers	Winter cold, some hot dry summers	Winter cold, often dry late summer
Precipitation						
10. —mean in (annual)	40–50	35–45	25–35	10–20	5–20	40–80
11. —C.V. % of Precipitation (annual)	10–15	15–20	15–25	20–25	15–25	10–15
12. Special Climate Features			Rainfall distribution favourable (large proportion of spring rain)	Rainfall distribution favourable. Dry harvest period	Winter and early spring rainfall	Cool summer due to maritime influence
13. Weather Reliability	4	3	3	2	2	4
Soil:						
14. —Zone/Group	Podsolic	Podsolic	Podsols and Chernozems	Chernozems, Lithosols, Planosols	Complex, mixed	Complex, mixed
15. —pH	Low	Low	Medium	High	—	Low
16. —Special Features		Some thin soils. Drainage problems in glaciated areas	Deep; good physical and chemical status	Shallow and imperfectly weathered in Badlands	Prairie-type soil in the Palouse	Leached. Some valley soils badly drained
17. Vegetation	Mixed forest	Mixed forest	Hardwood forest to tall grass prairie	Tall grass merging to short grass further west	Timber, grassland, tundra. Varies with altitude	Coniferous forest
18. Relief	Flat or rolling lowland	Mountains, hills, valleys	Flat or rolling plains	Plains with some broken areas	Mountains, plateaux valleys	Low mountains and valleys
19. Market Access	4	4	4	2	2	3
20. Farming Systems	Intensive tillage	Intensive grass and intensive alternating	Intensive tillage and intensive alternating	Extensive tillage and extensive grazing	Extensive grazing and intensive tillage in valleys	Intensive grass; intensive alternating; intensive tillage and tree crops
21. Major Enterprises	Horticulture; potatoes; level milk production; poultry	Level milk production; maize; small grains; hay. Cattle raising	Maize; small grains; soya-beans. Pigs; beef cattle; some milk	Spring cereals; cattle ranching; some irrigated field crops	Cattle and sheep ranching. Cereals, dry peas and beans in Palouse. Irrigated horticulture and field crops in western valleys	Level milk production; small grains; grass hay and seed. Irrigated horticultural crops

THE CORN BELT (*Table 2.3.2, Sector (c)*)

The Corn (maize) Belt stretches from western Ohio about 800 miles into Nebraska and South Dakota where small grains (wheat, barley, oats) gradually take over from the corn (maize). It is a unique, highly productive and very important area of world-wide technical interest. Despite hot summers and cold winters the climate is unusually favourable for an inland area. Mean January temperatures are 20–30°F, July averages about 75°F, so there may be some heat stresses in livestock, though high summer temperatures are favourable for maize. The thermal growing season averages about 7 months, but, as in the Canadian Prairies, has unreliable starting and ending dates; it has a mean P–E–T of 26–27 in and an A–E–T of 25 in. Annual precipitation averages 35 in in the east and 25 in in the west, with a coefficient of variation (C.V.) of 15–20% increasing to 25% in the west of the sector. The once glaciated land is flat or rolling and is well suited to mechanised farming. The soils are loamy to rather heavy in texture and derived from glacial drift and its cover of wind-deposited loess, and were sometimes badly drained. On the east, grey-brown podsolic soils developed under hardwood forest. In the west, the chernozems and black prairies, developed under tall grass, form one of the world's largest areas of fertile soils. They are deep, moisture retaining, usually easy to work and were high in organic matter and in nitrogen when first ploughed. (The main grass species, which still persist on poorly drained uncultivated soils, were bluejoint (*Calamagrostis canadensis*), Prairie cord-grass (*Spartina pectinata*), American slough grass (*Beckmannia syzigachne*), manna grass (*Glyceria* spp), reed canary-grass (*Phalaris arundinacea*) and common reed grass (*Phragmites communis*).) Early farmers could not resist the temptation to overcrop and, apart from the loss of soil nutrients, water erosion has been serious in places and the top-soil has disappeared. Despite counteraction, some Corn Belt top-soil still finds its way into the Gulf of Mexico.

Farming systems. Here then is a flattish area, with good soils but with a mean A–E–T in the T–G–S that is 10 in higher, and where rainfall is greater as the crops start to make leaf area than in the Spring Wheat Belt (Fig. 2.2.4). After spring thaw, temperatures rise rapidly and frost risks fall; maize can be safely planted in early May, and from then until August good light receipts and fairly good moisture favour high yields from the specially bred hybrid maizes, particularly now that weeds can be controlled both by herbicides and inter-row cultivations and now that fertiliser is heavily used. Further, maize for grain, unlike small grains, provided it has matured, need not be harvested until October or November; which eases the seasonal work load. Finally, much of the lower land has been drained; one-third of the land in U.S. drainage projects is in the Corn Belt. Not surprisingly, nearly half the world's maize is produced here and this crop, and the livestock that converts it to cash, dominate the Corn Belt. But soya-beans (a leguminous oil-seed row crop) now rival maize in acreage. Of the small grains, wheat is most important, followed by oats; the latter are often undersown. The former increases in

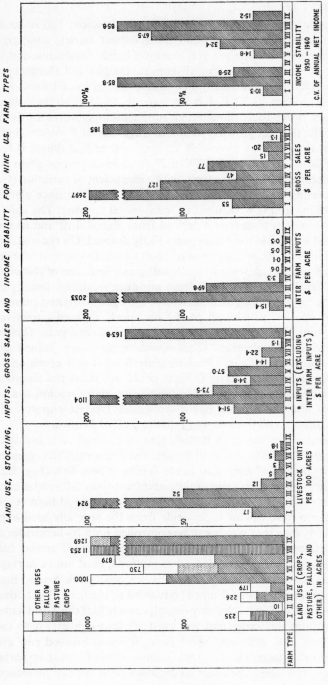

* TOTAL EXPENDITURE LESS INTER-FARM INPUTS.

Fig. 2.3.1 Land Use, Stocking, Inputs, Gross Sales and Income stability for Nine U.S. Farm Types. (Key on page 139)
Source: U.S. Dept. Agric.

Key to Nine U.S. Farm Types in Fig. 2.3.1

No.	Farm type	Location	Transect	Sector	h†	Transect table
I	Dairy Farm	Central North East	Northern	b	0·7 Wet	2.3.2
II	Egg Producing	New Jersey	Northern	a	0·6 Wet	2.3.2
III	Hog–Beef Fattening	Corn Belt	Northern	c	0·9 Wet	2.3.2
IV	Peanut, Cotton	South Coastal Plains	Southern	a	0·8 Wet	2.3.3
V	Cotton (Large)	Mississippi Delta	Southern	c	0·7 Wet	2.3.3
VI	Wheat, Small Grain	Northern Plains	Northern	d	0·6 Dry	2.3.2.
VII	Wheat, Grain Sorghum	Southern Plains	Southern	e	0·7 Wet	2.3.3
VIII	Cattle Ranch	South West	Southern	f	0·1 Dry	2.3.3
IX	Cotton (Large) General	California	Southern	h	0·6 Dry*	2.3.3

* Excluding irrigation. † h = hydrologic ratio (p. 462).

area as the Corn Belt blends into the eastern edge of the spring hard ¡wheat belt (Fig. 2.2.4) and in parts of Ohio where soft winter wheat yields well. Grassland (pasture and hay) often equals the area under maize, partly because the former is a useful break crop and provides bulky winter feed for cattle, and partly to check erosion, especially, for example, in the loessal areas bordering the Missouri and the Mississippi. Timothy (*Phleum pratense*) and Kentucky blue grass (*Poa pratensis*) are widely used; other introduced species include cocks-foot (*Dactylis glomerata*) and tall fescue (*Festuca arundinacea*), medium red, alsike and white clovers (*Trifolium pratense, T. hybridum, T. repens*) and, of course, lucerne (alfalfa, *Medicago sativa*).

The systems are mainly *intensive alternating* (Figs. 2.2.4 and 2.3.1 and Col. III of Tables 2.3.4 and 2.3.5). A common 4-year rotation is maize; maize; oats (undersown) or soya-beans (*Glycine max*); one year ley for hay. This is basically the same as rotations found in the better soiled, less humid parts of north-west Europe. *Intensive tillage* is found on cash-crop farms with good soils. Areas liable to water erosion have more grassland and raise more cattle. Fattened beef cattle and hogs (pigs) (mainly Landrace, Hampshire and Duroc Jerseys) are the main sources of gross sales in the Corn Belt. Many cattle and some sheep are imported from western and south-western ranges and fattened on feed-lots on own grown or locally bought concentrates. Egg production is secondary, but the total egg output from the area is large. Small dairy farms producing milk for manufacture are common nearer the northern Lake states. The 200–300-acre family farms use little hired labour and are highly mechanised, labour saving and mostly technically advanced. But capital requirements are high, and crop-share and cash tenancies are common; in western Iowa and Illinois 60% of farms are tenant operated.

A One-and-a-half Man Intensive Pig and Beef Farm in Central Iowa, U.S.A.

This 300-acre *tillage* farm is snow-covered or frozen for 3 or 4 months and free of frost from May to September inclusive; field operations are confined to

F

7 or 8 months and most livestock is housed for 6 or 7 months each year. The soils are fertile, deep, dark brown or black, slightly acid and tile drained. The mean annual precipitation is 36 in and much of it falls in the thermal growing season (April to September). Maize (corn) which yields over 4,500 lb/acre, occupies more than half the rotated acreage. One-fifth is in soya-beans (2,100 lb/acre) and oats, (2,100 lb/acre) which do not have very high yields. A sixth is in mixed lucerne (alfalfa) and brome grass leys for hay (which yields nearly 9,000 lb/acre) and for grazing. So, if there were a formal cropping system, it would be 5 years in cereals or oil-seeds and one year in ley. The maize gets 60 lb of elemental nitrogen per acre before, and 20 lb at planting, whilst nitrogen at 30–40 lb/acre may be side dressed at last cultivation, plus phosphate and potash as needed. All seed is dressed. Herbicides and pesticides control weeds, insects and diseases, including corn borers (*Ostrinia nubilalis*), the main danger. On such a farm the tall maize stalks are chopped up after harvest, partly to deprive corn borers of shelter and sustenance, and partly to reduce erosion when the spring thaw comes. For the same reason ploughing (5 furrow) is not started until late March, followed by discing across the furrows and planting in early May. This is done in one operation at up to 60 acres per day by one man on a large tractor pulling a corn planter depositing dressed seed and fertiliser, spaced in 3-ft rows, a sprayer applying pre-emergence herbicides and insecticides and a harrow to cover the seed. The crop is machine hoed twice before it gets too tall and then harvested by machine corn-picker in October.

There are 28 sows which farrow twice a year (February and August) and produce 420 fat pigs per year which are fed on maize and a protein and mineral supplement, and sold at about six months of age. Each year 130 half-grown (feeder or store) Herefords and Angus cattle are bought at 450 lb live-weight, fattened on maize, protein and mineral supplements, hay and grazing and sold a year later at about 1,050 lb live-weight each. The local transport and public services are good and the cattle, pigs and soya-beans are marketed locally or sent to central markets. Apart from the buildings to house the livestock and to store feed, etc., the equipment includes 3 tractors, a corn-picker (maize harvester), a forage harvester and ancillary machinery, so that all work can be done rapidly by the owner and his family, plus some hired labour for harvesting. The manpower used, however, equals only 1½ full-time workers, i.e. 200 acres per man. This is an average mid-western standard. It is higher than English and much higher than continental European countries. To this end, investment is heavy in labour-saving machines and devices of all kinds. (*Courtesy:* H. R. Rowell, Extension Economist, Iowa.)

NORTHERN GREAT PLAINS (*Table 2.3.2, Sector (d)*)

The Northern Great Plains rise gently from 1,000 ft, where they meet the Corn Belt, to 5,000 ft in the foothills of the Rockies, and are well suited to mechanisation, except the rough broken land of the Black Hills and Badlands of South Dakota and Wyoming, and the Nebraska Sandhills where cultivation would expose loose dune sands to erosion. The mainly grassland soils are easily worked, of high pH and in the east are mostly dark brown-black chernozems, originally under tall grasses (p. 141 above), grading into chestnut soils and brown soils in the west (see Canadian Prairies). On the short grass prairies in the eastern part, the original vegetation was mostly blue grama (*Bouteloua gracilis*), buffalo grass (*Buchloë dactyloides*), western wheat grass (*Agropyron smithii*) and needle grasses (*Stipa* spp). The early settlers made their 'houses' of grass sods because trees are rare except in 'bluffs' on the northern slopes of rough terrain or along streams where Great plains cottonwood (*Populus sargentii*), balsam poplars (*Populus balsamifera L.*) and willows (*Salix* spp) grow. The cool dry climate is like that of the Canadian Prairies (p. 129) but the thermal growing season, P–E–T and A–E–T are higher and the summer is hotter. Yearly precipitation means range from 25 in in the east to 13 in in the west. The reliability of rainfall, spring and autumn temperatures is low and there may be sudden high winds and storms or very severe winters. Yields and net incomes are unstable (Col. VI, Tables 2.3.4 and 2.3.5).

Farming systems. Extensive tillage (spring wheat), on the lines already outlined (p. 130) for Canada, dominates, with some *extensive grazing*. Stock numbers are kept low because otherwise pastures are overstocked in the frequently occurring drought years, and erosion results (Figs. 2.2.4 and 2.3.1 and Col. VI, Tables 2.3.4 and 2.3.5). To the north of the transect, for example in the Red River Valley, or where the rainfall is higher, the farms are not so highly specialised. On these *intensive tillage* farms small grains are grown in combination with flax-seed (linseed), maize and sometimes cash roots (all spring sown). Livestock enterprises are common on such farms, especially where the relief is accentuated. (For a useful study on the relation between tillage acreage and land-form, see Hidore 1963.)

A Farm in the Red River Valley (*of the North*) *on the north-east edge of the Interior Plains*

Moisture is major limitant. Hence 300 acres are fallowed to conserve moisture before sugar-beet.

Crops and stock. Sugar-beet, 300 acres; spring wheat and spring barley, 700 acres; lucerne (cut for hay and sold), 100 acres; fallow, 300 acres; a total of 1,400 acres. 600 cattle are fattened each winter.

Crop methods. Cereals are treated as in the Canadian Prairies. Sugar-beet: 11 lb N (maximum allowed by the factory); 150 lb P_2O_5/acre; no salt; K rarely; dung from feed-lot. Pelleted seed drilled in 22-in rows, thinned to 25,000 plants/acre. Harvested in October (as frozen up by November) by lifter and separate topper (each 3 row). Two lorries and men to take away 25 acres (350 ton) per day.

Beef. Calves are bought in the autumn from ranches at 450 lb live-weight (l.w.); sold the following April/June at 1,100 lb l.w. They are yarded in an open-front shed, with a roof over mangers. Mechanised feed (by auger) of barley (moist stored in tower silos); beet pulp; some maize; protein supplement.

Labour and machinery. This consists of father, son and one hired man, say 300 acre crops/man, together with big, powerful machines. (*Source: Farmers Weekly.*)

The western, drier parts of the Dakotas and most of Montana, together with the sandhills of Nebraska, are seasonal non-migrational short grass range lands used for cattle and sheep, which need only be fed for 2–4 months in the winter. The beautiful Sweet Grass Hills of Montana are well named. Calves or yearling cattle from this *extensive grassland* system are either sold as feeders or fattened on grain, silage and lucerne grown on near-by irrigated land.

THE ROCKY MOUNTAINS AND ADJOINING AREAS (*Table 2.3.2, Sector (e)*)

This sector of high plateaux, valley land and mountains, with peaks where arctic tundra provides some seasonal grazing for sheep, contains, some interesting farming systems in scenic locations and extends, for present purposes, from the end of the Interior Plains over to the coastal mountains. Above 2,000 ft the winters are long and cold and snowy. Summers are dry, and, in the mountains, cool (July mean temperatures 55–65°F, which is 10°F lower than in the Palouse). The thermal growing season may last 6–7 months in the lowlands, but only one in the highlands. Precipitation (mean annual) varies from 10 in to 40 in and is unreliable. In the lower rainfall areas it comes mostly in winter and early spring. The soils are varied and mixed. The vegetation is mostly coniferous forest. However, the Palouse (in south-east Washington on the west of the Rockies) developed under prairie *not* under forest, and its soils are mainly black, sandy or silty in the drier parts, but silt loams where precipitation is higher. It had some of the best natural grazing lands in the 'old west'. The main grass species were bluebunch and beardless wheat grasses (*Agropyron spicatum, A. inerme*), Idaho fescue (*Festuca idahoensis*), blue grass (*Poa* spp), giant wild rye (*Elymus condensatus*) and needle and thread grass (*Stipa comata*). However, overgrazing has encouraged wind and water erosion and the invasion of sagebrush (*Artemisia* spp).

Farming systems. Extensive grassland (cattle and sheep ranching) is the major land use. The grazing is mainly seasonal, the animals being fed for 3–5 months of the winter on hay from mountain meadows, from irrigated land near the ranch headquarters or bought from irrigated farms which are increasingly important. Most of the ranches are more than 2,000 acres. Increasing attention, in range management, is paid to ecology.

Two other systems, not in the Rockies proper, are technically interesting The mean annual precipitation in the *Palouse* ranges from 18 to 25 in. It falls mainly in the winter and is just enough for a *tillage* crop to be taken each year, as nearly all the moisture is retained by the 'grassland' soils, and is available for crop growth in the spring and early summer before temperatures are excessive and evapo-transpiration too great. The small areas of flat land in the valleys of the Palouse are badly drained and are in extensive grazing. Hence, in this area of strongly rolling hills with slopes up to 55%, it is the slopes that are in annual crops. Severe soil erosion has been avoided because the precipitation intensity per hour is low and the high organic matter retains water. However, after more than 80 years of continuous cropping, soil organic matter has fallen by half. Whereas, under virgin sod, total run-off was about 5% of the total precipitation, it now averages about 20%. More green manuring crops, grass/legume swards and livestock are indicated. The thermal growing season is 5 to 6 months and there is some risk of severe winter frost. Nevertheless, winter wheat is preferred as it yields better (over 2,300 lb/acre) and provides some winter cover to check erosion. If it fails, the land is often reseeded to spring wheat, barley or peas. The family farms in this *semi-extensive tillage system* are fully mechanised (as the slopes are steep, crawler tractors are used) and have 400 or more acres cropped each year in wheat, barley or peas.

In the *Yakima Valley*, which is in the Columbia basin and gets its water from the Cascade Mountains, favourable temperatures, soils and land slopes are the bases of *intensive irrigated tree fruits and tillage* systems on small family farms. Only 8% farms exceed 100 acres. In the spring the apple, pear and peach blossoms call for a Van Gogh to paint the scene. However, there are also many small tillage farms, producing hay, cereals, horticultural specialities, sugar-beet, potatoes, hops and mint. Local sheep ranchers utilise stubble, sugar-beet by-products and hay for autumn, winter and spring feeding, and the bordering mountains supply summer grazing. Feeder (store) cattle are fattened from crop by-products in the valley and small dairy herds are found throughout the area. Turkey production has expanded rapidly in the last few years.

NORTH-WEST PACIFIC COASTAL ZONE (*Table 2.3.2, Sector (f)*)

This zone is of alternate mountains and lowlands running parallel to the coast. A transect at constant latitude would pass from the Yakima Valley through limited summer grazing areas on the slopes of the Cascade Mountains and into the Puget Sound Lowland, with its dairying, poultry, berries, flower bulbs, flower and vegetable seeds, and vegetables for canning and freezing. The cool humid climate has high winter and moderate summer temperatures (January means 35–45°F, July means 55–70°F). There are over 200 frost-free days on the coast and 180–200 in the Willamette Valley (which lies south of the transect), but less than 120 as one ascends the Cascades. The mean lowland thermal growing season is about 11 months. The mean annual precipitation is high, 50–100 in or more on the coast and Coastal Ranges, and 30–50 in in the inland valleys. Half the precipitation comes in the winter and less than 5% in July and August, and this is the main difference between this region and north-west Europe and greatly influences yields and farming patterns. In midsummer, the Willamette Valley is as dry as the Central Valley of California. The mean P–E–T in the T–G–S is 26 in but the summer drought reduces the A–E–T to 21 in and favours grass seed production. The winter rain is of low intensity so erosion is slight. The annual C.V. is 15–20%. The soils are very mixed. They have been moderately or highly leached. The pH is 5·4 to 6·3 in the valleys and 5·0 on the coastal lowlands. About half the alluvial soils of the Willamette Valley and the coastal lowlands are poorly drained or water-logged. The natural vegetation is coniferous forest, notably Douglas fir (*Pseudotsuga taxifolia*).

Farming systems. Only a fifth of the land area is in farms and the upland regions are mostly in forest. On the deeper soils, deep rooting crops (tree fruits, alfalfa) can make use of stored soil moisture, but, on shallow soils, annuals with short life-cycles, such as small grains and vetches, which complete their cycle before the summer drought, produce well, but may be limited by soil acidity and deficiencies of available N, P_2O_5, K_2O, boron and sulphur.

Where the land is drained, limed and fertilised, the Willamette Valley, Oregon, which is not untypical of the interior lowlands, is in intensive *alternating* (*mainly dairying*) *or tillage systems*, in which sprinkler irrigation is increasingly important, or, where drainage is poor and winter rainfall very heavy, in *intensive grassland* for grass and legume seed production. In the wet coastal lowlands milk production from *intensive grassland* is virtually the only enterprise. The pastures are of cocks-foot, tall fescue, rye-grass and reed canary-grass (*Phalaris arundinacea*), with red alsike and Ladino clover, *Lotus major*, or New Zealand white clover; purchased concentrates are fed in winter to meet the year-round market for milk.

U.S.A.—Southern Transect (*see Fig. 2.2.4 and Table 2.3.3*)

The Southern Transect crosses the north of the Cotton Belt then passes through the southern Interior Plains (with their winter wheat region), the grazing regions of the Rocky Mountains and the arid regions of Arizona and California with their islands of intensive irrigated farming, before reaching the highly complex specialist farming of coastal California (Fig. 2.2.4).

The Cotton Belt. The Cotton Belt is a historical and social but no longer an agricultural description. It has a humid climate, suitable for cotton, with a long thermal growing season, high summer temperatures and more than 200 frost-free days per annum. It includes the first four sectors in Table 2.3.3 and stretches from the east coast almost to the 100°W meridian. It is by no means a continuous belt of specialised cotton-growing farms. Over a hundred years of overcropping with cotton, resulting in severe soil erosion and loss of soil fertility, and aggravated by economic pressures and social problems, led to a mass exodus of tenant farmers and share-croppers from the great plantations to the cities. Within the belt, cotton growing is now confined to a few favoured areas where mechanisation is feasible and there is less danger of soil erosion. This crop has been migrating westwards for 100 years. Now Texas has by far the greatest acreage, followed by Mississippi and Arkansas, whilst, in New Mexico and California, with irrigation, cotton yields are amongst the highest in the world. Most U.S. cotton varieties have short- or medium-staple length and collectively are referred to as the Stoneville varieties. (U.S.D.A. has a cotton-breeding station at Stoneville, Mississippi.) The production of American Egyptian cotton only amounts to 100,000 bales out of a total production of about 15,000,000 bales annually. Most of the American Egyptian cotton is produced in Texas, where some of the acreage is irrigated. California and Arizona produce, with irrigation, about 2,850,000 bales with an average yield of 2·4 bales of 480 lb net per acre. The other states produce the remainder with an average yield of only 0·85 bales per acre.

At the same time beef production has been moving east. Much abandoned cotton acreage is now in grass and legume pastures for cattle (beef cows have more than doubled in 20 years), whilst food crops and forestry are increasing. Soft woods do well and more than 55% of the land area of the south is forested, mainly with pine (loblolly—*P. taeda*, slash—*P. elliottii*, long-leaf —*P. australis*, short leaf—*P. echinata*). A well-managed pine forest producing pulping timber is profitable and a good inflationary hedge from which the potential yield/acre of cellulose for man-made fibres is as high as that of cotton.

THE SOUTHERN ATLANTIC COASTAL LOWLANDS
(Table 2.3.3, Sector (a))

Summarised in Table 2.3.3, Sector (*a*); one interesting system still here is a type of shifting cultivation; one-quarter to one-third of the land is cropped, with half the cropped acreage in cotton and the rest in winter wheat, oats, hay and tree fruit (peaches). When this land is exhausted, the land that has 'reverted' is recleared and again cropped.

THE SOUTHERN APPALACHIANS *(Table 2.3.3, Sector (b))*

The southern end of the Appalachians rise to over 5,000 ft in the lovely Blue Ridge country. Then, after the Tennessee and Coosa River Valleys, rolling uplands gradually decline in altitude to the Mississippi bluffs and delta.

The climate is warm humid with annual mean precipitation of 45–55 in of rain (C.V. 15–20%). Even in the relatively dry autumn, the rainfall is still 3 in a month. The thermal growing season is 9–12 months with 180–230 frost-free days and a mean P–E–T of 29–35 in and mean A–E–T of 28–33 in. Summer

Table 2.3.3

U.S.A. Southern Transect
Lat. 34°N ←——→ Lat. 37°N
Wilmington (N. Carolina) — Californian Coast

1. Sector	S (a)	S (b)	S (c)
2. Name	*Southern Coastal Lowlands*	*Southern Appalachians*	*Mississippi De*
3. From	Wilmington (N.C.)	Greenville	Corinth
4. To	Greenville (S.C)	Corinth (Mississippi)	Little Rock (A
5. Miles	260	350	210
6. Climate-type(s)	Warm Humid	Warm Humid	Warm Humid
Thermal Growing Season: 7a.—months at or >42½°F	12	9–12	11–12
7b.—months at or >72½°F	4	4	4
8a.—P–E–T (in) in T–G–S	36–38	29–35	36
8b.—A–E–T (in) in T–G–S	36	28–33	33
9. Growth limited by		Cold, some heat	Heat
Precipitation: 10.—mean in (annual)	40–50	50–70	45–55
11.—C.V.% of Precipitation (annual)	15–20	15–20	20–25
12. Special Climate Features			Hot humid sun
13. Weather Reliability	3	3	3
Soil: 14. Zone/Group	Podsolic–latosolic	Podsolic and podsolic–latosolic	Alluvial
15.—pH	Low	Low	Medium
16.—Special Features	Often light, leached	Light, water erosion	Deep, rich, nee drainage. Some hard-pan. Floo
17. Vegetation	Southern coniferous forest	Southern coniferous and mixed forest	River bottom h woods and cyp
18. Relief	Plains, flat to rolling	Mountains, hills and rolling	Floodplain, wi meander ridges
19. Market access	3	3	3
20. Farming Systems	Tillage, some alternating	Intensive tillage, some intensive alternating	Intensive tillag
21. Major Enterprises	Horticulture; maize; peanuts; soya-beans. Tobacco in north. Cotton in restricted areas. Some intensive livestock	Maize; forage crops; tobacco; cotton; horticulture. Beef and dairy cattle	Cotton; soya-maize; oats; fo crops. Specialis areas of sugar-and rice

S (d)	S (e)	S (f)	S (g)	S (h)
rkansas Valley and olands	*Southern Great Plains*	*Rocky and Western Mountains*	*Sierra Nevada*	*Pacific Coastal Zone*
ttle Rock	Oklahoma City	Santa Rosa	Wickenburg	Porterville
klahoma City okl.)	Santa Rosa (N.M.)	Wickenburg (Ariz.)	Porterville (Calif.)	Coast (Calif.)
300	400	440	380	130
arm Humid arm Dry	Warm Dry	Warm Dry (Desert)	Warm Dry (Desert)	Warm Dry (Medit.)
9–11	9	7–12	12	12
4	3–4	0–5	5–7	0–4
34–36	30–35	21–23	45–55	30–40
30–33	20–30	13–18	2–8	10–18
ld, heat	Cold, heat, drought liability	Cold, heat, drought	Heat, drought	Heat, drought
30–45	15–30	10–20 (more on high peaks)	0–10	5–35
25–30	30–35	35–40	30–40	30–50
	Precipitation is unreliable in West		Very dry, hot summers. Cold nights	Hot, dry summer. Cool, wet winter
2(?)	2	1	1	2
dsolic–latosolic d mixed	Chernozemic desertic	Lithosolic-desertic	Desertic	Mixed
ries	High	High	High	High
any cherty soils en with fragipan	High organic matter on chernozems. Wind erosion	Sierozems often have caliche. Saline or alkaline in places	Saline or alkaline in places	Some problems with alkali soils and panning
xed forest and l grass prairie	Tall grass prairie; short grass prairie in West	Coniferous forest; scrub; desert grass-land	Scrub or none	Mediterranean woodland, scrub, grass
l and Valley	Plains, rolling in places	Mountains, high plateaux, valleys	Mountains, valleys	Mountains, valleys
2–3	2	2	2	3
ries	Extensive tillage; extensive grazing	Extensive grazing, some intensive tillage	Some intensive tillage	Intensive and extensive tillage; tree fruits; extensive grazing; intensive alternating.
ize; hay crops; s; wheat; ghum; cotton. rticulture. Dairy l beef cattle; ne sheep; poultry	Winter wheat; grain sorghum. Cotton (some irrigated). Range cattle and some sheep, turkeys	Cattle and sheep ranching. In irrigated valleys, forage and horticultural crops, cotton. Dry-lot dairy farming	Isolated and small irrigated areas in desert produce horticultural crops; hay; poultry and dairy products, but mainly non-agricultural	Irrigated cotton horticultural crops; rice. Some extensive grain. Tree fruits. Extensive cattle or very intensive for beef and milk. Poultry

temperatures are high (July mean 75–80°F), and so, then, is evapo-transpiration. In the limestone valleys, there may be moisture stress in crops because
summer rainfall is erratic. Mixed oak and pine forests are the natural vegetation. The soils are podsolic/latosolic, pH *c.* 5·5, clays and silts in the valleys
and sandy or silty loams in the uplands and stony in the mountains. Water
erosion is a problem.

Farming systems. Systems depend mainly on the soil, aspect and degree of
slope. Much of the Blue Ridge is forested and used for recreation; it is too
steep and rugged for more than limited cropping. The few small farms usually
crop less than 20 acres and are often, as in other marginal areas, for example
Vermont in New England, residential or part-time homes for families who
work in near-by towns. Moderate slopes in the limestone valleys usually have
alternating systems, for example based on maize as a row crop, and erosion
reducing crops of forage (especially legumes) or small grains, e.g. winter
wheat. Lucerne yields well because of its longer growing season and its deep
root system which helps it get soil moisture in summer dry periods. But it is
expensive to establish and is liable to crown rot (*Sclerotinia*) in its first winter.
Hence, *Lespedeza striata* and *L. stipulacea* (annual types), which are usually
undersown in small grains in the spring, are important annual forage and anti-
erosion crops. When a forage crop sward is ploughed out the surface is left
rough until maize is drilled. This maize is harvested early if possible to enable
early seeding of winter wheat and thus provide some erosion check by late
autumn. Further, to reduce run-off, contour and strip-cropping are usual,
especially where sheet erosion is a potential danger. Livestock, especially
cattle, are increasing in numbers as farming systems improve.

The alluvial bottom lands where erosion is not serious are in *intensive tillage*
systems for maize, burley tobacco, and such crops as lettuce, green beans,
cabbages and strawberries. The average government tobacco allotment per
farm is only 1 acre, but its high gross sales per acre make it a worthwhile, if
difficult, crop. After the tobacco is harvested, a winter anti-erosion cover crop,
such as crimson clover (*Trifolium incarnatum*) and rye-grass, may be sown and
ploughed out in the spring. This procedure and heavy fertilisation enables
tobacco to be grown continuously on the same field, although for disease, etc.,
control tobacco is generally taken 'round the farm'.

THE MISSISSIPPI DELTA (*Table 2.3.3, Sector (c)*)

The 'Delta' consists of a floodplain with low meander ridges; it is bordered
on either side by the Mississippi 'Bluffs' of rolling uplands, which vary in
width from a few yards to 75 miles. The warm humid climate has a moderately
reliable evenly distributed rainfall averaging 45–55 in per annum (C.V. 15–
20%) although it may be rather low in late summer. January temperatures
average 40–45°F. The thermal growing season is about 11 months, with 220–
240 frost-free days and with mean P–E–T 36 in and A–E–T 33 in. Summer
temperatures are a trouble with livestock. (July averages over 80°F). The vege-

Farm Costs by Class of Input in the U.S.A. (Mean of Samples in 1961–63)
(Nine Farm Types)

Column number	I	II	III	IV	V	VI	VII	VIII	IX
Type of farming (Means of Samples)	Dairy Farm Central North East	Egg Producing New Jersey	Hog-Beef Fattening Corn Belt	Peanut-Cotton South Coastal Plains	Cotton (Large) Mississippi Delta	Wheat-Small Grain-Livestock Northern Plains	Wheat-Grain Sorghum Southern Plains	Cattle Ranch South West	Cotton (Large)-General California
Climate and System	Cool Humid Intensive Alternating or Grazing	Cool Humid Intensive (Landless)	Cool Humid Intensive Alternating or Tillage	Warm Humid Intensive Tillage	Warm Humid Intensive Tillage	Cool Dry Extensive Tillage	Warm Dry Extensive Tillage	Warm Dry Extensive Grazing	Warm Dry Intensive Tillage (irrigated)
U.S. Transects Sector (Tables 2.3.2 and 2.3.3)	N (b) Appalachians	N (a) Atlantic Coastal	N (c) Corn Belt	S (a) Atlantic Coastal	S (c) Mississippi Delta	N (d) Great Plains	S (e) Southern Plains	S (f) Rocky Western Mountains	S (h) Pacific Coastal
Summary per acre (in dollars)									
(1) Land and buildings	7·1	286·6	18·1	5·9	10·8	2·8	6·35	0·67	56·2
(2) Hired and family labour	17·4	496·8	22·4	8·8	14·8	3·8	4·15	0·22	25·2
(3) Inter-farm inputs	15·4	2032·7	69·8	3·3	0·6	0·1	0·51	0·34	
(4) Industrial inputs	19·4	206·6	22·5	18·7	29·0	6·5	9·77	0·28	69·1
(5) Miscellaneous crop and livestock expenses	2·6	38·2	4·9	0·7	0·5	0·4	0·56	0·15	
(6) Other expenses (and taxation)	4·5	75·5	5·6	0·7	1·9	0·9	1·61	0·14	13·3
(7) Total expenditure	66·4	3136·4	143·3	38·1	57·6	14·5	22·95	1·80	163·8
(8) Net farm income	17·1	313·3	31·8	25·5	34·9	8·2	10·49	0·64	67·8
Percentage Distribution of Expenditure	%	%	%	%	%	%	%	%	%
(1) Land and buildings	11	9	13	15	19	19	28	37	34
(2) Hired and family labour	26	16	15·5	23	26	26	18	12	16
(3) Inter-farm inputs	23	65	49	9	1	1	2	19	
(4) Industrial inputs	25	7	15·5	49	50	45	43	16	42
(5) Miscellaneous crop and livestock expenses	4	1	3	2	1	3	2	8	
(6) Other expenses (and taxation)	7	2	4	2	3	6	7	8	8
(7) Total expenditure	100	100	100	100	100	100	100	100	100
(8) Net farm income (as % of total expenditure)	26	10	22	67	61	56·5	46	35·5	41
Income Instability (1950–60) Coefficient of variation of annual net income	10·3	85·8	25·8	n.a.*	14·8†	32·4	67·5	85·8	15·2

Source: Calculated from U.S.D.A. 1964 and Hathaway 1963.
Details and origin of figures:
Capital charges: Calculated at 5% on Land and Buildings; Machinery and Equipment; Livestock and Crops. *'Per acre'* = Land in Farm.
Inter-farm inputs: Include purchase of Seed, Feed and Grazing Fees, Livestock.
Industrial inputs: Include expenditure on Machinery, Fertiliser, Fuel, Crop/Livestock Expenses, Poison, Ginning, Machinery Work Hired, Operating Expenses of Machinery.
Land and Buildings: Include Irrigation facilities. *Family Labour:* Calculated at same rate as Hired Labour, i.e. *Labour Expenditure* number of hours worked.
* n.a. = not available.
† Small cotton farms in Mississippi Delta.

153

Table 2.3.5

Cropping and Stocking (Nine Farm Types)—U.S.A.
(Mean of Samples in 1961–63)

Type of Farming (Means of Samples)	Unit	I Dairy Farm Central North East	II Egg Producing New Jersey	III Hog-Beef Fattening Corn Belt	IV Peanut-Cotton South Coastal Plains	V Cotton (Large) Mississippi Delta	VI Wheat-Small Grain-Livestock Northern Plains	VII Wheat-Grain Sorghum Southern Plains	VIII Cattle Ranch South West	IX Cotton (Large)-General California
I Land in farm	Acre	235	10	226	179	1,000	730	879	11,253	1,269
Cropland harvested	,,	86·3	0	147	73	611	347	362	10	1,039
Pasture (or range land)	,,	87·3	—	—	—	—	202	—	11,243	103
Fallow and idle	,,	—	—	—	—	—	—	—	—	—
II Crops harvested	Acre	—	—	8·1	32·7	19·3	—	—	—	20·6
Maize for grain	,,	11·5	—	—	—	—	—	—	—	—
Maize for silage	,,	13·5	—	30·2	—	—	—	—	—	—
Small grains	,,	—	—	—	—	—	111·1	230	—	—
Wheat	,,	—	—	—	—	29·6	127·8	—	—	—
Oats	,,	—	—	—	—	—	—	13	—	293
Barley	,,	—	—	—	—	—	—	94	—	—
Sorghum	,,	—	—	—	—	—	—	15	—	—
Sorghum for forage	,,	—	—	29·2	—	20·6	53·9	—	—	—
Hay (ALF = Lucerne)	,,	—	—	—	—	258	42·5	—	—	ALF: 293
Cotton	,,	—	—	—	16·1	283	—	—	—	432
Peanuts	,,	—	—	—	22·8	—	—	—	—	—
Peanut hay	,,	—	—	—	7·3	—	—	—	—	—
Soya-beans	,,	—	—	—	—	—	—	—	—	—
Flaxseed	,,	—	—	—	—	—	—	—	—	—
III Livestock on farm (In L.U.'s)	L.U.									
Cows and heifers 2 years and over	1	30·8	—	5·2	0·8	—	4·3	1·2	148·3	—
Beef cows	1	—	—	—	5·4	—	8·7	12·2	—	—
Purchased feeder cattle	0·8	14·9	—	71·0	—	76·3	12·0	50·2	83·6	—
Other cattle	0·6	—	—	6·2	5·8	—	—	—	5·9	—
Horses	1	—	—	—	—	—	4·5	5·1	1·8	—
Pigs raised	0·25	—	—	197·3	50·4	24·5	45·5	16	17	—
Chickens	0·02	—	—	91·0	—	—	—	—	—	—
Average number of layers on hand during year	0·02	—	4,619	—	—	—	—	—	—	—
Livestock units per 100 acres		17	924	52	12	5	3	5	1·8	—
IV Tractors on farm	No.	2·05	0	2·6	1·07	6·7	2·3	1·91	0·49	n.a.
V Total labour	Hour	4,540	5,023	4,407	3,777	28,397	2,568	3,203	3,677	29,750
Operator and family	,,	3,707	4,120	3,833	2,613	3,200	2,197	2,723	2,490	2,600
Hired	,,	833	903	573	1,164	20,920	366	480	1,187	27,150
Cropper	,,	—	—	—	—	4,277	—	—	—	—

Source: As for Table 2.3.4. Calculated by authors from source data.

L.U. = Livestock Units per Animal.

tation is a mixed forest of river bottom hardwoods and cypress. Variable in
texture, the alluviums of the floodplain are fine sandy loams or loams along
the old stream channels, clays in the broad flat basins and silt loams or silty
clays between the basins and the ridges. Good drainage is the key to full use
of, and reliable crops on, these soils. But, until recently, adequate drainage
projects have not been economically feasible in this region where *per capita*
incomes are the lowest in the U.S.A.

Farming systems. Cotton is still the dominant crop. It is grown on two types
of farms. The first consists of small family *tillage* farms of about 60 acres,
which have one-quarter of their total acreage in cotton which provides about
three-quarters of the total cash receipts, and a similar acreage in soya-beans
with which cotton is rotated. These farms are the last stronghold of the mule
and are poorly mechanised. They are either owner-occupied or held on 'share-
cropping' tenancies, which are today found only in the Mississippi 'Delta'
and on the coastal plains of the Carolinas. Despite the system's bad name,
'croppers' are often better off than the owners of small farms because share-
cropping allows more flexibility of cropping and some diversification, e.g.
through an added livestock enterprise. The landlord provides the land, house
and fertiliser (which is used heavily); the tenant supplies the labour; the gross
output is usually divided 50 : 50. Increased demand for labour in the towns
has reduced tenant numbers. Landlords have met this exodus by the mechanisation of, and the amalgamation of farms left on their hands, and by diversification. The result has been to raise living standards all round.

The second type, viz. large owner-operated cotton plantations of 1,000 acres
or more, is prominent west of the Mississippi (see Figs. 2.2.4 and 2.3.1 and
Col. V in Tables 2.3.4 and 2.3.5). These *tillage* farms are highly mechanised
and have a wage-earning labour force; some are technically advanced with
high inputs of good genotypes, herbicides and fertilisers including nitrogen
injected into the soil as liquid ammonia. More than one-third of the harvested
crop land is in cotton, which accounts for three-quarters of gross sales. Most
of the rest comes from soya-beans which cover half the crop acreage. But
maize, grain sorghums and oats (often harvested for hay) are also grown, and
wheat on the heavier land. Although less than 10% of the gross sales of such
farms comes from livestock (mainly cattle and pigs) and only 10–15% of the
acreage is in pasture and forage crops (cow-peas (*Vigna sinensis*), lucerne and
lespedeza), these are of growing importance. Abandoned, overcropped cotton
fields are being put into grass/legume swards for beef and dairy cattle. The
year-long grazing season makes it good pasture country, though livestock still
face some awkward pest, disease and heat stress problems.

Two *sub-tropical crops*, viz. sugar-cane and rice, are grown on the alluvial
soils of the lower Delta in Louisiana. Here three-quarters of the gross sales
are from sugar and the rest from rice and vegetables. Cotton does not ripen
well (insects and fungus) in areas where autumn rainfall exceeds 10 in; thus
cane is a better crop for some of these areas. Drainage is a major problem in the

low-lying areas and nitrogen has to be applied heavily to sugar. Rice is also grown just south-east of Little Rock. Potential is very large, but there is a surplus of rice in the U.S.A. The crop is highly mechanised (see California, p. 161, and France, p. 283). Nitrogen is heavily applied, but excess nitrogen causes lodging. P_2O_5 is needed on some soils. Weed control is the primary purpose of the rotation which is commonly: 1–2 years rice, then 1–3 years lespedeza, soya-beans or pasture. An unusual variant rotation is 2 years of rice then 1–2 years of water fallow for fish production.

THE ARKANSAS VALLEY AND UPLANDS (*Table 2.3.3, Sector (d)*)
    Summarised in Table 2.3.3, Sector (*d*).

THE SOUTHERN GREAT PLAINS (*Table 2.3.3, Sector (e)*)
    Rising gradually from 1,000 ft in the east to 5,000 ft in the west, the southern Interior Plains are broken in places by rolling or steep terrain. The climate changes from warm sub-humid in mid-Oklahoma to warm dry in the Texas 'panhandle', and in eastern New Mexico. The mean annual precipitation, which is unreliable in the west (C.V. 25–35%) declines from 35 in in the east to 15 in in the west. Much of the rain comes in May and September and there is a mid-summer drought hazard. The mean thermal growing season of 8 to 9 months, with a mean P–E–T of 30–35 in and A–E–T of 15–30 in, is often cut short by · the sharp frosts. Mean January temperatures are below freezing in places. Summer temperatures are high (July means 75–80°F) and, if combined with drought, result in much crop damage. The soils are generally southern chernozems or reddish brown prairie ranging in texture from heavy clay to sand on which wind erosion is a problem. Tall grasses dominate the natural vegetation in the east, but give way to short prairie grasses in the west. The most common of the short grasses are buffalo-grass (*Buchloë dactyloides*) and grama grasses (*Bouteloua* spp).
    Farming systems. This sector crosses the southern part of the most important winter wheat producing area in the country (Fig. 2.2.4). The east of the sector is transitional; in north-central Oklahoma, winter wheat is combined with general farming, whilst further west it is combined with *extensive grassland* ranches (mainly cattle). Next comes a band of cotton-producing farms extending from western Oklahoma north-east into the Texas High Plains. The 'Cotton Belt' and the 'Wheat Belt' are merging in this area, and the relative prices of these crops influence the extent to which each crop is grown on *semi-intensive tillage systems*.
    West and north of this transitional area, grain sorghum, which stands drought better than maize, is second in importance to wheat, and is widely grown on the lighter soils for feed grain. Forage sorghums or sorghum—Sudan grass hybrids are also grown for silage or hay, mainly for beef cattle. These *extensive tillage wheat-sorghum farms* (Fig. 2.2.4 and Column VII of Tables 2.3.4 and 2.3.5) are mostly mechanised family units of about

800–900 acres. Most of the gross sales come from beef and wheat. As there is
no satisfactory legume which is adapted to the limited rainfall and periodic
droughts, rotations are alternate winter wheat and sorghum, or wheat and
summer fallow. Winter wheat is drilled with some fertiliser in October and is
still sometimes grazed in winter by sheep moving from the far western ranches
to the Corn Belt for fattening; it provides a ground cover in winter and early
summer before it is harvested in June or July. In contrast with the spring-
sown crops such as sorghums, barley and oats, this cover helps to check wind
erosion. Care needs to be taken to prevent a recurrence of the 'Dust Bowl'
that, in the 1930s, covered parts of five states in this region.

The potential production of some semi-arid areas when water is less limiting can be seen in the large irrigation area of the High Plains round Lubbock
(Texas). Here *intensive tillage* farms of 400–500 acres may have 150 acres in
cotton, which provides more than two-thirds of the gross sales, and another
150 in grain sorghum. Yields of cotton are more than doubled and of grain
sorghum trebled by irrigation. Livestock are not important.

On the western edge of the sector, year-long range land (*extensive grassland*) predominates. The best range supports one beef cow or yearling to every
4–6 acres. Although some sheep are kept in eastern New Mexico, the biggest
'sheep country' lies in southern Texas on the Edwards Plateau, which also has
most of the goats (Angoras for mohair) in the U.S.A.

THE ROCKY AND WESTERN MOUNTAINS (*Table 2.3.3, Sector (f)*)

This sector includes, for present purposes, the Western Mountains and
Basin, the Western Interior, the Rockies and the Intermountain West, the
Western Highlands and Basins. High plateau country (5,000–9,000 ft) is intersected by valleys, canyons and some mountain ranges. The warm dry climate
is more erratic than that of the previous sector. The mean annual precipitation is 5–15 in, except on the high country where it may be 20 in or more.
This is reflected in the very high C.V. of net profit on south-western cattle
ranches (see Column VIII of Tables 2.3.4 and 2.3. Frequent summer
storms often cause dangerous flash floods or severe erosion, especially
where the range has been over-grazed. Mean January temperatures are,
in most parts, below freezing. Night frosts occur in all but the summer
months. Towards the desert, in the west, at lower altitudes, there is excessive summer heat (July averages round Phoenix 80–90°F), and dairy
cows', conception rates may then drop 20% (Roddick 1965). The mean
thermal growing season ranges from 7 to 12 months with P–E–T 21–23 in and
A–E–T 13–18 in. Winters are very dry and sunny. In the east there are chiefly
brown soils, varying in texture from sandy to clay silt-loams; whilst in the west
there are lighter-coloured chestnut soils. The indigenous plant cover, as modified by ranch livestock, remains on more than 90% of the land. On the plateaux open ponderosa pine forests predominate with a grass under-storey of
fescues. (*Festuca arizonica, F. idahoensis*), *Muhlenbergia montana*, pine

drop seed (*Blepharoneuron tricholepis*), blue grama (*Bouteloua gracilis*) and blue grasses (*Poa* spp). At the lower fringes of the mountain masses, there is a wider variety of trees and also bluebunch wheat grass (*Agropyron spicatum*), needle and thread grass (*Stipa comata*), blue grasses and Indian rice-grass (*Oryzopsis hymenoides*). In the south, mesquite (*Prosopis* spp), a small tree which shades out productive vegetation and thus lowers the carrying capacity of the range, has infiltrated where the limited grass has been overstocked.

Farming systems. Ranching. The role of the higher forests is water conservation and water erosion control, timber and hunting (shooting) and other recreation. Four-fifths of the land area is not used for farming. Agriculturally, however, New Mexico and Arizona are *extensive grassland* states with small and scattered, but important, irrigated areas of intensive tillage. The best range, which is privately owned, will carry one head of cattle unit *for a month* on 4–6 acres. This falls to 6–10 acres in the higher plateau regions and 10–16 acres on the highest land. About 15% of the land, mostly of low stocking capacity, is owned by the Federal Government but grazed by private ranchers. The average size livestock range unit is 11,000 acres and would have 10 acres in crops (for hay) and 150 Hereford cows and heifers. In the spring and summer the bulls (mainly Herefords) run with the cows, which calve in the early spring and are weaned in late summer if it is a summer grazing area. The calves are branded, innoculated against blackleg, castrated and dehorned at 3 to 4 months old; the cows are kept until they are 10–12 years old. The annual calf crop is about 80%. A few ranches have sheep but in these rugged conditions cattle do better. Mostly, the range is grazed year round, the best grazing being in autumn, winter and early spring. There is little feed in high summer. A little hay may be made to partially offset seasonal or drought shortages. The main sales are weaned (feeder) calves, but in the better areas these may be carried on for sale as 'long' yearling stores (feeders) in their second year, or if near an irrigated area, may be fattened on local hay, grain and arable by-products and sold at $2\frac{1}{2}$ years for slaughter (see Figs. 2.2.4 and 2.3.1 and Col. VIII of Tables 2.3.4 and 2.3.5).

Irrigated farming (intensive tillage and alternating systems). Less than 5% of the land is irrigated. But output from such land is at least 25 times that of range land, so intercommunity and interstate arguments about water supplies are not surprising. In the 'bad old days', most gun fights, it is said, were caused by women, whisky or water. The small scattered areas of irrigated land in this sector are centred mainly on the Rio Grande, San Juan and Pecos valleys in New Mexico and round Phoenix (Arizona). The latter, where 725,000 acres are irrigated from the local rivers and from the groundwater-table, is one of the most productive counties in the U.S.A. The water-table is, however, rapidly falling and the outlook is grave unless the project to bring in Colorado River water materialises. But irrigation in this climate brings salinity problems. Most irrigated soils in the west are underlain by strata of alkaline materials. If the drainage is poor, alkaline salts dissolve in the applied water and

rise by capillarity up the soil profile, adding to the salts left by the high evapo-
transpiration. Sometimes the pH goes beyond the tolerance of the majority of
plants and, unless the soil is thoroughly leached (which of course means more
applied water), crop growth ceases (see p. 56). Around Phoenix, important
irrigated crops are highly mechanised cotton, wheat, sorghum, as well as
citrus, apples, winter and spring lettuce, carrots, cabbage, cauliflower and
melons. Here, and the south of California, are the only places where dates are
grown in the U.S.A. Lucerne (alfalfa) is grown for hay and for seed; a common rotation in the area is 2 years flood-irrigated alfalfa, 2 years furrow-
irrigated row (vegetable) crops, but increasing use is made of sprinklers,
especially where the terrain is rough. Feeder cattle and some 'pasture fat'
animals are produced, whilst use is made of by-products from irrigated
crops. Increasing numbers of animals are being finished by dry-lot feeding
(see p. 163).

A Dairy Farm near Phoenix (Arizona)

Near Phoenix are 250 large-scale dairy farms averaging 150 cows each and
producing milk for liquid consumption but also some cream for butter. The
cows are held on small areas; for instance, one 'farm' has 800 milking Friesian
cows (averaging 14,500 lb at 3·45% fat), plus dry cows and calves—a total of
1,100 head, kept on 30 acres. Milking is done thrice daily in three 8-hour
shifts. The milk is piped away to a tank. In the stanchion-type barns, four
'herds' of 40–50 cows can be milked and fed at once. In eight hours one man
handles 6 herds, say, 250 cows. From two trucks moving up the central gang-
way, the cows are fed individually according to yield, etc., on concentrate
(barley and cotton-seed meal at 1 lb for 4 lb milk) whilst hay, maize silage,
chopped green sorghum are mechanically unloaded in the concrete outdoor-
laying yards, which have some shading against the sun. All the feed is bought
in. Each cow's expected and actual performance is checked on a punch-card
system. The bull calves are sold for veal at 250 lb live-weight and the heifers,
which, like the bull calves, get whole milk, are sent away at six months to a
ranch 1,000 miles away for further rearing. Artificial insemination only is
used but low conception rates and other heat stress problems are common in
very hot weather and this farm, like others in the south-west and California,
is experimenting with insulated buildings and cool air-conditioning for cows.
(*Source:* Roddick 1965.) This farm is not exceptional; 'town dairies' of this
type and size can be found near Los Angeles and elsewhere.

THE SIERRA NEVADA (*Table 2.3.3, Sector (g)*)
Summarised in Table 2.3.3, Sector (*g*).

SOUTHERN PACIFIC COASTAL ZONE (*Table 2.3.3, Sector (h)*)
The Central Valley of California lies between the Sierra Nevada and the
Coastal Range which rises to 7,000 ft and is separated from the sea by a nar-
row strip of coastal lowlands. The climate is warm dry (Mediterranean) with

up to 90 % of the precipitation between November and April, and with hot, dry cloudless summers. Both the sea and the mountains influence local climates markedly and partly account for the wide range of farming. The mean annual precipitation on the coast is between 15 and 20 in, rising, on the western slopes of the mountains, to over 30 in, but is below 10 in per annum in parts of the Central Valley in the rainshadow of the Coastal Range. The coefficient of variation is high, 35–45 % or even higher near Los Angeles. The thermal growing season is year long, except on the mountains, with a mean P–E–T ranging from 30–40 in, but A–E–T only 10–18 in. Inland, excessive heat may restrict plant growth and distress livestock in the summer. July temperatures average 80–85°F in the Central Valley, but are moderated by the sea on the coast and by altitude in the mountains. Winters are mild, viz. about 50°F on the coast (which may have over 300 frost-free days), 40–45°F in the Central Valley and 35–40°F in the Coastal Range. Coastal Range soils are dark brown prairie-type soils; in the bottom lands of the Central Valley are alluvial or extensive peat and muck (fen) soils, sometimes poorly drained, whilst in the rainshadow of the Coastal Range are sierozems. Where mean annual precipitation exceeds 25 in, the natural vegetation is woodland with grass and brush and includes bluebunch wheat grass (*Agropyron spicatum*), Idaho fescue, pine blue grass (*Poa scabrella*). Steeper, more rugged country may be covered by chaparral or Mediterranean scrub, mainly *Ceanothus* spp, manzanita (*Arctostaphylos* spp) and other shrubs plus scattered grasses and herbs.

Farming systems. Like Israel (p. 306), south-east France (p. 279), southern Australia (p. 176) and other areas with Mediterranean climates, California has a wide range of farming systems and farm products. *Intensive* irrigated *tillage* and *intensive* irrigated *tree fruit* systems are found in the southern (San Joaquin Valley) and northern (Sacramento Valley), arms of the Great Central Valley and on the coastal lowlands. *Extensive grassland* (*range land*) and *extensive tillage* are located on the unirrigated slopes and uplands. Intensive, almost landless, livestock 'dry-lot' farms rely on the tillage farms for concentrates and hay and on the ranges for feeder (store) cattle. Many, perhaps most, farms in California are highly specialised, and concentrate on one main product, or occasionally on two, and have no interest in others. This is true even in the market garden (truck farming) areas. In order to have some form of crop rotation, in some places it is still the custom for farmer *A* to grow break crop *X* for one year on the land of grower *B* who specialises in crop *Y*. The farmers as well as the crops are rotated.

Irrigated farming in the Central Valley. Economically, the most important single crop in California is surprisingly cotton, the production of which is only exceeded by Texas. Irrigated cotton farms are generally large and highly mechanised with mechanical cotton-pickers, etc. (see Column IX of Tables 2.3.4 and 2.3.5). Most are farmed by their owners, employing some full-time and considerable part-time migrant labour. A farm of 1,000 acres would be fully cropped with, on average, 400 acres in cotton (yield 1,000 lb—2 bales—of lint per acre), a few acres of lucerne for hay (11,000 lb/acre), 300 acres barley (2,700 lb/acre). This *intensive tillage* rotation is sometimes varied by a speciality crop (e.g. potatoes). Fertiliser and herbicide, etc. inputs are high. All crops are sold and no livestock is kept.

Other leading irrigated crops in the Central Valley are Irish potatoes, rice, beans, sugar-beet and wheat. Most potato crops are irrigated and are harvested in late spring, although there is some production throughout the rest of the year, including the winter (as in Florida). Flood-irrigated rice is grown in the Sacramento Valley on hard-pan soils, which are shallow and are of limited value for other crops. Average yields are about half those obtained in Europe (e.g. the Po Valley), but higher than in the tropics. Production is highly mechanised and the crop is often seeded from the air. The plants are not submerged until they are in need of water because soil moisture is generally sufficient for germination and initial growth. The water is drawn off just before harvesting, which is by specialised combines. Purple vetch is often sown in the standing rice just before the water is drawn off, to act as a green manuring crop. A common rotation is rice, beans, wheat, but often only the rice is irrigated. The Sacramento Valley is also the main sugar-beet growing area in California.

Winter wheat is grown on *extensive* methods and is the crop of the non-irrigated sides of the Central and Salinas Valleys where rainfall averages about 15 in yearly. Hay, mainly from lucerne, but also from oats, vetches, etc. and mainly irrigated, occupies a quarter of the crop land in California. It is vital as a 'break' crop on many kinds of specialised farms and is essential to the numerous large-scale and intensive dairy and beef units relying on bought-in feed.

More than a third of the state's gross sales comes from vegetables and fruit, of which grapes are the most important crop. Raisin grape (for drying) production is centred in those parts of the San Joaquin Valley where the summers are very hot and dry. Wine grapes, which yield many pleasant table wines usually produced by innoculated fermentation in highly aseptic factories are more widely distributed. Table grapes are found in the Central Valley and in the valleys of the desert regions, further south (which produces the earliest crop). Tree fruits grown in the Central Valley are navel oranges and deciduous fruits (particularly prunes, peaches and pears) and nuts, especially walnuts and almonds. Melons are locally important. Half of the United States' output of tomatoes, lettuce, asparagus and celery is in California, mainly in the Sacramento Valley (except lettuces which are concentrated on the coastal lowland round San Francisco and Santa Barbara).

Irrigated Farming on the Coastal Lowlands

As in the Central Valley, irrigation water is the key to the farming, which is mainly in one enterprise units (e.g. lemons or strawberries). Citrus and

truck crops are the chief sales, particularly lemons, valencia oranges, strawberries, tomatoes, broccoli, green lima-beans, lettuce, nuts (walnuts and
almonds), nursery products and cut flowers, avocados and chili peppers. The
only important field crops are potatoes, sugar-beets and lucerne hay.

A One-man Specialised Intensive Irrigated Lemon Farm in the Coastal Low-lands

This 40-acre farm lies north-west of Los Angeles in Santa Barbara county,
and has a typical warm dry Mediterranean climate. The average July temperature is about 67°F, which is 7°F higher than south-east England. The January temperature averages 53°F; in this part of California sea-bathing is not
uncomfortable on Christmas Day. The thermal growing season is more than
eleven months. There is no snowfall but frost protection is needed between
November and February with orchard heaters and wind machines. The annual
P–E–T is about 30 in and the average annual precipitation is $17\frac{1}{2}$ in, which
nearly all falls in the winter between November and April on the recent alluvial sandy loam soils. The summer is very sunny. The whole farm is in lemon
and avocado orchards. Nitrogen is broadcast in February at 200 lb/acre (as
urea or ammonium nitrate); trace elements (zinc and manganese) are sprayed
direct on to the lemon tree foliage in June or July, partly because high calcium
in the soil blocks uptake of other basic ions. The large variety of pests and
diseases are controlled by chemical sprays and plant hygiene including, for
virus diseases, careful bud selection.

The farming year. Frost protection is usually required during November–
February. Pruning by hand is a fill-in job often done in early spring; mechanised pruning and brush disposal is often done in later summer. Weed-spray-
ing is necessary throughout the year with one full coverage applied in
November. Gopher control throughout the year. Trees for replacement are
planted in May; fumigated the previous March. Irrigation usually from May–
December (once every 3–6 weeks; additional applications may be necessary
if it does not rain in winter). Pest-control sprays in May and November;
disease sprays in November. Picking is done year round; fruit picked selec-
tively, by colour (maturity) and size, 4–7 times; peak picks January–July. Fruit
stored in local packing houses up to 6 months then shipped to market.

The owner of this one-man farm is busy throughout the year with irrigation,
chemical weed control, applying nitrogen, controlling gophers (a small rodent
Citellus spp), etc. Pest and disease spraying and mechanical pruning are done
by a travelling contractor whilst the fruit is harvested throughout the year
by the crews of the lemon packing firms which buy the crop. Of the annual
input of 380 man hours/acre only 56 come from the owner. This farm, therefore, is an example not only of a highly specialised crop enterprise with a high
manpower efficiency but also of the highly specialised division of labour
characteristic of California and indeed of much of the American and Canadian

economies (*Courtesy:* Mr G. E. Goodall, Farm Adviser, Santa Barbara, California).

Livestock in California. In spite of the financial importance of crop production, more than one-third of farm sales in California come from livestock. Cattle are kept on range, on non-irrigated forage crops and natural pasture, on irrigated pastures or in dry-lots. Parts of the Central Valley and large sections of the Coastal Ranges up to 1,500 ft are open grasslands, which are *extensively grazed* at all seasons but are at their best in winter and spring. The higher land towards the Sierra Nevada, in the timber zone, is used for summer grazing. The more arid land is used only for winter grazing. The carrying capacity of range land varies from less than 0·1 A U M/acre* on the steeper, rougher, chaparral-covered land, where the sale product is weaned calves (for which there is a good demand by dry feed-lots) up to 2·0 A U M/acre on the best open grasslands in the Coastal Range where yearlings (feeders) may be the output. Although some beef cattle are slaughtered off grass, open-air *feed-lot fattening* is increasing on irrigated *semi-alternating* field crop farms. Beef cattle use pasture grown as a break-crop, crops grown in rotation with the main crop (e.g. hay and grain in the rice-growing areas in the Sacramento Valley) and also crop by-products such as cotton-seed hulls and sugar-beet pulp. These bulky feeds are supplemented by barley and 2 lb/head/day of a mixture of cotton-seed, meal, molasses, minerals and vitamin A. Most *feed-lots*, however, are not on farms but are open-air factory units, often operated by companies (*The Times* (London), June 1965). 70% of the feed-lots in California accommodate more than 10,000 head (for example, on a feed-lot near Bakersfield 40,000 cattle are fattened on 600 acres with a labour force of 35 men). The feeding operation is often automated. In this dry climate dung dries quickly and can be scraped off the dry ground or concrete easily.

Dairying is important throughout the state (Shultis *et al.* 1963). Round San Francisco and Los Angeles, where land is very expensive, there are large dry-lot dairies (200–800 cows) producing milk for liquid consumption (see p. 159). In the northern end of the San Joaquin Valley, land is cheaper and herds are smaller, averaging 80 cows per farm and per man. Cows are kept on irrigated pasture in the summer and on alfalfa hay and silage in the winter plus concentrates as needed. Dairymen usually rear their own replacement heifers, which, as on dry-lot dairies, are mainly Friesians. The milk goes mostly for liquid consumption, but some is manufactured.

Poultry farms are numerous round the two major cities. Eggs, and to a lesser extent turkeys and broilers, have (like beef) a ready market in this rapidly urbanising state. Some units such as Egg City, Moorpark, are very large.

* A U M (Animal Unit Month) = 1 mature cow or equivalent (e.g. 5 sheep) for one month. In other words, if a pasture could carry 1 cow per acre the entire year then the carrying capacity would be 12 A U M/acre.

This one has 750,000 battery-caged layers in houses with adjustable louvred sides, etc., which control ventilation, light and, except in very hot weather, temperature. Each can hold 70,000 hens. Feeding, collection, cooling, grading, packing, etc. are fully automated where possible. There is one poultry man per 17,000 birds. In numbers, if not relatively, pigs and sheep are important in California. The main sheep areas are in the northern part of the state, where range grazing is on the foothills bordering the Sacramento Valley, whence some sheepmen move sheep to the Sierra Nevada for summer grazing. Fattening is concentrated in the Sacramento–San Joaquin Delta area where grain stubble, volunteer grain and beet tops are common sheep feeds. Though lamb is still popular in California, sheep are declining.

United States — Stability and Efficiency

Stability. Thirty-five years ago the combination of the depression of the early thirties, droughts, erosion and belief in *laissez-faire*, made farming in the U.S.A. very unstable, particularly west of 100°W. But since then there has been a remarkable improvement, as shown by Fig. 2.3.2. This has been due to the application by government of Keynesian economics, to ever-increasing domestic markets, to the use of support prices, to surplus storage and export programmes, to some quantitative capacity control (e.g. of rice, cotton, liquid milk) and other measures of the Central government. These have been augmented by the growth of the food processing and storage industries which increasingly contract with farmers, by active government measures of soil and water conservation and action to get marginal yield-variable land out of production, by pest disease and weed control, by fertiliser use, by mechanisation which reduces the weather sensitivity of field operations, by the extension of irrigation (especially of sprinklers), and by the growth of grain-consuming livestock enterprises which have more 'built-in' stabilisers than grazing systems. In fact, the application of science and technology has probably equalled government economic action in reducing income instability. But, over large sections of the country, especially the arid parts with low mean precipitation, the sometimes savage climate and weather variability remain sources of instability. This is shown by the greater C.V. of net income in extensive grazing and extensive tillage systems than of more intensive systems located in more reliable climates (Table 2.3.4, bottom row). This trend towards less instability will no doubt continue as greater control over both economic and ecological uncertainties is achieved.

Efficiency. There are areas of poverty, low yields and poor farming in pockets along the Appalachians (especially in the south), the Ozarks and the 'old' plantation states, especially Alabama, Mississippi, Louisiana. In these 'poverty pockets' there are still too many people on farms, and output per farm man is, therefore, even by European standards, low. The T.V.A. (Tennessee Valley Authority) has, however, helped in the Appalachians. Further,

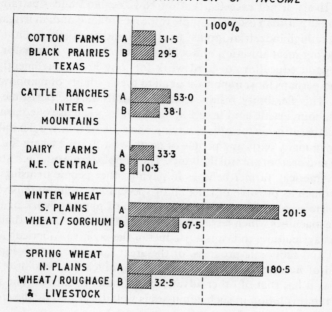

U.S.A. INCOME INSTABILITY BY TYPE OF
FARM AND DECADE
CO-EFFICIENT OF VARIATION OF INCOME

COTTON FARMS BLACK PRAIRIES TEXAS	A	31·5
	B	29·5
CATTLE RANCHES INTER- MOUNTAINS	A	53·0
	B	38·1
DAIRY FARMS N.E. CENTRAL	A	33·3
	B	10·3
WINTER WHEAT S. PLAINS WHEAT / SORGHUM	A	201·5
	B	67·5
SPRING WHEAT N. PLAINS WHEAT / ROUGHAGE & LIVESTOCK	A	180·5
	B	32·5

A = 1930–1939
B = 1950–1960

Fig. 2.3.2 U.S.A. Income Instability by Type of Farm and Decade (Coefficient of Variation of Income). *Source:* Hathaway, 1963

in the extensive tillage and extensive grazing areas, yields per acre are limited by low mean A–E–T and other climatic constraints which do not justify 'high farming'. Thus overall mean U.S. yields per acre (Table 2.2.1) are not high. But comparing like with like (as Menzies Kitchin (1951) did in his classic study in which he compared Illinois with East Anglia in the United Kingdom), it is clear that, on *intensive tillage* or *intensive alternating* farms, yields per acre and per animal are as high in the U.S.A. as they are in most parts of the north-west of Europe. However, yields of livestock products from *intensive grassland* are not as good in the U.S.A. as on comparable farms in the Netherlands, the United Kingdom or New Zealand. This is because concentrates are cheap and because there are, in North America, no good grassland climates north of about 43°N. South of this line, despite the great potential of the south-east and, if irrigated, of the south-western states, intensive grassland is too new to have commonly reached high mean livestock output per acre.

However, 'added value' per man in the U.S.A is about 50% above that of north-western Europe (France, p. 290), and is, on ecologically comparable farms, twice as great as in those parts of Europe with highest labour productivity and rivals, if not exceeds, that on New Zealand and Australian farms. This high output/man is due to ample resources of land and, in irrigated areas, of water, to high industrial inputs, to skill in and competitive enthusiasm for labour-saving mechanisation and willingness to borrow money for this purpose, to good education, to hard work for long hours intelligently at peak times, and perhaps most important of all to the flexibility of manpower supply and use. This flexibility results from the family farms, part-time farming, migrant labour, elastic land-tenure systems, widespread use of contractors and a constructive attitude towards changing one's job or home, a flexibility which is lacking in many parts and people of north-west Europe where the inertia of tradition and custom are still limitants. The sample farms outlined above show that the American farmer believes in paying other people or using machines to do as much as possible of the dull or specialised capital-intensive work of farming. He sees no virtue in drudgery, in hard physical work or in owning expensive machines which he rarely uses. But all this goes deeper than farming. American agriculture, today, is a product of the positive American attitude to technical and economic progress, of the great mass-production input industries, and of modern science and technology which has revolutionised U.S. farming as it has that of other advanced industrial countries. The speed and effect of the revolution in such countries is well illustrated by the story of the last 50 or 60 years in the U.S.A. shown in Fig. 2.3.3.

Conclusions and Prospects

The most impressive aspects of American agriculture are its dynamism and its use of capital and technology. Indeed, the technology of transport, production and processing as they affect the location of farm systems, as well as changing markets and government policy are just as important as physical resources (if not more so) in determining present-day farm activity (Anderson, J. 1967). Can this dynamism, and the rate of progress shown in Fig. 2.3.3 continue, or will the curve flatten out in the U.S.A. and elsewhere? Prediction is dangerous. But there seems to be no major biological or industrial resource reasons why the curve should flatten, unless the crop acreage re-expands to ecologically less productive areas, or unless market and export problems become critical. Energy, steel and chemical (including fertiliser) resources appear adequate for 30–40 years ahead. The most likely limitant is water. The water-table is falling and irrigation supplies are not unlimited; much water is unwisely used. Fortunately desalination, possibly by nuclear power, seems a likely economic starter. Given adequate, scientifically used water to exploit the high P–E–T of the southern half of the U.S.A., food production could greatly increase. The current tillage crop output of the U.S.A. could, if sup-

CANADA AND THE U.S.A.

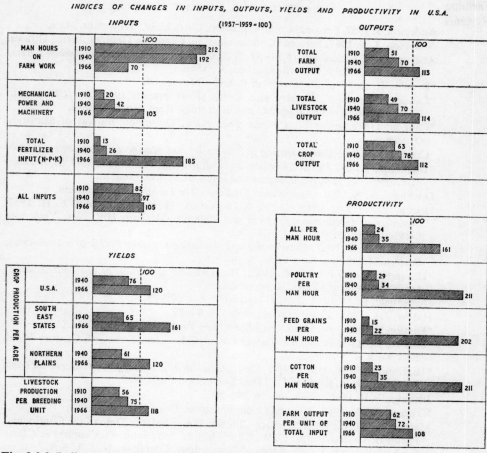

Fig. 2.3.3 Indices of Changes in Inputs, Outputs, Yields and Productivity in the U.S.A. *Source:* U.S. Dept. Agric.

plemented by synthetic amino acids, synthetic vitamins and minerals, make a good, if unpalatable human diet, and if fed direct to human beings would probably feed twice or three times as many people as today. There seems no reason why the number fed on the present diet should not be doubled by the first decade of the 21st century.

Acknowledgements and thanks for research assistance to Kathleen Down, Janet Z. Foot and for comments and criticisms to J. A. Anderson, J. Anderson, W. R. Bailey, Joan Bostock, H. J. Critchfield, M. D'Aoust, B. A. Eagles, J. Eaton, R. E. Hodgson, C. Hudson, W. K. Kennedy, J. D. MacQuigg, G. S. Mehren, L. W. Osborne and W. Wilson.

167

References and Further Reading

Anderson, J. 1967. Personal Communication, Clark Univ., Worcester, Mass.

Anderson, J. A. 1967. Personal Communication. Canadian Dept. of Agric., Ottawa.

Association of Agriculture. *Farm Study Scheme. Sample Studies of Individual Farms: Canada.* (Manitoba, Ontario, Quebec.) London.

*Bennett, R. H. 1946. *The Compleat Rancher.* New York.

*Canadian Dept. of Forestry and Rural Development. 1966. *The Climates of Canada for Agriculture.* Canada Land Inventory. Rept. No. 3. Ottawa.

*Canada, Government of. *Canada Year Book.* Dominion Bureau of Statistics, Ottawa.

*Crisp, D. J. (Ed.) 1964. *Grazing in Terrestrial and Marine Environments.* Oxford.

*Duckham, A. N. 1952. *American Agriculture: Its Background and its Lessons.* Ministry of Agriculture, Fisheries and Food. H.M.S.O. London.

Duckham, A. N. 1963. *Agricultural Synthesis: The Farming Year.* London.

*Finch, V. C., Trewartha, G. T., Robinson, A. H. and Hammond, E. H. 1957. *Elements of Geography: Physical and Cultural.* McGraw-Hill. New York and London. 4th edn.

Hathaway, D. E. 1963. *Government and Agriculture: Public Policy in a Democratic Society.* Macmillan. New York.

*Haystead, L. and Fite, G. C. 1955. *The Agricultural Regions of The United States.* Univ. of Oklahoma Press. Norman, Okla.

Herschfield, D. M. 1962. *J. App. Meteorology.* 4. 575–78.

Hidore, J. L. 1963. *Economic Geography.* 39. 84–89.

*Higbee, E. 1958. *American Agriculture.* New York and London.

*Higbee, E. 1963. *Farms and Farmers in an Urban Age.* Twentieth Century Fund. New York.

*Highsmith, R. M. (Ed.) 1964. *Case Studies in World Geography, Occupance and Economic Types.* Prentice-Hall Inc., Englewood Cliffs, N. J.

Jennings, P. R. 1966. *Econ. Bot.* 20. 396–402.

Kennedy, W. K. 1967. Personal Communication. Cornell Univ.

Mather, J. R. (Ed.) 1964. 'Average Climatic Water Balance Data of the Continents', Part VII, United States. C. W. Thornthwaite Associates, *Publications in Climatology*, XVII, No. 3, Laboratory of Climatology, Centerton, N. J., U.S.A.

Menzies–Kitchin, A. W. 1951. *Labour Use in Agriculture.* Farm Econ. Branch, Dept. of Agric., Univ. of Cambridge. Rept. No. 36.

*Nelson, L. 1954. *American Farm Life.* Cambridge, Mass.

Roddick, N. 1965. *Farmer's Weekly*, London. 14 May.

Sharp, G. and Capstick, C. W. 1966. *J. Agric. Econ.* 17. 1. 2–16.

Shultis, A., Forker, O. D. and Appleman, R. D. 1963. *California Dairy Farm Management.* California Agricultural Experimental Station. Extension Service Circular No. 417.

*U.S.D.A. (United States Department of Agriculture). *Yearbooks*—1941, *Climate and Man;* 1948, *Grass;* 1955, *Water;* 1957, *Soil;* 1958, *Land;* 1964, *Farmer's World.*

*U.S.D.A. *Agricultural Statistics* (Annual).

U.S.D.A 1950. *Generalised Types of Farming in the United States.* Agriculture Information Bulletin No. 3.

U.S.D.A. 1955. *Guide to Agriculture, U.S.A.* Agriculture Information Bulletin No. 30.

U.S.D.A. 1964. *Farm Costs and Returns. Commercial farms by type, size and location.* Agriculture Information Bulletin No. 230 (and later issues).

Van den Noort, P. C. 1968. *J. Agric. Econ.* **19**. 1.97–103.

*Van Royen, W. 1954. *The Agricultural Resources of the World.* Prentice–Hall Inc., N.Y.

Ware, D. W. 1963. *The Variability of and the Sources of Farm Income 1926–1960.* Canadian Dept. of Agric., Economics Div. 63/2 March, Ottawa.

*Wilsie, C. P. 1962. *Crop Adaptation and Distribution* San Francisco and London.

* Suggested further reading. These sources have been freely used in the above text. Additional sources for areas or subjects are the bulletins, etc. of the U.S.D.A. and of the Land Grant Universities and such professional periodicals or annuals as *Advances in Agronomy*, *J. Animal Science*, *J. Farm Economics*, and many others.

CHAPTER 2.4

Australia and New Zealand

(a) Australia — Southern Half

Introduction

Australia has roughly the same land area as the continental U.S.A., but only 6% of its human population, of which more than half are in the six major cities on the seaboard. It is, therefore, owing to population distribution, basically an urban country with large distances and scattered farming communities between major cities. Though the resultant problem of communication has been largely solved by broadcasting and an excellent air network, the transport of supplies to and produce from farming units remains a costly and difficult business. In one way it is fortunate that the really arid areas are in large blocks, viz. the central 'dead heart', the Nullarbor Plain in the south and the middle sector of the western coast. For this, at least, helps to concentrate the social and economic capital invested in farming in the more humid regions, many of which are served by sea transport. To this extent, Australian agriculture is better off than Canadian, where the Prairies are cut off by the Canadian Shield or the Rocky Mountains and have no waterway connections.

Mineral deposits such as gold, silver, zinc, lead and copper are of economic importance; there are also huge reserves of bauxite, particularly in the Northern Territory, black coal in New South Wales, brown coal in Victoria and in Western Australia there are some of the largest deposits of iron-ore in the world. Oil has been found in Queensland and Western Australia and natural gas in Victoria and Queensland. Thus, Australia is rich in natural resources. Her population is able to enjoy one of the highest standards of living in the world (see Tables 1.5.1 and 1.5.2). This is maintained by well-protected local industries, carefully regulated immigration, an ebullient and vigorous working population, rapidly developing manufacturing industries and the ability of her agricultural exports (which form 70% of her total exports)—'soft' wheat, meat, dairy products, dried fruit and especially wool—to earn foreign exchange to pay for her imports of manufactured goods for consumption and industrialisation (U.S.D.A 1967). Traditionally, these exports went to Europe and particularly to the United Kingdom, but since the Second World War Asian and North American markets have become major outlets. Over one-third of Australian wool exports go to Japan. The U.S.A. is a growing market for meat, whilst China has become a big purchaser of Australian wheat.

Table 2.4.1 indicates the relative importance of the main farm enterprises which are sheep for wool and fat lamb, wheat, beef raising, fruits, vegetables and milk.

170

AUSTRALIA

Table 2.4.1

Crop	AUSTRALIA Area (000's acres)	AUSTRALIA Yield (average for 1963–65) (lb per acre)	NEW ZEALAND Area (000's acres)	NEW ZEALAND Yield (average for 1963–65) (lb per acre)
Wheat	17,515	1,106	184	2,798
Oats	3,768	752	31	2,289
Barley	2,298	1,061	85	2,721
Rye	64	386	—	—
Maize	198	1,611	10	4,662
Sorghum	432	1,174	—	—
Rice	62	5,400	—	—
All cereals	24,455	1,060	310	2,785
Sugar-cane	511	66,305	—	—
Potatoes	89	12,785	26	19,000
Vegetables (a)	119		24	
Dry peas	47	940	30	1,362
Oil-seeds (b)	654	679	16	1,258
Tobacco	27	1,145	5	1,653
Cotton (lint)	62	452	—	—
N.Z. flax	—	—	40	240
		Production 000's lb		Production 000's lb
Tree crops (fruit) (c)		3,797,010		317,520

Livestock	Numbers 000's	AUSTRALIA Yield (average for 1963–65) (lb per animal)	Numbers 000's	NEW ZEALAND Yield (average for 1963–65) (lb per animal)
Total cattle	18,816	(e) 437	6,801	(d) 882
Dairy cows	4,057	(f) 4,831	3,174	(f) 6,520
Pigs	1,660	(e) 108	716	(d) 150
Sheep	170,621	(e) 42 (lambs 35)	53,748	(d) 101 (lambs 64)
Wool		(g) 10·4		12·2
Goats	80		33	
Poultry (chickens)	21,500		4,527	

(a) Onions, tomatoes, cabbages, cauliflowers, green beans, green peas.
(b) Groundnuts, cotton-seed, linseed and sunflower seed (N.Z. linseed only).
(c) Apples, pears, plums, prunes, cherries, peaches, nectarines, apricots, oranges, tangerines, grapefruit, lemons, limes, bananas, pineapples and grapes (grape yield—Australia, 9,973 lb per acre).
(d) Live-weight of animals slaughtered.
(e) Dressed weight of animals slaughtered. Approximate conversion:
Cattle and sheep: live-weight = dressed weight × 2
Pigs: live-weight = dressed weight × ⁴⁄₃
(f) Milk per milking cow.
(g) Excluding lambs.

Agrarian Structure

The first attempts at farm settlement were made well over 150 years ago. Since then, land tenure has had a rather stormy social and legislative history. The original wool-growing 'squatters', who initially made, out of virgin land, large ranch-type holdings of thousands of acres of poor carrying capacity, resented the intrusive 'smaller' would-be producers of wheat, milk, dried fruit, etc. The land tenure situation is now complicated. Over half the land in New South Wales, South Australia and the Northern Territory, and four-fifths in

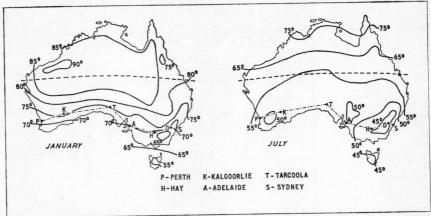

Fig. 2.4.1 Australia. Mean temperatures for January and July. *Source:* C.S.I.R.O. 1960

Queensland, is held under some form of lease or license from the Government whereas in Victoria and Tasmania most farmland is owned freehold. In Western Australia about one-third of the land is on lease and the remainder unoccupied. Leasehold property is almost as marketable as freehold land. It can be resumed by the Government(s) if it decides to split up any holding into smaller 'living areas' under the policy of closer settlement. Long-term leases are mainly for large (beef or wool) semi-arid grazing properties, whilst small farms are mostly privately owned. Today, the large wool-grower is losing his hold on the top of the (social) pyramid, but the 'cow cocky', i.e. a milk producer with less than 50 cows, remains near the bottom with wheat men and the larger fruit-growers in between.

The family-sized unit predominates. But the larger wool-growers and cattle ranchers may have 6 to 12 regular hired hands and smaller farmers may employ itinerant casual or contract workers, for example, for shearing or sugarcane harvesting. (The play *The Summer of the Seventeenth Doll* dramatised one aspect of the life of the latter workers.) Manpower is very short and regular hired labour not easy to obtain. Although, by European standards, the proportions of farmers with public electricity and water supply may be low, farmhouses almost always have their own generating sets and modern amenities. Sound radio plays a large part in the life of the more remote farms in the 'outback', providing education for the children, access to medical advice and services (e.g. the flying doctor) whilst leaving time for plenty of farm-to-farm radio gossip.

As there are few areas where the mean temperature in any month is less than $42\frac{1}{2}°$F, house heating is not a difficult problem nor is the housing of livestock which need protection from the sun and from heat stress more than from the cold, wet or wind. Capital invested in fixed equipment at the farmstead itself is thus rather low, but, for example on wool-growing holdings, the

money sunk in well-bores and fencing is substantial. Mechanisation is highly
developed. Most farms have one or more 'utes' (i.e. a pick-up 'utility' car or
truck), whilst the shearing or milking-shed or the drilling and harvesting
equipment for grain, as the case may be, are more advanced than in Europe
and rival North American standards of labour saving. Several implements,
for example, the 'stump-jump' plough were invented in Australia, and Australian farmers and their 'off-the-farm' colleagues show great ingenuity in
tackling working problems, even if they have been laggard, until recently, in
implementing some ecological advances. However, during the last ten years
the sheep population has increased by 38 million, largely as a consequence of
pasture improvement, so advances are now being made, although the shortage
of labour is still a major obstacle to better husbandry.

The primary role in basic agricultural research is played by the Commonwealth Scientific and Industrial Research Organisation (C.S.I.R.O.). It has,
with the departments of Agriculture of the six states and the Australian universities, made great scientific contributions of a high standard to Australian
farm problems. Unfortunately, the machinery for imparting these results to
the farmer, which is the responsibility of the State governments, is not yet as
well developed as it needs to be. Both Federal and State agencies are now
trying hard to remedy this defect and the farmers themselves are solving the
problem in some areas. The farm management club system is highly developed
in Western Australia and, to a lesser extent, in the eastern states. This is a
growing movement. The principle is that a group of agriculturalists, usually
in a fairly confined area, employ their own advisory officer by payment of
subscriptions to the club.

Ecological Background

The central desert (Fig. 2.4.2) is cut by the Tropic of Capricorn. Thus, to
the north of the Tropic ($23\frac{1}{2}°$S), the climate is tropical savanna (summer rain
and winter drought); to the south, it is either warm dry Mediterranean (winter
rain and summer drought) or warm humid (see p. 42). Though the tropical
sections, with their sugar-cane, tobacco, tropical fruits, beef ranching, etc.,
are of economic importance to Australia and of interest to tropical agriculturalists in so far as the shortage of labour and high-level technology has
forced mechanisation further than in other areas with comparable climates,
they are not considered further in this work, which is limited to the southern,
temperate parts of the continent.

Along a southern transect from Sydney to Perth (Table 2.4.3 and Fig. 2.4.1),
and on each side of it, the dominant ecological factor is rainfall (Table 2.4.2
and Fig. 2.4.2). This is unreliable and, in many areas, low when the high loss
by evapo-transpiration is taken into account. Temperatures are relatively
high, averaging about 50°F in winter and 70°F in summer. The concept of a
thermal or geographic 'growing season' (p. 34) based on the days or months
that are at or above $42\frac{1}{2}°$F is almost invalid, except in Tasmania. When an

Table 2.4.2

Climatic Data—Australia

Station	Bright Sunshine hours/day (mean)	Thermal Growing Season				Annual Precipitation		Moisture Index Im	Max. recorded rain/day inches (month of occurrence)	Mean months >72·5°F	Hottest month			
		Length in months	Mean Temp °F	Mean P–E–T (inches)	Mean A–E–T (inches)	Mean (inches)	Coefficient of variation				Month	Mean Max. Temp °F	Mean Min. Temp °F	Mean Midday Humidity
Sydney	6·8	12	63	33·0	32·4	47·0	15–20%	40 humid	11·1 (March)	0	January	95	58	64%
Wagga-Wagga Dubbo*	—	12	63·5*	34·1	21·1	21·1	15–20%	−38 semi-arid	5·1 (January)	4*	January*	93*	64*	32%*
Deniliquin Hay*	—	12	62·5*	34·6	15·8	15·8	20%	−54 arid	4·4 (November)	3*	January	91*	61*	28%*
Adelaide	7·0	12	63	34·8	20·9	20·9	15%	−40 semi-arid	5·6 (February)	2	January	108	51	31%
Tarcoola	—	12	65·5	38·2	6·0	6·0	25%	−84 arid	5·5 (February)	4	January	110	51	24%
Kalgoorlie	—	12	66	38·9	9·3	9·3	20%	−76 arid	7·0 (February)	5	January	108	53	25%
Perth	7·7	12	64	35·1	24·6	35·7	15%	1 moist sub-humid	3·9 (June)	2	January	102	54	44%

* Nearest alternative station.

Sources: See p. 107.

Australian talks about the growing season he usually means the months that
are *wet* enough for plant growth. A growing month used to be defined as one
in which rainfall was greater than one-third of the evaporation loss (E) from
an open water surface. This has now been replaced by the definition that, in a
growing month Precipitation $\geqslant 0.4E^{0.75}$. On this criterion, the arid centre
has no growing season, the sparsely stocked wool-growing areas have 1–5

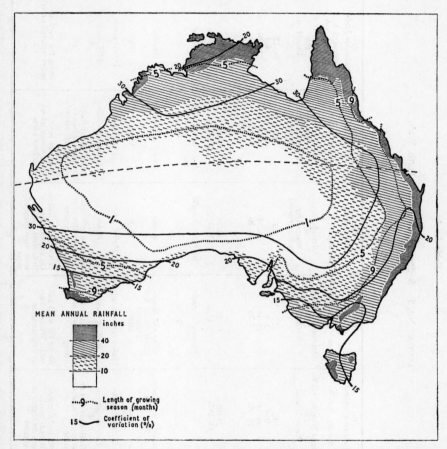

Fig. 2.4.2 Australia. Mean Annual Rainfall, its variability and length of the
hydrologic growing season. *Source:* C.S.I.R.O. 1960 and Tulloh *et al.* 1962

months (Sector (*e*) in Table 2.4.3), the wheat or wheat and sheep lands have
more than 5 months (Sectors (*b*), (*c*), (*d*)), and dairy and mixed farms have
more than 9 months growing season (Sector (*a*)) (Fig. 2.4.2), (Aitken *et al.*
1962, p. 189). The relatively short hydrologic growing season in the drier area
is aggravated by its unreliability (Fig. 2.4.2).

Australia has many cloudless days and adequate sunshine with tempera-

G   175

Table 2.4.3

Transect No. 5 — Sydney N.S.W. to Perth, W.A.

(Lat. 34°S) (Lat. 32°S)

	(a) Eastern Highlands	*(b)* Riverina and North Victoria	*(c)* Murray Basin	*(d)* S. Australia	*(e)* Nullarbor Plain	*(f)* W. Australia
1. Sector / 2. Name	Eastern Highlands	Riverina and North Victoria	Murray Basin	S. Australia	Nullarbor Plain	W. Australia
3. From	Sydney, N.S.W.	Wagga Wagga, N.S.W.	Deniliquin, N.S.W.	Adelaide, S.A.	{Elliston, S.A. / Tarcoola, S.A.}	Esperance, W.A.
4. To	Wagga Wagga, N.S.W.	Deniliquin, N.S.W.	Adelaide, S.A.	Elliston, S.A.	Esperance, W.A.	Perth, W.A.
5. Miles	250	150	400	250	800	400
6. Climate-type(s)	Warm Humid	Warm Dry	Warm Dry (Mediterranean)	Warm Dry and Desert	Desert	Warm Dry (Mediterranean)
Thermal Growing Season:						
7a.—months at or > $42\frac{1}{2}$°F	12	12	12	12	12	12
7b.—months at or > $72\frac{1}{2}$°F	0–4	3–4	2	2–4	4–5	2–4
8a.—P–E–T (in) in T–G–S	30–35	33–38	33–38	33–38	36–40	33–38
8b.—A–E–T (in) in T–G–S	20–35	15–25	14–20	14–22	4–9	10–25
9. Growth limited by	—	—	—	—	—	—
Precipitation:						
10.—mean in (annual)	25–40	15–25	10–20	10–20	5–10	10–40
11.—C.V. % of Precipitation (annual)	15–20%	20–25%	15–25%	15–25%	15–30%	15–20%
12. Special Climate Features	—	Summer drought	Summer drought	Summer drought moister on coast	Drought	Summer drought
13. Weather Reliability	2	2	2	1	1	2
Soil:						
14.—Zone/Group	Podsolic or Solodic or Red Brown	Grey and Brown Chernozemic and Deseric	Brown	Brown or Deseric	Deseric	Podsolic, Red-Brown, Brown
15.—pH	Low	High	High	High	High	High (mostly)
16. Special Features						
17. Vegetation	Eucalypt forest	Mallee scrub/grassland	Mallee scrub	Mallee scrub	Desert scrub	Scrub/grassland some Eucalypt forest
18. Relief	Low mountains	Rolling plains	Plains	Plains with hills	Flat plain	Plains with hills
19. Market Access	2		1	1	0	1
20. Farming Systems	Intensive grazing or alternating; extensive grazing	Extensive grazing; semi-intensive alternating	Intensive tree fruits; semi-intensive alternating; extensive grazing	Intensive tree fruits; extensive tillage or alternating; extensive grazing		Intensive tree fruits; extensive tillage; extensive grazing
21. Major enterprises	Level and summer milk near coast. Sheep on higher or drier land	Sheep for wool and meat. Winter wheat in moister areas	Horticulture, including grapes; rice (irrigated); Level milk from irrigated grass	Sheep for wool. Some winter wheat in moister areas	Sheep for wool in better watered areas	Sheep for wool. Winter wheat. Horticulture, including grapes (irrigated or in moister coastal areas)

tures (Table 2.4.2, Fig. 2.4.1) which may reach traumatic levels in summer.
This, combined with low, unreliable rainfall, means that her husbandry
problems have more in common with those of south-western U.S.A. and
Israel than those of Europe or of the northern half of North America. There
is great interest in the strategy of drought resistance, in fodder and water conservation against drought risk, in 'water harvesting', in irrigation and in the

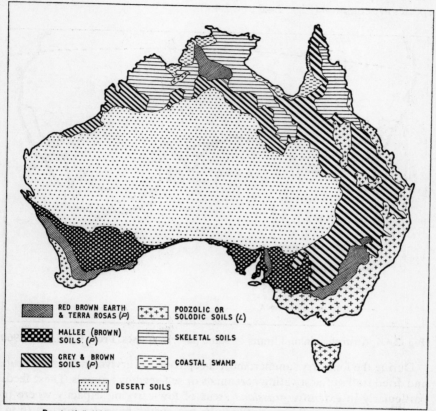

RED BROWN EARTH & TERRA ROSAS (P)

MALLEE (BROWN) SOILS. (P)

GREY & BROWN SOILS (P)

DESERT SOILS

PODZOLIC OR SOLODIC SOILS (L)

SKELETAL SOILS

COASTAL SWAMP

P = pedocal (soil which has calcium carbonate in its profile)   L = leached of main nutrients especially calcium and potassium

Fig. 2.4.3 Australia. Principal Soil Zones. *Source:* Wadham *et al.* 1957

feeding of herbage that has 'cured on the stalk', as well as in the possibilities
of transpiration-reducing chemicals (C.S.I.R.O. 1967). Indeed, in pasture
improvement under semi-arid conditions, or where summer drought is a
major risk, Australia probably leads the world. Summer drought can be
'dodged' *either* by the aestivation of some perennials (e.g. as in *Phalaris
tuberosa* or *Danthonia semiannularis* and other native perennial species) *or* by
self-seeding species with short annual life-cycles (e.g. Wimmera rye-grass
(*Lolium rigidum*), subterranean clover (*Trifolium subterraneum*), *Medicago*

spp, etc., (Wadham 1966). In the moist areas with more than 9 months
hydrologic growing season, perennial species such as perennial rye-grass
(*L. perenne*), white clover (*T. repens*), cock's-foot (*Dactylis glomerata*),
Paspalum dilatatum and Rhodes grass (*Chloris gayana*) may be used. In drier
areas with hydrologic growing seasons down to 5 months, self-seeding annual
mixtures (mostly *L. rigidum* and *T. subterraneum*) are practicable. But where
pasture improvement has not been or cannot be attempted, aestivating 'native'
species are common (Figs. 2.4.2 and 2.4.4).

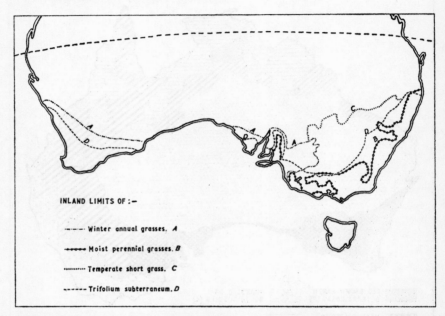

Fig. 2.4.4 Australia. Inland limits of some pasture grasses. From C.S.I.R.O. 1960

During the long, dry summer, most sheep have to survive on the seed heads
and dried stalks of aestivating perennials or self-seeding annuals. These feeds,
particularly in *extensive grassland* areas of low-carrying capacity where re-
seeding with improved varieties may be too costly, have dry matters (d.m.)
which are high in fibre (low digestibility) and low in protein (the latter may
fall as low as 2·5% or lower, compared with 8–10% in poor English hay).
Low digestibility and, in Australia, perhaps low protein, limit intake by sheep.
Animals may lose weight on poor quality roughage which is available in quantities of more than 2–3,000 lb d.m. per acre. Indeed, sheep may continue to
grow wool, although at a lower rate and with a smaller fibre diameter, whilst
starving to death in a drought. This problem is, perhaps, more serious in the
tropical grasslands of the north, but on the temperate grasslands (natural and
sown) it can also be critical. Research has shown that the feeding of supplementary non-protein nitrogen (e.g. urea) to animals grazing low-quality

178

roughage results in an increased rate of passage of food residue through the digestive tract; in consequence there is an increased intake of roughage and therefore of animal production (Coombe and Tribe 1962, Coombe and Tribe 1963, Vercoa, J. E. *et al.* 1961). Licking blocks of urea and molasses are better than herbage sprayed with urea.

In both grass and wooded areas, as in other warm dry climates when temperatures are high and vegetation dry, there is always the danger of bush fires which are assisted by winds and the oil content of the eucalypts. Such fires can move very rapidly and are a major and, to those affected, invariably a terrifying hazard.

Transect (*Fig. 2.4.1 and Table 2.4.3*)

EASTERN HIGHLANDS (*Sector (a)*)

The Eastern Highlands sector south-west of Sydney consists of a coastal plain and low mountains rising up to 3,000 ft and dissected by river valleys. The area is covered with leached soils of the podsolic–latosolic type (Fig. 2.4.3). These are slightly acid, deficient in lime and phosphates, respond to nitrogen, sulphates and molybdenum and are well drained and have a loam texture. There are patches of fertile alluvial soils on the coastal plain. The vegetation is an open sclerophyll forest with a shrub understorey. Species of *Eucalyptus* are the dominant trees. The climate is warm humid, with a mean annual temperature of 63°F at Sydney. At Dubbo, on the edge of the western slope, the annual mean temperature is almost the same but, whereas in Sydney no month in the year has a mean temperature of 72·5°F or above, at Dubbo four months are above this temperature and heat stress in breeding livestock can be a problem, but feed is of good quality and appetite not depressed. Sydney has an average of 6·8 hours of bright sunshine per day and receives an average of 47 in of rain a year with rather more falling during the autumn and winter, although each month receives a mean of not less than 3 in. At Wagga Wagga (Dubbo) (Table 2.4.2 and (D) on Fig. 2.4.1), the mean annual precipitation of 21 in is more strongly concentrated between May and October (winter and spring). The coefficient of variation of the annual rainfall is 15–20% throughout the region (Fig. 2.4.2). The eastern edge is liable to have very intensive rainfall, 11·1 in being recorded at Sydney one March day. Annual mean precipitation ranges from 40 in on the coast down to 25 in inland, with P–E–T at 30–35 in and A–E–T at 20–35 in (Table 2.4.3). Wadham 1966 notes, however, that Thornthwaite's data (p. 22) do not really fit Australian conditions.

Farming systems (Fig. 2.4.5). Farms are mainly of two types—*intensive grassland* on the coastal plains and *extensive grassland* in the mountains behind, though many Tableland sheep farmers have become much more intensive and go in for prime lamb production as pastures improve. In the mountains, the farms are largely wool producing, but some specialise in beef cattle breeding. From the climatic data, it would seem that intensive alternating or tillage farms would be possible, at least on the coastal plains, but their occurrence is rare. This may be due to the natural infertility of most of the soil, high costs of grain production and the pressure on the better areas to produce milk for the populations of Sydney and other towns. However, mixed

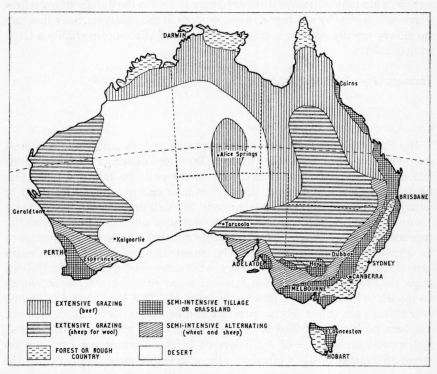

Fig. 2.4.5 Farming Systems

Land Use. Australia. Note: semi-intensive tillage and grassland includes beef or dairy cattle, tree fruits, vegetables or sugar-cane. From Tulloh *et al.* 1962

farms with the accent on potatoes, for instance, do occur and are increasing in numbers.

Intensive grassland—summer and level milk production. The typical dairy farms in Australia are of 150–400 acres with an average herd of 60 cows; the farms in coastal New South Wales are mostly under 200 acres. Because of the proximity to Sydney, where one-fifth of the country's population lives, most dairy farms produce for the liquid market. With natural rainfall, the pastures can support about 1 cow in milk to 1 acre during the spring flush. The usual practice has been to stock at lower rates than this and to graze some 'native' pastures only lightly during the spring. The ungrazed grasses run to seed and

180

remain as dry roughage of low digestibility but with patches of green growth of higher digestibility. Nevertheless, large quantities of concentrates have to be fed to maintain level milk output (Geddes 1964). In the past, the autumn and winter feed was provided by growing maize, sorghum, oats and other crops which were either fed off in the fields or cut green or made into silage. But since the advent of superphosphate and since 'introduced' perennial or annual species of clovers and grasses have become widespread, there is now a spring surplus which is made into hay or silage (using labour-saving forage harvesting attachments to small tractors) for late summer and winter feed, though concentrate use is still heavy.

The steadiness of the Sydney milk market has also encouraged 'water harvesting'. Farmers build ingenious storage devices to collect run-off water during storms and then sprinkler* irrigate their best pastures, offsetting rainfall uncertainty in spring, summer and autumn, so that milk production can be maintained at a high level throughout the year and stocking rates increased. Geddes pioneered this system at the Camden farms of the University of Sydney, which have annual rainfall of irregular arrival of 26–28 in and high insolation and summer temperatures. Now *Paspalum dilatatum* and Ladino (white) clover (which is more heat-tolerant than western European and New Zealand cultivars and is widely used in the warmer parts of the U.S.A.) provide irrigated summer grazing that can, if given nitrogen, feed up to 3 cows to the acre in the warmer months.† Winter bulky feed is provided by using a sod-seeder to drill oats, with fertiliser, into the pastures which are, thus, irrigated grass/clover grazing in the summer and oats for grazing in winter. The aim is to supply 30 lb dry matter (at 15% protein) of grazing per cow every day of the year. Concentrate consumption is much reduced. Fifteen years ago only less than 2% of the milk farms in the Camden district irrigated, today 75% do so (Geddes 1964).

However, elsewhere, many of the pastures still consist largely of native or volunteer species, which established themselves after the forest and shrub had been cleared. But in the moist, cooler areas 'made' perennial 'summer' pastures are increasingly important with, perhaps, perennial rye-grass (*Lolium perenne*), and white clover (*T. repens*) (Wadham *et al.* 1957, p. 174; C.S.I.R.O. 1960, p. 91) the most common species. If the land is too dry for perennials, annual pastures based on Wimmera rye-grass (*L. rigidum*) and subterranean clover (*T. subterraneum*) which, as noted, self-seed in the summer, commonly provide 'winter' pastures. Where possible, irrigation is practised, especially of 'summer' pastures, which require 2 or 3 times more water than 'winter' pastures.

Farmers supplying manufacturing milk (for butter, cheese and other products) generally do not receive a high enough price to justify much pasture

* Sprinkler or other overhead irrigation is used in Australia for private schemes for vegetables and intensive grassland; flood irrigation in official irrigation schemes.
† At Sonning Farm, Reading University, we run about 3 cows to the irrigated acre in summer, but the effective A–E–T is lower than in New South Wales.

improvement or supplementary bulky feeds or concentrates. They aim to calve their herds when grazing starts and dry off 8–9 months later. So the milk output is very seasonal. In many cases the milk is still separated on the farm and cream supplied to the factory. Payment is on a butter-fat basis, so high butter-fat breeds are favoured. Although Jersey cows are most abundant where conditions are good and a local breed, the Illawarra Shorthorn, is popular, Friesians are steadily increasing, especially for the liquid milk market. The average production per cow in New South Wales is 3,640 lb per year. This is lower than all other states except Queensland, and is probably because much of the dairying in the state is north of the transect, i.e. in the summer rainfall area where *inter alia* summer temperatures are too hot for both the pastures and the cattle (C.S.I.R.O. 1960, p. 149). Diseases, for instance tuberculosis, are not so serious as in countries where the stock has to be housed during the winter. However, contagious abortion and mastitis do cause serious losses.

Extensive grazing (*wool-growing*). This type occurs on the rolling open plateau and valleys of the Southern Tablelands of New South Wales. In Australia, sheep population/acre is greatest in districts with between 20 and 30 in of precipitation and A–E–T per year, where the original vegetation was open forest. This particular area, following pasture improvement, has some of the highest sheep stocking rates in the country. A. G. Lloyd (in Barnard 1962, p. 359) estimates that, under Australian conditions, 3 sheep/acre is the optimum economic level, but some wool producers who practice rotation grazing carry 8 or 10 sheep/acre with apparent economic and technical success. However, few farmers, even in the best areas with adequate and reliable rainfall, stock above 3 sheep/acre as, at higher rates, profitability rises only slowly and the problems of management and of feed shortage risks from drought increase. Mechanised forage harvesting has encouraged hay and silage making from surplus pasture during the spring flush and, on these Tablelands, has permitted increased stocking rates; stored fodder offsets the risk of overstocking during low rainfall seasons; as the coefficient of variation of the annual rainfall is 20% or more, this risk is considerable.

The district is still largely in 'native' pasture, including *Danthonia* spp (Wallaby grass) and tussocks of *Poa caespitosa*; but on about half the farms small areas are under crops (mainly oats) for grazing or conservation, whilst nearly all the farms, which are mainly between 1,000 and 5,000 acres with flocks averaging 1,500 sheep, have some pastures, top-dressed with superphosphate, in short cycle annuals, e.g. subterranean clover and Wimmera rye-grass. Most farmers breed some, at least, of the replacements for their predominantly Merino or Merino cross flocks (which largely consist of wethers). But the risk of sharp frosts, snow and late cold spells during lambing, force many small wool-growers to rely on bought-in wethers. A few farms in better-watered parts concentrate on breeding crossbred flocks (usually long-woolled Border Leicester rams on Merino ewes). The ewe progeny are then crossed with Southdown, Romney Marsh or Dorset Horn rams to produce prime lambs. Lambing is generally in the late winter or early spring (July–October) and shearing between September and November. Though here, as elsewhere in Australia, shearing is becoming less seasonal. Shearing is usually done on contract or by the owner himself engaging shearers living locally. The average fleece yield of 8·4 lb/sheep is below the country's average of 10 lb/sheep. All farmers dip regularly and also crutch the sheep to control flystrike (*Myiasis*) due to blowfly (*Lucilia* spp). Some have cattle as well as sheep, either breeding and selling young store cattle and fat steers or purchasing and fattening store (feeder) cattle.

RIVERINA AND NORTHERN VICTORIA (*Sector (b)*)

Here rolling plains slope gently westward from the Eastern Highlands. The soils, as one moves west, are of three main types, red-brown earths with a vegetation cover of savanna woodland grading into short grassland on the grey soils and mallee scrub and salt bush on the solonised brown soils (Fig. 2.4.3). The climate is warm dry. The mean annual temperature is $62\frac{1}{2}°F$ at Hay, on the very western edge, where there are on average three months (which coincide with the time of lowest rainfall in midsummer) when the mean temperature is $72·5°F$ or more. Mean annual P–E–T is about 35 in. Precipitation averages 14 in in the west to 25 in/year in the east, with A–E–T from 15 in to 25 in per annum. Rain falls mainly from May to October (winter/spring); it is not generally very intensive but summer storm rains do cause sheet and gully erosion of the gently sloping Riverina. The annual coefficient of variation is 20–25%. High temperatures, low rainfall, high P–E–T and feed shortage, combine to make high summer a period of stress.

Farming systems (Fig. 2.4.5). The wetter east is mainly in *semi-intensive alternating* wheat–sheep farms, whilst the poorer soils and the drier western parts of the region have *extensive grazing* (wool-growing) similar to those already described. There are also some irrigated lands which support a Mediterranean type of *intensive tree fruit and alternating agriculture* and some specialised farms with Merino and Corriedale sheep studs.

Semi-intensive and extensive alternating (*wheat and sheep farms*). These wheat–sheep farms are essentially the same all over the country; they hold about 40% of the national sheep flock and produce the greater part of the wheat crop. This important system evolved out of continuous wheat production as the need to maintain soil fertility was realised. This led first to an extension of the traditional half-year fallow and to the introduction of superphosphate and such annual legumes as subterranean clover, medics (*Medicago* spp) and some grasses. The resultant reduced acreage under wheat was balanced by higher grain yields and by increased livestock and wool sales. The two enterprises dovetail well together and are an excellent example of alternate husbandry. This partnership between P_2O_5, annual legumes and annual grasses and wheat in semi-arid areas is the Australian counterpart of the P_2O_5,

white clover, perennial rye-grass combination which has proved so successful
in New Zealand (p. 199). Australian wheat production is characterised by low
and variable yields per acre but high production per man. New South Wales
had an average yield of about 1,000 lb grain per acre for the years 1957/58–
1961/62. However, within this period the annual yield varied from 280 to
1,250 lb per acre. The annual coefficient of variation of wheat yields is over
25%. Similar average yields are recorded in southern Australia and Western
Australia and about 1,250 lb/acre in Victoria. (For a classical analysis of yield
variability in southern Australia, see Cornish 1949, 1950.)

The farms are family units of 500–2,500 acres, of which about one-fifth is
under wheat at any one time, and with a flock of sheep averaging 1,000 head.
The rest of the land is under varying proportions of (1) 'natural' or 'invader'
pastures, e.g. *Bromus* spp, *Vulpia myuros*, *Medicago* spp, *Trifolium glomera-
tum*; (2) sown pastures, often of mixed grasses, especially Wimmera rye-grass
and special strains of subterranean clover or of *Medicago* spp; (3) fallow and
(4) oats for forage. A common rotation is, firstly, sown pasture, preferably
for several years, then, secondly, oats to provide stock feed and to clean the
soil of cereal footrot, take-all (*Ophiobolus graminis*) and other fungal diseases
of wheat, followed by, thirdly, several wheat crops before undersowing the
last crop to the grass and clover ley. In drier areas the rotation is, firstly,
wheat; secondly, two years 'volunteer' pasture (weeds and invader grasses)
and, thirdly, one year fallow and/or sown annual legumes. The ground is prepared for the wheat crop by rough discing in February to April (autumn) and
then using a 'combine' spring-tine cultivator and seeder which gives a very
fine seed-bed and dislodges the weeds; the 'combine' is later used for drilling
in early May (early winter). In the same operation, it applies superphosphate
and, if ley is to follow, it broadcasts the grass or clover seed mixture. Herbicides are increasingly sprayed for weed control. The major weeds are skeleton
weeds (*Chondrilla juncea*), wild mustard (*Sisymbrium* spp), wild turnip (*Brassica tournefortii*). The growing wheat is sometimes grazed during June and
July (mid-winter) to prevent rank growth. In October, a fire break is cleared;
this involves cutting a strip of the crop 18 ft wide which is made into wheat
hay. This strip is then ploughed as well as a further 10 ft round the outside of
the field. Harvesting is done by headers which cut the grain high up leaving
most of the straw standing. (The latter, with its weeds and some fallen grain,
is cleaned by sheep and then ploughed in.) Though bulk tank and farm-storage
systems are increasing rapidly, some grain is still fed into a bushel box and
then down chutes into bags. These are taken by lorry to the local state-owned
silo. The Australian Wheat Board (a statutory Commonwealth body) markets
the country's crop.

In wheat–sheep areas with better rainfall or irrigation, farmers often have
flocks of crossbred ewes for wool and prime lamb production. On such farms,
all-the-year-round grazing is provided by lucerne, native pasture, stubble and
clover/medics in the summer, and subterranean clover/Wimmera rye-grass

(or annual medics in the drier areas) in the autumn, winter and spring. However, where the moisture is lower and less reliable, or the farmer likes a more
leisurely life, pure Merinos for wool are kept and pasture management is less
advanced. Culling and classing are considered very important for maintaining
flock standards, especially of wool quality and yield. Ideally, the ewes are
culled before the first shearing and then reviewed annually. Ewe replacements
are bred on the farms or bought-in (if for prime lamb production). The wool
yield from Merino wethers falls off after about four years, at which time the
animals are fattened and sold as mutton for the home market.

A Wheat–Sheep Farm

The farm of W. H. Boothby of Pechelba East in north-east Victoria is an
example of a semi-intensive wheat–sheep farm and lies south of Transect
Sector (*b*). Temperatures here rarely fall below 40°F and the mean annual
precipitation is 22 in. Summer drought extends from mid-November to mid-
March, but some spray irrigation is carried out. The farm consists of 1,000
acres of flat, heavyish land. The rotation is 3 years cereals (250 acres wheat,
100 acres oats) followed by 6–7 years of ley undersown at a bushel of oats (as
a starter) plus *Phalaris* spp with subterranean clover *or* Wimmera rye-grass
with subterranean clover. The ley or stubble is broken with a chisel plough
and the land is cultivated in the 'early wet' (March/April). Drilling is in May/
June and the crop is harvested at Christmas (equivalent of late June in the
United Kingdom). Superphosphates at 112 lb /acre are applied each year and
the wheat, which is winter wheat with a spring habit of growth, yields 2,800
lb/acre. This is nearly three times the Australian average. Silage from the leys
is stacked with a forage harvester and buck-rake and the sheep can be self-fed
at any season of the year. There is now some (strategic) drought reserve silage,
which is 16 years old (one of his neighbours has 'mislaid' or forgotten the
whereabouts of 1,600 tons of pit silage!). The Border Leicester/Merino cross
ewes, of which there are 800, are put to Dorset Horn rams to lamb in late
autumn (May/June). There is a 90–100% crop of lambs, which are sold at 17
weeks at 36–40 lb dead-weight, just before 'the dry'. There are also 200 pedigree Aberdeen Angus cattle. As Mr Boothby himself is now semi-retired, the
farm work is done by his son, a hired man and his wife. Farm machinery
includes combine harvesters, chisel ploughs, drills, tractors, etc.

Murrumbidgee: intensive irrigated alternating and tree fruits. Here P–E–T
averages about twice the precipitation and A–E–T 15 in to 20 in per year.
Fortunately the Murrumbidgee irrigation scheme supplies about 100,000
acres with water during August (early spring) to April (autumn). Allocations
of water depend on the kind of land use, but are generally 1 acre foot per acre
per year for horticulture and 1 acre foot per 4–10 acres for other purposes.
(Wadham *et. al.* 1957, p. 276; C.S.I.R.O. 1960, p. 63). By U.S. standards these
rates are low for such a climate; they would not be very high even in north-
west Europe.

185

The area has two kinds of farms. The first are highly mechanised units of 600–700 acres each, which specialise in rice and meat production. The soils (grey-brown clays) are too heavy for fruit-growing and often very difficult to manage for other crops. One-seventh of the land in any one year is planted with rice, which forms part of a three- or five-year rotation with pasture. In midsummer, some of the highest Net Assimilation Rates in the world have been recorded in this area, and rice yields very well (about 4,200 lb/acre; the world average is about one-fifth of this). Californian cultivars of rice are sown in the late spring in specially levelled and graded bays, in which water is maintained from October to the end of February (early autumn). Then the ground is dried out before harvesting with combine harvesters. The Rice Marketing Board limits the acreage that can be grown by the individual farmer, but as rice needs about 6 acre feet of water, the area drilled is partly determined by water availability. The irrigated pastures are mostly used for fat lamb production although cattle are also fattened on some farms.

The second type is much smaller (10–50 acres), on lighter soils, devoted to tree fruits and to vegetables; they are family farms only employing Italian and Greek hired labour for harvesting. The tree crops are citrus fruit, vines for winemaking, apricots and peaches for canning, and prunes (dried plums). On the heavier soils, tomatoes, beans and green peas are grown for canning, and on sandier soils, carrots and rock melons. The vines and tree fruit are weeded mechanically during the summer; in the winter a leguminous cover crop is grown between the rows and later ploughed in. The vines are dressed with P_2O_5 and K_2O; nitrogen is applied to citrus.

Semi-intensive grassland. To the south of the transect, between the Riverina Plains and the Great Dividing Range, lie some better-watered, rolling hills, many of which are used for fat lamb or beef production. The original vegetation of gum-trees (*Eucalyptus* spp) has been cleared by ringing, burning and stumping and native grasses have invaded and taken over. If P and Ca plus trace elements are applied, subterranean clover will invade (volunteer) or may be sown. Mean annual precipitation for the area is about 27 in and there are 3–4 months of summer drought. Soil erosion occurs, especially when the autumn rains hit the dry, hard-grazed soil. There is some pasture growth all through the winter, but the peak growth comes in early spring.

On one farm, paddocks of 100 acres were well wired and watered and on the lower, wetter land there are some pastures of perennial rye-grass or *Phalaris tuberosa*, with subterranean clover. Superphosphate (112 lb) and 2–3 oz of molybdenum salts per acre are often applied annually. Border Leicester cross Merino ewes were put to polled Dorset rams to lamb in mid-winter (July). There was a 90% crop of lambs, which grew quickly and were sold fat in November/December at 40 lb dead-weight at 4–5 months. Twins were not wanted because all lambs, except own replacements, must be sold before the dry weather (cf. drier fat lamb farms in New Zealand). Drenching was carried out with thio-benzidrole for worm control. Hay making took place in late

spring, using string—not wire—for tying the bales. These were left where
they fell and grazed *in situ* by sheep and cattle during the summer drought.

MURRAY BASIN (*Sector* (*c*))

The area is a flat artesian basin formed from sedimentary rocks which yield pressure-water. The Murray is regulated by a system of dams, locks and weirs to provide a year-round flow for irrigation. The desertic and solodic soils are very variable, from deep sandy brown soils to shallow clay soils; they are all very alkaline with lime deposits and often saline layers. They are low in phosphorus and nitrogen and, locally, in zinc and copper. Zinc deficiency can usually be rectified by a dressing of 5 lb zinc salts/acre, repeated every five years. The natural vegetation is often mallee and thus, these soils are often known as mallee soils (Fig. 2.4.3). The mallee consists of low, much-branched eucalypts 10–20 ft high rising from a basal woody mass and is so dense that there is little or no useful growth of grasses. But much of the region is semi-desert. The figures given for Hay in Table 2.4.2, describe the climate in the centre of this region, whilst the climate near Adelaide is discussed in the next sector.

Farming systems (Fig. 2.4.5). Excluding the area round the north of Adelaide, wool-growing (*extensive grazing*) is almost the only type of farming that is ecologically feasible without added water. However, irrigation schemes at several points along the Murray River have opened up very productive areas for intensive farming, mainly *tree fruits*, but some *grassland* dairying. The very high annual P–E–T– (e.g. 35·4 in annual mean at Euston on Murray River) indicates the agricultural potential, provided that water can be applied to augment the low precipitation, that the soils are suitable for irrigation, that fertilisers are used where necessary, and that soil salinity problems can be avoided.

Extensive grazing (*wool-growing*). These unirrigated farms, or stations as they are called in Australia, differ from those in the N.S.W. Tablelands (p. 182) owing largely to the drier and hotter summers. Only Merino sheep (mainly wethers) which are able to withstand the rugged conditions of heat stress and poor feed, are carried. (Wethers are hardier and need less attention than ewes.) Pasture improvement is rarely practised because of the high cost of clearing mallee vegetation and because of the low rainfall and infertile soils. The average size of stations in the area is 20,000 acres, with an average flock of 3,200 sheep, i.e. an average of one sheep to just over 6 acres. However, this carrying capacity varies from year to year and falls from east to west as the rainfall decreases.

Most stations are divided by fences into huge paddocks and the sheep are rotated from one to the other. For a month or two after rains the sheep feed on the ephemeral 'native' ground vegetation and grasses such as Wallaby grass (*Danthonia* spp). After this has been exhausted they feed on the shrub vegetation including saltbush (*Atriplex vesicarium*) and bluebush (*Kochia*

spp). If a prolonged drought develops and the flocks are not hand-fed with
purchased feed or the numbers are not reduced, then the shrubs may be eaten
beyond the point at which they are able to recover when the rains come. If this
occurs, then wind erosion results, and large areas are lost to agriculture.
Water from constructed catchment basins or artesian wells is provided for
the sheep in each paddock.

Stations usually have one man to 3,000 sheep and his time is spent moving
the sheep from paddock to paddock, repairing fencing, maintaining water
supplies and controlling blowfly (*Lucilia* spp) strike. Lack of manpower precludes fodder conservation; in any case the natural vegetation is not suitable
or in sufficient quantity to conserve; conservation would only help in a dry
spell and would make no appreciable difference in a prolonged drought. The
stations often carry enough ewes to breed their own replacements, but some
may buy-in wethers from better-watered districts, whilst others cross their
older Merino ewes with Border Leicester rams to produce the crossbred
progeny wanted by the meat producers in the wheat–sheep and high rainfall
areas, or produce surplus lambs for sale. Lambing takes place either in the
autumn or the spring. Besides normal sheep pests and diseases such as blowfly,
worms and trace element deficiencies, sheep stations have to contend with
mammalian and avian predators and competitors. Dingos are no longer common, except on the fringes of the deserts and in rough country. However,
fencing helps to keep them out, but they can still be a problem on outlying
stations, especially further north. (In 'outback' parts of Australia feral camels,
wombats and kangaroos of various types, may damage dingo fences. Kangaroos, rabbits (much diminished by myxomatosis), hares and even wild
donkeys and goats, amongst mammals, together with various seed-eating
and grazing birds (wild duck, galah (*Kakatoe roseicapilla*), compete with the
sheep for grazing in varying degrees in different regions. The same species
may also feed on wheat and other crops. The dingo, the beautiful wedge-
tailed eagle (*Aquila audax*), crows, and, amongst introduced mammals which
have become feral, wild pigs and foxes are alleged, from time to time, to prey
on sheep, especially on lambs. (B. V. Fennesey in Barnard 1962.))

Intensive tree fruit (*dried fruit production*). The flood-irrigated areas along
this section of the Murray are chiefly famous for their dried fruit, but, wine,
apricots, citrus fruits and sprinkler-irrigated vegetables are also produced.
The dried fruit industry is centred on Mildura and, helped by the hot, very
dry period between November and March, and the rarity of severe frosts,
produces, perhaps, the best quality sultanas in the world. During the ripening
and drying period (the latter is done in the open air) rain can, however, cause
serious damage to the grapes. Most of the dried vine fruits are produced on
small farms of 15 to 40 acres owned and worked by one man, but casual
workers are taken on for harvesting. The vines are pruned after leaf fall in
May (autumn) and then the ground is alternately cultivated and irrigated
during the winter. The ground (between the rows) may also be sown with

leguminous cover crops during the autumn and winter. The vines are sprayed against fungal diseases from September to harvest time. Harvest starts at the beginning of February (late summer) and by the end of March most of the fruit is off the vines. The fruit is dipped in hot soda or cold potash solutions before drying in order to improve its colour. Vines for winemaking are grown on large farms owned by the winemaking companies, which chiefly produce a sweet fortified wine of the Spanish and Portuguese type. There are, however, some excellent light clarets and dry white wines made from selected vineyards.

SOUTH AUSTRALIA (*Sector (d)*)

The South Australian climate is warm dry, becoming rapidly drier as one moves inland from the long coastline. The topography is plains with low mountains. Most of the soils (Fig. 2.4.3) are solonised (mallee) soils lacking adequate nitrogen and phosphorus and with copper, zinc or molybdenum deficiencies in particular areas. Much of the region carries typical mallee vegetation with a dense shrub layer; this has been cleared for wheat production. Where the natural vegetation is sclerophyllous shrub and ericaceous heath, some of it has been cleared for pastures based on subterranean or medic clovers, and various grasses, plus trace elements and superphosphates, and supports a substantial population of sheep, mainly for wool. The mean annual temperature is 63°F at Adelaide and 65½°F at Tarcoola (inland). The days are much sunnier than in north-western Europe with an average of 7 hours of bright sunshine per day. Adelaide has an average of 2 months a year when the temperature is above 72½°F and Tarcoola has 4. Annual P–E–T is about 35 in. The mean annual precipitation in Adelaide is 21·1 in but in Tarcoola, about 130 miles from the nearest coast, it is only 6·6 in. However, most of the sector has between 10 and 20 in per year and A–E–T is also in this range. Inland, what little rain there is, occurs at any time of the year, but near the coast precipitation is typically Mediterranean, falling mainly in the winter months of May to September. The coefficient of variation of annual precipitation is 15–25%, increasing from the coast, inland. Rainfall is often intensive and water erosion can, therefore, be a difficulty on sloping, overgrazed pastures which are bare when the autumn rains break. In the coast area, storms carry salt spray inland, and, as the rain fails to wash the salt right out of the soil profile, it tends to accumulate to dangerous levels.

Farming systems (Fig. 2.4.5). *Extensive tillage, extensive grazing* and *tree fruit* systems are important, whilst dairying for urban milk supply is also practised round Adelaide and in a few other areas.

Extensive tillage (*wheat*). On the rolling country and plains north of Adelaide and on the Eyre and Yorke peninsulas, the system of grain production is much the same as in northern Victoria (Sector (b)). However, farmers especially on the Yorke peninsula, have an alternative cash crop in high-grade malting barley. This sector has a particular problem due to the shedding of grain, both wheat and barley, a few days before harvest, as a result of strong, hot winds which blow from the deserts of the north, and which, on occasions, have caused extensive damage. The shedding is now controlled by rolling the grain with a heavy roller up to 4 days before the anticipated harvest date. This, apparently, enables most of the crop to be saved, whatever the wind conditions.

Extensive grazing (*wool-growing*). This is found where precipitation and A–E–T (6–15 in per annum) are too low for tillage, in country that has mulga as its natural vegetation. This is desert scrub with hardy *Acacia* spp trees (e.g. *A. aneura, A. loderi*). The 'leaves' and twigs and especially the pods and seeds are readily eaten by livestock. During period of drought branches may be lopped or even whole trees felled to increase feed for the sheep. However, this practice and overgrazing of the interspersed volunteer grasses have tended to increase wind erosion. A sheep station at Salt Creek in South Australia lies 120 miles south-east of Adelaide in rolling country with a vegetative cover of mallee scrub (not mulga) and roughly half of the 14,000 acres have recently been cleared and seeded to improved pastures. Clearing has been carried out at the rate of 120 acres per day, working day- and night-shifts with the aid of modern heavy machinery. The soil is deficient in phosphate, zinc and copper. The general procedure has been to disc-plough twice with D.7 caterpillar tractors, roll with a 16-ton roller and then drill with 28-row drills. Super-phosphates, copper sulphate (7 lb/acre) and zinc sulphate (7 lb/acre) have then been applied. After the pasture is established, the land is regularly top-dressed. As the farm has its own landing strip the top-dressing is done by aeroplane (cf. New Zealand) at the rate of 250 tons of superphosphate spread over 2,500 acres in $2\frac{1}{2}$ days. Subterranean clover and veld grass (*Ehrharta* spp) are allowed to seed each summer and the effect of the clover is very apparent as the phosphate builds up nitrogen in the soil which feeds the veld grass. This is particularly apparent where the clover is established in the same tuft as grasses. The stock, which are fenced into 500/600-acre paddocks, graze on the autumn and spring flushes of green, but in the summer, when few perennial grasses can survive, they feed on 'cured on the stalk' herbage and seed-heads. The stocking in October 1962, was one Merino sheep for wool production per cleared acre. The station, which is run by the owner's son, does not carry much machinery as shearing, well drilling, superphosphate-spreading, fencing, etc. are done on contract.

Intensive tree crops (*wine production*). The Barossa Valley, north-east of Adelaide, is the best-known winemaking area in Australia. Here, the rainfall is more reliable, between 20 and 35 in, per annum, and the vines are grown without irrigation. The soils are clay loams and this, together with the climatic conditions, produces grapes with a high sugar content suitable for the production of full-bodied wines.

NULLARBOR PLAIN (*Sector* (*e*))

This plain in western South Australia and eastern West Australia is agriculturally useless.

WESTERN AUSTRALIA (*Sector (f)*)

This region consists basically of the centrally placed Darling plateau which ends abruptly with an escarpment leaving coastal plains 10–15 miles wide. Bordering the scarps are flat alluvial plains, built by the streams coming off the plateau and having naturally fertile soils. The soil types follow the rainfall with leached podsolic–latosolic soils in the extreme south-west and cherno-zemic–desertic soils giving way north-eastwards to desertic mallee soils. The vegetation is the same on each class of soil as in the eastern states except that heavier mallee soils carrying sclerophyll woodland with salmon gum (*E. salmonophloia*) are common. All soils are deficient in phosphates and other minerals. It was, in fact, in this region about 35 years ago that Underwood and others did their well-known research on trace element deficiencies, particularly on cobalt, copper and zinc. This, together with Marsden's work in South Australia, opened new areas for farming not only in Western Australia but also in many other parts of the world (see E. J. Underwood in Barnard 1962 for a summary).

Western Australia has a warm dry Mediterranean climate and abundant sunshine; Perth has an average of 7·7 hours of bright sunshine a day, a mean annual temperature of 64°F, and mean precipitation of 35·7 in a year, whilst inland, Kalgoorlie has only 9·3 in. Perth has a mean annual P–E–T of 35·1 in and a mean A–E–T of 24·6 in whilst the corresponding values for Kalgoorlie (inland) are 38·9 and 9·3 in. Rain falls mostly in the autumn and winter months (April to September). The coefficient of variation of the rainfall increases from 15 to 25% with decreasing rainfall and, as in most comparable parts of Australia, in inland areas, it is considerably more variable than the world mean C.V. for its particular mean annual precipitation.

Farming systems (Fig. 2.4.5). In the moister areas round Perth and other coastal towns, intensive irrigated dairying and fruit and vegetables, sometimes on the same farms, are found. This sector also produces some excellent light wines. But the major land uses are extensive tillage and extensive grazing.

Extensive tillage and grazing (*wheat and sheep farms*). These farms lie in a broad belt from Geraldton in the north to Albany in the south. The annual precipitation is between 10 and 25 in (mean). The holdings are mainly between 1,500 and 2,500 acres and carry an average of 1,000 sheep and 500–1,000 acres of winter wheat. Some, however, are bigger (10,000 acres or more), particularly in areas where the soils are 'patched' by areas of ironstone and saline spots; much of such farms may be uncultivable, but can be grazed by sheep. In West Australia varieties of wheat with a fairly short growing season and which can be harvested early, are grown. This is necessitated by the concentration of rainfall in April to September, and the importance of good rains during autumn drilling and at flowering in August/September. Harvest starts in November and December (early summer) and safely continues in the 'dry' for 10 weeks or so.

191

On a 10,000-acre holding, about 160 miles north of Perth, with a mean annual precipitation of 13 in, the cultivated land is divided equally between winter wheat, stubble (one year) and fallow (one year) although the area under fallow is steadily being reduced as pastures are improved with early maturing legumes (*Medicago* spp) and an alternating system is introduced. Three brothers run the holding with one hired man and his wife. The main work peak is at seed-time, which begins with the autumn rains, when 6 casual workers are hired. 3,000 acres are then drilled in one month, when each tractor and seeder may put in 100 acres of wheat in 24 hours. Harvesting is more leisurely as just noted; one combine harvester may do 600 acres per season compared with about 150 acres in the United Kingdom (cf. Duckham 1963, p. 344). Yields average around 1,100–1,200 lb per acre. The sheep which graze the improved pastures and the weeds in the stubble are shorn on contract.

Extensive grazing (*wool-growing*). Further inland with precipitation averaging 10–15 in per year, stations run from 2,000 to 50,000 acres in size and support 500–5,000 sheep. Due to very high temperatures 'summer sterility' occurs, fertility rates of rams are low and lambing percentages are very poor. In other respects, the wool-growing is not unlike that described earlier (p. 190).

Efficiency and Stability

The dominant agricultural problems of Australia are the relative sparseness of farms and farm people with the adverse effect this lack of density has on infra-structure and services generally; the uncertain and low rainfall in inland areas; and the difficulty of finding stable and profitable export markets for wheat, meat, dried fruits and wool which is experiencing severe competition from man-made fibres. These factors combine to lower overall efficiency, because uncertainty is always expensive and often retards progress. The advanced and very ingenious levels of technology exhibited by some research stations and progressive farmers are hard to apply generally in a country of low-density rural services where uncertainties of climate and markets encourage one to 'play safe' and avoid the risk of over-investment, both in variable inputs and in durable resources. One could hardly expect the Australian fat lamb or milk producer to be as biologically efficient as the New Zealander or the Englishman. On the other hand, in the irrigated areas, where services are ample and water adequate, there are some enterprises, for example, dried fruit, fruit for canning, rice-growing, which are very efficient. (Australian rice yields are amongst the world's highest.) Further, the wheat–ley–sheep system (p. 183) is obviously a sound method of land and labour use whilst, though yields/acre may be low, the extensive wool-growing and wheat–meat production areas achieve a very high output per man. Lastly, it is noteworthy that, in irrigated areas, the Australians, despite high evapo-transpiration losses, do not seem to have created the problem of soil salinity on the same scale as have the Americans. Possibly the enforced economy of irrigation water use—due to generally high water prices* or small water allocations—is a blessing in disguise.

The problems of finding export markets and, in particular, the competition of man-made fibres with wool, largely depend on political and economic developments, especially in the Far East (and to a lesser extent in Europe and the U.S.A.) and are not likely to yield to applied science except in so far as its results help to reduce costs of wool production and so enable wool to compete with man-made fibres.

Conclusions and Prospects

In the long run, the growing industrialisation of a country, rich in mineral resources and human enterprise, should offer outlets for both surplus farm population and for farm produce—though this will hardly help the wool-growing areas. At the same time, it should provide some surplus capital for investment in rural services—though low farm population density will probably remain. The future of the water problem is hard to foresee. On the one hand, improved application of statistical method to farm management and operation should permit decisions to be more scientifically based on rainfall probabilities. On the other, in the very long run, the diversion of rivers and possibly the development of sea-water desalting may both take the risk out of and increase yields in a climate with a high P–E–T which is, given water, admirably suited to advanced intensive farming. The ambition of one leading Australian agricultural scientist is that not a drop of the country's river waters should reach the sea. Indeed, the vision and large-scale thinking of Australian scientists, engineers and economists, suggests that by the end of the century many of her now chronic ecological problems will have been solved by technology, provided investment capital is available.

Acknowledgements. For research assistance to Kathleen Down and Sarah Rowe-Jones, and for comments and criticisms to R. Elphick, Janet Foot, H. J. Geddes, A. J. Macfarlane, J. Melville, D. E. Tribe, E. J. Underwood and S. M. Wadham.

References

*Aitken, Y., Tribe., D. E., Tulloh, N. M. and Wilson, J. H. 1962. *Agricultural Science*. Melbourne.

*Barnard, A. (Ed.) 1962. *The Simple Fleece*. Melbourne.

*C.S.I.R.O. 1960. *The Australian Environment*. Commonwealth Ind. and Sci. Res. Organisation, Melbourne.

C.S.I.R.O. 1967. *Div. of Meteorological Physics. Ann. Rept. 1966–67*. Commonwealth Ind. and Sci. Res. Organisation, Melbourne.

Coombe, J. B. and Tribe, D. E. 1962. *J. Agric. Sci.* Camb. **59**. 125.

* Variations in water prices are, however, wide and so influence land use (Wadham 1966).

Farming Coombe, J. B. and Tribe, D. E. 1963. *Aust. J. Agric. Res.* **14**. 70.
Systems Cornish, E. A. 1949, 1950. 'Yield trends in the wheat belt of S. Australia 1896–1941',
          Aust. J. Sci. Res. Ser. B. **2** (2), **3** (2), 178, 218.
        Duckham, A. N. 1963. *Agricultural Synthesis; The Farming Year*. London.
        Geddes, H. J. 1964. *Outlook on Agriculture*. **4**, 4. 182.
        U.S.D.A. 1967. *Agricultural Policies in the Far East and Oceania*. Foreign Agricul-
          tural Economic Report No. 37. U.S.D.A. Washington, D.C.
        Vercoe, J. E., Tribe, D. E. and Pearce, A. L. 1961. *Aust. J. Agric. Res.* **12**. 689.
        *Wadham, S. M., Wilson, R. K. and Wood, J. 1957. *Land Utilisation in Australia*.
          Melbourne, 3rd edn.
        Wadham, S. M. 1966. Personal Communication. Melbourne, Australia.

Further Reading
(SEE ALSO ASTERISKED ITEMS)*
        Bureau of Agricultural Economics, 1957. *Australia Sheep Industry Survey, 1943.
          High Rainfall Zone*.
        Robinson, K. W. 1960. *Australia, New Zealand and the South-west Pacific*. London.
        Spate, O. H. K. 1956. *Australia, New Zealand and the Pacific*. London.
        Wilsie, C. P. 1962. *Crop Adaptation and Distribution*. San Francisco and London.
        Yearbook of the Commonwealth of Australia.

CHAPTER 2.4

Australia and New Zealand

(b) New Zealand. BY R. W. WILLEY AND A. N. DUCKHAM

Introduction—Economic and Social Structure

New Zealand lies between latitudes 34°S–47°S 'in the bottom right-hand corner of the world', and covers an area roughly equal to that of the United Kingdom. It consists of two main islands (North Island and South Island) and a few minor ones. The North Island is rather smaller than the South but it supports about two-thirds of New Zealand's 2½ million population. The climate, New Zealand's major physical asset, is cool to warm temperate humid with abundant sunshine, but rather windy. Much of the terrain is broken and mountainous (Fig. 2.4.6), creating problems of field operation, erosion and transport. The transport problem is further aggravated by the rather elongated shape of the country.

New Zealand's agricultural history, of any importance, began in the mid-19th century with wool farming. However, the early prosperity which this created was short-lived as over-production severely lowered returns. Fortunately, the advent of refrigerated shipping in the 1880s opened the way for the export of frozen lamb and, later, of dairy products to the rapidly industrialising countries of Europe. This was the birth of the country's present-day agriculture; for today, wool, meat and dairy products provide over 80 % of the gross farming income and the greater part of these are still exported to the United Kingdom and Europe. Thus, New Zealand is extremely dependent on distant and, at the present time, rather unreliable markets, and in recent years she has made considerable efforts to tap the great potential of the Asian markets.

New Zealand has considerable resources of water power, and some steam power from hot geysers, but her mineral resources are limited though oil or gas is a possibility, and as yet she has not been able to build up any major manufacturing industry. She therefore relies heavily on imports of machinery, fertilisers, chemicals, fuels and manufactured goods as well as imports of high-calorie foods such as sugar. Even tourism, which potentially could earn these lovely islands some badly needed foreign exchange, is severely limited because of the geographical isolation. The country's chronic balance of payments problems (U.S.D.A. 1967) have been particularly accentuated by the competition of man-made fibres which can substitute the low-quality 'carpet' wool which she produces.

There is an acute shortage of manpower which has resulted in high farm wages and a high degree of mechanisation. The latter, aided by industrial and

195

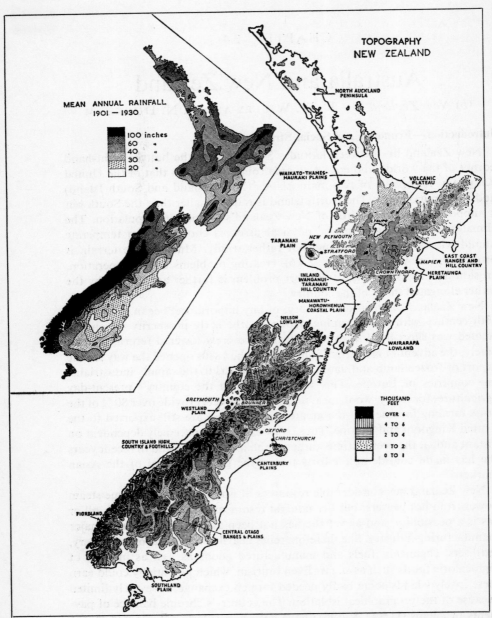

Fig. 2.4.6 Relief and inset of mean annual rainfall. (From N.Z. Lands and Surveys and Meteorological Service)

scientific inputs, has raised farm output *per man* to a level as high as anywhere in the world. The people enjoy an excellent nutritional level and have a high standard of living from the purely material viewpoint (Beckerman Index, 58·6, p. 77). They are highly literate, very democratic and, as a result of excellent social welfare services, look forward to the longest life expectancy in the world.

Agrarian Structure

Of 66·4m acres, about two-thirds are actually farmed. Of the unfarmed land, 4·5m acres are reserved for the Maoris and the remainder is mountain, government-owned forest, or unsettled scrub and tussock grassland. Much of the unsettled land is gradually being brought into use by the Government's Lands and Surveys Department, which is responsible for determining the problems of particular areas (e.g. mineral deficiencies), alleviating these where possible, and then handing over already viable units to farmers. Rather less than half of New Zealand's 27,000 farms are held on perpetually renewable leaseholds, whilst the remainder are mostly held freehold, often with a mortgage. The New Zealand farmer has little prejudice against borrowing money. Long-term credit for buying land is available from Government bodies and life insurance companies, and short-term credit is available from banks or 'stock and station agents'.

About an eighth of the population lives on the farms, which are usually run as family units with perhaps a 'boy' to help. These farm people are sober, very hard working and mostly Protestants. Because of their limited contacts with the rest of the world their outlook is inclined to be conservative, yet they have a keen appetite for technical progress. Many of the ideas on intensive grassland and livestock management now at work in north-west Europe and parts of North America originated in New Zealand. In most areas services to the farming community, including education, are excellent. An intricate system of roads, railways and coastal seaways provides adequate transport in the intensively farmed areas such as Taranaki and Waikato. But the rough terrain of the hilly 'sheep' country makes access to the large scattered farms more difficult. The rural electricity and telephone services are good and domestic water supplies, though usually private, are adequate. Rural housing standards are good; simple but uncrowded. Most kitchens have electric cookers and refrigerators. Marketing is highly organised. Butter and cheese factories are largely co-operative, fat lamb freezing works are mainly in commercial hands, and stock and station agents handle much of the wool. The quality, and sometimes the quantity, of farm products for export is controlled by Government and Statutory Boards, but attempts at fixing support prices and the like are meeting with difficulties. Research is advanced and centres such as Lincoln College in the South Island, and Massey University, the Ruakura Research Centre and the Grassland Division of the Department of Scientific and Industrial Research at Palmerston North in the North Island have international reputations. Technical services such as milk and butter-fat recording, seed

FARMING SYSTEMS OF THE WORLD

certification and plant and animal health control are good. Advisory work by Government and Co-operatives, although perhaps not notable in the past, has greatly improved in recent years. In some areas groups of farmers also employ their own advisers (as in Australia, p. 173), whilst stock and station agents frequently act as unofficial consultants.

Technical aptitude for labour saving is seen in the stress placed on the efficient layout of farm and buildings. The land is divided into blocks or paddocks of convenient size and access from these to the buildings is usually good. Round the buildings are pens, races and crushes for efficient handling of livestock. The buildings themselves are few as the mild climate makes it unnecessary to house stock during the winter. On dairy farms, there is usually an open-fronted milking shed (abreast or herring-bone), an implement shed and a hay barn: on sheep farms a large wool shed with a shearing stand and an implement shed. Labour-saving devices such as milking and shearing machines, electric motors for pumping cooling water, etc. are used wherever possible. Tractors and Land Rovers are ubiquitous and in the extensive farming areas aircraft are widely used for top-dressing, re-seeding, spraying or dropping fencing materials. In brief, because of the labour shortage and high wages, the New Zealand farmer has to pay much greater attention to his output *per man* than many of his European counterparts.

Climate, Vegetation and Soil

New Zealand's latitude suggests a warm temperate climate, but her maritime position in the path of the westerly winds and her high mean altitude combine to make the climate wetter and cooler than expected. Air mixing over the mountain backbone plus high winds reduce cloud cover, so that high levels of sunshine are general. This humid, sunny climate with a thermal growing season of at least 10 months is ideal for crop growth. Except in the drier eastern areas where late summer droughts and hot winds can be troublesome, it is particularly suited to grassland production. Although seasonal grass peaks are still marked, the long growing season for pasture gives the New Zealander a great advantage over the United Kingdom or Danish farmer. The very noticeable winds in New Zealand, although not particularly cold, do present a livestock protection problem. In Scotland, Blaxter (1964) reported that the maintenance requirement of outdoor stock was 10–15% greater than that of stock in indoor calorimeters. But it has been suggested that comparable New Zealand figures are anything up to 50–80%. Windchill, particularly in association with rain-chill (Chap. 1.2. p. 70), may be responsible for their apparently higher maintenance needs.

The location of New Zealand, stretching from the sub-tropical north in the Auckland peninsula where citrus fruits are grown, to the sterner, colder climate of the extreme south, permits a diversity of farming. (The names of the towns in the south, e.g. Invercargill, Dunedin, and the fact that swedes and oats are grown, give clues not only to the Scottish origins of some of the settlers, but to the climate and country.) The 'natural' vegetation shows a similar range. To the west of the mountainous spine, the wet climate supported evergreen rain forest, forming leached podsolised soils. In the drier east and in the mountains, tussock grassland, with lighter coloured soils, or forest was probably the original climax. Certainly the early Maoris fed, until they exterminated them, on the large Moa birds which grazed these lands (Duft 1957) just as the buffalo grazed the North American plains and the kangaroo their Australian equivalent. The main soil problems are deficiencies of lime, phosphate and trace elements such as molybdenum, copper, cobalt, boron and magnesium. At one time, raising the pH with lime was chiefly relied on for combating the trace element problem, but nowadays lime is used rather more judiciously ($1\frac{3}{4}$m acres are dressed annually) and trace elements are frequently added directly in 'fortified' phosphate fertilisers. These phosphate fertilisers, half of which are applied by aircraft, are the main key to New Zealand farming and they constitute about 80% of all the fertiliser used. They are particularly important for stimulating the growth of clover to provide the legume nitrogen on which the New Zealand farmer relies so heavily. The rich perennial clover/perennial rye-grass sward allows a high stocking rate which in turn, ensures a short sward to preserve clover, maintain digestibility and provide a high organic matter return and hence fertility for the rye-grass. This fertility cycle has become a classic case of scientific farm ecology comparable to the Australian wheat/annual clover and annual rye-grass/sheep ley story (p. 183).

Land Use, Crops and Grass

Of the 44m acres of farmed land in New Zealand, 20m acres consist of sown pastures and 'permanently improved' grassland (i.e. 'natural' grassland improved by oversowing with clovers and grasses). These 20m acres of intensive grassland are economically the most important and technically the most interesting: their management is outlined below. 'Unimproved' land, which includes large areas in the central mountains of the North Island and in the South Alps, is used largely for extensive grassland (sheep ranching) producing wool and store sheep. There is little cattle ranching in New Zealand although some animals are kept on the better sheep land to 'top' the pastures.

Only 1·1m acres (Table 2.4.1) are in tillage, two-thirds of which are green fodder or roots (oats, swedes, turnips, rape and kale). These are fed mainly to sheep but are also important for winter feed to dairy stock on farms producing liquid milk all the year round for town supply. The remaining tillage acreage is largely in cereal grains and peas. These are grown mainly on the Canterbury Plains to the east of the mountains in the South Island (i.e. in the rainshadow of the prevailing north-westerly winds) under a system of intensive alternating farming in which there is usually a longish ley of 6–8 years then 2–3 years tillage. Wheat is the most important cereal and recently production has increased to the point where New Zealand is almost self-sufficient. A little maize is grown but this is almost entirely confined to the east coast of the North Island. Potatoes are a fairly important crop, grown in the Canterbury Plains, around Wellington and in Central Auckland, this last area being particularly suitable for 'earlies'. Some potatoes — 'Pacific Smalls' — are exported to the Pacific Islands and the New Zealand Potato Board would like to send some to Australia if they could meet the quarantine regulations. Seed crops of grain and clover are also important, mainly in the Canterbury Plains. New Zealand also has market gardens (13,000 acres) and intensive commercial orchards (tree fruits) (18,000 acres). Of the market gardens, in the North Island, the Pukekohe district supplies Auckland with a wide variety of vegetables and is particularly noted for onion-growing. The Otaki and Levin districts supply Wellington; Ohakune supplies a wide area with cabbages and broccoli and Hastings has become a centre for canning and freezing. In the South Island the market gardening is concentrated round the cities; for example on the rich soils round Christchurch, and to the south of Oamaru and Dunedin. Nelson is also an important area, much of the produce from here being shipped across to Wellington. Tree fruits are largely confined to areas with particularly suitable climate and soil. The Nelson and Hawke's Bay districts are notable for peaches, apples and pears, a high percentage of the two latter fruits going for export. Central Otago is particularly suited to apricots, and Kerikeri (North Island), Tauranga (Bay of Plenty) and Gisborne to citrus.

Finally, land use in New Zealand cannot be discussed without referring to forest reversion and invader pests. Gorse (*Ulex europaeus*), blackberry (*Rubus fruticosus*) and ragwort (*Senecio jacobaea*), all introduced from Britain, are major weed-infestations, which are liable to take over grassland and tussock country unless controlled by physical means (e.g. burning) or by herbicides. (The introduction of gorse has produced the same problem in the coastal areas of Oregon (U.S.A.), though there, as in New Zealand, it adds considerably to the beauty of the countryside.) Gorse and blackberry often 'move in' on land which has been cleared of the bushy weed Manuka (*Leptospermum scoparium*). But perhaps a greater problem is that of reversion to forest or 'second growth' on established pastures. Avoidance of reversion demands, in some areas, great attention to detailed grassland management (p. 202 below). Rabbits, which were also introduced from Great Britain, were a major problem until very recently, not least because of the erosion danger caused by their sward destruction in hilly regions. Attempts were made to introduce myxomatosis but these failed because of the lack of a suitable flea or mosquito to act as a vector. However, the introduction of the Rabbit Destruction Act, which encouraged the large-scale use of poison, has proved very successful and rabbits are no longer considered a serious problem.

Livestock

Nine-tenths of the 7·3m cattle and more than half the 54m sheep are in the North Island (Table 2.4.1). The majority of the cattle, and certainly of the dairy cows, are on the wetter western side of a line drawn along the mountain backbone of New Zealand (Fig. 2.4.6). The majority of the sheep are on the eastern, drier side of this line. The emphasis on cattle on the flat terrain west of the central mountains in the North Island is partly associated with the greater capacity of cattle to produce and remain healthy under wet conditions, whereas the breeds of sheep used in New Zealand are less tolerant than say the Scottish Blackface or Welsh breeds. But the shorter lactation period of the ewe also make it better fitted to the drier east with its higher C.V. of annual precipitation and its tendency to late summer drought. The lactating cow, milking perhaps 7 to 9 months, cannot withstand even a few weeks under-nutrition without the remainder of her lactation being seriously prejudiced (Blaxter 1956).

The *dairy cows* are mainly Jerseys, which are well adapted to warm conditions (compare the dominance of Friesians in the northern U.S.A. with the former frequency of Jerseys in the Southern States) and whose high butter-fat milk is particularly well suited to butter if not to cheese production. However, Friesians are popular on farms selling liquid milk all the year round to urban areas. On the farms producing milk for manufacture, calving occurs in the spring and starts at 2 years old (younger than in the United Kingdom). The heifer calves are often weaned early and grazed in front of the milch cows. Grazing, with hay or grass silage when the cows are dry in winter, is usually the only feed. Annual yields per cow at 8,000 lb of milk from grassland alone are of the same order as, for example, in the intensive, tillage based, high labour cost dairying in Denmark. But in New Zealand the emphasis is on yield of butter-fat per man and per acre. For the latter 300–320 lb per acre is now fairly common and some farms have reached the 500 lb/acre level. (At the Ruakura Research Centre a record figure of 626 lb/acre has been achieved purely from grass (Campbell 1966).) Output per man may reach the remarkably high level of 35,000 lb butter-fat per year. About 90% of the milk output is manufactured, mainly into butter (80%), cheese (15%) and whole milk products. Also some of the butter-milk and skim-milk is dried to be used for calf rearing. *Beef cattle* are mainly of Aberdeen Angus origin though there are some Herefords; they are largely used, as noted, as 'mowing machines' to top pastures to keep them at sheeplength and at high digestibility for lamb production.

Cattle health is, on the whole, good. There are no epizootics, such as foot and mouth, thanks to good veterinary controls on imported stock. A tuber-culosis eradication scheme is nearly complete and a brucellosis-control eradication scheme introduced in 1966 now requires both beef and dairy heifers to be vaccinated. Mineral deficiency diseases are now less common but New Zealand still has her share of mastitis, sterility and abortion problems.

On the extensive grassland on the high country in the South Island, where the main product is fine- to medium-grade wool, the *sheep* are the hardy and agile Merinos (60+ wool count) and Merino crosses (56–58 count). The rough-terrained high country can be used by sheep but not by milk cows which are not agile, cannot survive the climate and are nutritionally too demanding. This extensive sheep system allegedly avoids the metabolic strain of intensive systems. Long-wool British breeds (Border Leicesters and Lincolns) and a local derivative, the Corriedale, came with the introduction of refrigeration in 1882 and are still important. But another long-wool, from the flat alluvial marshlands of east Kent, England, viz. the Romney, with its well-spaced grazing habits, is now, together with Romney crosses, the basis of the fat lamb industry in the North Island. (This breed is also locally important in some of the higher rainfall areas in the south of South Island.) Southdown rams on Romney-type ewes give excellent small fat lamb at 25 to 30 lb carcass-weight well suited to the U.K. market. These are slaughtered and frozen at the well-distributed privately-owned or co-operative 'freezing works'. In the high country, wethers usually survive 7–8 annual wool clips of about 10 lb each, but lambing percentage is usually well below 100%. In the lowlands this may be a little higher, at 120%, but this rather low percentage can be offset by high stocking rates of 5 ewes to the acre; yields are up to 300 lb carcass-weight per acre. Ewe clips on lowland flocks average about 10–11 lb, usually of 40's to 50's count. Though dipping is compulsory, there is no sheep scab (*Psoroptes ovis*) in New Zealand and, as in the United Kingdom, the main disease problems (apart from deficiency disease) are those of salmonella and clostridial infections, pregnancy toxaemia, internal and external parasites and footrot.

Pigs were of limited importance in the past on manufacturing milk dairy farms where they were fed on skim-milk or whey bought back from the factories. But they have never been very popular, partly because of the high cost of cereals and, more recently, because of bulk collection of milk. In some cases interest in pigs has revived and some large-scale units have developed where up to 300 sows are grazed outside and the fattening is done intensively.

Grassland

Particularly in the North Island, grassland management has been raised to a level that is world famous. It has been fully analysed and described (Brougham 1959, 1966; Campbell 1966; Curry 1958, 1962, 1963; Hamill 1958; Hamilton 1944, 1954; Hendrie 1952, 1954; Hulton 1954) and can only be summarised here.

Skilled sward management and the well-integrated fertility cycle (noted above) are the core of New Zealand's economy, which is founded on a climate that usually provides rain and sunshine when they are most needed. Rain is reliable at least in the west and north of the North Island, viz. a C.V. of annual precipitation 12–14% rising to 15–20% on the eastern side of the country. But at high stocking rates, e.g. five ewes or a cow to the acre, minor short-term falls in A–E–T can substantially affect product output, especially as the grassland (Chain C) is the most weather sensitive of the four food chains. To be on the safe side farmers tend to stock at lower rates than the weather probabilities justify (Curry 1963). Most of the 'sown' grass acreage is in permanent pasture or long leys, based on locally evolved ecotypes of seed introduced from England, viz. perennial rye-grass (*Lolium perenne*) and perennial white clover (*Trifolium repens*) mixtures (called 'rye/white') and some timothy (*Phleum pratense*). Some short leys are grown, for which H I (short rotation) rye-grass (*L. perenne* × *multiflorum*) and red clover (*T. pratense*) are often used.

On lowland farms dairy pastures are usually paddocks of about 5 acres, which are well fenced and grazed rotationally, often in strips with electric fences in the early part of the season. Thus, cattle only remain on the same strip for one or two days. But, as in Britain, interest in paddock grazing is increasing. The spring surplus is either saved for seed or shut up for hay or silage, which is made in November, for feeding in the winter when the cows are dry, or in a late summer drought when the cows are still milking. New Zealand farmers believe in keeping grass very short for sheep and, on fat lamb farms, cattle are used for topping the grass and thus maintaining digestibility. This helps to check invasion by blackberry and other shrub weeds and as suggested by Hendrie (1952, 1954) it may reduce the incidence of worms in sheep. Recently a good deal of pasture has been renovated by burning off the old sward with paraquat and then disc-drilling new seed. In higher country, rough grassland is improved by the use of aerial top-dressing and aerial re-seeding, as indicated earlier, and by feeding hay from good genotypes to stock on poor pastures and letting the seed in the dung germinate and become established. Or a small area is sown with a good mixture of grasses and allowed to seed; then the stock are fed by day on the good grass and by night on the poor pasture — an effective and labour-saving renovating technique.

North Island Transect (*Table 2.4.5 and Figs. 2.4.7, 3.1.4*)

TARANAKI (*Sector (a)*)

This area lies to the west of the central volcanic plateau and consists of plains, hills to 2,000 ft (much dissected by rivers) and the extinct volcano, Mount Egmont (8,260 ft). Much of the hill country has very steep slopes with shallow, unstable soil and has remained undeveloped. The soils (Fig. 2.4.7) are (*a*) yellow-brown loams and clay loams derived from volcanic ash and (*b*) loamy yellow-brown earths derived from siliceous parent material and formed under a mild, humid or superhumid climate and forest vegetation. Both respond to phosphate, lime and, in some cases, potash or molybdenum; they are easily worked and make good pastures. However, on slopes overgrazing must be avoided if water erosion is to be prevented. Aerial top-dressing, etc. has extended the area in intensive grassland, but, for the control of rivers and protection of the lower country, much of eastern Taranaki must be left as

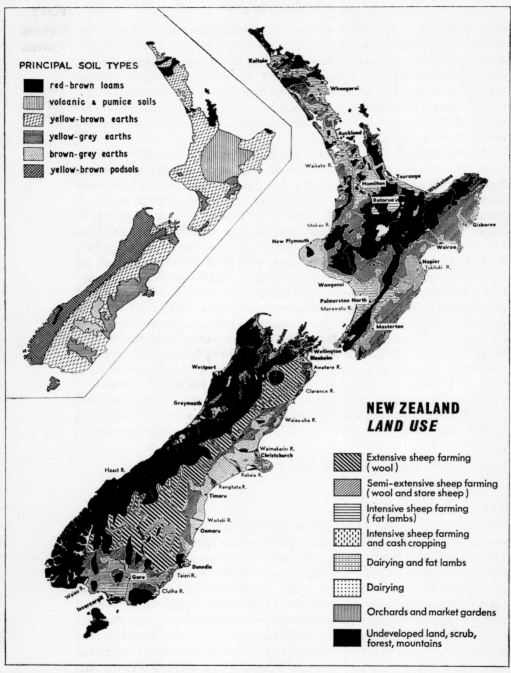

PRINCIPAL SOIL TYPES

- ◼ red-brown loams
- ▥ volcanic & pumice soils
- ▨ yellow-brown earths
- ▦ yellow-grey earths
- ▧ brown-grey earths
- ▩ yellow-brown podsols

NEW ZEALAND
LAND USE

- ◩ Extensive sheep farming (wool)
- ◪ Semi-extensive sheep farming (wool and store sheep)
- ▤ Intensive sheep farming (fat lambs)
- ▦ Intensive sheep farming and cash cropping
- ░ Dairying and fat lambs
- ⋮ Dairying
- ▥ Orchards and market gardens
- ◼ Undeveloped land, scrub, forest, mountains

Fig. 2.4.7 Farming Systems and Principal Soil Types. (From N.Z. Dept. Agric. and Geol. Surveys)

204

forest. The natural vegetation is mainly podocarp/mixed hardwood forest with softwood timber trees of the Podocarpaceae (rimu (*Dacrydium cupressinum*) miro (*Podocarpus ferrugineus*) etc.) and hardwood species of tropical and sub-tropical families, but a narrow strip on the coast is manuka scrub (*Leptospermum* spp) and fernland (mainly bracken fern, *Pteridium aquilinum*).

The climate is mild, sunny and wet with plentiful rainfall and frequent strong winds. The climatic changes between seasons are small but the daily weather variations are considerable. The mean annual temperature at New Plymouth (Table 2.4.4) is 56·5°F and diurnal range is low on the coast but rather greater inland. However, like most of New Zealand, this district is occasionally subject to very hot or very cold weather. Snow is rare but ground frosts occur about twenty times a year. The mean annual precipitation ranges between 60 and 100 in and is evenly spread through the year, with a slight peak in the winter and a slight trough in the summer. The coefficient of variation of annual precipitation is low at 12–14%. This is due to the influence of westerlies and north-westerlies and the very regular cyclic quality of much of New Zealand's weather. Cloud cover, as noted p. 198, does not remain long. Thus rainfall, although intense, is generally short-lived and there are fewer wet days for the same annual precipitation than in, for example, the United Kingdom. As in most areas of New Zealand the amount of sunshine per unit of precipitation is, therefore, high. Thornthwaite's figures (Table 2.4.4) show that at New Plymouth the mean A–E–T in the T–G–S is good at 27·6 in and is equal to the P–E–T, emphasising the fact that there is no water deficit. In fact, mean annual precipitation exceeds mean A–E–T by 20 in or more.

Farming systems (Fig. 2.4.7). There are three types of farming in Taranaki: (i) intensive grassland dairying; (ii) intensive grassland fat lamb; (iii) semi-extensive grassland sheep farming.

(i) *Intensive Grassland Dairying*

About 90% of New Zealand's dairy farming is in the North Island. It is centred round Auckland, but the Taranaki area is still important because it now produces over half the cheese for export. Dairy farms range from 100–200 acres, with stocking rates up to 1 cow per acre or even higher. With the advent of the herring-bone parlour it is not uncommon for one man to handle 100 cows, and these he will milk twice daily at the rate of about 60 per hour. Dairy farming is thus hard work, at least for the 10 months or so of the year that the cows are in milk. The farms are divided into 5–10 acre paddocks separated by wire fences, 'quick' hedges (usually barberry or boxthorn) or sometimes pines or macrocarpa which also act as wind-breaks. There are usually concrete 'races' from the paddock to prevent poaching ('pugging', 'treading'). Poaching of paddocks themselves is a very serious problem and in the wetter areas concrete or sawdust 'pads', where the cows may be kept in wet weather, are becoming common.

Table 2.4.4

Climatic Data — New Zealand

Station	Bright Sunshine hours/day	Thermal Growing Season length in months	Thermal Growing Season Mean Temp °F	Mean Months >72½°F	Thermal Growing Season Mean P-E-T (inches)	Mean A-E-T (inches)	Mean Annual Precipitation (inches)	Moisture Index Im	Annual Precipitation Coeff. Var. (C.V.)	Max. recorded rain/day inches (month of occurrence)
Auckland	5·6	12	59·0	0	28·8	28·8	49·9	73 humid	12-14%	6·4 (February)
New Plymouth	6·1	12	56·5	0	27·6	27·6	60·5	119 per humid	12-14%	7·3 (February)
Karioi	4·7	9	51·8	0	21·0	21·0	48·2	103 per humid	12-14%	8·0 (March)
Napier (Hastings)	6·6	12	57·5	0	27·9	26·5	31·8	14 moist	20-24%	8·0 (March)
Hokitika	5·2	12	53·0	0	25·7	25·7	114·4	344 per humid	10-12%	9·2 (February)
Lake Coleridge	5·5	9	53·6	0	23·3	22·5	31·6	24 humid	14%	(Arthur's Pass) 12·7 (November)
Ashburton	5·0	10	53·8	0	24·1	23·6	29·4	16 moist sub-humid	16%	
Christchurch	5·4	12	52·5	0	26·0	26·0	25·8	−8 dry sub-humid	16%	4·7 (April)

The New Zealand dairy farmers in this area seldom grow tillage cash crops and only the very intensively stocked farms feed any concentrates. However, a few forage or root crops may be grown, particularly on 'town supply' farms, because of greater winter feed requirements; these are grown after ploughing old leys prior to re-seeding. Up to one-fifth of the farm may be shut up in spring for 'saving' as hay or silage, although the latter is losing popularity. The rest of the farm is grazed rotationally, nowadays more often on a paddock system than strip-grazed with electric fences. Some of the paddocks may be shut up in autumn to preserve end of season growth for winter or early spring grazing. The grass receives about 550–700 lb/acre of superphosphate per year, or even more on the very intensive farms, and this is usually split between spring and autumn applications.

Winter housing is not usually needed, but occasionally some farmers fit jute jackets to their stock to keep out the wind. About one-fifth of the herd is replaced each year and the culled cows are sold in autumn as the herd is drying off. Except on the 'town supply' farms, where calving has to be both spring and autumn, calving has traditionally been concentrated in August (late winter). However, it has been found that September–October (early spring) calving with peak milk output in November, does not affect yields and helps to avoid expensive late winter and early spring feeding. New-born calves may be fed on whole milk for 10 days, then on skimmed milk and cereal meal, and are weaned at 2–2½ months. Sometimes they are grazed ahead of the cows so that they receive the best grass. Below, an example 'town supply' dairy farm is briefly described. This type is less common than that producing milk for manufacture, but most aspects of general management are identical apart from the need to produce milk all year round.

A 'town supply' dairy farm, Palmerston North (1962). This farm of 94 acres lies south of the Transect and is not in fact in Taranaki. The land is flat and rather sandy and adjoins a river, but no irrigation is practised. The farm carries 94 milking Friesian and Jersey cows, which are managed by the owner and a boy. Eighteen young stock are also kept, and two bulls; no A.I. is used. The grass is 'rye-white', which is paddock or strip-grazed and also produces some hay or silage.

The equipment is simple and includes an open bail milk shed, piped water to all fields, 2 tractors and simple machinery; the owner is thinking of buying a forage harvester. The milk output is about 25,000 lb per day in summer, falling to about 16,000–17,000 in winter. Output per acre, therefore, runs at about 8,000–10,000 lb per year, a level which is not uncommon from grass alone.

(ii) *Intensive Fat Lamb from Grassland*

Fat lamb farms are interspersed with dairy farms in Taranaki and are usually a little larger than the latter. The pastures are mostly sown with 'rye-white' but may include crested dog's-tail (*Cynosurus cristatus*) cock's-foot

H

(*Dactylis glomerata*) or timothy (*Phleum pratense*). Special-purpose pastures
for winter and early spring grazing may be sown to short-rotation rye-grass
(*Lolium multiflorum* var), Prairie grass (*Bromus unioloides*) and white clover.
The farms are divided into paddocks of 10–20 acres which are sometimes
rotationally'grazed but are more often 'set-stocked'. Stocking is at a rate of 4–8
ewes per acre, and a few wether sheep or beef cattle are also bought in for
fattening. The pastures receive about 100–220 lb/acre of superphosphate per
year and produce 6,000–10,000 lb dry matter per acre, yielding (at these stocking rates) about 250 lb carcass meat and 40–50 lb wool per acre. The flocks
average about 1,000 pure or crossbred Romney ewes, which are good grazers,
hardy and resistant to footrot, but have a low lambing percentage (80 to
120%). They are put to Southdown or Romney rams, and give lambs with a
very good carcass. Ewes are bought from hill farms as two-tooths, or as four-
or five-year-old culled ewes which are kept for one or two lamb crops and then
sold off fat in midsummer before feed is in short supply. The ewes arrive in
February, are dipped, fed well and put to the ram in March. During April,
May and June part of the farm is shut up to provide feed for the start of lambing which begins in August (i.e. before the spring growth has got really under
way) and goes on to the end of September. Ewes, wethers and sometimes
lambs are shorn in November. On the best farms lambs may reach a carcass-
weight of 30–32 lb in as little as 8–10 weeks. As many as possible are sold
straight off the ewes as prime milk-fattened lambs; the remainder are usually
weaned in about December and fattened, together with the culled ewes, on a
catch crop such as rape (*Brassica napus*).

(iii) *Semi-extensive Grassland Sheep Farming*

These farms provide store sheep and ewes for the lowland fattening farms.
Such farms are situated on low hills in both the North and South Islands and
in this sector are on the land east of the Taranaki lowlands. Here the natural
vegetation was forest which was cleared by felling and then burning although
large trees were left standing. Rye-grass, cock's-foot, timothy and clover seeds
were sown by hand in the ashes and, ideally, established a closed community
which excluded secondary growth. But, in practice the burn was often not hot
enough to kill the seeds of the forest trees, shrubs and ferns and these soon
re-established themselves. This has resulted in large tracts being covered with
shrubs such as blackberry, hard fern (*Paesia scaberula*), manuka, pinpiri
(*Acaena* spp), gorse and tutsan (*Hypericum androsaemum*). This country can
only be fully recovered by further burning and re-seeding, often with 'brown-
top'(*Agrostis tenuis*), *Lotus major*, Yorkshire fog (*Holcus lanatus*),sweet vernal
grass (*Anthoxanthum odoratum*) and some clovers, perennial rye-grass and
Danthonia pilosa.

Aerial top-dressing with superphosphate has increased the carrying capacity of these hills to one or two sheep per acre. However, skilled management
is needed if reversion (secondary growth) is to be checked and for this, inter-

mittent grazing is much better than continuous grazing, especially if no cattle are kept as 'toppers'. The sheep are mainly crossbreds, with breeding ewes on the better land and wethers, dry sheep and cattle on the poorer sections. The latter, mainly Aberdeen Angus, Hereford or crosses thereof, are sold as stores (feeder cattle) to lowland farms. Usually the only animals bought in from outside are rams, selected carefully for their constitution and wool. The output of these farms is therefore wool, store cattle, wethers, store lambs, cast-forage ewes and young ewes culled on wool quality or lambing performance. Wool production is about 17 to 25 lb per acre. Flocks are usually 1,500 to 2,500 strong.

CENTRAL NORTH ISLAND (*Sector (b)*)

This sector is summarised in Table 2.4.5. Rugged relief, cool summers and frost risks restrict farming, but some easily cleared former forest land and the tussock grassland now support semi-extensive livestock farming (as in eastern Taranaki), though cock's-foot and red clover are more noticeable in the swards than in the warmer area.

HAWKES BAY (*Sector (c)*)

The region has mountains on the west and then hills sloping gently down to the alluvial coastal plains. The volcanic yellow-brown pumice soils of the mountains give way to yellow-grey earths on the coastal plain. These are seasonally drier than similar soils in other parts of New Zealand and are therefore less leached and less acid. They are often heavy but, if drained and kept clear of rushes (*Juncaceae*) and buttercups (*Ranunculus* spp) they make good fattening pasture. The natural vegetation of the whole region was scrub (mainly manuka, bracken fern, New Zealand flax (*Phormium tenax*) and sphagnum moss). Firing, and trampling by livestock, allowed the development of an open turf of *Danthonia pilosa*, sweet vernal grass (*Anthoxanthum odoratum*) and New Zealand rice-grass (*Microlaena stipoides*), but the dry summers precluded a full turf cover and permitted secondary growth. The climate of Hawkes Bay is complex. On the east side of the central mountains the winds contain less moisture and are associated with temperature extremes, moisture deficiencies and unreliability; the country's highest coefficient of variation of annual precipitation (24%) is recorded at Sherendon, just west of Napier in this sector. Rainfall is also more intense than in other parts of the country, falling in short, sharp showers; so on average there are only 115 rainy days a year and it is one of the sunniest areas in the country (Table 2.4.4). The clear skies result in relatively high diurnal ranges of temperature, a greater seasonal temperature range (18·3°F) and more frosts than on the west coast. Mean annual precipitation of 30–40 in exceeds the P–E–T of 26–28 in and the A–E–T of 25–27 in in the 12 months T–G–S, but as the C.V. of precipitation is high, late summer moisture deficits are frequent.

Table 2.4.

New Zealand Transects at Lat. 39°S. and Lat. 42°S. ←→ 44°S.

	(a) Taranaki	(b) Central Mountains	(c) Hawkes Bay	(d) Westland	(e) Southern Alps	(f) Canterbury
1. Sector						
2. Name	N. Is.	N. Is.	N. Is.	S. Is.	S. Is.	S. Is.
3. From	New Plymouth	Stratford	Crownthorpe	Greymouth	Lake Brunner	Oxford
4. To	Stratford	Crownthorpe	Napier	Lake Brunner	Oxford	Christchurch
5. Miles	40	100	30	20	60	40
6. Climate-type(s)	Cool Humid Temperate	Cool Humid Temperate	Cool Humid Temperate	Cool Humid Temperate	Cold Temperate	Cool Temperate
Thermal Growing Season:						
7a—months at or > 42½°F	12	9	12	12	0–9	10
7b—months at or > 72½°F	0	0	0	0	0	0
8a—P-E-T (in) in T-G-S	26–28	20–22	26–28	25–26	21–23	24–26
8b—A-E-T (in) in T-G-S	26–28	20–22	25–27	25–26	20–22	23–25
9. Growth limited by	—	Cold	Some drought	—	Cold	Drought
Precipitation						
10.—mean in (annual)	40–80	40–100	30–40	100–120	Up to 200	20–35
11.—C.V. % of Precipitation (annual)	12–14	14	20–24	10–15	15–20	15–20
12. Special Climate Features	Sunny	—	Very sunny, dry winds	—	—	Hot, dry winds, some frost
13. Weather Reliability	3	?	3	3	0	3
Soil:						
14.—Zone/Group	Podsolic/Latosolic	Podsolic/Latosolic	Podsolic/Latosolic	Mountain Complex	Mountain Complex	Podsolic
15.—pH	Low		Medium	Low		Medium
16.—Special Features	Volcanic leached	Mixed forest tussock grassland	Scrub and fern	Heavily leached	Forest, alpine and tussock grassland	Tussock grassland
17. Vegetation	Mixed forest	Mountains	Rolling plains	Mixed forest	Mountains	
18. Relief	Plains and hills	0		Coastal plains	0	Flat plains
19. Market Access	1	Extensive grassland	1	1	Extensive grassland	1
20. Farming Systems	Intensive grassland		Intensive and extensive grassland and alternating	Intensive grassland		Intensive alternating
21. Major Enterprises	Milk from grass. Fat lamb	Semi-extensive sheep in suitable areas (stores, wool)	Intensive fat lamb. Some horticulture, some grass, etc. seeds	Milk from grass in suitable areas	Extensive sheep for wool and stores in suitable areas	Intensive fat lamb. Grass seeds and other cash crops. Some intensive milk from grass with irrigation

210

Farming systems (Fig. 2.4.7). The plains of Hawkes Bay are devoted to intensive fat lamb production, whilst the surrounding hills are covered with semi-extensive sheep farms.

(i) *Fat Lamb Grassland Farms* (Plains)

These farms differ from those in the Taranaki in that they also produce grass and clover seed crops, and that lambing is about a month earlier so that the lambs can mostly be sold off before grass shortages caused by the late summer moisture deficits (Jan.–Feb.). Lambs which do not go fat off the ewes by December are finished with the culled ewes on rape or kale ('chou moellier'), which may also be grazed by the breeding ewes in the autumn. After these forage crops, a rye-grass/white clover sward is sown which gives a crop of rye-grass seed in the first harvest year and a crop of clover seed in the second. It may then be grazed for a further 8 years or so before being ploughed out and sown to winter forage for the ewes, and is probably followed by summer forage again.

A fat lamb and wheat farm near Holcombe, North Island. This farm of 600 acres of heavy clay in rolling country with some steep-sided but drilled and harvested slopes, averages about 38 in precipitation annually. Even so, it is often short of moisture in late summer when, however, the weather is good for grass seed and wheat harvesting. The farming system is very similar to that outlined above, except that after ploughing up the old leys (about 60–70 acres per year) and taking the winter forage crops, a spring wheat crop is grown. This yields up to 2,800 lb per acre. 2,000 to 2,500 Romney or Border Leicester ewes are carried, which are put to Romney or Southdown rams. The ewes and lambs are usually set stocked on large paddocks until weaning. Lambing is intentionally spread over several months and yields 100% lambs sold fat at 16 weeks or less of age. A lambing percentage of over 100% would mean more lambs on the land in the dryish late summer and hence finishing problems as well as a work clash with grass seed and cereal harvest and other late summer work on crops. The ewes are occasionally drenched for worm control but disease problems do not seem to be marked. Superphosphate is applied each year, usually all over the farm. Lime and potash are applied occasionally. The equipment is shared with the owner's sons (who farm 700 acres near by) and includes a crawler tractor for heavy work, 3 other tractors, a 'header' combine harvester and an excavator for drains which are laid when time permits. There is no hired labour.

(ii) *Semi-extensive Grassland Sheep Farms* (Hills)

These farms are similar to those described for Taranaki, but, owing to the intensive rainfall and the weak-structured rock, soil erosion problems are more serious, particularly where overgrazing occurs. In many areas *Danthonia* sward has been improved to dominantly rye-grass and clover (subterranean

South Island Transect (*Table 2.4.5 and Figs. 2.4.7, 3.1.4*)

WESTLAND (*Sector* (*d*))

   This is a hilly region of very high rainfall (100–120 in) and infertile podsolised soils; much of the podocarp/hardwood forest cannot be cleared because
of the erosion danger. Even where it can be cleared, burnt and seeded, it
frequently reverts to useless scrub because the high rainfall precludes a sufficiently hot burn. However, on rolling country near Lake Brunner and on
swamp soils near Hokitika, pasture has been established and some butter-
producing dairy farms, and a few beef cattle, are found. The dairy farms are
similar to those described in the Taranaki section but are much less intensive,
stocking only about 1 cow to 3 acres (Fig. 2.4.7).

SOUTHERN ALPS (*Sector* (*e*))

   The transect crosses the mountains at 8,000 ft but further south altitude
rises to 12,000 ft (Mount Cook). The 50-mile wide range rises more steeply
on the west than on the east and is deeply dissected by wide and steep-sided
valleys; many of them are blocked by glacial moraine and are occupied by
lakes. In the wetter parts the soils are podsolised yellow-brown earths, but in
the drier, high country they are moderately acid with brown loamy top-soils.
They have now been improved by aerial top-dressing with superphosphate
fortified with molybdenum and, in some cases, sulphur. The higher tussock
grassland (over 3,000 ft) is dominated by snow grass (*Danthonia flavescens*)
and the lower tussocks by *Poa* spp and *Festuca novae zelandiae*. On the wetter
western flanks, the forest is podocarp/mixed hardwood, but on the eastern
slopes it is beech (*Nothofagus* spp) and beech/podocarp. Tussock grassland
occurs in areas with 80 in annual rainfall and forest in places with only 30 in;
there is no clear correlation between the forest/tussock grassland interface
and climatic conditions, possibly because the early moa-hunting Polynesians
(Duff 1957) burnt the forest in search of their game and grass species were
allowed to establish. The climate is Alpine with mean annual precipitation
anything from 30–300 in and there is a permanent snowline. However, at, for
example, Lake Coleridge (Table 2.4.4) which has a precipitation of 32 in, there
is still 5·5 hours of sunshine per day and a 9 month thermal growing season
with an A–E–T of 22·5 in. The farms are extensive sheep 'stations' (Fig. 2.4.7),
covering thousands of acres of the 'natural' tussock grassland and producing
wool and some store sheep, from Merinos and Merino crosses. A typical
station is described below.

   '*Cecil Peak*' *extensive sheep station* (Collins Bay, Queenstown). 'Cecil
Peak' Station is south of the transect line and is a 'high country run' of 34,000
acres, most of which has to be mustered on foot, as it is too steep for horses.

Rising to 6,500 ft, it is snow-capped through the winter months. Precipitation is 28 to 32 in a year and stock watering is no problem. Vegetation at the higher levels is tussock, but lower down, after aerial top-dressing, cock's-foot and red and white clover have been established amongst the tussock grass. Scrubby manuka, thorny matagouri, native fuchsia, sweet-briar and bush lawyer are weeds which are hard to keep in check. Keas (a native bird) attack sheep, pecking into the kidneys, sometimes panicking sheep headlong over a bluff. Other pests are red deer, goats and opossum. The Merino flock is 8,000 strong, i.e. 1 to 4 acres. The old 'culled' sheep are fed as 'dog-tucker' to the 50 or 60 sheep-dogs used at musters. Lambing percentage is about 75%. Snow losses, fly-strike, keas and other hazards, make it difficult to keep up replacements. There are three main musters during the year. The autumn muster brings the flock on to the 'winter country'. The shearing muster starts with rams, wethers and hoggets in September, and the ewes are shorn after lambing which is usually about mid-November. Shearing is by contract and the annual wool clip is about 60–70,000 lb. Later there is a 'tailing' muster when the lambs are tailed and ear-marked. At other times there are the routine jobs of fencing, crutching, foot-rotting and, after the not infrequent sudden heavy snowfalls, finding and digging out snow-bound sheep. January is hay making time and between 3,000 and 4,000 bales of hay are made and stacked. This, together with 30 acres of lucerne, supplies winter feed for the sheep, cattle and horses kept in the valley. There are also 200 breeding cows. The steer calves are kept for 18 months and then sold for fattening. There are 15 to 20 horses born and bred on the property and the same number of 'permanent' dogs. Power is generated by a diesel lighting plant. The station is nearly self-supporting in food with mutton, beef, pork, poultry, milk, turkey, eggs, garden vegetables, etc. Schooling of the children is by correspondence until they go away to boarding school.

CANTERBURY PLAINS (*Sector (f)*)

This is a broad coastal plain, averaging 30 miles wide, formed from material washed down by fast rivers from the Southern Alps. The soils are mostly light, stony, seasonally dry and yellow-grey with gravelly or stony subsoils; there is also some recent alluvium near the river mouths. Both soils respond well to irrigation. In the west the natural vegetation was beech forest; elsewhere it was short *Poa* spp or *Festuca* spp tussock grassland, originally grazed by the Moa birds. The climate is cool humid with mean annual temperatures of 51·7°F and 52·5°F at Ashburton and Christchurch respectively. The region is drier than any sector so far mentioned, firstly, because it lies in the rainshadow of the Southern Alps, and secondly, because of the frequent 'north-westers',—winds which have blown over the Alps and are hot and dry because of adiabatic effects (compare the 'Chinook' off the Rockies in the north-west of the U.S.A.). These 'north-westers' can be very strong to gale force, they lower humidity, raise human irritability and cause very high temperatures; they can, however, bring suitable weather for pasture seed and cereal harvesting. They are often followed by rather unpredictable invasions of cold air which bring rain and contribute to the high diurnal ranges of temperature. Shelter belts of *Cupressus macrocarpa* or *Pinus radiata* are commonly found. Mean annual precipitation is 20–35 in and, like other parts of eastern New Zealand, has a rather higher coefficient of variation (16%). Rainfall is not so intense as elsewhere but may fall for long periods, resulting in flooding; it has a summer maximum and a spring minimum. Lack of local relief causes a rather cloudy climate and rather less light energy receipts than elsewhere. In the T–G–S, A–E–T is less than the P–E–T (Table 2.4.4). In spring, soils are usually at field capacity but later in the season moisture deficits build up, aggravated by the 'north-westers'. Rickard *et al.* (1960, 1961) and others have stressed the irrigation potential of this area, where the soils are not moisture retentive, where snow-melt from the Alps provides a fairly reliable water source without need for storage dams, and where the natural slope of the terrain allows gravimetric transport and application of water.

Farming systems (Fig. 2.4.7). In the mid-19th century the Canterbury Plains, like the other farmed areas of New Zealand, were devoted to wool production. The slump in wool prices caused a change to extensive wheat production, and then the advent of refrigerated shipping resulted in the development of *intensive alternating farms* producing cash crops and fat lambs. This farming system now dominates the Canterbury Plains, although near urban areas '*town supply*' *dairy farms* are also found.

(i) Intensive Alternating Farms

On these farms the emphasis is on the grass and the production of the world famous Canterbury lambs. The leys are rye-grass, cock's-foot and red or white clover; they usually provide a grass seed crop in the first year and a clover seed crop in the second, the threshed stalks making useful hay. After 5–7 years they are ploughed up and followed by 2–3 years of tillage. This includes cash crops of wheat, barley, oats, potatoes and vining peas, and forage crops of rape, kale, swedes, turnips and lucerne for silage or hay. Rainfall variability makes it difficult to judge stocking rates but cash crops provide flexibility; for example, the pasture seeds or oat crops may be grazed if necessary. Lime, at about 2,200 lb to the acre, is often applied to the new leys. All fields receive 110–220 lb per acre of superphosphates, with higher rates for new leys and lucerne. Until recently the superphosphates were fortified with DDT to control grass grubs (*Costelytra zealandica*) and porina caterpillars (*Oxycanus cervinatus*). Now, only prilled DDT* is allowed. This is not proving effective and these pests can be a very serious problem, particularly in dry years when leys can be completely killed out. Cereal yields average about 3,200 lb for wheat, 3,100 lb for barley and 2,400 lb for oats.

The average farm is about 300 acres. This will carry about 1,000 half-bred

---

* In an attempt to reduce the DDT content of the resultant animal products.

or Corriedale ewes, which are either cast-for-age from the hill farms, in which
case they are kept for two seasons before selling off fat, or own-bred replacements. The ewes are put to Romney or Southdown rams in March, for lambing in August and September. The stocking rate is about 5 ewes to the acre
of pasture and the weaned lambing percentage, at over 100%, is high for New
Zealand. Over 50% lambs are sold 'milk-fat' straight off the ewes; the remainder are finished on grass or lucerne, and all are sold at 25–30 lb carcass-
weight. Lamb sales at about 5 to the acre make a total carcass yield of about
150 lb/acre including old ewes. Many farms have some irrigated acreage with
border-dyke or flood irrigation. But irrigation involves heavy capital investment and has, at present, high labour needs. In theory, the reduced drought
risk, which would allow increased stocking rates and give higher crop yields,
should more than offset the greater financial inputs. However, irrigation is
financially doubtful (Stewart 1963). But technological changes that can reduce
labour costs, e.g. automatic irrigation, may, in the future, make irrigation
more attractive in a region which, as noted, is well suited for it. The labour
input on these farms is very small, often only the occupier, or with just one
other man. Such jobs as shearing, sheep crutching, hay-baling and harvesting
are often done on contract, which reduces the need for expensive machines on
each farm. Two example farms are briefly described below.

*An intensive alternating commercial farm at Lincoln College, near Christ-
church* (1962). The annual precipitation is 24 in with a high C.V. The farm is
400 acres of heavy loam, supporting 940 breeding ewes 4–5 years old, 400
crossbred fattening wethers and some Southdown rams. Lambing starts in
early August (averaging 112%) and as many lambs as possible are sold
straight off the ewe at 30 lb carcass-weight before 'the dry' comes in November–December. The ewes are stocked at 8 to the acre on 'rye-white' pasture
until weaning. During the winter months of June to August the sheep get
supplementary feed of lucerne hay, rye-grass, straw and green oats. The rotation is 4–5 year ley; peas or potatoes; wheat; barley, then back to ley. Manurial
policy and the taking of pasture seed crops are as indicated above. Yields are
wheat 2,500 lb/acre; barley 3,100 lb; grass seed 800–1000 lb; white clover 500
lb plus. Machinery is partly shared with other farms. Manpower is a working-
manager, a regular hired man, a student for 5 months, plus contract shearing,
baling, etc. (*Courtesy:* J. D. Stewart, Lincoln College.)

A mixed farm near Ashburton (1962). This farm, of 323 acres with mean 25
in rainfall and long dry periods in summer, is run by the owner with some help
from his wife. The soil is stony silt loam over shingle with large boulders up
to 8 or 10 inches diameter; when these are brought up by the plough they are
forced back by a heavy roller and seem to do little damage to field machinery.
There are also some pockets of clay soils. Rotation is grass; swedes or
linseed; grain crop; new grass (8 to 10 years). 200 acres can be irrigated at
4/6d. (half a dollar) per acre foot for water, but high labour needs preclude
full use of irrigation on a one-man farm. The 1,080 ewes (600 to Romney and

480 to Southdown rams) are kept at four or five to the acre and yield 112%
lambs, of which half are sold straight off the ewe at 28–30 lb carcass-weight
'before the dry comes'. Machinery consists of 1 tractor, cultivation equipment,
1 truck and a half share in a hay baler.

(ii) *Intensive Dairy Farms*

These farms are largely for 'town supply' or liquid milk. They differ from
comparable farms described in the Taranaki section in that there is usually
an acreage of cash crops, and cows are housed for the three winter months
during which time they are fed concentrates.

An example farm is described below:

A 'town supply' dairy farm near Christchurch. The annual rainfall is 23–26
in with high summer evaporation. The farm consists of 98 acres of medium
to heavy silt loam over a retentive subsoil. 75 acres can be sprinkler irrigated
from an artesian bore. 55 milking cows are run on about 70 acres of pasture,
the rest of the farm being in lucerne for hay, fodder beet and peas. The pastures
are based on perennial and short rotation rye-grass, timothy, cock's-foot and
red and white clovers. Calving is both spring and autumn, but largely the
latter. In the three winter months, besides the usual forage, they receive 4 lb
barley per day. 530,000 lb of milk are produced annually: liquid milk quota
is 1,600 lb per day, the rest is sold for manufacture. The labour on the farm is
the owner and a boy.

Stability and Efficiency

Ecologically, New Zealand farming systems enjoy long-term stability
except in erodable areas; any short-term instability is more likely to be due to
intensive stocking under weather-sensitive conditions than to major pests or
diseases. Economically the vagaries of the international markets, to which
New Zealand is decidedly vulnerable, are more important than indigenous
ecological variabilities. Efficiency is high by almost every standard. Ecologically the farming is generally excellent. Yields per acre and per animal are
relatively high and still rising. Industrial inputs are widely used and research
is highly developed. The farmers are technically keen, hard-working and have
a very high output per man. But two questions may be posed. First, how far
can New Zealand continue to raise her output from grass whilst still relying
entirely on legume nitrogen? It seems that legume N cannot contribute much
more than 250 lb N per acre/year to the sward; yet under her well-watered
sunny conditions swards would probably respond up to 400–500 lb of fertiliser
N. Fertiliser N is expensive in New Zealand but if production of grassland
products is to be raised still further, can New Zealand afford to be limited by
the capacity of the clover plant to fix atmospheric nitrogen? Second, has
specialisation been carried too far? In New Zealand's largely favourable
climate (p. 461) a wide range of enterprises would be practicable at least on
the smoother lowlands. Specialisation undoubtedly permits competitive

marketing, efficient capital use and high output per man; but it tends to make
farmers prisoners of their own system and to rob them of the mental, technical and capital flexibility they need to meet the changing world conditions
(e.g. the formation of the E.E.C., the fall in demand for 'carpet' wool and the
growing markets of the Far East). Is New Zealand in danger of becoming a
victim of its own specialised production efficiency and market ties?

Conclusions and Prospects

The typical New Zealander is now a city-dweller. But the minority who
comprise the declining manpower on farms, produce, aided by ancillary industries, some 90% of the country's export earnings. The farming industry is undoubtedly economically efficient, but, as stressed above, herein may lie the
weakness as well as the strength of the country's economy. For the high standard of living largely depends on imports of manufactured goods, of raw
materials for urban industry and of carbohydrate foods, which can at present
only be paid for by exports of wool, lamb, butter and cheese to markets which
are by no means guaranteed and which are subject to political as well as
economic fluctuations. The country is industrialising rapidly to save imports
and in the hope of exporting manufactures. Meanwhile, despite falling manpower, farming has in recent years been increasing both total farm output
and agricultural exports. But this is 'in a world trading context which, in all
probability, will not favour New Zealand's traditional products' (Franklin
1967). Nearly half (45%) of her trade is still with Britain, 16% is with the
E.E.C., 15% with the U.S.A. and Canada (mainly the former), $7\frac{1}{2}$% with
Japan and $4\frac{1}{2}$% with Australia. Much depends on whether or not the United
Kingdom joins the E.E.C. and to what extent the south-east Asian markets
can be developed, but in all probability New Zealand faces the prospect of
drastic economic change. And she, like the United Kingdom, is rather conservative and perhaps too proud of her undoubted achievements in agriculture, social welfare and social institutions. But given the will to change, New
Zealand has, agriculturally, great potentialities. For, at least in the drier but
potentially irrigable lowlands of the east, a wide range of enterprise choice in
intensive tillage and horticultural crops exists, whilst, on the wetter west, intensive grass beef should be able to rival dairying (or even, looking ahead,
grass protein compounds might find a market in the protein-starved countries
of south-east Asia and the less-developed world generally).

Acknowledgements. For research assistance to Sarah Rowe-Jones and for
comment and criticism to R. W. Brougham, A. G. Campbell, D. J. Collyns,
T. I. Cox, T. G. Sewell and J. D. Stewart.

References

Blaxter, K. L. 1956. *Proc. Brit. Soc. Animal Prod.*, p. 3.

Blaxter, K. L. 1964. *J. Brit. Grassl. Soc.* **19**. 90–99.

Brougham, R. W. 1959. *N.Z. J. Agric. Res.* **2**. 6. 283–96 and **2**. 6. 1232–48.

Brougham, R. W. 1966. *N.Z. Agric. Sci.* **2**. 19–22.

Campbell, A. G. 1966. *Proc. Massey Univ. Dairy farmers week.* Palmerston North, N.Z.

Curry, L. 1958. 'Canterbury's grassland climate', *Proc. N.Z. Geography Conference*, Christchurch, August 1958.

Curry, L. 1962. *Geog. Rev.* **52**. 2. 174–94.

Curry, L. 1963. *Econ. Geog.* **39**. 2. 95–118.

Duff, R. 1957. *Moas and Moa-hunters.* Govt. Printer, Wellington, N.Z.

Franklin, S. H. 1967. *Geography.* **52**. 1. 1–11.

Hamilton, W. M. 1944. *The dairy industry in New Zealand.* Council for Sci. and Indust. Res. Bull. No. 89. Wellington, N.Z.

Hamilton, W. M. 1954. *Proc. N.Z. Soc. Animal Prod.* **14**. 14–26.

*Hammill, W. 1958. *Report on the dairy industry and grassland husbandry in New Zealand.* Govt. Printer, Wellington, N.Z.

Hendrie, 1952. *Agriculture.* **59**. 324. London.

Hendrie, 1954. *Agriculture.* **61**. 338. London.

Hulton, J. B. 1954. *Dairy-farm survey of Waipa County 1940–41 to 1949–50.* N.Z. D.S.I.R. Bull. No. 112. Wellington, N.Z.

Rickard, D. S. 1960. *N.Z. J. Agric. Res.* **3**. 3. 431–41 and **3**. 5. 820–28.

Rickard, D. S. 1961. *N.Z. J. Agric. Res.* **4**. 5 and **6**. 667–75.

Stewart, J. D. 1963. *The comparative profitability and productivity of a sample of irrigated and non-irrigated farms in the Ashburton–Lyndhurst area of mid-Canterbury, New Zealand.* Lincoln College Publication No. 1. Lincoln, N.Z.

U.S.D.A. 1967. *Agricultural Policies in the Far East and Oceania.* Foreign Agric Econ. Rept. No. 37. U.S.D.A., Washington, D.C.

Further Reading

(SEE ALSO ASTERISKED ITEMS)*

Dept. of Agric. 1957 (or later issue). *Primary Production in New Zealand.* Wellington, N.Z.

Mackenzie, D. W. (various dates). *Man, Map and Landscape.* (A series of booklets.) Reed. Wellington, N.Z.

McLintock, A. H. 1959 (Ed.) *A descriptive atlas of New Zealand.* Govt. Printer, Wellington, N.Z.

Oliver, W. H. 1960. *The Story of New Zealand.* London.

Robinson, K. W. 1960. *Australia, New Zealand and the South-west Pacific.* London.

The *New Zealand Year Book* and the (annual) *Pocket Digest of N.Z. Statistics* (Govt. Printer, Wellington, N.Z.) have much statistical information. The bulletins and research station reports of the N.Z. Dept. of Agriculture and the Annual and other Reports of the various Divisions of the N.Z. Dept. of Scientific and Industrial Research, are useful sources of technical information. The N.Z. Dept. of Education Primary School Bulletins about farming are elementary but technically informative.

CHAPTER 2.5

Sweden and the Netherlands

BY R. W. WILLEY AND A. N. DUCKHAM

(a) Sweden

Agriculturally *Sweden* has much in common with Norway and Finland. Each runs south to north through more than 10° of latitude well into the Arctic Circle. Each has considerable areas of mountainous or otherwise operationally unsuitable land, and the soils are often shallow or very acid. This type of terrain, and the long-thin shape of these countries, has resulted in the farming areas, and often the individual farms themselves, being widely scattered and relatively isolated. This, as in New Zealand and Canada, adds to transport costs and handicaps the provision of a full infra-structure and social amenities. Until 100 years ago the Scandinavian countries were unindustrialised ones of small and often fragmented farms, with patches of forest usually providing a cash crop and earning some foreign exchange. Population then pressed heavily on resources. Substantial emigration eased this problem in all three countries but its solution lay largely in the growth of fishing and shipping in Norway, of forest industries in Finland and of industrialisation in Sweden. In this last country in 1880, 75% of the population of $4\frac{1}{2}$ million gained their living directly from agriculture, compared with 10% of $7\frac{1}{2}$ million today. At the same time, agrarian structure has been rationalised and Sweden has achieved a very high mean standard of living (see Tables 1.5.1 and 1.5.2) even though her net output (added value) per man on the land is not notably high.

Denmark also was, 100 years ago, unindustrialised and had a relatively undeveloped agriculture. Following her war with Prussia in 1861, she remodelled her agrarian structure and converted much forested sandy land in Jutland to intensive farming. In Denmark, small owner-occupied farms practice intensive tillage and livestock farming. The successful pioneering of the agricultural co-operative movement; the installation of processing plants (bacon factories, creameries); the importation for livestock feed of cheap cereals from the New World and of oil-seeds from the tropics; and an aggressive export policy of marketing high-quality standardised animal products (bacon, eggs, butter, etc.) in the United Kingdom and Germany, have all combined to make her an international model of early 20th-century agricultural reform. She has been able to expand her farming output and usefully re-deploy her population without recourse to emigration or, until the last 30 years or so, to industrialisation. Today, however, though agricultural exports remain vital to her economy, Denmark is rapidly industrialising; but, her adherence, for social reasons, to an agrarian structure, which has become somewhat outdated by modern technology, is becoming a constraint on her continued agricultural progress.

To represent Scandinavia, Sweden has been chosen for three reasons: (*a*) the success with which she has integrated technological advance with the reform of agrarian structure; (*b*) she provides good material on the effect of latitude and altitude on cool temperate farming systems; (*c*) she exhibits, in her 1,000 miles length, most of the farming systems found in Scandinavia. For example, in the far south in Scania, the farming is in effect an eastward extension of Danish farming, whilst in the centre and north the ecological background has many homologues in Norway and Finland; it is indeed permissible to call on data from Norway or Finland to illustrate particular facets of Swedish agriculture. Much of the ecological picture which emerges for northern and central Sweden is also partly repeated in northern Quebec and northern Ontario, in the higher land of the Maritime provinces of Canada and northeastern U.S.A., and in the far southern tip of New Zealand. These also have cool or cold climates, where high latitude and/or altitude leave little margin of safety to meet adverse weather, and where the inclemency and uncertainty of the climate rather than the rugged relief or the poor soil often make such land marginal for farming.

Economic and Social Background

Sweden is a highly advanced welfare state with marked social and economic egalitarianism. There are few very rich people and practically none that are very poor. Economic planning, a large and strong co-operative movement, vigorous and powerful trade unions, advanced technology and progressive capitalism all appear to be successfully blended. Education, scientific research and medical services are all at a high level. Exports of timber and industrial products provide foreign exchange for imports of cereals and protein feeding stuffs (and for the summer holidays on the Mediterranean which so many Swedes seem to need after their long, dark, cold winter). With these exceptions, Sweden has been about 95% self-sufficient agriculturally for the last 15–20 years. However, considerable changes are likely in the near future, for in 1966 the Government declared its intention of substantially decreasing the level of self-sufficiency by the late 1970s.

Agrarian Structure

The self-sufficiency policy was adopted in 1947, after the wartime food shortages, out of a desire to secure an independent neutrality. The programme

had three prongs. First, a system of price fixing, including import fees within
certain limits, combined with an income goal for farmers.* Secondly, governmental measures to subsidise farm amalgamations designed to increase farm
size, to modernise farm buildings, to reduce fragmentation (both of farms and
of forest units) and to integrate farming with forestry. (Half the area of Sweden
is in forest whilst only 10 % is in farms.) For example, at Ekeby, 7 holdings
with a mean of 38·8 acres of crops and grass and 11·7 acres of forest have been
merged into 3 holdings averaging 88·9 acres crops and grass and 42 acres of
forest. These are not large by North American or even English standards for
mechanised family farms, but they are a great improvement. The aim is to
create units of 170–250 acres each operated by 1 or 2 men. The total number
of farms in 1961 (230,000) was 25 % less than in 1939–45. This farm rationalisation has been supported by substantial Government aid for farm buildings,
for better infra-structure (land drainage schemes, roads, etc.) and controlled
or 'guided' credit.

The third prong has been the drive to maintain national food output whilst
releasing manpower intentionally—by land reform and mechanisation—to
manufacturing industry. Today, only 9 % of the national manpower is in
farming and 2 % in forestry. In 15 years productivity per man has risen 60 %
(compared to about 50 % in the U.K.) *without* increasing total farm output.
This is a considerable achievement (Meissner 1956); for it is much harder to
save labour without expansion of output than it is to use higher industrial
inputs to increase the output of men already 'trapped' on the land and so
raise total output.

Other aspects of infra-structure are good. Despite the length of the country
and the dispersal of farming areas, transport facilities by road, rail and air are
effective, electric power is cheap and almost every farm has mains electricity.
Technical, research and advisory services are highly developed. The international reputation of the Agricultural High School at Uppsala and the
Svalöf Plant Breeding Station is well established, and Swedish research work
in livestock housing, e.g. at Lünd, and in the application of linear and other
advanced programming to whole farming areas is also well known (Birowo
1965, Folkesson and Renborg 1965). Input services and supplies of fertilisers
are good and are helped by the geographical distribution of manufacturing
and forestry. Agricultural marketing is highly organised and is largely under
the control of farmers' co-operatives, which form a powerful central federation and which administer much of the Government's price support and control policy. The Ministry of Agriculture is a large, powerful body with wide
ramifications and a philosophy of close co-operation with, or of indirect control of, agricultural institutions. Family farms predominate and, despite

* The farmer's income (including that from forestry, haulage and other part-time occupations) on 'model' farms of 37 acres *arable* land, is linked to industrial wages. If the latter
rise, the farmer's 'basic' income (which is recognised to be lower than that of industrial
workers) goes up proportionately. Import tariffs and home market conditions are manipu-
lated accordingly (Holmström 1961).

Table 2.5.1

Crop	SWEDEN Area (000's acres)	SWEDEN Yield (average for 1963–65) (lb per acre)	NETHERLANDS Area (000's acres)	NETHERLANDS Yield (average for 1963–65) (lb per acre)	FRANCE Area (000's acres)	FRANCE Yield (average for 1963–65) (lb per acre)
Wheat	684	3,235	390	3,946	11,169	2,700
Oats	1,100	2,515	247	3,416	2,644	1,986
Barley	1,149	2,524	244	3,547	6,005	2,625
Rye	148	2,328	242	2,643	546	1,505
Mixed grain	358	2,379	79	3,018	420	1,974
Maize	—	—	—	3,696	2,152	3,089
Sorghum	—	—	—	—	82	2,509
Rice	—	—	—	—	74	3,369
All cereals	3,440	2,661	1,196	3,414	23,178	2,588
Sugar-beet	104	32,826	225	37,851	976	35,293
Potatoes	227	20,635	306	26,165	1,421	16,383
Vegetables (a)	25		85		376	
Dry peas and beans	27	1,355	50	2,531	197	1,479
Oil-seeds (b)	210	1,459	66	1,603	601	1,140
Tobacco	—		—		52	1,852
Flax fibre	2	862	54	1,127	126	1,076
		Production 000's lb		Production 000's lb		Production 000's lb
Tree crops (fruit) (c)		712,215		1,133,370		29,449,980

Livestock	Numbers 000's	Yield (average for 1963–65) (lb per animal)	Numbers 000's	Yield (average for 1963–65) (lb per animal)	Numbers 000's	Yield (average for 1963–65) (lb per animal)
Total cattle	2,261	(e) 493	3,751	(d) 1,051	20,244	(d) 622*
Dairy cows	994	(f) 7,509	1,723	(f) 9,164	9,624	(f) 6,089
Pigs	1,893	(e) 104	3,752	(d) 236	9,043	(d) 173
Sheep	225	(e) 36 (sheep and goats) n.a.	484	(d) 110	8,821	(d) 35 (sheep and goats)
Wool	15		92		1,041	6
Goats	15		12			
Poultry (chickens)	9,099		42,279		75,000	

(a) Onions, tomatoes (not in Sweden), cabbages, cauliflowers, green beans and green peas.
(b) Linseed, rape-seed and (sunflower seed, in France only).
(c) Apples, pears, plums, prunes, cherries (peaches, apricots, grapes, olives in France (grape yield—France, 6,125 lb per acre)).
(d) Live-weight of animals slaughtered.
(e) Carcass weight of animals slaughtered. Approximate conversion:
    Cattle and sheep: live-weight = carcass weight × 2
    Pigs: live-weight = carcass weight × $\frac{4}{3}$
(f) Milk per milking cow.
* Excluding calves.
Source: F.A.O. *Production Yearbook*, 1966.

rationalisation, are still small. Over 80% are still less than 50 acres (excluding forest) and only 3% are more than 125 acres. This often means that capital is insufficient and labour supply excessive for optimum economic efficiency. For the small man who combines farming with forestry the problem of deciding the best use of his labour is often chronic.

Crops and Livestock

About 40% of the agricultural land is in grass, and a similar area is in cereals (Table 2.5.1). The oat crop has in the past been the most important cereal but recently barley has been steadily increasing (largely replacing mixed corn) until in 1966 it occupied the greater acreage. Of the other important crops, potatoes are grown in most farming areas, oil-seeds are important in central and southern Sweden, and sugar-beet is restricted to the south. The average fertiliser usage of 41 lb N, 32 lb P_2O_5 and 28 lb K_2O per acre (1965) and the average crop yields (Table 2.5.1) are not high by north-west European standards; but, in its better farming areas Sweden compares very favourably with many European countries.

Other technological aspects of crop production, such as breeding, seed certification, the use of chemicals, etc., are well advanced. The major biological problems are probably those associated with the cereal crop—for example, cereal root eel-worm (*Heterodera avenae*), eye-spot (*Cercosporella herpotrichoides*) and *Fusarium* footrot, and Wild Oat (*Avena fatua*) and couch-grass (*Agropyron repens*). The pollen beetle (*Meligethes aeneus*) is also a serious pest of the oil-seed. A crop hazard of a different kind is the risk of crop losses in severe winters, for example, in the winter of 1965–66, 90% of the winter oil-seed rape crop (*Brassica napus*) was killed out in the Uppsala district. To

Table 2.5.2

Agricultural Land Use (1964)

Agricultural Land (%)		Agricultural Land (%)	
Winter wheat	3·7	Potatoes	2·2
Spring wheat	3·4	Sugar-beet	1·2
Barley	12·4		
Oats	13·4	Ley	26·0
Mixed barley and oats	4·7	Pasture	14·0
Rye	1·1	(Total grass)	(40·0)
(Total cereals)	(38·7)		
		Fallow	9·0
Oil-seeds	3·0	Fodder crops	2·0
		Miscellaneous	3·9

compensate farmers for this type of loss there is a national Crop Failure Insurance, which covers all the major crops. In practice a farmer receives compensation if the average yield of his area falls below the 'normal' yield by a certain amount (usually 10–15%). Half the cost of compensation is actually paid by the farmers themselves via crop levies, whilst the Government pays the other half plus administration costs.

About three-quarters of the total output of crops is fed to livestock. Nationally, milk accounts for 40% of gross sales from farms and livestock sales for a further 40%. The main cattle breeds are the Swedish Friesian (S.L.B.) in the south, the Red and White (S.R.B.) in central Sweden and the smaller, lower yielding Swedish Polled in the north. In 1965, the yields for recorded herds of these breeds were, respectively, 11,920 lb (at 4·02% fat), 10,660 lb (at 4·14% fat) and 8,150 lb (at 4·42% fat) (Roos 1966); however, only a little over a third of the Swedish herds are in fact recorded, and the average yield over all herds and all breeds was only about 8,000 lb. Artificial insemination is used in over 80% of the recorded herds and in about half the non-recorded herds. Herd size is very small (over 80% have less than 10 cows), but machine milking is usually employed. The long lean bacon sides from the Swedish Land-

race and Swedish Large White pigs are less well known in the United Kingdom than they used to be but internal consumption of bacon is high and the
pig industry is up to date. Poultry keeping (mainly White Leghorns) is rather
a side-line on most farms, as in other Western countries, but there is a growing
segment of large specialised broiler and egg-producing units. Though the
Swedish Landrace ewe is said to average 2 weaned lambs, the long cold winters
preclude sheep being important, but in the far north 275,000 reindeer are
economically significant.

Official livestock (genetic) improvement schemes including artificial insemination are noteworthy, whilst on the disease side tuberculosis and contagious abortion have been eradicated, leaving mastitis, fowl pest and the usual
plethora of pig diseases as the main animal health problems. The long winter
emphasises the need not only for animal health precautions but also for good
insulation of livestock, which has to be housed for 7 months or more (compare Canada), and for labour-saving livestock mechanisation. As to crop
mechanisation, almost all farms have tractors and, with the exception of
mixed corn and to a less extent oats, most of the cereal acreage is combine
harvested. The Swedes love labour-saving gadgets; in this, as in some other
respects, they are the Americans of Europe and they rival the United Kingdom as leaders of European farm mechanisation.

Climate and Soils

Climate (Table 2.5.3) and soils and farming systems are discussed in more
detail below for each sector in a South to North transect (Table 2.5.4). The
points to be stressed here are the saving graces of the Gulf Stream and the
westerly winds which combine to raise winter temperatures well above the
latitude mean; the wide range of climates, with latitudes ranging from 56°N
(compare Edinburgh) to 69°N (in the Arctic Circle) and altitudes approaching
6,000 ft (Fig. 2.5.1); and the fact that over much of the country the soils have
been heavily glaciated and are thin. There are, however, good lowland soils
in Scania in the far south, and in the central plains; together these account for
substantially more than half the total agricultural output. Mean annual precipitation in most stations exceeds the total P–E–T in the thermal growing
season (Table 2.5.3) and has a fairly low variability (Table 2.5.4). Light energy
and the uncertainties of temperature (in air and soil) at the beginning and end
of the growing season are more important constraints than precipitation.
Energy receipts, as measured by P–E–T are not high and A–E–T does not
exceed 20in in the T–G–S. The thermal growing season (T–G–S) is short (for
example, 4–5 months in mid-Sweden) but, though the solar altitude is low,
the summer days are long. It has been shown that light energy is, photo-
synthetically, more efficiently used by grass at low temperatures. Stanhill
(1960) has shown that growth per unit of water transpired is greater at
high than at low latitudes and suggests that potential growth is as great at
60°N as it is at 40°N and greater than at 50°N (e.g. London). Although his

Table 2.5.3

Climatic Data — Sweden

Station	Bright Sunshine hours/day (mean annual)	Thermal Growing Season (>42½°F) length in months	Thermal Growing Season Mean Temp °F	Mean Months >72½°F	Mean Annual Precipitation (inches)	Annual Precipitation Coeff. Var. (C.V.)	Evapo-transpiration in Thermal Growing Season		Moisture Index Im	Max. recorded rain/day inches (month of occurrence)
							Potential (mean inches)	Actual (mean inches)		
Malmö (Copenhagen)	4·0	7	54·8	0	20·7	10%	21·6	19·1	−10 dry sub-humid	1·6 (July/August)
Cotëborg		6	55·7	0	29·0	10-15%	20·5	19·4	26 humid	2·9 (August)
Jönköping		6	53·7	0	21·1	10-15%	19·5	18·0	−1 dry sub-humid	3·3 (June)
Uppsala	4·9 (Stockholm)	5	55·4	0	21·5	10-15%	18·3	16·6	4 moist sub-humid	1·7 (August/September)
Härnösand		5	52·8	0	24·9	10-15%	17·2	16·1	32 humid	2·9 (August)
Östersund		5	51·5	0	19·6	10-15%	16·5	15·6	9 moist sub-humid	2·1 (September)
Gällivare		4	52·0	0	18·2	15%	14·1	13·0	14 moist sub-humid	1·7 October
Abisko		4	47·6	0	10·5	15%	12·9	9·4	−26 semi-arid	1·0 (October)

225

research was not experimental, it is clear from Swedish results that high latitudes are not inconsistent with good yields, provided one can get the crop in
and harvested in the short working season whilst the ground is not frozen.
This difficulty, of course, does not apply to the grass leys which tend to
dominate the north of Sweden, but it does restrain tillage cropping, particularly as late summer weather may be unfavourable for cereal harvesting.
Nevertheless, although there is no wheat in Sweden north of 61°N, rye, oats,
six-row barley and potatoes are practicable at higher latitudes. Potatoes, if
sprouted in a warm place, do well on peaty soils (compare the garden potatoes
of northern Newfoundland). In this context Varja's 1965 map of the northern
boundaries of various crops in Finland is illuminating.

South to North Transect (*See Table 2.5.4 and Fig. 2.5.1.*)

Scania (Sector (a))

Over one-fifth of the country's agricultural output is grown on the rolling
plains of the southern tip of Sweden. On the coast the soils are calcareous,
morainic clays, fertile and readily worked, whilst just to the north they are
sandy, lower in pH and less fertile. The natural vegetation of the area is
mixed beech and oak forests. The thermal growing season in Scania is 7
months, from April to October, and during this period the mean temperature
is 54·8°F, with a July mean of 63·5°F. Lowest temperatures are in February
(mean 32°F). Night frosts are a considerable risk during the spring and early
summer. Mean annual precipitation in the area ranges from 19 in in the east
to 27 in in the west, with coefficients of variation under 10%. P–E–T and
A–E–T in the T–G–S average 20–22 in and 19–20 in. Lowest monthly precipitation is from January to May and most rain is received in July and August.
Thus, the distribution of precipitation hinders cereal production in two ways
—during spring and early summer there may be insufficient moisture* for
optimum growth and heavy rainfall during August may hamper harvest
operations.

Farming systems—intensive tillage and alternating with livestock. This sector, like most of Denmark, is too dry for intensive grassland, but favours intensive tillage. The highly mechanised farms have similar cropping and stocking programmes to those in Denmark but are larger in area, although over
40% are still under 25 acres and 30% are between 25 and 50 acres. Most of
the soils are easily worked, for the winter frosts break down even the heavier
clays; 80% of the area is in arable crops, and very little of the grass is permanent pasture. Cereal growing is the main tillage enterprise (roughly 60% of

---

* Black 1960 has found a region of high radiation receipts (especially during the summer)
centred over eastern Denmark and southern Sweden and suggests that, given adequate
moisture, the region should be more suited climatically to agriculture than other areas of
northern Europe. There are, in fact, many parts of eastern Denmark (especially the light
soils of east Jutland) and parts of Scania which have a high irrigation potential.

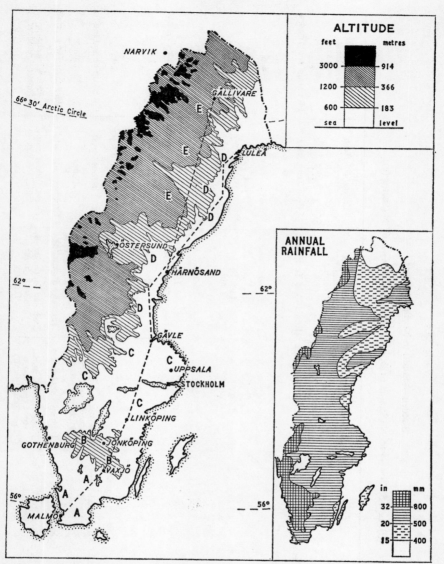

Fig. 2.5.1 Relief of Sweden and transects with inset of Mean Annual Rainfall

the arable acreage*), barley and wheat being the main crops. Spring wheat is almost as important as winter wheat, a reflection of the relatively long growing season and the fact that even in this southernmost area winter wheat may still be killed out in severe winters. Oats, rye and mixed corn are of less importance although there are considerable acreages of these on the lighter,

* Arable = tillage plus temporary leys.

227

Table 2.5.4

Sweden — South to North Transects

(i). Lat. 55°N ⟶ 66°N (Sectors (a, b, c, d₁) (ii) Lat. 63°N ⟷ 67°N (Sector (e))

	(a) Scania	(b) S. Swedish Highlands	(c) Central Swedish Lowlands	(d) Coastal Northern Sweden	(e) Upper Swedish Highlands
1. Sector					
2. Name	Malmo	Vaxjö	Linköping	Gävle	Härmösand
3. From	Vaxjö	Linköping	Gävle	Lulea	Gällivare
4. To					
5. Miles	120	105	150	410	340
6. Climate-type(s)	Humid Cool Temperate	Humid Cool Temperate	Humid Cold Temperate	Cold Temperate	Mountain and Tundra
Thermal Growing Season:					
7a.—months at or >42½°F	7	6	5–6	5	4–5
7b.—months at or >72½°F	0	0	0	0	0
8a.—P–E–T(in) in T–G–S	20–22	19–20	18–20	16·5–18·5	14–16·5
8b.—A–E–T(in) in T–G–S	19–20	18–19·5	16·5–18	15–16·5	13–16
9. Growth limited by	Cold	Cold	Cold	Cold	Cold
Precipitation:					
10.—mean in (annual)	19–27	19–23	19–22	17–25	17–20
11.—C.V. % of Precipitation (annual)	<10%	10–15%	10–15%	10–15%	15%
12. Special Climate Features			Long summer daylight	Long summer daylight	Long summer daylight
13. Weather Reliability	4	3	3	2	1
Soil:					
14.—Zone/Group	Podsolic	Podsolic	Podsolic	Podsols	Podsols
15.—pH	Low or Neutral	Low	Low or Neutral	Low or Neutral	Low
16.—Special Features					
17.—Vegetation	Mixed deciduous forest	Mixed coniferous/deciduous forest	Mixed deciduous forest	Coniferous forest	Birch, alpine and tundra
18. Relief	Plains	Hills and valleys	Plains	Plains	Mountains
19. Market Access	4	3	4	1	1
20. Farming Systems	Intensive tillage and alternating	Intensive alternating	Intensive alternating	Extensive alternating	Very extensive grazing
21. Major Enterprises	Cash and fodder crops, livestock	Fodder crops, grass, some cash crops Livestock, forestry	Cash and fodder crops, livestock. Some forestry	Grass, fodder crops, milk, forestry	Extensive sheep, forestry, reindeer

228

more acid soils of the north and east of the sector. The very important oil-seed crops, mainly rapes, constitute over 10% of the arable cropping. These are potentially high-value crops, but high yields are not easily obtained; winter crops may be killed out by frosts or by waterlogging in early spring, and the rape pollen beetle (*Meligethes aeneus*) can be a serious hazard. The potato crop is largely grown for alcohol or starch, but the household trade is important in the more fertile and more heavily populated south-west. Many of the farmers growing household varieties also have a profitable 'early' market in the populated district round Stockholm where harvesting is appreciably later. Very good crops of sugar-beet are grown and again these are centred on the more fertile soils of the south-west. However, if the suggested policy of decreasing self-sufficiency is going to be implemented in the near future, it seems fairly certain that sugar-beet is one of the crops that will suffer.

In Scania, crop technology is particularly good (the mean fertiliser usage per acre in 1964–65 was 69 lb N, 55 lb P_2O_5 and 53 lb K_2O) and yields are extremely high for this latitude; for example, compared with the national figures, cereals average 700 lb, potatoes 2,250 lb and winter rape 400 lb higher per acre. The traditional rotation is that associated with Danish farming, that is, an 8-year one in which cereal runs of 1–2 years are alternated with ley, oil-seeds, roots, etc. (often regarded as a double Norfolk 4-course). However, there is a recent trend towards rotations more typical of the United Kingdom to give a longer, cleaner break from cereal pests and diseases. Thus 5–6 year rotations are now something like: 2–3 years non-cereals; wheat; barley; oats. This typical order of cereals after the break emphasises the particular importance in this area of the soil-borne diseases Take-all (*Ophiobolus graminis*) and Eye-spot (*Cercosporella herpotrichoides*), for this is the order of decreasing susceptibility to these diseases.

In Scania the main livestock enterprise (dairying) is declining. This trend clearly indicates the relative profitability of tillage crops versus milk, especially as most of the milk goes for manufacture and therefore commands a price only about 60–70% of the liquid market price. Beef cattle have not proved a paying alternative, so livestock numbers continue to fall. The associated decrease in grass acreage is causing concern about soil structure, particularly with the ever-increasing traffic of heavy machinery on the land. However, the relatively slow rate of organic matter breakdown in these higher latitudes helps to alleviate soil structure problems, and, to judge from U.K. experience, some of these Swedish farmers are perhaps worrying unnecessarily.

Example farm. This farm, lying in a small area of fertile coastal plains around Kristianstad, is an owner-occupied one of 175 acres of farmed land and 52 acres of forest. It is mainly boulder clay, but with some patches of acid heath soils which are stony and difficult to work in places. Mean annual precipitation is 21·5 in with a late summer peak which is a problem at harvest time. Only 25 acres are in permanent grass, but there is a further 18 acres in a 2-year ley and 16 acres in a longer ley. The permanent grass receives 50–70 lb/

acre of nitrogen, whilst the leys receive up to 180 lb/acre/year. The leys are
mainly Timothy (*Phleum pratense*) Meadow Fescue (*Festuca pratensis*) and
Red Clover. The farm carries 40 Red and White milking cows, averaging
11,000 lb, plus followers. In winter they are fed hay, silage and concentrates
according to yield, which is recorded. They are housed in a traditional cowshed
with slatted dung passage and central feeding passage. Calving is mainly in
autumn because of the price differential. The tillage consists of 39 acres barley, 15 acres rye and 12 acres oats. There are no potatoes or sugar-beet
although these are grown locally. The rotation is ley; rye or barley; barley;
oats. The main problem is frit fly (*Oscinella frit*). All grains are sold to the
co-operative and concentrates are bought back ready mixed. There are also
700 hens on deep litter. The labour consists of the owner himself, his son, wife
and one hired man. There is an extensive range of field machinery; cows are
machine milked on a pipeline system to a bulk tank. (*Courtesy:* Mr Ohlson
and D. T. Pritchard.)

SOUTHERN SWEDISH HIGHLANDS (*Sector (b)*)

In the west of this sector annual precipitation is as high as 29 in, but moving
east this decreases to 18–20 in with increasing risk of early summer drought.
Rainfall can be intense, with up to 3 in recorded in 24 hours in August. The
thermal growing season averages 6 months, with mean P–E–T and A–E–T of
19–20 in and 18–19 in respectively. The soils are essentially coarse glacial
moraines with some patches of peat bog. Most of the area is covered with
mixed or coniferous forest; the soft woods are mainly Scots pine and spruce,
the hard woods include oak, ash, lime and elm.

Farming systems—intensive alternating. Only 20% of the area is farmed,
i.e. the finer sands and, in some cases, the better-drained peat land. The
farms are badly fragmented and very small; 80% still have less than 25 acres,
although most have substantial areas of forest as a source of added income.
Poor soils and small farm size (and in the west the high rainfall) mean that
only half the farmed land is in tillage crops. These are almost entirely cereals,
with small acreages of potatoes and fodder crops of roots, peas and vetches.
Oats and barley are the main crops, the former being particularly important
in the wetter west. Very little wheat is grown, but rye is found in the drier
areas of the east. Potatoes are usually grown for stock feed; these yield surprisingly well, largely because spring frosts do not seem to be a serious hazard
in this area and therefore this crop is usually the first one planted in spring.
However, there is a danger of the crop being frozen in the ground before
lifting. Most of the grass is in short leys, based on Timothy and Meadow
Fescue with perhaps a little Red Clover, whilst the longer ones include *Poa
pratensis*. Timothy and Meadow Fescue soon give way to *Poa annua* and then
to other poorer species in this climate, and the problem is to maintain the
better species in the sward. Silage and hay are made and small quantities of
the latter are usually fed right through the summer. Dairying, with Red and

White cattle, is the major livestock enterprise, but, as milk goes largely for
manufacture, pig fattening on the skimmed milk is also fairly common. As in
most of Scania, the low price for manufacturing milk is rapidly making dairying financially unattractive. But, whereas in Scania the farmers have been able
to decrease dairying and place more emphasis on tillage crops, in this sector
dairying is the most ecologically suitable enterprise and so, rather ironically,
farmers have been forced to increase their grass acreage in recent years to try
to improve the efficiency of milk production. However, small farm size, poor
layout and the traditional cowshed system make low-cost milk production
difficult. The high labour demands of the cowshed system make it increasingly
costly, and yet money spent on milking parlours or better buildings may increase overheads too much.

Faced with these problems farmers have been paying more attention to
their forestry enterprises. In Sweden forestry has long acted as a buffer
against the changing fortunes of agriculture. However, the extent of this
buffering is limited and it seems inevitable that much of the land in this area
may soon cease to be farmed. As much of it is marginal, it might well be
argued that from the national viewpoint this might not be a bad move.

Central Swedish Lowlands (*Sector (c)*)

This sector is second only to Scania in agricultural importance. It covers a
belt of lowland, including 4 large lakes and many small ones, which stretches
the entire width of Sweden but is broken by the numerous eskers, i.e. low ridges
of gravelly moraine. The transect (Table 2.5.4) from Linköping to Gävle
crosses 2 fertile plains divided by an area of higher ground of stony, infertile
morainic soils, of which only 15% is farmed. The thermal growing season is
5–6 months during which temperatures differ little from those in Scania. The
shorter growing season is partially compensated by more daylight hours and
less cloud cover so that, especially in the north, vegetative growth is rapid.
However, there is a greater risk of night frost at the start and end of the growing season. Annual precipitation in this sector is about the same as at Malmö.
P–E–T and A–E–T during the thermal growing season decrease steadily with
increasing latitude; mean A–E–T is about 17 in.

The Östergötland plain at the south of the sector has fertile clay soils which
need little liming and are as good as the best soils in Scania, especially when
they are tile drained. The plains surrounding Lakes Mälaren and Hjälmaren
and further north round Uppsala are mostly heavier than in Östergötland,
and harder to work. However, winter freezing and thawing alleviates the
problem. Most of the soils are rich in potash and sometimes also in phosphate.

Farming systems—intensive alternating. The farms in this area are basically
similar to those in Scania; they combine cash crops with the production of
milk from grain and fodder crops. Farm size is also comparable; for although
farms in this sector were very small, consolidation schemes have been success-

ful. In Uppsala County for example, mean farm size is now 120 acres. However, the higher latitude of this sector is evident in a more restricted range of
tillage crops, a rather greater emphasis on grass, and lower, more variable
crop yields. Cereals are the most important tillage crop and now occupy more
than half the farmed area. Because of the severe winters, only a small proportion is winter crops (largely wheat with a little rye). Barley, oats and mixed
corn are the main spring cereals, in order of importance; spring wheat is only
grown in particularly favourable conditions, for the short growing season
frequently prevents satisfactory ripening. Oil-seeds are important, although
in this latitude, yields are low and unreliable. These are mainly winter sown,
partly because of the higher yield potential and partly because, like wheat, late
ripening spring crops have harvesting problems. However, winter rape has
the disadvantage that it should be sown by early August which limits the
cropping situations it may follow, and even then there is still a risk of winter
kill. About one-quarter of the farmed land is in grass, and as in the southern
Swedish Highlands this is largely short leys. But poor returns from dairying
are decreasing the grass acreage, as in Scania. Some farmers have omitted
grass entirely and are now growing all tillage crops.

Rotations are traditionally of the type: 3–4 years ley; 2 years cereal; fallow;
winter rape; 2 years cereals. However, the increasing emphasis on cereals has
led to longer runs of these and the rotation now frequently takes the form:
ley; winter rape; winter wheat; oats; barley; then possibly up to two more
cereals. Or, where grass is omitted, the break is usually fallow; winter rape and
then the cereals. The taking of oats before barley (compared to the Scania
order of wheat, barley, etc.), whilst perhaps partly traditional, indicates the
importance of cereal root eel-worm (*Heterodera avenae*); oats grown later in
the rotation would be more susceptible to losses. A one-year ley ploughed out
after hay time or a fallow, fits into the farming system, for it allows for sowing
of winter oil-seeds. A fallow also allows cultivations to control wild oat (*Avena
fatua*) and couch grass (*Agropyron repens*) which constitute major problems.
Fertiliser usage is considerably lower than in Scania. However, crop yields are
surprisingly high considering the latitude, for example in Uppsala County the
mean yields of winter wheat, barley and oats for the three years 1963–65 were
respectively 3,500 lb, 3,000 lb and 2,900 lb.

Example farm. This farm, near Nyköping, is basically like that at Kristianstad (p. 229), although it is 300 miles further north. It consists of 190 acres
of farmed land plus 116 acres of forest, and lies in a small boulder clay plain.
Annual precipitation is 20 in with a July/August peak. There are 17 acres of
permanent grass and a further 17 acres of ley, some of which is cut for seed.
Species are again Timothy, Meadow Fescue and Red Clover. There are 23
autumn calving Red and White milking cows, yielding 12,000 lb each and 14
Landrace sows producing mainly weaners for sale. Tillage is divided more or
less equally between winter rape, winter wheat, barley and oats. The rotation
is ley; rape; wheat; barley; oats. There is a good range of machinery. Labour

consists of the farmer himself and one semi-retired man. The farm is typical of many in Sweden in that expansion of the dairy herd is hampered by traditional buildings. Some buildings were being modified for the pigs, but the farmer is considering going out of milk rather than investing in an expensive new cowhouse. (*Courtesy:* Äke Karlsson and D. T. Pritchard.)

COASTAL NORRLAND (*Sector (d)*)

The thermal growing season of 4–5 months is some 2 or 3 months shorter than in Scania. The winter average temperature is over 20°F lower. The mean temperature in the T–G–S is about 5°F lower but the risk of frost in the T–G–S is less. As in central Sweden, this shorter T–G–S is partly compensated by the great length of cloud-free daylight in the summer (for example Jokkmökk, on the Arctic Circle, has more hours of sunshine in June than Rome or Madrid). Nevertheless, the range of crops and enterprises is limited by the early autumn frosts and the unfavourable pattern of rainfall (the seasonal maximum is in the autumn when light and temperature are too low for plant growth). As in Finland the higher the latitude, the smaller the range of crops that can be grown. Here, as elsewhere in the world, the more marginal the climate, the fewer are the choices open to the farmer.

The mean annual precipitation, and its coefficient of variation, is similar to that of central Sweden. However, P–E–T during the thermal growing season declines to 15 in at the higher latitudes and is a major limiting factor. The natural vegetation is coniferous forest and the rugged relief precludes much farming. The only areas of significance are those in the lower reaches of the larger river valleys where the climate is less extreme and where the soils are derived from recent marine deposits.

Farming systems—intensive alternating. Each farm has a considerable acreage of forest which provides an important part of the farmer's income. Because of the short growing season, there is little emphasis on annual crops, and very little of the land feels the plough. Fully acclimatised six-row barley, with its very short growing season requirement, is the only important cereal. Oats are sometimes grown, but largely for feeding green. Potatoes yield higher than the national average, because of the absence of eel-worms and virus diseases, and because this crop is only grown in the most favourable conditions. The grassland is mainly in long leys or permanent pasture: the species are similar to those used in the southern Swedish Highlands. The farms (using the hardy Swedish Polled cattle which have low yields of high butter-fat content) concentrate on liquid milk for the local market; production is based on grazing, hay and silage, fodder crops and some concentrates.

UPPER SWEDISH HIGHLANDS (*Sector (e)*)

This final sector has both high altitude and high latitude and is unimportant agriculturally. It is summarised in Table 2.5.4. The farming system is very extensive grazing of the forest pastures and moorlands by sheep of the

improved Landrace breed and occasionally a few Swedish Polled cattle. Where
the land is cultivable, crops of six-row barley, green oats or potatoes are
sometimes grown. Most of the area lies within the Lappmark (a demarcation
line to check colonisation of Lappland by Swedish farmers) and many of the
Lapps still lead a semi-nomadic existence, hunting, fishing and raising reindeer.

Efficiency and Stability

In ecological terms, the efficiency of land use in Sweden seems to be high.
The transect has shown how, at the different latitudes, the farming systems
are well adapted to climate, relief and soil type, and that yields are somewhat
higher than might be expected. Possible climatic reasons for this were suggested on p. 224, but undoubtedly it must also be due to the genetic improvements, high inputs of fertilisers and protective chemicals, good technical services and the careful husbandry so characteristic of Swedish farming. It is
difficult, in a country in which the farming is substantially insulated from
world prices, to assess overall economic efficiency. But two questions may be
asked. First, despite the planned reduction in farm manpower, is not the net
output per man on the farm rather lower than one would expect in an industrialised country with such a high standard of living? Second, although the
policy of farm amalgamation is going some way to achieve economies of scale
and of specialisation, is it going far enough and is it sufficiently flexible to
meet changing technological needs? May it not be 'trapping' on farms invest-
ments in fixed resources which may become obsolescent too rapidly? These
questions are perhaps partly answered by the Government's recent change of
attitude over self-sufficiency; for this change suggests that, despite the major
achievements of the last 10–15 years, progress in the rationalisation of agriculture is not fast enough.

The considerable support given to farmers to achieve and maintain self-
sufficiency has produced substantial income stability. But a major difficulty
in these latitudes is yield variability. For example, Renborg (1962) gives coefficients of variation for cereal yields in Scania and the plains of Östergötland
—two of the most favoured parts of the country—as 22% and 23% respectively.* And, quoting from Finnish experience (Pessi 1960) 'in every decade
occur two years of crop failure and three years of poor crops' and 'of the
fifteen years of crop failure during the past one hundred years, the cause in six
years was frost, in three years excessive rainfall'. In other words, in countries
such as Sweden, particularly in the more marginal areas of high latitude and/
or high altitude, tillage crop production is frequently limited by the uncertainty of climatic factors, such as severity of the winter, the onset and end of
the growing season, late spring and early autumn frosts and rainfall at cereal
harvest. Though the climatic constraints may be different, the situation is not

* Renborg also gives some useful C.V.'s for major farm operations and for labour inputs
for various crops.

unlike those found in the Welsh uplands in the United Kingdom, on the
northern edge of the Canadian Prairies, and, at the arid end of the spectrum
(p. 481), at the frontier between the wheat and sheep belt and the purely
wool-growing districts of Australia.

Conclusion

Sweden illustrates very well the climatic influence of latitude and altitude
on cool and cold temperate farming systems when the economic and social
factors (other than access to market) are held constant. The substantial
changes in agrarian structure and the high degree of market organisation
provide good examples of what can be done by planning in a democracy. Her
self-sufficiency policy (though now less marked) has shown that it is possible,
no doubt at some cost to the consumer, for an industrialised country with a
high level of agricultural technology, but without great economic assets, to
achieve an independent food supply at the same time as it achieves a remarkably high standard of living.

References

Birowo, A. T. 1965. '*Lantbrükskögskolans*', *Annaler*. **31**.

Black, J. N. 1960. *Archiv für Meteorologie, Geophysik u. Bioklimatologie*, Ser. **B. 10**.
2. 182–91.

Folkesson, L. and Renborg, U. 1965. *Optimal Inter-Regional Allocation of Agricultural Production and Resources in Sweden*. (Paper presented at Symposium on
Operational Research. Dublin, 1965. Mimeo.

Holmström, S. 1961. *The Basis for Agricultural Products Prices in Sweden*. Agric.
Econ. Res. Inst., Stockholm.

Meissner, D. 1956. *J. Agric. Econ*. **11**. 4. 444–56.

O.E.E.C. 1954. *Grassland: Seed rates and seed mixtures*. O.E.C.D. Paris.

Pessi, Y. 1960. *Majaniemi*. **1**. 1. Sorvoo, Helsinki.

Renborg, U. 1962. *Studies on the Planning Environment of the Agricultural Farm*.
Agric. Econ. Dept., Agric. Coll. of Sweden, Uppsala.

Roos, A. 1965. 'Sweden Breeds for Yield', *Farmer's Weekly*, 15 July. 81.

Stanhill, 1960. *Proc. 8th Int. Grassl. Congr*., 293–96.

Varja, U. 1965. 'The Finnish farm: seen from the viewpoint of geographical
typology of agriculture', *Fennia*. **92**. 1. Helsinki.

Further Reading

Agricultural Economics Research Institute, 1964. *Swedish Farming and forestry in
figures and charts*. 2nd English edn. Stockholm, 1.

Jonasson, O., Höijer, E. and Biorkman, T. 1963. *Agricultural Atlas of Sweden*.
Royal Swedish Academy of Agriculture, Stockholm.

Ministry of Agriculture and National Board of Agriculture. 1964. *Swedish Agriculture*. Stockholm.

Sweden and the Netherlands

By R. W. WILLEY and A. N. DUCKHAM

(b) The Netherlands

Introduction

The Netherlands is studied because, firstly, socio-economic, altitude and climatic factors can be regarded as held constant; so one can examine the influence of soil structure and of soil/water relationships on location and intensity of farming systems. Secondly, pressure of population has been relatively high for many centuries. Thirdly, because farming yields are amongst the highest in the world and standards of living (Beckerman Index 45) and nutrition (Table 1.5.1) are both good.

National survival depends on successful control over a relatively unfavourable ecological inheritance (viz. sea-water and salinity, flood risks, sandy soils, clay soils reclaimed from the sea, and rather poor working weather) with, however, a mean actual evapo-transpiration (A–E–T) in the thermal growing season (T–G–S) of 20–22 in. The centuries old fight to check invasion by the sea and by armed forces has produced a sturdy national temperament, and a lively democracy in which pragmatic economic and ecological control and forward planning play a prominent role in land and water utilisation. Thus plans for the creation of new polders, and for schemes to protect against flood, stretch several decades ahead. 'God created the world, but the Dutch the Netherlands.' This aphorism is at least half true; for more than 50% of what is today land surface was once under the sea, but is now protected from flooding and soil salinity by an elaborate system of built-up dykes (embankments, bunds, levees) and drainage canals. The latter are used, in addition, to control the level of the water-table for crop production and, aided by the great delta of the Rijn (Rhine), Waal and Maas (Meuse) rivers, to provide cheap water-barge traffic. This water network is supplemented by a well-developed system of railways and roads. The high bicycle population bespeaks the flat terrain, but its suitability for mechanisation is limited by the small area of the average farm, the fragmentation of fields, the frequency of ditches and canals and the large areas of low-lying heavy, more or less untillable, clay soils. (Fig. 2.5.2.)

The Netherlands is one of the most densely populated countries in the world, averaging 930/sq mile, rising in the western provinces to 2,000 head/sq mile. The total population of 12 million, of which about 40% are Roman Catholics and the rest Protestants, is rising at about 1·1% per annum. Both infant mortality and overall death-rate are low. Medical and social services,

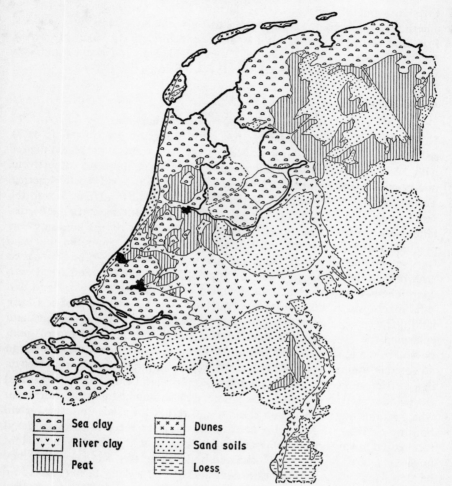

Sea clay
River clay
Peat
Dunes
Sand soils
Loess

Fig. 2.5.2 Soils Map of the Netherlands

including education, are advanced; there is little poverty. One in 6 houses have telephones, and practically every home has mains electricity, whilst 4 out of 5 rural dwellings have piped water supplies (Table 1.5.2). Holland is essentially a vigorous trading and manufacturing country based on growing energy sources (coal, oil and natural gas) and a wide range of industries—many of them sophisticated (e.g. electronics). Energy and steel consumption are high (Table 1.5.1). Raw materials and about 4·0m tons per annum of bread grain and feeding stuffs are imported and exports are mainly of high value industrial, agricultural and horticultural products (eggs, milk products, fruit, bulbs). Many of these have a high labour requirement, a reflection of the high rural population. Ethnically the two basic peoples of the Netherlands were the

Friesians and the Saxons. The Saxons held tenaciously to a rule of
inheritance, whereby the farm passed to only one son. Thus, in north-east
Friesland, settled by the Saxons, the farms are still relatively large (120 acres).
But, in Friesian areas the holdings were divided on inheritance and now
average about 30 acres, often fragmented (see below).

Agrarian Structure

Farm structure, land and water policy. More than half the farms in the
Netherlands are under 25 acres, often with too many enterprises. In 1960,
the area per farmer and farm worker on the land was 15·5 acres compared
with 27·5 acres in Denmark and about 100 acres and 300 acres respectively,
in ecologically comparable parts of the United Kingdom and North America.
Technologically, Netherlands agriculture is decidedly advanced but the
small mean size of farm (30 acres), the severity of fragmentation, the distance between fragments, and the excess of manpower on the land not only
preclude realisation of the full technical and economic potential, but also
trap, in agriculture, resources such as manpower, which could be better used
elsewhere in the economy. The Government have, therefore, embarked on a
comprehensive plan for the redeployment of existing resources of land and
rural manpower and the creation, in the Issel Meer (Zuider Zee), of new farm
land. The programme has two main prongs—first, consolidating and
enlarging the areas of individual farmers; secondly, the diversion of manpower from and rising specialisation in, farming. Consolidation schemes can
be voted in by a majority of the farmers or by a majority of the 'farm acreage'
in a district, subject to provincial and central government approval. The land
is then surveyed and valued; the whole area is bought by the Government
which retains not more than 5% for use or resale for roads, industry, urban
housing or recreation. In due course, each farmer receives a consolidated
farm equal in value, but not necessarily in area, to that he has given up. This
farm has a centrally situated house and buildings and, where necessary, may
be 'de-mixed', e.g. a specialist horticultural unit or a dairy farm (Fig. 2.5.3).
The new farm may be in the consolidated district or one of the new polders.
By the end of 1964 applications had been received for the consolidation of
3·0m out of the 3·7m acres requiring redevelopment, and actual consolidation
is 100,000 acres per year.

The traditional fight against water, mainly sea-water, is spreading to new
fronts. For not only are there problems, in coastal areas, of the infiltration of
salts (see Israel, p. 317), but also of industrial and human effluents from urban
Holland and from higher up the Rijn (Rhine), Maas (Meuse) and Waal in
Germany, Belgium and France. Although many horticulturalists can afford
to buy drinking water for irrigation, farmers on the higher, lighter soils, or in
the areas where the water-table has been lowered to reduce the need for innumerable drainage ditches, cannot afford such water for sprinkler irrigation
of grassland, sugar-beet and potatoes, and may not dare to use river water.

1 (*a*) Dairying on a big modern intensive grassland farm in the Netherlands. Table 2.5.6, Sector (*b*). (*With acknowledgements to the Royal Netherlands Embassy.*)

1 (*b*) Fat lamb production on intensive grassland in the North Island, New Zealand. Table 2.4.5, Sectors (*a*) and (*c*). (*With acknowledgements to the High Commissioner for New Zealand*).

2 (a) Flood irrigation of a strawberry crop in California, U.S.A. Table 2.3.3, Sector (h). (*With acknowledgements to the U.S. Feed Grain Council.*)

2 (b) Intensive horticulture under glass in the Westland district, Netherlands. Table 2.5.6, Sector (a). (*With acknowledgements to the Royal Netherlands Embassy.*)

3 (*a*) Intensive tillage in the U.S.A. Corn Belt—maize and alfalfa (lucerne) on the contour for control of water erosion. Table 2.3.2, Sector (*c*). (*With acknowledgements to the U.S. Feed Grain Council.*)

3 (*b*) Extensive tillage on the North American interior plains—strips of wheat and fallow on the contour for control of water and wind erosion. Table 2.3.2, Sector (*d*). (*With acknowledgements to the U.S. Feed Grain Council.*)

4 (*a*) Extensive grazing in New Zealand—a 'high country' sheep station. Table 2.4.5, Sector (*e*). (*With acknowledgements to the High Commissioner for New Zealand.*)

4 (*b*) Extensive grazing on sown land, South Australia. Table 2.4.3, Sectors (*d*) and (*e*). (*With acknowledgements to the High Commissioner for Australia.*)

5 (*a*) A typical piece of savanna, the most widespread vegetation type of the tropics. The photograph was taken in northern Uganda; the tree in the centre is a shea butter-nut (*Butyrospermum parkii*).

5 (*b*) A peasant farm in Zambia under the system of shifting cultivation. The crop in the foreground is finger millet. The trees at the back were too large for the farmer to fell, so have been ring-barked and left to die standing.

6 (*a*) An area denuded by over-grazing and eroded by gullying in Tanzania. (*Photo A. H. Savile.*)

6 (*b*) A soil conservation measure used in Tanzania is to make tied ridges (also known as 'basin listing'); in this case two rows of cotton have been planted on each ridge. (*Photo H. Gillman.*)

7 (*a*) A homestead in the dry area of northern Ghana. Crops are grown on ridges made with the ridging plough seen in the picture. (*With acknowledgements to Regional Information Office.*)

7 (*b*) Ploughing in south India. The leading pair of animals are buffaloes, the others bullocks. Planted areca palms in the background.

8 (*a*) A typical rice-growing landscape in Malaya. The fields are divided into small compartments by 'bunds' (earth ridges) to control the water level. An irrigation canal is seen at the right. (*With acknowledgements to Malayan Information Agency.*)

8 (*b*) In this photograph in a Trinidad village, the Indian bull of the Ongole breed, introduced for draft purposes, typifies the influence of East Indian immigrants, while the church behind represents the Christian tradition in settlement.

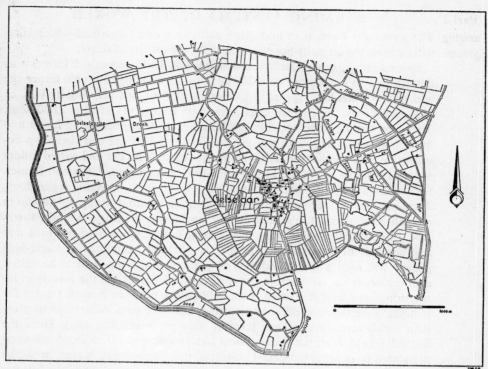

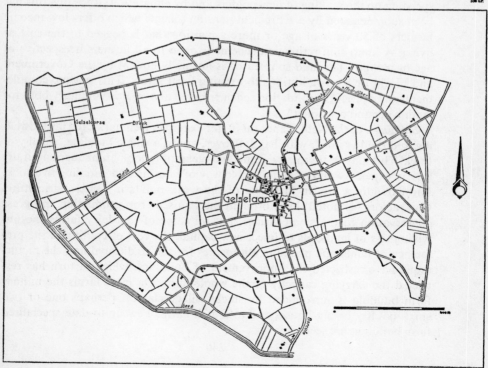

Fig. 2.5.3 Example of the effects of land consolidation in the Netherlands Above, before consolidation; below, after. *Source:* Min. Agric. Hague

The ambitious Delta plan and other plans to create large fresh-water lakes
will increase the quantity but not, some fear, the quality of water.

Specialisation, mechanisation and manpower. Until recently a farmer was
not allowed to keep more than 200 hens on his holding, but now larger egg
and broiler units have sprung up. Though they are not very big by U.S. or
U.K. standards, 25% of the eggs now come from units of over 1,000 hens.
The same trend can be noted in pig keeping; units are becoming larger and
more frequently divorced from general farming. In Friesland there is a co-
operative milk production unit which relies entirely on cut and carted fodder
supplies from its members, i.e. co-operative zero grazing. The small farmer,
now often with no children to help him, has to meet changing technology,
decreasing manpower and the need to specialise without 'trapping' capital in
specialised equipment, e.g. feed-mixing machinery. Wide use is made, therefore, of contractors who bring to the farm both a specialised machine and
another pair of hands on the man who works it. A potato or sugar-beet
grower will own a specialised drill or planter, and a mechanical harvester,
etc., but relies on contractors for spraying and harvesting the cereals in his
rotation. Grassland dairymen often only have a horse or a small tractor for
ploughing, seed-bed making and carting, and rely on a contractor to grow
their small acreage of tillage (cereals, turnips, mangolds, etc.). Since the
Second World War, the Government has encouraged migration from over-
populated farm areas by special industrial training colleges in rural areas, by
speeding up the making of new polders and by emigration. Recently, this has
been supplemented by a retirement pension scheme which offers low-income
farmers of 50 years of age or more a pension that is pegged to the cost of
living. A lump sum retirement scheme is open to all farmers irrespective of
age or income. The land so released is normally bought by the Government
and resold or let as additions to existing local farms. This scheme comple-
ments the district consolidation schemes noted above. It is voluntary both for
landowner and pensioner.

Buildings. In the Netherlands (as in, for example, Denmark), the farm dwelling and the livestock, with hay-loft over, are in a single building, usually of
brick. This may be thermally sound and pleasant to look at, but agriculturally
it is an expensive and often inflexible way of housing and handling livestock.
Farm buildings are, in fact, one of the weak points in Dutch agriculture.
Contractors (as noted) have brought mechanised crop production and grass
conservation to the small farmer, but on some of the dairy farms the unwillingness of landlords to inject new capital, the cost of credit and the preference of landscape planning authorities for the traditional and the picturesque have restricted the improvement of buildings, which in turn has restricted the carrying capacity of the whole farm. On most farms the unitary
farm building is surrounded by stacks, silage clamps, perhaps one or two
small poultry or pig houses and a 'Dutch barn'. Few up-to-date specialised
farm buildings are seen.

Services to farmers. Government control of genotype use, of plant and animal health, of marketing and of product quality is strong. Education is well organised. The research and teaching at the Government research institutes and at the Agricultural University at Wageningen enjoy world recognition. The farmers are organised denominationally; there is a Protestant, a Catholic and a small 'neutral' (Liberal) farmers' union. To some extent the advisory service has to follow a religious pattern, but this does not seem to impair its marked efficiency. In both input supply and product marketing, as well as credit facilities, co-operation plays a large part. Thus, over four-fifths of the milk marketed is handled by co-operatives, though for other products the proportion is lower. Government price guarantees for milk, sugar-beet and potatoes are limited to specific quantities; any excess has to be sold at world, or at least E.E.C., prices. The Government also guarantees certain types of agricultural loans, and, as noted above, largely finances the consolidation and the rural depopulation schemes. This long-sighted rural reconstruction should, in the long run, so raise efficiency that it will reduce the need for subsidies.

Climate, Relief, Vegetation and Soils

The climate is cool humid. The mean annual precipitation is mostly between 25 in and 35 in and is markedly reliable as its C.V. is less than 10%. The Netherlands has more 'rainy days' than any other country in Europe. This is an operational constraint. Mean annual sunshine hours per day are between 4·0 and 4·6 (Table 2.5.5) and the thermal growing season (T–G–S) is 9 months in the south-west and 7 months in the north-east. In the T–G–S the mean temperature is about 54°F, the mean P–E–T is 23 in or 24 in and mean A–E–T ranges from 21 in in the north-east to 23 in in the south-west. The annual temperature range is rather greater than in the United Kingdom; the July mean at Groningen is 61°F and the January mean is 34°F. Summer temperatures make maize for grain feasible in parts of the south, but the fairly cold winters demand full winter housing of cattle. Frost is a problem inland. The relief is flat except for the gentle hills towards the east. The natural vegetation on the land not reclaimed from the sea was mixed forest or, on the sandy 'uplands', heath. The soils fall into six main groups. Fig. 2.5.2 shows the large area of sea and river clays, often heavy, in the reclaimed areas of the north and west and near the river valleys, and the dominance of sandy or peat soils elsewhere. There are few loamy soils. Fig. 2.5.4 shows that large parts of the sands and some of the peats are not farmed; that most of the grassland (which is used mainly for intensive milk production) is on those clays that are not really suitable for tillage, or are liable to flood; and that tillage, horticultural and alternating (mixed) farming systems are found on the sands, the less-waterlogged and workable peats and the more workable clays of the west and the north. (Visser, W. C. in Rutter and Whitehead 1963 establishes an interesting relationship between groundwater depth and cropping pattern.)

About $2\frac{1}{4}$m acres, or 40% of the farmed land, are in tillage crops. Cereals

Table 2.5.5

Climatic Data — The Netherlands

Station	Bright Sunshine hours/day (mean annual)	Thermal Growing Season (>42½°F) length in months	Thermal Growing Season Mean Temp °F	Mean Months >72½°F	Mean Annual Precipitation (inches)	Annual Precipitation Coeff. Var. (C.V.)	Evapo-transpiration in Thermal Growing Season		Moisture Index Im	Max. recorded rain/day inches (month of occurrence)
							Potential mean (inches)	Actual mean (inches)		
Rotterdam Katwijkaan de Rijn	4·3 (De Bilt)	9	52·4	0	29·1	<10%	24·7	23·3	12 moist sub-humid	2·0 (September)
Winterswijk	4·0	7	55·9	0	29·8	<10%	23·9	22·2	13 moist sub-humid	2·5 (July)
Groningen	4·0	7	56·3	0	28·3	<10%	23·1	21·7	11 moist sub-humid	3·0 (July)
Den Helder	4·6	8	53·8	0	25·9	10%	22·9	20·9	3 moist sub-humid	2·7 (September)

occupy about 1¼m acres. Wheat and barley are the main crops, being grown largely on the clays of the north and west. Almost all the wheat is milled, and about 15% of the barley crop (exclusively from the south-west) is sold for malting. Oats and rye are important on the acid sands of the south-east, but these cereals are gradually being replaced by barley. Despite the high proportion of cereals grown, 60% of the required feed cereals and two-thirds of the required bread grains are imported. Potatoes (325,000 acres) are an important crop on most soils except the very heavy clays and bog peat land. About 25–30% of this crop is grown for starch manufacture, an industry which is centred on the large mechanised farms of the north-east; of the remainder, a further 25–30% is for home domestic consumption, up to 15% is exported and the rest is used largely for stock feed. Sugar-beet (225,000 acres) is grown mainly on the more workable soils of the west. Sugar-beet tops are highly valued as a stock feed and the desire to keep them free from soil has tended to slow down the mechanisation of this crop. Amongst fodder crops, the acreage of fodder beet, which is grown on the clays, peats and sands and fed chopped to cows after the New Year, is noteworthy. Turnips as catch crops and small areas of red clover, lucerne and fodder lupins (on sandy soils) complete the list. Grassland is discussed below.

Most of the viral, bacterial and fungal diseases are under effective control, though with the tendency to specialisation (e.g. in cereals), and the relaxation of rotations, fungal diseases in particular may increase. Insect pest problems appear to be minor, and the main problems appear to be yellow rust of cereals (*Puccinia glumatum*), cereal root eel-worm (*Heterodera avenae*), sugar-beet eel-worm (*H. schachtii*) and potato eel-worm (*H. rostochiensis*). The last is probably the most serious and by law potatoes cannot be grown more often than one year in three and may even be forbidden for up to, say, 10 years on land that is heavily infested. Almost all technological aspects of crop production are very highly developed, for example, fertiliser usage in 1963–64 reached the extremely high national level of 82 lb of N, 45 lb of P_2O_5 and 59 lb of K_2O per acre. Coupled with a reliable, though admittedly not very high A–E–T it is therefore not surprising that this very high standard of farming achieves some of the highest crop yields in the world (Table 2.5.1).

Horticulture is very important in the Netherlands, which has made use of technical skill and an excess of manpower to specialise in crops with a high labour requirement and a growing consumer demand from her industrial neighbours. The industry has several sections: in the west, vegetable growing, especially the south Holland glasshouse district; top (tree) fruit and soft fruit again in the west; flower bulbs in the north-west, especially on chalky, sandy soils but increasingly on clays; flower crops under glass and in the open; woody nursery stock, especially rose-trees and fruit-trees; horticultural seeds and medicinal and aromatic herbs.

Livestock: sheep, pigs and poultry. Sheep are not of national importance but are quite a common sight, particularly in the coastal areas, grazing dykes which have been grassed down. The only breed, the Texel, is very prolific and fast growing. A large percentage of ewe lambs can successfully conceive at 6–8 months and the carcasses average about 66 lb dead-weight at 6 months. Pigs (3m) and/or poultry (50m) are found on most farms on the sandy soils of the east and south. This is perhaps partly because this was traditionally a feed-grain producing area; partly because the farms are small and these enterprises provide a valuable addition to gross farm sales; and partly because family labour is plentiful as the people are mainly Catholics. But the old combination of home-grown grains and milk by-products as a basis for feed has been replaced by reliance on purchased compound feeds ready fortified with vitamins and minerals. (The cereal acreage on such farms has largely given way to grass for milk.) The Netherlands Landrace pig (which has good length and a good lean to fat ratio) is used for making Wiltshire bacon for local use and for export, whilst the Large White, which is found mostly in the industrial west, is used primarily for pork. Poultry are mainly crossbred, with Rhode Island Red genes in many crosses. Feed conversion efficiency of both pigs and poultry is good and is encouraged by strict health and genotype control, for example, through litter testing of pigs. Farm herds and flocks predominate, but specialist units, including broilers, are increasing. Large quantities of eggs, broilers and day-old chicks are exported.

Cattle, milk production and grassland management. Dairying is the most important farm enterprise in the Netherlands. Historically, the production of liquid milk was centred on the heavy clays near the large seaport cities of the west, where there was a ready market and easy access to imported concentrate feeds; milk for manufacture was concentrated on the low-lying clays of the north and north-east. This pattern still exists today, and the only real change on many of these farms has been increasing intensification, particularly in the traditional liquid (market milk) area. Cattle total 3·8m. About one-quarter of the cows are of the red-and-white Meuse–Rhine–Yssel (M.R.Y.) breed, which used to be dual purpose but has been made a milk breed; it is found on the banks of, and between, these three rivers and on the sandy soils of the east. The main breed is the black-and-white Friesian, however, which is found in all districts but especially in the west and north. The Netherlands strains, like the British, are more dual purpose in conformation and purpose than the Friesian (Holsteins) of Canada and the U.S.A. Yields are almost the highest in the world and average 9,160 lb per cow per annum at 3·75% butter-fat for all breeds. Progeny testing is compulsory. There is now little difference between the performance of pedigree and commercial herds. (Israel, which produces an even higher mean milk yield, uses Friesian cows under controlled environments and on expensive feeding regimes (p. 310). By contrast, the Netherlands relies less on controlled environment and much more on intensive grassland grazed *in situ*. Her output is, therefore, more seasonal than in Israel, which makes the high Dutch yield the more remarkable.)

Seventy per cent of the dairy herd is artificially inseminated, and peak calving occurs in mid to late winter. Little use is made of milk substitutes and most calves are reared on whole milk on the farm of their birth until weaned at 4 or 5 weeks. Calf mortality is low. Most cows are individually recorded for milk yield, fat and protein percentage. Bull calves are not castrated if they are to be killed as baby beef (which is treated much as barley beef in the U.K.), but they are castrated where bullocks are needed for converting tillage surpluses, e.g. sugar-beet tops, or surplus grass into beef sold at 20 months or more. A programme completed in 1956 successfully eradicated bovine tuberculosis throughout the country and this situation is maintained by testing all cattle annually. A similar project to control brucellosis (*Brucella abortus*) is proving successful. Until recently foot-and-mouth disease was controlled by vaccination backed by rules banning the movement of un-vaccinated stock; to ease the exporting of live cattle, a slaughter policy on the British model has now been adopted. Probably the most serious ailments of cattle in the Netherlands are those caused by parasitic worms, largely because of the predominance of dairying on heavy, wet soils. Grass tetany (hypomagnesaemia), a problem found elsewhere in north-west Europe and on the Pacific North West of the U.S.A., may be associated with certain aspects of fertiliser use (Van den Molen 1964), but can be controlled by administering magnesium in suitable form.

Increased fertiliser use combined with improved grass management, regulation of the water-table, intensive grazing regimes and better methods of hay and silage making have raised the national average grass yield to 2,900 lb starch equivalent/acre—a very high figure. The stocking intensity in livestock units (1 unit = feed needed by a milch cow) varies from district to district. In typical all-grassland districts, it is as high as 80 units per 100 acres; on sand, where part of the tilled land produces fodder crops, the rate may rise to 97 units per 100 acres grassland. Of the national grass crop 70% is grazed, 20% is made into hay and 10% into silage. However, on the permanent pastures, such as in west Holland, where tillage crops for fodder are impracticable, much more of the grass crop is made into hay and silage.

The pastures are almost invariably simple mixtures or single species *without* clover (except in certain favoured situations) and nitrogen application is usually 125 lb or more per acre. The lack of clover (in contrast to New Zealand (p. 199, for example) can be attributed to relatively cheap fertiliser nitrogen, to the low pH and winter flooding of the heavy soils, to the lack of sunshine and relatively low temperatures, and to the ease of management of grass-only swards. Grazing stocking rates are high, often reaching $1-1\frac{1}{4}$ cows/acre. Rotational grazing of small paddocks at perhaps only 5–8-day intervals is usually practised, a system which helps to avoid poaching on the heavy soils and 'soiling' of the pastures with dung and urine. Most farmers make both hay and silage, and some practise grass drying on a small scale. In the north, particularly in Friesland, so-called warm silage is made which, although perhaps allowing high losses in dry matter, has the advantage that it can be handled by the farmer without outside help. In other areas, the silage is usually wilted and carefully sealed in clamps or low-concrete silos; here great attention is paid to quality. The main problems of conservation are inadequate buildings and machinery on the small farms. The machinery contractor is increasingly important for supplying transport, a forage harvester or baler, and even manpower.

Although the thermal growing season ranges from 9 months in the south and west to 7 months in the north-east, the danger of poaching the wet autumn and spring pastures, and the need to protect livestock against cold winds and rain, mean that, in practice, the grazing season is limited to little over 6 months. Winter housing is usually in traditional cowsheds, although occasionally covered yards are used. In the permanent grassland areas the only winter fodders which the farmers can produce themselves are those based on grass, but, depending on the distance to the arable areas, there is often an opportunity of buying sugar-beet tops, chat potatoes, lucerne hay, etc. Bulky feeds such as sugar-beet tops or pulled and carted turnips are usually fed before the New Year, with more emphasis on silage afterwards. As the good quality silage is better feeding value than turnips, the cows are often in better condition after the New Year than they are before it—a reversal of the late winter malnutrition found in many temperate countries. As mid-winter is a popular calving date, this autumn underfeeding can adversely affect 'steaming-up'. Concentrates for cows, and indeed all livestock, are generally bought from the co-operatives ready mixed and fortified with protein, minerals and vitamins. (Even cereal growers raising pigs or poultry sell their grain and buy back compounds.) Only about 30% of the total milk output of the Netherlands is consumed as liquid milk. The rest is manufactured into butter, cheese, condensed milk and milk powder. Small-scale butter or cheese making on the farms themselves is not very common today, although a notable exception is the production of the famous full-cream cheese in the Gouda district. About half the butter and cheese, and nearly all the other dairy products, are exported.

Trace elements. Apart from biological pests and diseases of crops and livestock, trace element deficiencies and toxicities are economically important. These imbalances occur principally on the river clays and on the sandy soils of low pH; soil type and pH are the main contributory factors, but in some instances high fertiliser inputs have aggravated the problem. The major deficiencies are of copper, cobalt, magnesium, manganese, molybdenum and zinc. These seem to be corrected mainly by direct addition of the appropriate elements to the soil rather than by the use of 'fortified' fertilisers. The major toxicities are manganese, molybdenum and zinc, the latter two due to industrial pollution. These occur very locally. In one district in south Holland zinc toxicity is so bad that farmers have been forced to abandon tillage crops in favour of grass.

Table 2.5.6

The Netherlands Transects

Transects (see Fig. 2.5.4) lie between 52°N and 54.4°N

	(a) West Holland	(b) Rhine, Waal, Maas Valleys	(c) Guelderland and Overysell	(d) N.E. Holland	(e) Polders and Friesland
1. Sector	(a)	(b)	(c)	(d)	(e)
2. Name	West Holland	Rhine, Waal, Maas Valleys	Guelderland and Overysell	N.E. Holland	Polders and Friesland
3. From	Hook of Holland	Gouda	Arnhem — Winterswijk	Den Ham	Amsterdam
4. To	Gouda	Arnhem	Den Ham	Groningen Coast	Groningen
5. Miles	25	55	40	70	90
6. Climate-type(s)	Cool Humid	Cool Humid	Cool Humid	Cool Humid	Cool Humid
Thermal Growing Season:					
7a.—months at or > 42¼°F	9	8	7	7	7–8
7b.—months at or > 72¼°F	0	0	0	0	0
8a.—P-E-T (in) in T-G-S	24-25	24-25	23-24	22-5-23-5	22-5-23-5
8b.—A-E-T (in) in T-G-S	23-24	22-23	21-5-22-5	21-22	20-5-22
9. Growth limited by	Cold	Cold	Cold	Cold	Cold
Precipitation					
10.—mean in (annual)	25-35	25-35	25-35	25-35	20-30
11.—C.V. % of Precipitation (annual)	<10%	<10%	<10%	<10%	<10%
12.—Special Climate Features	Water-table control	Water-table control		Some parts have water-table control	Water-table control
13. Weather Reliability	4	4	4	4	4
Soil:					
14.—Zone/Group	Podsolic	Podsolic	Podsolic	Podsolic	Podsolic
15.—pH	Neutral	Neutral	Low	Low	Neutral
16.—Special Features	Dunes and marine clay	River clay	Sands	Peat bogs and peat reclamation soils, marine clays	Marine clays
17. Vegetation	Sea	Flooded	Mixed forest, heath	Mixed forest, heath	Sea or flooded
18. Relief	Flat land below sea-level	Floodplains, much of it below sea-level	Low hills	Low hills and below sea-level plains	Flat land below sea-level
19. Market Access	4	4	4	4	4
20. Farming Systems	Intensive tillage. Intensive grass	Intensive tillage. Intensive grass	Intensive alternating. Forestry	Intensive tillage. Intensive grass	Intensive tillage. Intensive grass
21. Major Enterprises	Horticulture, wheat, dairying	Dairying, some orchards	Pigs, poultry, rye and potatoes, and increasing grassland dairying	Cash and fodder crops, commercial potatoes, dairying	Dairying, cash and fodder crops

247

Transects (*Table 2.5.6 and Fig. 2.5.4*)

WEST HOLLAND (*Sector (a)*)

This sector of flat country below sea-level, with a ridge of sand-dunes along the coast, lies just north of the Rhine estuary. It is protected from flooding by the ridge of dunes, by sea walls, and along the tidal and other waterways by dykes. The soils, mainly reclaimed from the sea, are young sea-clay (rather low in humus), old sea-clays (low in humus and calcium) and peaty clays or shallow clays over peat. The thermal growing season (T–G–S) is 9 months and in general the winters are mild and the summers cool. The mean temperature during the T–G–S is 52·4°F with average daily maximum temperatures reaching 70°F in July and August. Bright sunshine averages 4·3 hours per day. Precipitation is reasonably evenly distributed but there is a slight peak between August and October, so that rain at cereal harvest is a hazard, and a slight trough in February to May. Mean annual precipitation at Rotterdam is 29·1 in. On the soils that have not been reclaimed from the sea, the original vegetation was forest.

Farming systems. This sector has *intensive grassland* dairying on the heavy marine clays and peats, and *intensive tillage* farming (including intensive horticulture under glass) on the lighter soils and reclaimed peats. Soil type, and therefore the farming system, may change very rapidly over short distances.

Intensive grassland dairying. These farms are mainly concentrated on the east of the sector, in the low-lying river valleys of the Rhine, Waal and Maas. A general description of this type of farming was given earlier, but a 'model' farm here would be about 35 acres. Although consolidation schemes are making great strides, fragmentation is a particular problem here and the average farm could still have 10–20 fields. These fields could be widely separated with few, if any, hard roads linking them. The soil type is clay overlying peat. The water-table is frequently too high and permanent grass is the only feasible cropping. This farm would support about 26 milk cows and their followers, perhaps a few sheep, and up to 30 pigs. The cows would be Friesians, yielding about 11,000 lb at 3·75% fat. Hypomagnesaemia and parasitic worms would be the main health problems. The farm buildings would be old and dilapidated and the cows may be housed in more than one cowshed. (The pigs may also be housed in 2–3 separate units in unsatisfactory buildings.) Cheese may be produced on the farm itself, in which case only the late spring and summer milk is used; winter milk is sold to the dairy factory. About two-thirds of the pastures would be cut for conservation, mainly hay with some silage. The farmer may still dry hay on racks in the fields; on the other hand, he may make 'barn-dried' hay with cold air ventilation, but out-of-date layout of his buildings could make this difficult. Silage would probably be stored in above-ground concrete silos because of the high water-table. Mechanisation would almost certainly be poorly developed; for example the farmer may

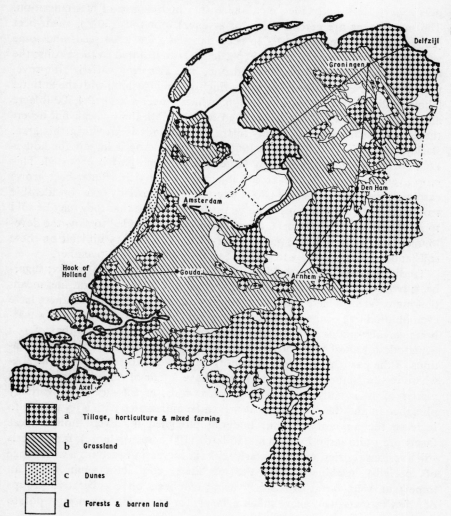

Fig. 2.5.4 Farming Systems of the Netherlands. *Source:* Min. Agric. Hague

not even have his own tractor. In any case, however well mechanisation was developed by local standards, the farmer would still rely fairly heavily on a machinery contractor.

Intensive tillage farming (excluding intensive horticulture). In west Holland and Zeeland (south of the Hook) intensive tillage farms are found on the lighter morainic soils and the sandy clays. These soils tend to be rather acid and need regular liming; they are also rather deficient in P_2O_5 and K_2O. The level of the local water-table is controlled by the Drainage Board and all fields are tile drained. The farming system is intensive cash cropping with high inputs of fertilisers, herbicides, etc., and with a high degree of mechanisation. The main crops are cereals (35% of acreage), potatoes (30%), sugar-beet (25%), with the remaining 10% in peas, flax and a few grass seeds and some vegetables. Of the cereals, wheat is particularly well suited to these soils; the autumn-sown crop is the traditional one, but, as spring varieties improve, there is greater emphasis on spring sowing to avoid clashing with the autumn labour peak of root harvesting. In 1965 winter wheat averaged 4,500 lb/acre and spring wheat 4,150 lb. Barley is grown less often than wheat, but nevertheless all the the malting barley in the Netherlands is grown in this area. Barley averaged 4,032 lb/acre in 1965. The potato crop is largely for household consumption in the near-by large cities, though part is exported. This crop averages about 24,600–26,800 lb/acre of total yield. Sugar-beet grows extremely well and averages 44,800 lb/acre. Farmers usually take considerable trouble to keep the beet tops clean whilst harvesting so that they may be sold to neighbouring dairy farmers, a practice which has tended to slow the development of mechanised harvesting. Harvesting can be very difficult on these soils and the aim is usually to get the beet crop off by mid-November.

Traditional rotations are of the type: potatoes; wheat; peas or flax; sugar-beet; barley. However, increasing intensification of the rotation has meant that many farmers have dropped the peas or flax, and in some instances have even omitted one of the cereals as well, giving a very short rotation of sugar-beet; cereals; potatoes. The main problem which this raises is that of soil nematodes, both of beet and potatoes, although thanks to the balanced rotations of the past this problem does not, as yet, seem to be as serious as elsewhere in the Netherlands. Farmers also have a fear of reducing the organic matter of their soils too much; to offset this, cereal straw is invariably ploughed in and the cereal crop may even be undersown with a ley merely to increase the amount of plant material ploughed in. Even though these farms are quite large by Dutch standards (100 acres or more), farm size is still a serious restriction, particularly with the increasing need for a wide range of specialist machinery. Thus the machinery contractor again plays an important role, but there is increasing emphasis on co-operative groups of a few farmers who share machinery or, for example, have a joint potato store.

Horticulture. An area particularly well suited to intensive horticulture is found along a narrow coastal strip in the west of this sector (Westland). It is sheltered from the prevailing westerly winds by the coastal dunes, it has a higher P–E–T, very fertile medium clay or peat soils, a well-controlled water-table and easy access to markets both at home and abroad. In addition, the more general factors of land shortage, a plentiful supply of labour and cheap water transport have aided the development of high-value, labour-demanding, cash crops. Westland is, in fact, one of the most intensively cultivated regions of the world (Russell 1954). There are about 10,400 acres, of which more than half is under glass, both heated and cold frames. The main glasshouse crops are tomatoes, cucumbers, lettuce, grapes and flowers. The sheltered areas between the glasshouses are ideal for high-value vegetable crops.

RIJN, WAAL, MAAS VALLEYS (*Sector (b)*)

The climate of this sector is like that of west Holland but with a slightly shorter T–G–S. The mean temperature in the T–G–S is a little higher owing to its inland position, but the P–E–T and A–E–T are in the same range. This sector lies, as a band about 20 miles wide, between the Rijn (Rhine) and Maas. It is flat and low lying with dykes to prevent flooding by the rivers in the east and by the sea in the west. The soils, formerly flooded, are recent clays laid down by the rivers, and are, on the whole, workable with adequate drainage. But there are some 'basin soils', i.e. very heavy intractable clays that were waterlogged for much of the year and then dried out and cracked in the summer. Until 1955 these only yielded poor grazing and poor quality hay, which was sometimes made useless by the presence of the poisonous bog horse-tail (*Equisetum palustre*). But now drainage ditches, pumping stations and new roads are being installed, and the land is being reallocated into small compact holdings (25–50 acres) with modern new farmhouses. Grassland yields have doubled and standards of farm living are greatly improved. (De Jonge 1955, 1957).

Farming systems. On the 'basin' clay soils only *intensive* permanent *grassland*, mainly for milk production (as on p. 248 above) is economically feasible. On the lighter and the topographically higher clays, *intensive tillage* farming and fruit production is common. Formerly orchards were all in grass, but newer orchards are now bush trees grown clean cultivated; apples, pears, and some plums and cherries are grown.

SOUTH-EAST HOLLAND, GUELDERLAND AND OVERYSSEL (*Sector (c)*)

The climate is more continental than in previous sectors and has a higher summer mean maximum temperature (75°F in July at Winterswijk) and a greater diurnal range of 19°F in July as against 12°F at Rotterdam. Average winter temperatures are 3–4°F lower than near the western coast. Mean annual precipitation of 30 in is as at Rotterdam but with a more marked peak in July and August. The higher summer temperatures are often offset by a shorter thermal growing season (7 months), resulting in slightly lower P–E–T and A–E–T in the T–G–S. The 'natural' vegetation is mixed forest and heathland. The soils are glacier-deposited Ice Age Pleistocene sands and are extremely complex. They have a low pH, generally few minerals and often suffer from lack of water. Push moraines, formed as the ice-sheet moved south, remain today as low ridges and hills. The first 2 or 3 ft of many of the older tillage lands are entirely man-made; centuries of applying farmyard manure has raised not only the organic matter content and the quality of the humus,

but also the level of the fields by several feet. The mainly *alternating farming*
systems depend on the varied relief, the pH of the soil and the relative economic advantages of grassland or tillage, the former being currently more profitable.

Farming systems. These are *intensive grassland and alternating* family farms between 25 and 50 acres. The grassland is permanent on the brook-valley soils but on the podsols and 'old arable' land, pastures need re-establishing after 5 to 10 years and are either directly resown after ploughing up or taken into the rotation as a 5 to 10 years ley. But ley farming as such is not widespread. Only a relatively small proportion of the grass crop is conserved because tillage crops provide winter feed. Accordingly, summer stocking rates per acre of grass are higher than on all-grass farms.

In the southern part of this sector, M.R.Y. cattle are more common than Friesian–Holland cows and the surplus milk is used for butter making. However, pigs and poultry (see below) can be quite important in this sector. The tillage crops here are rye, oats, mixed corn (oats and barley), potatoes and fodder beet. With the exception of the potato crop and some rye exported for seed, the tillage crops are grown for livestock feed on the farm. Turnips are grown in the rye stubble; they are fed green when the cattle are first housed in the autumn or are made into silage. Rye is still the dominant cereal grown on these sandy, inherently acid soils, but the acreage has now fallen to less than 247,000 owing to the low rye price, an increase in grassland, the reduced need for bedding straw and the increasing competition from barley as the pH of these soils is raised by liming. The average yield of rye is about 2,700 lb/acre, and of oats about 3,100 lb/acre. As on all tillage soils in the Netherlands (with the exception of the stiff, sticky clays in the northern arable districts) potatoes are an important crop; due to the nearness of the peat colonies (see Sector (*d*)), production of seed potatoes and also of farina is important.

Pigs are important on these small farms. In Guelderland, both the Large White and Landrace breeds are kept, the former being fattened to about 275 lb live-weight for pork and the latter making a very good 'Wiltshire' bacon pig. Many local pig-breeding societies keep good boars for breeding. Pig testing stations, run by the herdbook society in most provinces, compare different feeding rations as well as selecting better genotypes. Poultry are kept for egg production, but broilers and other table poultry are rapidly increasing. About 80% of all Dutch hens are crossbreeds, usually with Rhode Island Red blood. Legislation has resulted in very high standards of poultry breeding and all flocks ultimately stem from registered pedigree breeders, who supply chicks and breeding cocks to licensed producers of hatching eggs, who, in turn, supply registered hatcheries. All farmers *must* buy their future laying stock as day-old chicks from these hatcheries. The small flocks are kept in wooden houses with thatched or tiled roofs which house up to 100 hens. In summer these have access to grass, but during the winter they are kept in their houses which usually have electric light and are well ventilated. 'Colony-houses' are used for rearing young birds; they are small simple structures, usually placed on, and moved over, ground not previously used for hens, to cut down internal parasite infection. Most poultry farms buy compound feed (concentrates) and add it to their home-grown grain, which comprises about 40% of the birds' intake. Egg marketing is highly organised and, aided by good quality control, about half the total production is exported, mainly to West Germany. About 40% of eggs are marketed by producer co-operative organisations. Exports of poultry meat and day-old chicks are growing. As in other countries and enterprises, technology and the need to economise on capital are stimulating specialisation in pigs and poultry. Fewer farms keep pigs, whilst those that do have larger herds and tend to concentrate on breeding or fattening. Poultry keeping and pig production are becoming separate activities.

Example mixed farm (near E. Drenthe). This farm of 27 acres is split into eleven fields which average 1½ miles from the farm buildings. There are three soil types: low-lying, poorly drained soil in permanent pasture; humus-rich 'man-made' soil in tillage cropping; and reclaimed sandy soil in a ley/tillage rotation. The permanent pasture is prone to winter flooding and the reclaimed soil to summer drought. Sixteen acres are in grassland and the rest in potatoes (for seed), sugar-beet and cereals. The grass receives 180 lb or more of nitrogen per acre and conservation is largely as hay. This grassland supports 16 Dutch–Friesian dairy cows and 20 young stock. Milk yield averages about 8,800 lb/cow at 3·8% fat. The herd is housed in an out-of-date cowshed which has to be cleaned out twice daily. The main health problems are lungworm and grass tetany (hypmagnesaemia). Other livestock consist of 50 laying hens and 20 fattening pigs which are fed on purchased concentrates in poor buildings. The farmer owns a tractor and some hay-making machinery; he shares a liquid manure tank and a binder with other farmers. Drilling, spraying and root harvesting are done by contract. The manpower is the farmer and his son.*

NORTH EAST NETHERLANDS (*Sector* (*d*))

This sector, north through the provinces of Drenthe and Groningen, embraces first the so-called 'peat colonies' and then, to the north, the marine clay area. These 'colonies' were formed after peat had been excavated for fuel and peat moss. The top layer of peat (containing *Sphagnum*) was retained after excavation and mixed with the underlying sand or with the sand excavated when canals and ditches were built to drain the area and transport the peat

* An interesting feature of this farm, as for many other farms of this type, is that there will not be a successor. The farm buildings are owned by the farmer in an as yet undivided inheritance with his brothers and sisters. Part of the land is similarly owned, but the main part is rented. Within a few years the estate must be divided and the farmer does not have the capital to buy it all. He is now fifty-five and he will certainly stop farming when this division occurs.

turf. An area of well-drained soils of good structure but low fertility was thus created. This has been improved by high fertiliser use as well as compost (from town refuse) and farmyard manure. North of the peat colonies the soil is marine clay, which is younger towards the coast. The peat colonies are notorious for late killing spring frosts. August is the only month in which frosts have not been recorded. On the clays nearer the coast, the nearness to the sea delays the spring rise in temperature and affects the acreage of spring-sown crops. In this flat, open country, wind is also a problem. The natural vegetation was mixed forest or heath.

Farming systems. The peat colonies and the lighter marine clays support *intensive tillage* farming; the heavier clays near Groningen *intensive grassland* for milk. Mixed, but not alternating, farms are found on sandy soils near the peat colonies, intensive tillage crops being grown on higher land and pasture on the low-lying ground. These mixed farms are otherwise very similar to those in Sector (*c*).

Intensive tillage. The farms in the peat colonies are very long and narrow, but unlike other parts of Holland, individual fields are well shaped and average 1·75 to 2·5 acres. Canals border every field and create the problem of maintaining a water regime favourable for both farming and transport. Although spring wheat and oats are important, manufacturing potatoes are the chief crop. Special 'farina' varieties are grown and transported by canal and by road to the co-operative starch-mills where potato farina is made. Cereal straw is sold for manufacture into strawboard and allied products.

Tillage farming on the marine clays in north Groningen is quite different. The soils are reasonably workable with a good lime and P_2O_5 status, but are low in K_2O and organic matter. The farms are much larger (100 acres or more); mechanisation and modern techniques are well developed. The 15% of the area in grass is steadily giving way to more tillage. Traditionally most of the grass was in permanent pasture near the farm buildings, but as tillage crop-ping has increased, leys have joined the rotation to make an *alternating system*. The grass receives about 160–180 lb/acre of nitrogen, but this high rate is more a reflection of the high rates applied here to tillage crops than an indication of the importance attached to ruminant livestock. The grass is utilised by dairy herds. The white-faced black 'Groningen' breed, which has long been so characteristic of this area, now accounts for only 40% of all herds; the remainder are Dutch Friesians. Average yields of all herds (including those on the heavy clays round Groningen) are about 9,700 lb of milk with a fat content of 4%. Cereals are the most important tillage crop (65% of the area) with wheat, oats, then barley in decreasing order of importance. This order to some extent reflects the decreasing value of the straw, which is sold to the strawboard factories; this provides a very useful additional income from the cereal crops, for wheat straw may fetch up to £9 per short ton. The oat crop is still important here (relative to south Holland for example) because of its

resistance to *Ophiobolus graminis* and *Cercosporella herpotrichoides* in the predominantly cereal rotations. Yields of the three cereals may average up to
4,000 lb/acre. Most of the farms now own a combine harvester, but mechanisation developed slowly because until recently the straw factories would not
accept small pick-up bales. An average fertiliser dressing for cereals is about
70 lb N, 40 lb P_2O_5 and 80 lb K_2O per acre; the high K_2O emphasises the low
potash status of the soil. Although potatoes only occupy 5% of the area, a
large proportion of all Dutch seed potatoes are produced in this part of the
Netherlands. The sector is particularly suited to potato seed production because aphid* attack is slight, due to the persistent winds, and only comes late
in the season. An average fertiliser dressing is 100 lb N, 100 lb P_2O_5 and 200 lb
K_2O per acre. Harvesting of 60% of the crop is completely mechanised. A
high proportion of the seed tubers is exported, for example, to Italy and the
U.S.A.

Sugar-beet is also important (5% of the area), again particularly for seed
for export. A difficulty which is developing in the seed crop is the increasingly
high cost of labour for drying the crop on tripods. To avoid this some
farmers are trying to treat the crop like, for example, a grass seed crop by
attempting to combine it directly from wind rows. Two other cash crops of
interest are flax (for retting for linen production) and oil-seed rape. In the past
flax has been a popular crop because it provides such a good break for cereal
diseases and because it increases soil organic matter, especially when undersown with a legume to provide some stock feed between harvest and winter
and again the following spring. However, it is rapidly disappearing, partly because of the general competition from man-made fibres, but more specifically
because retting proves so expensive for the Dutch farmer. 'Natural' retting as
practised in France for example (i.e. cutting the crop and leaving it in the field
to be retted by the weather) is not usually possible by reason of the low temperatures and because the Dutch farmer has always liked to undersow the
crop. Today most of the Dutch crop is sent to Belgium for chemical retting.
Oil-seed rape is likely to increase in the future, particularly with the decline in
flax acreage. This crop could prove just as satisfactory as flax from the cereal
disease and organic matter aspects; also it seems to offer much better future
prospects because the initial E.E.C. price was a good deal higher than that
previously received by the Dutch farmer, and the Dutch Government accordingly gave substantial acreage subsidy in 1965.

Traditional rotations are of the type: 2 years ley; winter wheat; potatoes,
or possibly sugar-beet; oats or sugar-beet seed or flax, etc. However, some
farmers are trying to omit the grass break and are moving towards a 3-year
rotation such as cereals; cereals; sugar-beet seed or potatoes or oil-seeds,
etc. This trend is hardly surprising as the average herd size is only 7 cows, but
it does aggravate the cereal disease and organic matter problems. Undersowing the cereal crops helps to maintain the organic matter (as, for example, in

* Aphids are the vectors of certain virus diseases and are discouraged by wind.

FARMING SYSTEMS OF THE WORLD

south Holland sector), but the farmers' concern over this problem is illustrated by the fact that some are now even ploughing in their potentially valuable straw.

THE POLDERS AND FRIESLAND (*Sector* (*e*))

Polder is a Dutch word meaning an area of low-lying land with its own individually controlled water system and water-table. It is enclosed by dykes which are often surrounded in turn by ring-canals, frequently with a higher water level. Excess water from the polder is pumped into the ring-canal, which may also supply water in time of drought. The main Polders are being made out of the former Zuider Zee, which is being drained and turned into productive farm land. So far three of the projected five polders have been drained and the North-East polder (the first of the large polders) was occupied by individual farmers from 1948. Reclamation involves draining with surface ditches, deep ploughing and levelling, and the building of canals and roads. In the third polder (East Flevoland) the whole area at this point was sown (by helicopter and aeroplane) with reeds; these grew quickly, prevented the establishment of other weeds and helped to dry the soil out by evapo-transpiration. When the soil was dry the reeds were either harvested, burned or chopped up and ploughed in and the land was ready for farming.

For the first few years, the polder land is operated by the State as development farms of about 1,250 acres each. Labour requirements are kept to a minimum by concentrating on those crops which are amenable to large-scale mechanisation, e.g. cereals, oil-seed rape, lucerne for drying, etc. Also, towards the end of this period, the land is tile-drained. Meanwhile, farm buildings and social amenities are erected so that communities are complete as soon as the farms are ready. Farms are then leased to farmers of proven ability and who have sufficient capital. Farm size varies from 30 to 120 acres depending on the farming system for which the land is suitable. Quite a wide range of soils, not all of which are initially suitable for farming, are found on the polders. However, poor soils are often improved by very deep subsoiling in the early stages of reclamation, and, in the more recent polders, a greater proportion of the area is being treated in this way. Nutrient status of the soil has changed quite markedly with successive polders. In the Weiringermeer, the first and smallest polder, P_2O_5 was very deficient and sometimes K_2O also. In East Flevoland on the other hand, neither P_2O_5 nor K_2O have been found necessary in the early years of cropping.

Friesland, which is north-east of the polders, has flat, heavy marine clay soils. However, on the eastern side of Friesland is a wide belt of peat bogs and 'peat reclamation' sandy soils interspersed with stretches of open water; this area was formerly poorly drained but big schemes of land reallocation and drainage, together with new mains water pumping facilities, have greatly improved its potential. The climate of Friesland and the polders is influenced by maritime rather than continental conditions and, therefore, has a slightly

longer thermal growing season (8 months), a lower mean temperature for the **Farming**
same period and higher mean annual precipitation than further east.   **Systems**

Farming systems. The north-east polder has mainly *intensive tillage* farms,
but also some mixed alternating farms and a few market gardens and orchards
Friesland is largely under permanent *intensive pastures* except for a coastal
strip of tillage farms, where the chief crop is potatoes for consumption and
for seed potatoes for other parts of Holland and for export, but cereals are
also of importance.

Farming in the North-East polder. During the period of 3 or 4 years when the
State farms the land, experiments with fertilisers, crops and machinery are
carried out so that, when the land is handed over to tenants, many of the risks
of pioneer farming have been eliminated. A small acreage is usually retained
as experimental farms and information centres. The farmers supply their own
fertilisers, stock, machinery, capital, etc., while the State provides the land
and all the farm buildings at a rent intended to be comparable to other parts
of the Netherlands. Certain cropping restrictions are stipulated in the contracts; for example it may be required that a proportion of the farm is in
grass, or that certain crops must not be grown too frequently. In east Flevoland, the farmers have been given much greater freedom over their cropping
policies, and in this polder, the rotations which are developing are often very
short and extremely intensive (e.g. sugar-beet; potatoes; onions is not uncommon).

The chief tillage crops grown in the North-East polder are cereals (33% of
the tillage area), sugar-beet (26%), potatoes (24%), oil-seeds, fibre and seed
crops (13%), pulses (3%) and lucerne (1%), whilst some 16% of the total
polder is under leys. The level of technical inputs is extremely high and, not
surprisingly yields are amongst the highest in the country; for example, winter
wheat 4,150 lb/acre, oats 4,150 lb/acre, potatoes 30,250 lb/acre, sugar-beet
44,800 lb/acre.

A 'model' 59-acre mixed farm on the North-East polder. A farm of this size
would consist of only one 'field' which would be ringed with drainage ditches.
The land would be tile-drained, and control of the water-table would be
excellent. Soil type would be somewhere between a light to sandy clay. The
grass acreage would be stipulated in the farmer's tenancy agreement and
could range from one-sixth to five-sixths of the whole farm. The grass would
be rotated round the farm and would be utilised by a dairy herd. Stock would
be carried at the rate of just under one cow per acre, so a farm with, say, one-
third in grass would have about 15 milking cows. The grass would receive
about 200 lb/acre of nitrogen. Conservation would be largely as hay. Cows
would receive between 1,300–1,500 lb of concentrates per lactation and average milk yields would be about 9,700 lb at 4% fat content. The buildings,
being relatively new, would be in good condition, but could well be proving
too inflexible for up to-date dairying practice. On a farm with relatively little
grass, potatoes and sugar-beet would probably be the most important crops.

But, it would be stipulated that potatoes should not be grown more often than 1 year in 3, nor sugar-beet more often than 1 year in 4. The manpower would usually consist of the farmer himself and one full-time worker; additional labour might be employed seasonally. Operations such as hay baling, silage making, spraying, etc., would be done on contract or the farmer might share some machinery with a few neighbours.

Grassland dairy farming in Friesland. Farms are small to medium sized with most of their land under permanent pasture, although fodder crops are grown on some of them. Many farms specialise in breeding Friesian–Holland cattle. Eighty-five per cent of Friesland cattle are milk-recorded. The province has its own Herdbook Society in Leeuwarden, and cattle for export are an important source of income. Most of the milk in the area is supplied to co-operative factories for butter and cheese making, but some goes by road across the Lake Yssel dam as liquid milk to population centres in the provinces of North and South Holland.

Efficiency and Stability

The success of the Netherlands in its struggle against the sea and river floods is, of course, one of the world's great engineering and social achievements. The fact that this country also is amongst the world leaders in mean yields of crops and milk is a measure of the ecological efficiency of its farming. These high yields are attributable to a useful bio-climate, to the use of well-adapted local crop genotypes, to high fertiliser consumption, to strict pest and disease control and to the great emphasis placed on attention to detail, whether this is aimed at individual animals or at individual cropping situations. Thus, even on small farms one sees that the slightly higher, drier fields are in tillage crops, whilst the adjoining, slightly lower fields with higher water-tables and more liable to flood, are in grassland. (This practise does, of course, add to the enterprises on, and therefore to the management and equipment problems of, small farms.) Against these advantages must be set the facts that although bio-climate is reliable, the working climate is not favourable for operations, especially on the low-lying wet clays. This tends to limit the full exploitation of the thermal growing season, both in tillage cropping and in grassland management (by reason, for example, of the 'poaching' problems).

The farming structure, on the other hand, is not well adapted to modern technology. This is recognised by the strong policies for land consolidation (in order to get rid of fragmentation and increase farm size) and for reducing manpower on farms. Further, slow progress in modernising farm buildings is a major problem.

Ecologically, and to a lesser extent economically, stability is relatively high, but a major short-term uncertainty is the risk of severe river floods and sea inundation—the memory of the 1953 disaster is still fresh. In the longer run, the fight for adequate water of good quality may become important and, when all the possible new polders have been created, the pressure of the rising population on a decreasing supply of farm land, may become more critical. But if yields per acre can continue to rise as more scientific and industrial inputs are used on farms, then total food output may rise at a rate which both offsets the loss of farm land to urban development and meets the food needs of the increasing population.

Conclusions

The Netherlands is a classic example of what man can do to master his environment by the control of water, by the creation of land and by the application of science and technology in a densely populated industrial country, in which there is, in contrast to, say, the U.S.A., not much room for manoeuvre. At the same time, it illustrates well, within one socio-economic and climatic framework, the influence of soil and water relationships on the location of farm enterprises and systems. It shows too (for example in the sand lands) the effect of modern technology in smoothing out edaphic and soil moisture differences. There is no reason to doubt that this country's marked technical progress in food production will continue for many future decades.

Acknowledgements. For research assistance to Kathleen Down and Sarah Rowe-Jones, and for comments and criticisms to M. L. 't Hart, E. W. Hofster, C. H. Henkens, M. A. Denig, and in particular to L. J. P. Kupers who also provided much detailed information on farming systems.

References and Further Reading

De Jonge, L. J. A. 1955, 1957. *Fatis Rev.* **2**. 4. 120–23 and **4**. 1. 1–5 O.E.C.D. Paris.
*Ede, R. 1955. *Agriculture*, **61**. 11. 534–37.
*'t Hart, M. L. 1955. *Fatis Rev.* Grassl. Suppl. 20–22. O.E.C.D. Paris.
*Highsmith, R. M. 1961. *Case Studies in World Geography: Occupance and Economy Types.* New York, pp. 55–59.
*Horring, J. 1958. *Dutch agriculture, its economic structures and problems.* Univ. London. Wye College. Occasional Pub. No. 10. Wye, Kent.
*Netherlands Foreign Agricultural Information Service 1962 and later editions. *Dutch Agriculture: Facts.*
*Netherlands Foreign Agricultural Information Service n.d. *Agriculture in the Netherlands.* The Hague.
*O.E.E.C. 1960. *Agricultural Regions in the E.E.C.* Documentation in Agriculture and Food No. 27. O.E.C.D. Paris.
Penders, J. M. A. 1957. *Fatis Rev.* 4 5. 142–44.
*Russell, E. J. 1954. *World Population and World Food Supplies.* London, pp. 83–96.
Rutter, A. J. and Whitehead, S. H. (Eds.) 1963. *Water relations of Plants.* Oxford.
Van der Molen, H. 1964. *Outlook on Agric.* 4. 2. 55–63.
*Visser, W. C. 1963 in Rutter, A. J. and Whitehead S. H., pp. 326–365, q.v.

* Asterisked items are for further reading.

CHAPTER 2.6

France

Introduction

French agriculture presents a confusing picture which is easier to comprehend if five factors are borne in mind. Firstly, France is a country of 136m acres, with 2·9 times as much land per head as the United Kingdom, but less total population, much of which is concentrated in the industrial regions of the north-east, of the Paris basin, and increasingly of the Rhône basin. The farm population, which is, at 18% (1963 Census) high for an industrialised country, is thus less subject to the impact of industry and its technological, social and intellectual influences than that of, say, the United Kingdom or Holland. Secondly, this large, rather isolated, rural population, though declining in numbers at 150,000 per year, is still politically powerful and is inclined to resist governmental attempts to modernise the structure and the technology of agriculture. Thirdly, the small mean size of farms, the fragmentation and the tendency for even 'ring-fence' farms to have many small parcels in different crops, hinders technological advance especially in specialised capital-intensive enterprises. Fourthly, comes the great climatic, altitudinal and geological range of the country which is divided from south-west to north-east by the Massif Central and bordered on the south-west and east by mountains. The northern half is, in effect, part of the North Sea basin and has crops, livestock and farming systems familiar in England or Holland. The southern half is warm temperate and the characteristic, and indeed the almost ubiquitous crop, is the vine for wine. But whilst the south-east has a warm, dry (Mediterranean) climate, the south-west has one which is more humid in summer and has some resemblance to that of the south-east of the U.S.A. In these circumstances, particularly as each region was originally largely self-supporting and has strong local traditions, it is not surprising that there is a galaxy of different farming systems and that the mean gross output per man and per acre varies greatly from district to district (Fig. 2.6.4). Finally, the European Common Market, with a common organised market, common prices and no significant intra-community barriers, is already quantitatively affecting French agricultural production, French exports and imports of agricultural produce: it is also changing the pattern of this trade. Historical influences have preserved much of French agriculture in its basically medieval form. The best French farms are, however, as good as any in Europe, and economic pressures, Government action and television are rapidly modernising the most competitive areas; other areas may have little future as farming districts (Beresford 1967). The sums being spent by the French Government and the grants available from E.E.C. sources for the improvement and modernisation of French agricultural structure (including amalgamation and reconstruction of farms and the pensioning off of those over 65 years of age) will show their effects in the years ahead.

Economic Background

France is an economically developed country (Beckerman Index 54), with a mean high standard of living comparable to that of other western European countries (see Tables 1.5.1 and 1.5.2). Her population has a high level of literacy, is well educated and enjoys a wide range of social and welfare services. She is largely self-sufficient in food. In 1963, food imports, mainly of tropical products, equalled 12% of household food consumption, whilst exports (cereals, sugar, wool, fruit and vegetables) equalled 11%. France, like the Netherlands, is a member of the European Economic Community and her agriculture is protected against non-European but not, in future, against intra-community competition.

Agrarian Structure

France has great problems of farm structure, for 60% of the farms covering 34% of the farmed area are between 5 and 50 acres. On the northern plains, the typical farm may be well over 100 acres, but nationally, and especially in the south, farms are too small in area. The smaller farms often consist of fragments of land which are often several kilometres apart, and which may be broken up into small parcels or fields which are, in turn, split into several patches of different crops or different varieties of the same crop. Out of 1·9m holdings, 1·2m are owner-occupied, 0·6m are cash rented, and the remainder are run by a manager or a share-cropper (*métayer*) (Severac 1961). The average farm has two workers—the farmer and a relative or an unrelated employee. Part-time and subsistence farmers are not uncommon. The Government is tackling these agrarian problems vigorously. Indeed, it is well ahead of farming opinion. Firstly, there are the schemes (S.A.F.E.R.) for the mutually agreed consolidation of fragmented holdings. Secondly, the Government has early retirement schemes for farmers and farm workers, and is, for example, encouraging the provision of alternative work in new industries as a means of depopulating farms, which, as noted above, is going on rapidly of its own accord.

Despite these structural problems, the physical living standards and agrarian services of country areas are on the whole good. On average there is one person per room whilst 9 out of 10 houses have electricity and 7 out of 10 piped water. Banking and credit services (some of them co-operative, others Government supported and some supervised and conditional on the execution of a pre-agreed farming plan) are well developed. Input and marketing services are also well developed both by private industry and by the co-operatives. The Government operates various price support schemes and market subsidies as well as some production subsidies, e.g. on the purchase of

FARMING SYSTEMS OF THE WORLD

machinery. It is promoting efficient marketing of the many small lots of a very wide variety of produce sold by hundreds of thousands of small farmers from week to week. Large 'central' markets have been set up in producing areas; refrigerated transport and storage are increasingly important. The French realise these marketing devices secure some of the advantages of scale to small farmers and must supersede the picturesque haggling over a few chickens or a bag of artichokes in the market place.

Fig. 2.6.1 Agricultural Regions of France. *Source:* O.E.C.D. Paris

Agricultural education extends from day-release classes, or their equivalent, through agricultural high schools to the regional agricultural colleges, and finally to the Institut National Agronomique in Paris. Some of these are at degree level. Responsibility for research lies with the I.N.R.A. (National Institute for Agricultural Research) centred at Versailles and Paris, but with regional and local out-stations. The technical services of the Ministry of Agriculture (e.g. plant protection, animal health, husbandry improvement, marketing, etc.) have advisory as well as regulatory functions, but advice on production matters has been, in part, transferred to local and commodity (e.g. tobacco) producers' organisations. Although mutual farm accounting schemes and management advisory services (Centres de Gestion) are common, and although academic interest in, for example, linear programming, is growing, 'farm management' is perhaps less developed than in most advanced countries. Nevertheless, despite the above rationalisation schemes and these services, gross product per man in farming (Sharp and Capstick 1966), net productivity per man and net factor productivity (i.e. value added per unit of capital *and* labour) (Van den Noort 1968) is substantially less in France than in the Netherlands, Denmark, Canada and the U.S.A.

Climate, Relief, Soils, Vegetation

The climate ranges from cool humid in the north-west with an annual mean precipitation of 25–30 in, to warm semi-humid in the south-west (27–32 in) and warm dry (Mediterranean) in the south-east (20–25 in). It is cold or cool humid in the mountainous Massif Central, the north French Alps and the Pyrenees (25–50 in or more) (Table 2.6.1, Fig. 2.6.2). The mean thermal growing season temperatures vary from 55°F in the north to 57°F in the south where, however, the mean summer (June–September) temperatures are about 7°F higher than in the north. Over most of the country the annual P–E–T is between 25 in and 30 in. but falls to just over 20 in in the high mountains and rises to 32 in in the Rhône delta. The mountains have excess moisture and the north-west coast has little deficit; but much of the farmland, particularly along the south-east (Mediterranean) coast, suffers from drought in the summer months. This deficit is aggravated by moisture loss attributable to wind—which does not feature in Thornthwaite's formula. (Thus it has been estimated that the energy in the wind (*apportée par le vent*) is 5 to 6 kWh/m²/year on the coasts of Normandy and Brittany and in the Camargue and Languedoc in the south, compared with 1 to 2 in the northern plains and 3 in Massif Central and the high Alps.) Wind-breaks such as are found in Brittany and in Provence (see p. 280), can (apart from preventing mechanical damage to crops) substantially reduce moisture loss and thereby increase the energetic efficiency of a cropping system. But a more serious problem, in the areas most liable to moisture deficit, is the uncertainty of the summer rainfall. The C.V.* of mean annual precipitation ranges from 10–15% in the north to 20% in the south-east. More important is the very high C.V. in July and August; this is one of the main weather hazards in the south for such late-maturing cereals as maize and grain sorghum and for vineyards, tobacco, some tree fruits and, of course, for vegetables. Irrigation with subsidised or zero-cost water is a well-established safeguard in such areas. Frost is another important hazard, even in the south-east; it attracts considerable research and various frost-control measures, e.g. amongst vineyards.

Only two-thirds of the rural land surface of France is farmed. Most of the remainder is in forest, xerophytic scrub or alpine vegetation. The great

---

* Standard deviation as percentage of the mean.

Table 2.6.1

Climatic Data — France

Station	Bright Sunshine hours/day	Thermal Growing Season length in months (> 42½ °F)	Thermal Growing Season Mean Temp °F	Mean Months > 72½ °F	Thermal Growing Season P–E–T (mean inches)	Thermal Growing Season A–E–T (mean inches)	Mean Annual Precipitation (inches)	Moisture Index (Im)	Annual Precipitation Coeff.Var. (C.V.)	Max. recorded rain/day inches (month of occurrence)
Lille	—	9	55·3	0	25·2	23·3	30·3	1 moist sub-humid	10–15%	1·9 (September)
Reims	4·9 (Paris)	7	58·5	0	23·3	20·7	25·2	−3 dry sub-humid	10–15%	2·6 (June)
Dijon	—	8	57·9	0	25·2	22·8	27·4	2 moist sub-humid	10–15%	2·5 (June/July)
Lyon	4·9	9	57·6	0	27·4	26·1	32·6	14 moist sub-humid	10–15%	4·3 (October)
Marseilles	7·2 (Nice)	12	57·6	0	30·2	22·5	22·5	−25 semi-arid	10–15% (Montpellier 18%)	8·7 (October)
Bordeaux	—	11	56·4	0	27·9	25·1	32·8	15 moist sub-humid	10–15%	2·1 (June)
Toulouse	—	11	56·7	0	28·0	25·2	26·9	−5 dry sub-humid	10–15%	5·5 (September)
Le Puy St Etienne	—	7	57·2	0	23·4	23·0	32·9	25 humid	10–15%	2·3 (November)

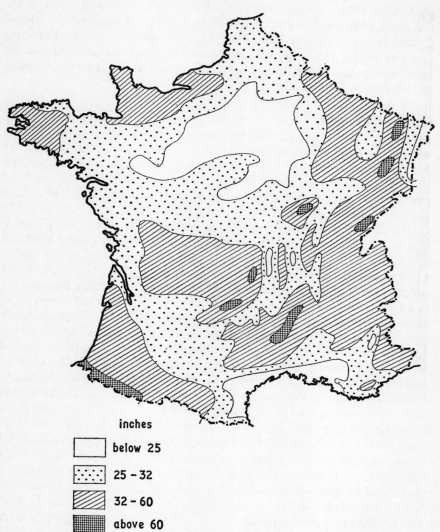

inches

☐ below 25

▦ 25 – 32

▨ 32 – 60

▦ above 60

Fig. 2.6.2 France. Mean Annual Precipitation. *Source:* Min. Agric. Paris

expanse of the Massif Central, with its very old formations, and the mountains of the south and the east not only break up the country, and thus hinder transport, but mean that much of the country lies above 1,500 ft, where A–E–T is often too low for anything except extensive grazing. Moreover, the great limestone plateaux which abut on the Massif Central are, except in the alluvial river valleys, often too permeable to retain enough water in the surface layer for cropping even if they are operationally suitable; whilst the thin

sand-dunes of the Landes are not only drought-prone but acid. Most of the
unfarmed areas are found in regions *H* and *J*, viz. the Jura and the Massif
Central in Fig. 2.6.1 and Table 2.6.2. They are naturally marginal and not
very productive. This is reflected in the crop and livestock statistics and the
value of outputs per acre, Table 2.6.2. Unfarmed areas are, however, found in
regions *D*, *G*, *I* and *K* as well.

Crops and Livestock

Table 2.5.1 summarises the acreage, numbers and yields of crops and livestock of the farming area of 95m acres. Nearly half are in tillage (field) crops
and temporary leys, more than a third in permanent grass and rough grazing
and nearly 10% in vineyards, tree fruits, vegetables, flowers, etc. Wheat is the
most important single crop and, like barley, is grown in all tillage areas. Maize
and sorghum, though sometimes grown north of Paris, are mainly found in
the warmer south-west where the summer rainfall is better and more reliable
than in the south-east, where, in the Camargue, rice is grown under flood irrigation. Yields have been increasing rapidly, particularly those of maize,
which on good farms in irrigated areas run at 4,500–5,500 lb/acre, whilst
wheat gives, in the better areas, 5,000 lb/acre. Cereal production has doubled
in 10 years. Potatoes are found on most farms but the chief 'main' crop
areas are in the north (14,500 lb/acre) whilst earlies come from Britanny
(compare Cornwall in the U.K.) and the south. Sugar-beet is popular on the
northern plains where it averages around 45,000 lb/acre, as do mangolds and
fodder beet which are widely grown (except in the south) and fodder cabbage
(usually cut and carted) which is a feature of French livestock feeding in
winter, especially in Brittany. Grassland is more important in the wetter,
cooler western or upland regions (see Tables 2.6.2, 2.6.3 and 2.6.4) such as the
Limousin and Charollais, though one- or two-year leys or lucerne or sainfoin,
often for hay or silage only, are frequently taken in the tillage districts of the
northern plains and sometimes, as break crops, in the irrigated farms in the
south. Grassland species, except in the southern lowlands, are the same as
those used in north-west Europe, but grassland management, though improving rapidly, has not yet reached Dutch or even English standards. With
such notable exceptions as the farms of André Voisin, cow-days per acre seem
to be only moderate, though 2,500 to 3,200 fodder units/acre are said to be
common on well-managed pastures, and live-weight gains of 355–535 lb/acre
reported. (One fodder unit = 1 kg barley.)

Though the emphasis is, in the acreage statistics (Table 2.5.1), on tillage
crops and tree products (including vines), in fact livestock products account
for the greater part of the gross sales of French farming. But tillage rather than
grassland is the basis of the French livestock industry. Until recently, cattle
were either dual purpose or even triple purpose—meat, milk and work—and
although in, for example, the south-west, one still sees draught cattle at work,
specialisation in meat or milk is rapidly increasing. Apart from liquid con-

Table 2.6.2

France — Summary of Farming Regions

Region	Description	Climate type	Farming systems	Mean yields		Index of gross sales (Region A = 100)	
				Wheat lb/acre	Milk lb/cow	Per unit area	Per active person on farms
A	*Channel* coast area from Belgian frontier to Seine estuary and inland to the foothills of Ardennes	Cool Humid (Maritime)	Intensive tillage and intensive alternating both with livestock	2,800	6,218	100	100
B	*Normandy* coastal region with its hinterland	Cool Humid (Maritime)	Intensive grazing. Very high livestock density	2,050	5,138	54	46
C	*Paris Basin* with the Central Region (Île de France) and outlying districts	Cool Humid/ Cool Dry	Intensive tillage. Few livestock	2,800	5,711	64	94
D	*North-eastern* Plateaux and hills of Lorraine and Alsace	Cool Humid (Continental)	Semi-intensive alternating with livestock but more intensive in Alsace	2,050	5,270	50	66
E	*Brittany*	Cool Humid (Maritime)	Intensive tillage and intensive alternating. High livestock density on arable by-products	1,800	3,550	70	48
F	Region south of Loire, with *Anjou*	Cool Humid	Less intensive tillage and alternating. Lower livestock density	1,900	4,256	43	53
G	Region of broken contour with *Burgundy* and adjoining terraced plateaux	Cool Humid	Less intensive alternating and grazing. Livestock. Vines on Burgundy plateaux and valleys	1,800	3,682	43	52
H	Mountains of the French *Jura* and northern Alps	Cool/Cold Humid (Alpine)	High proportion of uncultivated land, woods and forests. Extensive livestock. Some intensive trees (vines)	1,650	5,182	54	50
I	South-west France: Plains and valleys *Aquitaine Basin*	Warm Humid	Intensive and semi-intensive tillage, alternating or grazing. Horticulture in valleys including vines	1,450	3,153	50	40
J	*Massif Central* mountain region	Cool/Cold Humid	Extensive grazing	1,350	3,065	41	41
K	The French *Mediterranean* region from the foot of the Pyrenees across the Rhône delta, inland to the foothills of the Alps	Warm Dry (Mediterranean)	Extensive grazing on mountains. Intensive horticulture and tillage on plains. Vines very important. Low livestock density	1,350	4,542	69	62

Sources: Condensed from O.E.C.D. 1960 and other sources

sumption (about one-fifth of milk output) and butter and cream, the 300 varieties of French cheese provide an important outlet for milk, whilst 3-months-
old milk-fed veal is very popular and very good. The beef, on the other hand,
is not remarkable, though the specialist beef breeds like the Charollais make
excellent eating and trace back to the 'Durham' breed introduced from the
United Kingdom in 1825 (Beresford 1967). This breed and the Limousin are
favoured for meat whilst the Frisian (French Friesian), Normandy and
Tachetée de l'Est are most popular for milking.

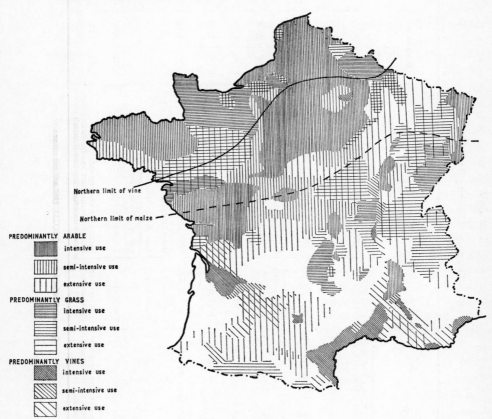

Northern limit of vine

Northern limit of maize —

PREDOMINANTLY ARABLE
     intensive use
     semi-intensive use
     extensive use

PREDOMINANTLY GRASS
     intensive use
     semi-intensive use
     extensive use

PREDOMINANTLY VINES
     intensive use
     semi-intensive use
     extensive use

Fig. 2.6.3 France. Farming Systems. *Source:* Min. Agric. Paris

Cattle are densest in the wetter north-west and in the uplands; in the south
they are often replaced by sheep and goats. Though sheep numbers and lamb
consumption are low, the richer French are very fond of young lamb—
wool is of some importance; and, in specialised areas, milk from ewes is
produced, mainly for the manufacture of Roquefort cheese. As in the United
Kingdom, there are many local breeds but it is worth noting that some of
these can conceive all through the year even north of Paris. The pig industry

is very important, especially in Brittany and central France. (There are few pigs in the south). The two main breeds are the Large White and the Western White. Though modern, large-scale, specialised and highly efficient units are on the increase, most farmers in the north keep a few sows or buy in weaners and feed them out by rather old-fashioned methods. Poultry, with the accent on meat rather than eggs, are widespread, even in the south. Though the barn-yard fowl is still in evidence, specialist broiler or egg units are on the increase. Minor livestock include milking goats for cheese, rabbits for meat, bees and, where the mulberry grows in the south-east, the remnants of the once great silkworm industry. In brief, the French livestock industry is very, almost too, varied, is not yet fully developed and is mostly in the north. Unlike Israel, the U.S.A. and Australia, there is little animal production in its Mediterranean regions, although, especially with irrigation and greater knowledge and con-trol of heat stress, there is plenty of room for expansion in the south and in fact some beef lots, *à l'Americaine* (p. 163) are appearing there.

Fruit and Vegetables

The south makes up in vines, tree fruits and vegetables what it lacks in live-stock. Although *vineyards* for wine are to be found in most of France except the north-west and the mountains, they are concentrated in the south-west, round Bordeaux and in the south, especially in Languedoc and in the south-east. There is, in fact, an overproduction of wine which is exacerbated by a tradition of imported Algerian wine which is cheaper and which 'finds its way' into French 'Beaujolais' bottled wine, etc. This has created an awkward politi-cal situation and the Government is trying to reduce output which averages about 3,500–4,500 lb/acre. Irrigation of vines and new planting are forbidden; and the grubbing up or abandonment of old vineyards and their replacement by tall tree fruits, vegetables and flowers is being encouraged on the lowlands as the higher quality, lower yielding wine grapes can be grown well on the poorer soils of hill-sides. *Tree fruits*, especially apples, of many types are important, but frost risk (see Israel, p. 311) precludes citrus. Fresh vegetables are fea-tured much in French diets and a great range of them is grown, mainly in the south (often irrigated), in Brittany and in market gardens round large cities. To the French farmer, however, horticulture or truck farming is not, as in the United Kingdom, a separate industry, but part of a farming system as several of the sample farms described below will illustrate. This integration has many rotational and labour-spreading advantages which go some way to offset lack of specialisation.

Crops: Weeds, Pests and Diseases

The broad-leaved weeds are mostly those found in northern Europe though naturally, in the south, Mediterranean species are more abundant. Although, possibly because of the importance of row crops and the greater availability of manpower, weed control by cultivation is emphasised, grass weeds such as

269

couch (*Agropyron repens*) in the north and hairy finger grass (*Digitaria*
sanguinalis) in the south are also controlled chemically when control by cultivation (e.g. before spring sowing) is not practicable. But a general dependence
on weed-killers is not so obvious as it is in the United Kingdom or the
U.S.A.

Diseases of small grains include bunt (*Tilletia* spp), smuts (*Ustilago* spp),
(*Fusarium* spp), leaf spots (*Septoria* spp), eye-spot (*Cercosporella herpotrichoides*), rusts (*Puccinia* spp), (*Helminthosporium* spp) and take-all (*Ophiobolus graminis*). Maize is subject to smut and a form of *Fusarium* infection
called Gibberella; sugar-beet to heart-rot (*Mycosphaerella tabifica*), leaf-spot
(*Cercospora beticola*), and virus yellows; potatoes to blight (*Phytophthora
infestans*), mosaic and other virus diseases, dry rot (*Fusarium coeruleum*),
common scab (*Actinomyces scabies*) and bacterial ring rot (*Corynebacterium
and Bacterium sepedonicum*) though this last is not common. Black scab, i.e.
wart disease (*Synchytrium endobioticum*) is apparently under control. Tobacco
diseases include wild fire (*Pseudomonas tabaci*), mosaic virus, blue mould
(*Peronospora tabacina*), whilst amongst weeds broom rape (*Orobanche* spp)
is a parasite of this crop and of some forage legumes. Amongst the latter, root
rots (e.g. *Rhizoctonia violaceae*) are troublesome. Control of crop diseases is
by familiar methods, e.g. by organo-mercury seed dressings, clonal hygiene,
genetic selection and choice of variety, spraying fungicides (where appropriate), and by husbandry means (e.g. dating of seeding, rotations). Vine
diseases are vine mildew (*Uncinula necator*), downy mildew (*Plasmopara
viticola*), black rot (*Guignardia bidwellii*), Anthracnose (*Gloeosporium ampelo-
phagum*), white root rot (*Rosellinia necatrix*) and a virus disease called
'Court-noue' (Gory and Gauthier 1963, p. 229).

Cereal root eel-worm is apparently absent but cereal pests include wire-
worm (*Elateridae*), in maize, small grain and sugar-beet; wheat bulb fly
(*Leptohylemyia coarctata*); the European corn-borer (*Ostrinia nubilalis*) in
maize; mangold-fly (*Pegomya betae*), and cockchafers (*Melolontha melolontha*)
in beet. There is no need to remind farmers and officials in south-east
England that across the Channel the Colorado beetle (*Leptinotarsa decemline-
ata*) is still a potato pest in France as are wire-worm, cockchafers and the
common silvery moth (*Plusia gamma*). Wire-worm and various caterpillars
attack tobacco. In vines mites (*Acarina*) produce blisters, whilst damage by
Sparganothis pilleriana, *Clysia ambiguella* (vine moth), *Polychrosis botrana*
(grape-fruit moth) and other pests are encountered. Weevils (*Phytonomus* spp)
and other small insects attack forage legumes. Pests of grasses do not seem to
be a problem, or if they are have not been diagnosed. *Phylloxera* destroyed
the greater part of the French wine industry between 1870 and 1900. Attempts
to control the pest, which attacked the roots of the vines, were by gassing (e.g.
with CS_2 (*sulfure de carbon*)) and by drowning (e.g. winter flooding), but today the industry relies on the grafting of high quality French varieties (genotypes) on *Phylloxera*-resistant stocks of American origin. Some damage to

leaves can still be caused by *Phylloxera* in its winged form but this can easily be chemically controlled.

Livestock: Pests and Diseases

Foot-and-mouth is endemic but its treatment is interesting; the vaccination policy appears to be effective to date (Beresford 1967). Amongst cattle, tuberculosis has been eradicated to a large extent and brucellosis is being vigorously tackled. Mastitis, warble-fly (*Hypoderma* spp), contagious vaginitis and trichomoniasis are also troublesome. Amongst metabolic diseases, milk fever is the most reported. In pigs, apart from internal parasites, swine-fever (hog cholera), erysipelas, contagious pneumonia and various *Salmonella* infections cause trouble. In the northern half of the country, footrot and internal parasites are common in sheep whilst in the warmer parts ticks are important amongst external parasites (see also Israel, p. 308) because they may be carriers of piro-plasmosis. Pullorum and coccidiosis call for precautionary counteraction in young chicks; in adult birds avian cholera, tuberculosis, fowl pest and variola call for attention as do internal parasites, fleas and mites. Mineral deficiency diseases are not widely discussed, but this may be because they have yet to be diagnosed and accepted by farmers or because the intensity of stock (e.g. sheep) or of production generally is lower than in say the United Kingdom or the Netherlands.

FARMING REGIONS

Fig. 2.6.1 and Table 2.6.2 show the locations of and summarise the characteristics of the main farming regions. Regions *A*, *C*, *G*, and *K* are discussed in more detail in the North to South Transect below (Table 2.6.3) and *I*, *J* and *K* in the West to East Transect (Table 2.6.4).

Of the remaining regions, *D* (North-eastern) and *F* (West centre) do not call for special comment. The latter is a southward but perhaps inferior extension of the Paris basin (Region *C*). Regions *B* (Normandy) and *E* (Brittany) have low proportions of large farms, high livestock densities and moderate product yields and gross sales per man. But they differ markedly in farming systems. Normandy, with a higher rainfall, has lower moisture deficits on Thornthwaite's formula (Arlery *et al.* 1954) and perhaps somewhat higher A–E–T, perhaps somewhat heavier soils and rougher, less cultivable terrain than Brittany and the Vendée (Region *E*). Normally it is largely in intensive grassland or alternating systems with milk and its products, veal and beef as major enterprises. Brittany has slightly less precipitation but more moisture loss through wind. Nearly three-quarters of its farmed area is in arable (as against one-third in Normandy) and there is much emphasis on fodder cabbage and other tillage fodder crops. A typical 3-year arable rotation is wheat; oats or barley; fodder crops. Cattle for milk, pigs and poultry are significant. The relatively frostless oceanic climate favours early potatoes and intensive vegetables, especially cauliflowers.

NORD (*Sector* (*i*))

This sector has a mild, cool humid climate with a thermal growing season of 9 months, a reliable mean annual rainfall of about 30 in (C.V. 10–15%). The mean growing season temperature is 55·3°F at Lille which has a mean P–E–T of 25·2 in and an A–E–T of 23·3 in; there are about 60 days of frost and 175 days of rain. (January mean is 37·4°F, July mean is 64·4°F.) The climate is somewhat more continental than that of south-east England and summer drought is more of a problem than winter cold. A region of gently rolling plains, much of it is covered by up to 3 ft of loam overlying permeable chalk. This soil is high in K_2O and organic matter but sometimes deficient in P_2O_5 and lime. Stones are few and cultivation is easy. It is well suited to intensive tillage, except on the heavy clays of north Flanders. The 'natural' vegetation in this and the next two sectors is mixed forest. Though there are still too many small farms, there are many large enterprises (for France), mostly in cash tenancies, on which mechanisation is advanced. Manpower is relatively scarce owing to widespread industrialisation, which, however, brings a good infra-structure; education, communications, rural electrification (97% of farms are on the mains) etc. are good though both urban and rural housing standards are rather low. Gross output per acre and output per man are relatively high (see Table 2.6.2, Region *A*, Fig. 2.6.4).

Farming systems. Intensive tillage or *intensive alternating* systems predominate. A typical rotation is wheat; barley or oats; cash roots (sugar-beet, potatoes) or fodder crops. Beet is processed locally but, as there is an excess of sugar production, and casual labour is short at thinning and harvest time, the acreage is fluctuating and the factories being rationalised. Most of the chicory (for coffee flavouring) in France is grown in this sector, but this crop, like flax, a traditional crop of Flanders, is declining. There is some market gardening round urban centres. Livestock (mainly cattle) density is high with a growing emphasis on liquid milk, though beef and pigs are also important. Yields (see Table 2.6.2, Region *A*) are amongst the highest in France.

PARIS BASIN (*Sector* (*ii*))

This sector has a lower annual precipitation than sector (i). The thermal growing season (7 months) is somewhat shorter and both the P–E–T and the A–E–T are less. The mean growing season temperature is higher, e.g. 58·5°F at Reims which has a January mean of 35·6°F, a July mean of 66·2°F, 74 days of frost and 151 of rainfall. It is thus, at least in its eastern half, more continental than the previous sector. The gently rolling plains are mostly covered, as in sector (i), with thick loams overlying chalks, but there are areas of clay with flints. Permeable underlying strata, a lower rainfall and higher summer temperatures combine to make this sector more prone to drought than sector (i). Farms are mostly large (60% are over 120 acres) and tenanted. Though its

Table 2.6.3

France — North to South Transect

Transect (a) Long. 2°E, Lat. 51°N ⟷ Long. 4°E, Lat. 43° 20′ N

	(i) North	(ii) Paris Basin	(iii) Burgundy	(iv) Mediterranean
1. Sector	(i)	(ii)	(iii)	(iv)
2. Name	North	Paris Basin	Burgundy	Mediterranean
3. From	Dunkirk	Cambrai	Joigny	Vienne (via Villefort)
4. To	Cambrai	Joigny	Vienne	Camargue Coast
5. Miles	75	150	185	150
6. Climate-type(s)	Cool Humid	Cool Humid	Transitional	Warm Dry (Mediterranean)
Thermal Growing Season:				
7a. — months at or > 42½°F	8–9	7–9	8–9	9–12
7b. — months at or > 72½°F	0	0	0	0
8a. — P–E–T (in) in T–G–S	24–26	22–24	25–27	27–30
8b. — A–E–T (in) in T–G–S	22–24	20–21	22–26	22–26
9. Growth limited by	Winter cold	Winter cold	Rigorous winters. Sudden cold spells	Summer droughts
Precipitation:				
10. — mean in (annual)	20–40	20–40	20–40	20–40
11. — C.V. % of Precipitation	10–15%	10–15%	15–20%	20–25%
12. Special Climate Features				Summer droughts. Cold, dry winds
13. Weather Reliability	3	3	3	3
Soil:				
14. — Zone/Group	Podsolic	Podsolic	Podsolic	Latosolic
15. — pH	Medium	Medium	Low	High
16. — Special Features	Heavy clays. Loams	Loams, sands, clays with flints	Chalkland. Sometimes poor, thin soils derived from deposits from Massif Central	Alluvium. Loamy dry soils
17. Vegetation	Mixed forest	Mixed forest	Mixed forest	Mediterranean woodland and scrub
18. Relief	Plains	Plains and hills	Hills and rolling plains	River valleys, plains, hills
19. Market Access	4	4	3	4
20. Farming Systems	Intensive tillage; intensive alternating	Intensive tillage; intensive alternating	Semi-intensive grazing; intensive alternating	Intensive tree crops; intensive tillage; extensive grazing
21. Major Enterprises	Wheat, barley, sugar-beet, potatoes, milk, pigs, poultry	Wheat, sugar-beet, vegetables, tree fruits, milk, pigs, poultry	Beef, sheep, pigs, poultry, wheat, barley, potatoes, vineyards	Vines, fruits and vegetables, cereals (rice (irrigated)), sheep and goats

273

presence ensures a good infra-structure, Paris is a great magnet for labour and manpower shortage is increasingly determining farming systems. Thus, one notes many small family-run market gardens (truck farms) on the one hand and on the other large fully mechanised farms of 500–800 acres employing hired labour. To save work many of the latter have eliminated livestock and replaced sugar-beet (which is still, however, a key crop in this sector) by hybrid maize or by colza (rape-seed for oil). Fertiliser usage is high and modern chemical control of weeds and pests is normal. Though livestock units per acre average only half that of sector (*i*) (Nord), the gross sales per acre are still (for France) fairly high whilst sales per man rival those of sector (*i*).

Farming systems. The systems in this rapidly changing and progressive area are *intensive tillage* with some *intensive alternating*. On the large farms, a common 3-year rotation is wheat; barley; sugar-beet or fodder crops (e.g. lucerne for hay) or sometimes maize for silage or grain. Where all the crops are not sold away, this rotation may be combined with winter fattening of cattle or with pig or poultry keeping or sometimes sheep—now mainly kept indoors as there are few fences and shepherding is too expensive. Liquid milk-producing farms may have longer rotations including leys or, to save fencing costs, they may keep their tillage and grassland (resown leys or permanent pasture) separate. Fruit, vegetables and broiler (chicken) production are also significant.

A tillage and fat-stock farm near Pontoise. This farm of 1,100 acres lies in an open district of gentle slopes with wooded crests, large arable farms and no fences or hedges. The soil is a heavy chalky loam of pH 7 to 7·2. There are 480 acres of winter wheat, 220 of spring barley, 120 of sugar-beet, 50 of potatoes—half of which are earlies—and the balance in lucerne. The 6–7-year rotation is, roughly, two wheat crops, followed by cash roots then barley and finally lucerne for 2 or 3 years. Wheat is drilled from mid-November to mid-December, barley in early March, beet from the end of March to mid-April. Barley is harvested by combine from mid-July, wheat in August and beet in October and November. The beet is machine thinned and mechanically lifted, but casual labour is also used as well as for the potatoes. Wheat receives 36 lb N/acre, 44 lb P_2O_5/acre and 44 lb K_2O/acre—part of the N is supplied as top-dressing. Barley receives somewhat less and sugar-beet about twice as much P_2O_5 and K_2O. Farmyard manure is mainly applied to potatoes. Herbicides are always applied to the cereals, whilst pre-emergence spraying is used for weed control in beet. Maize for grain has been tried but it ripens late and, apart from getting lodged, may contain up to 50% of moisture which makes drying costs excessive. Wheat and barley generally come in at about 20–23% moisture which is reduced to 15–16%. Pest and disease trouble includes '*Mouche du blé*' (*Leptohylemyia coarctata*) and '*Pietin*' (Take-all) (*Ophiobolus graminis*), '*Charbon*' (Smut) (*Ustilago* spp) in barley, and various insect pests; virus yellow in the beet is often sprayed with systemic insecticides two or three times to control the vector aphids. Yields in a good year are 4,500 lb/acre of wheat, 3,500 lb/acre of barley, 45,000 lb/acre of beet and 20,000 lb/acre of potatoes; the early potatoes are irrigated to get higher yield and earlier bulk-ing. A new crop enterprise, introduced to increase gross output, is heated glasshouses which are thermostatically controlled and sprinkler irrigated from about 5 ft above soil level. One crop of tomatoes and two crops of lettuce are taken each year.

The main livestock enterprise is a flock of 400 housed ewes of the Île de France breed. This was derived from a cross between the Leicester (Dishley) and the Merino and the ewes are said to take the ram at any time of the year. Twice-a-year lambing is the aim and with about 10% of doubles it is possible to sell two milk fat lambs per ewe per year. The lack of fences and the shortage of sheep-watchers preclude grazing, so the sheep spend their whole lives on deep straw litter in the ample farm buildings. The ewes receive lucerne, hay, cut and carted beet tops, beet pulp or beet pulp silage (which is made in towers), grain and protein supplements. In addition, 100 or so Charollais store cattle are purchased each autumn, tied up in byres and lie on plenty of straw. Their diet is much the same as that of the ewes though dried lucerne pellets may replace oil cake (for protein).

The equipment includes two combine harvesters, a potato harvester, a beet harvester, hay-making machinery, a large grain-dryer, grain storage and a small irrigation unit for the early potatoes. The farm is impressively run by two brothers with a regular work force of 15 men plus casual Spanish labour for potato, beet and glasshouse work peaks. By U.K. standards, the work force is on the high side and the indoor livestock systems are labour consum-ing; but the cattle and sheep provide a winter work load, an outlet for straw and a supply of dung. In brief, a rather old-fashioned system but one which uses modern technology and could be adapted (for example, by loose housing the cattle and reducing the root acreage) to a tighter labour situation.

A dairy farm in the Oise valley (L'Isle d'Adam). This farm of 265 acres of heavyish soil has a herd of 55 pedigree Jersey cows (which is unusual) and a herd of 50–60 commercial French Frisians. There are two large 'haylage' tower silos for maize and lucerne–grass silage, respectively. These feed automatic-ally into an auger-operated feeding trough which runs the length of the partly covered yard where the Jerseys and some Frisians spend most of the year. A.I. is not used. Calving takes place throughout the year with a September–Octo-ber peak. The calves are reared on dried-milk based concentrate until 3 months. Through most of the year the cows' feed is maize silage in the morn-ing and lucerne-grass silage in the afternoon. Concentrates, grown and ground on the farm, are fed except from the end of April till the end of September when the herd is rotationally grazed on 160 acres of permanent pastures or long-term grass/clover leys. The two-level milking parlour has 4-in-tandem stalls, with direct delivery to a bulk milk tank which is emptied daily for the liquid market. The Jerseys average 7,700 lb per lactation. The intensive alter-nating system is on a 6–7 year rotation. Normally, there are 30–37 acres of

275

lucerne or lucerne/grass which is left down for 2 or 3 years then ploughed up
and followed by maize (35 acres, half for grain, half for silage), wheat (winter
or spring) (49 acres), or spring barley (62 acres) for 3 years before reseeding.
N at 36 lb/acre on wheat and 65 lb/acre on maize, supported by K_2O and
P_2O_5, are applied, and typical crop yields are maize (for grain) 4,500 lb/acre,
wheat 4,000 lb/acre and barley 3,600 lb/acre. The maize, which receives pre-
emergence weed treatment with Simazine, is planted in early May and cut for
silage in late September or harvested for grain by a contractor's corn-picker in
October.

Some lucerne/grass (mostly cock's-foot (*Dactylis glomerata*)) hay is made,
but most of the temporary leys and the excess grazing on the permanent pastures are cut for silage, using an American cutter-bar-type forage harvester
which gives very short material suitable for feeding in and out of the mechanically discharged American-type forage wagons and thence to the tower silos.
Large quantities of haulms of vining peas are purchased from a near-by 'petit
pois' canning factory and made into excellent pit silage which is fed to the
young stock and dry cows, etc. 2,800 Rhode Island Crosses are kept on a semi-
deep litter system with high dropping boards along which runs a conveyor-
belt (controlled by a time-switch) bringing a high energy meal ration. In addition, a scratch feed of kibbled maize and black oats are fed. Yields are said to
be over 250 eggs per annum per hen housed. In addition to the working
manager, there are 4 regular men. The farm is heavily mechanised and has, in
addition to the equipment mentioned, a combine harvester, a grain-drier, a
cold air hay-drier, sprinkler irrigation and 5 tractors (one of them a caterpillar), which are also, however, used by the forestry enterprise with which
this farm is combined. About half the land in grass is irrigated 3 to 5 times per
season at 1 in to $1\frac{1}{2}$ in per application. In brief, a large but somewhat over-
equipped dairy farm well up on modern techniques and with, through winter
work in the forest, a well-spread work load.

BURGUNDY REGION (*Sector (iii)*)

This sector runs across Burgundy. Its transitional climate is compounded
of maritime, continental and Mediterranean influences. Precipitation is fairly
abundant but temperature fluctuations are large. Rigorous winters and sudden
cold spells are partly offset by warmer springs and mild autumns of Mediterranean origin. The thermal growing season is 8 to 9 months with a mean temperature of about 58°F and a mean P–E–T of 26 in and a mean A–E–T of 23
in. The mean annual precipitation on the lower lands is about 30 in but is not
very reliable. At Lyon the mean January temperature is 37°F and July is 70°F;
there are, on average, 62 days of frost and 150 of precipitation. Winter cold,
summer heat stress and moisture deficits are problems. Apart from the Seine
and Rhône valleys, and small districts which have good deep soils high in
K_2O, P_2O_5 and organic matter, most of the terrain is hilly, 600 to 1,500 ft in
altitude, and has poor thin soils often overlying chalk. The famous Burgundy

wines of the Côte d'Or are grown on a chalk outcrop. More than half the farmed land is in holdings of 25–120 acres and 40% are over 120 acres. Nearly half this area is owner-occupied but *métayage* (share-cropping) is still significant. Manufacturing industry is increasingly important. The schemes to provide navigable waterways, irrigation water and electric power from the Rhône and Durance rivers and a local source of natural gas will speed industrialisation. This, whilst providing local markets for milk and vegetables, will draw manpower from farms at the same time that it improves the local infrastructure and access to industrial inputs for farms. Hence, although gross sales per acre and per man and yields are low or mediocre (Table 2.6.2, Region *G*) they will no doubt rise.

Farming systems. Although some favoured hill-sides of the Saône valley (e.g. Côte d'Or) are famous for their wines and although tillage crops are grown in the valleys, the relief, difficult communications and heavier rainfall make *semi-intensive grazing* or *alternating* systems common on the higher hills and plateaux. About three-fifths of the farm acreage in this region is devoted to grass and fodder crops. The Charollais, Nivernais and the Morvan districts—the latter dotted with turreted chateaux—concentrate on livestock. In the Charollais, more than one beast per acre can be fattened in the 8-month growing season, but the typical farmer in this district also grows some tillage and forage crops and has pigs, poultry and perhaps some sheep. But there are many farms which are not good enough to fatten off grass and there is a considerable sale of store (feeder) cattle to, for instance, the Paris basin for finishing.

MEDITERRANEAN (*Sector* (*iv*))

The climate is typically Mediterranean with mild winters, heavy autumn rain (much of which comes in a few intense soil-eroding storms), light spring rains and warm dry summers with little or no precipitation. Even in the southern foothills of the Massif Central and in the High Alps to the east, this winter rainfall pattern prevails. In the lowlands during the thermal growing season of 12 months, the P–E–T is 27–30 in, the A–E–T 22–26 in and the mean air temperature is 57·6°F. But these figures are misleading because, except when irrigation is practised, the low and highly variable summer rainfall effectively limits crop production, though the strong insolation and the deep roots of the vine favour wine production. At Marseilles where the mean annual precipitation is 22·5 in (C.V. 20%), the January mean temperature is 45°F, the July mean 68°F, and there are, on average, 37 days of frost and 82 rainy days. Compared with Israel, south and west Australia and southern California (pp. 296, 173, 159), all of which have Mediterranean climates, summer heat stress and drought are less acute in southern France. The frequency of frosts which, if severe, can destroy olive trees and vines, accounts for the absence of a citrus industry. The cold wind from the north, i.e. the *mistral*, has profound effects on farming practices, especially west of the line Vienne to Marseilles. Here, the Rhône valley and delta and the Bas-Rhône and Languedoc areas are mainly flat with loamy soils, giving way as one moves towards the Massif Central to gravel and rocky terraces and uplands. To the east, in Provence and the high (southern) Alps, the relief is accented, the *mistral* less of a problem, the geological formations are often calcareous and the soils thin, though there are alluvial deposits in some valleys. The 'natural' vegetation in this sector is mostly Mediterranean woodland and scrub.

The infra-structure in the lowlands is good and improving, especially now that malarial mosquitoes have almost been eliminated, and as the Rhône–Durance multi-purpose river control scheme and the Bas-Rhône–Languedoc plan outlined below stimulate industrialisation. Four-fifths of the farmland is owner-occupied. Although 70% of the farms are less than 120 acres, there are some large semi-intensive sheep farms in the uplands, some big vineyards and fruit farms in the lowlands, and some big rice-growing enterprises in the Camargue. The gross sales per acre and per man and also yields are only moderate (Table 2.6.2 Region *K* and Fig. 2.6.4), but these are likely to rise as the area in vines is reduced, and as irrigation and industrialisation grow.

Farming systems. To illustrate the wide variety of farming systems we take a sub-transect north-west to south in the Department of Gard (to the west of the Rhône); and then two villages in the eastern half, one near the coast in the Department of Var, and the other in the Hautes Alpes.

The Gard farming ranges from *extensive mountain grazing* through *semi-extensive tillage* (cereals) to *intensive irrigated tillage*. The lowlands are dominated by *intensive fruit and vegetables* and, in particular, vines which occupy 25% of the tilled land and provide half the gross sales. Apart from the fact that inferior wine is in excess supply, vineyards have many weather hazards (frost, hail, heavy showers) and disease problems. Most lowland farms, however, have several other enterprises as well, e.g. vegetables, tree fruits or cereals. Except for sheep and goats, livestock are not important. The difficulty of providing summer keep, the hot days of high summer (see Israel), the *mistral* and the lack of large local consuming populations nearby combine to discourage high-value crops (such as artichokes), The Department of Gard runs from the edge of the Cevennes in the north (a part of the Massif Central up to 4,500 ft) with heavy precipitation, through the limestone plateaux and slopes of the Garrigues, with 23·5 in to 39·0 in of precipitation, to the coastal plain where, at Aigues Mortes, the mean annual precipitation is only 21 in, mainly in the autumn. At Montpellier in 3 years out of 10 the precipitation in July is less than 0·4 in. The C.V. of annual precipitation is about 18%. Evapo-transpiration at these latitudes is of course high in mid-summer when, usually, the sky is cloudless. For instance Vernet and Marger (1946) record that in May a wheat crop transpired more than 7·8 in. Hence, without irrigation, only deep-rooted crops such as vines or drought-resistant species such as sorghum are in the land in midsummer. Wheat is harvested in June and the sheep are in the hills by May. Another problem (Godard and

Nigond 1954) is spring frosts, sometimes severe as in 1956 and 1962 when many olive trees were killed. In the plains, sudden floods are occasional dangers. But the main hazard is the *mistral*. This strong, dry, cold wind comes mainly from the north and the north-west off the Massif Central and lasts for 3–5 days at a time. At Montpellier, winds blow from these directions on a third of the days in the year. *The mistral* is almost as important as the hot, dry summer in determining not so much farming systems as farm practices. Market gardeners and fruit growers with young trees use protecting windbreaks of cypress trees or canes; rice growers hasten their harvest in case the *mistral* shakes the grain out of the ripe crop; sheep keepers house their flocks partly to collect sheep manure for which there is a ready market and partly to save shepherding costs but also to shelter both sheep and shepherd from the *mistral*. Dry, hot winds from Africa are not troublesome as they generally pick up moisture on their way across the Mediterranean and usually bring rain with them.

The Cévennes in the north of Gard, are well-watered, rough, granite mountains. The sweet chestnuts of the hills are said to give a piquant flavour to pigs fattened on them, but there is little farming and this is on the light, acid soils of the slopes and valleys, where potatoes, apples and other hardy tree fruits are grown. A typical 2-man farm of 86 acres would have 76 acres in pasturage (volunteer invaders), 0·5 acre vines, 0·5 acre vegetables, 1·8 acres apples and 7 acres in cereals and forage crops. Some market gardening, e.g. onion growing, is done on man-made terraces. There is much rough grazing of an almost alpine type used by sheep and goats, mainly for milk for cheese. In the Causses (to the west of the Cévennes) milking-sheep for Roquefort cheese are, indeed, the major enterprise. Hand milking is usual though on some farms machine milking is practised.

The Garrigues lies south-west of the Cévennes and are rough, rock-strewn plateaux and hills with a xerophytic scrub vegetation dominated by evergreen oak and aromatic shrubs, with some grass clearings. Formerly these were grazed by sheep, but shepherds will no longer live and work in this lonely inhospitable country. Farming is confined to occasional patches of cultivable land on sheltered, less stony slopes and to the valley bottoms, some of which have good deep soils. A one-man farm of 20 acres would be half in cereals, $7\frac{1}{2}$ acres in vines, $2\frac{1}{2}$ acres in olives plus some rough grazing for the sheep. Although there are some larger farms specialising in cows' milk production or asparagus, *semi-intensive tillage* (dry farming), is more common on the bigger units.

A tillage farm near Nîmes. One such 250-acre farm north-west of Nîmes has 50 acres in vines, 80 acres on a cereal rotation, and a further 25 acres being bulldozed free of scrub vegetation for more cereals. The rotation is 1st year: winter wheat drilled from mid-October to end December and harvested in June; 2nd year: grain sorghum or sunflower for seed or vetches (*Vicia* spp) for green manure. The sorghum (U.S. varieties) is drilled 23·5 in between and

5·5 in within rows in April after autumn ploughing and weed-killing cultivation in spring. It is harvested in August before the grape harvest in September. Sunflowers (Russian varieties) for oil and protein feed, are drilled at 29·5 in between and 10 in within rows in March/April and harvested in late August or early September. Ploughing after wheat used to be started on the hard ground in July/August but the need for powerful tractors, the high rate of breakages of ploughshares, etc., and the annual August holiday not only of the farm workers but of the suppliers of spare parts, forced reversion to autumn ploughing after the rain had softened the rather heavy alluvial soil. This has poor structure which could not be improved without leys for livestock for both of which the climate is not very suitable. Weed control is largely by cultivation—sunflower and sorghum being treated as cleaning crops in preparation for wheat. Although herbicidal sprays are occasionally used to control grass weeds in the spring cereal crops, herbicides are not habitually used. The farmer claims that he has no disease or pest problems in his cereals. Fertiliser usage is fairly high at, for example, 36 lb P_2O_5/acre, 114 lb K_2O/acre and 50 lb N/acre for sorghum. The cereals and the sunflowers are harvested by one combine suitably set for each crop. Yields, which vary greatly with rainfall which ranges from 27·5 in to 55 in per year, are, in a good year, 4,000 lb/acre of soft wheat, 2,250 lb/acre of hard wheat and 3,500 to 5,500 lb/acre of grain sorghum. The long-term stability of this farming system is perhaps uncertain, as, to maintain yields, inputs of fertiliser are having to be raised. Apart from the combine and usual cultivating and other machinery, there are 6 tractors and 4 men, including the farmer.

Lowland plains. The lowland plains are mostly accessible to sprinkler or even flood irrigation but there is not enough irrigation water for salinity to be a real problem, though some areas have to be flooded in winter to leach out salt. There are large areas of good, if somewhat sandy, alluvial soils, but there are also districts (e.g. the Costières, see below) where, as in the Canterbury Plains of New Zealand, large, round, water-deposited pebbles make cultivation difficult. If they have access to irrigation and can put up wind-breaks, some farmers, however, collect the stones in piles (as in Quebec) or in rows at the field edge. On a small, one-man farm of say 12 acres, vines for wine will often be the only crop, but on larger farms market gardening and tree fruits (peaches, cherries) complement the vineyards. Since, in order to reduce the excess of poor quality wine, the irrigation of vineyards is not allowed, some smaller farmers are now diversifying, though they protest that the price of irrigation water is too high. Many such farms have a high gross output per acre of labour-intensive crops, and two men working a mixed farm of 50 acres can expect gross sales of say £150 an acre and, after meeting all costs, have £750–£1,000 each for their work and skill. In the Bas-Rhône or Languedoc area such a farm,* if irrigated, might have a choice of garlic (yield 4,500 lb/

* Société du Canal de Provence et d'Aménagement de la Région Provençale. September 1964.

acre), artichokes 4,500 lb/acre, beetroot (*Betterave potagère*) 22,500 lb/acre, melon 6,500 lb/acre, potatoes 22,500 lb/acre, grossing overall £300 to £500 per acre. If 18 acres were devoted to these high-value crops, the same acreage to cereals or lucerne for hay (grossing £35/40 per acre) and there were 12 acres of vines (grossing £80/90 per acre) there would be well-spread risks and a good rotation on an intensive tillage system. Contrast this highly mixed cropping with the single fruit or vegetable farms of California (p. 162). (Note that here, as elsewhere in France, there is little distinction between horticulture and agriculture as understood in the United Kingdom, they are both farming; the term horticulture is often restricted to flower growing.) Especially on the larger lowland farms, migrant labour, mainly from Spain, is very important from April to September for transplanting rice and harvesting fresh vegetables, then picking tree-fruits, table grapes and finally grapes for wine. A large farmer contracts with a gang and then 'lets it out' to his smaller neighbours as the work peak varies on his own farm.

An intensive tree fruit and grassland farm on the Costières. The Costières are terraces of large pebbles up to 3 or 4 lb in weight (deposited originally by the Rhône together with sand and finer particles), on which intensive farming is made possible by the sprinkler irrigation and land settlement schemes of the Compagnie Nationale d'Aménagement de Région du Bas-Rhône et de Languedoc. This company is jointly financed and controlled by Government and private capital. It is, as it were, a cross between the Tennessee Valley Authority in the U.S.A. and the Kiryat Gat scheme in Israel (p. 315). It has wide powers needed to revitalise the whole economy of the area by river and other water control works and by the encouragement of outside industry, tourism and factories to process farm products. In agriculture, its chief activities are the amalgamation and reallocation of small or fragmented farms, the supply of irrigation water, advisory and other technical services and the development of abandoned or virgin land.

The present farm of 124 acres, half of which is being planted to peaches whilst half is in leys for sheep, comes into the last category. The scrub is first bulldozed away (though suitable wind-break trees are left untouched) and the larger pebbles worked to the surface and then gradually moved to the perimeter of the fields. Peaches are planted (with suitable shelter belts) after a large dressing of Ca, K_2O and P_2O_5 (N is applied later to the leys). Between the peach rows, Italian rye-grass (I.R.G.) (*Lolium multiflorum*) is drilled into irrigated soil at the end of July and then grazed by the flock of 400 ewes from October through to January or February when it is allowed to grow to a foot or so; it is then ploughed in as green manure. The unplanted parts of the farm are in I.R.G. and/or lucerne for grazing and hay making. The ewes, a local breed (Merinos des Arles), will take the ram at any time of year. Although many sheep keepers allow lambing at any time, this farmer concentrates on two periods, viz. October when the ewes have got back from the Alps, and early spring (January/February) when the grazing is ample and there is time to market the lambs at 2–3 months old at 55 lb live-weight before the ewes leave the farm in June. This transhumance is done on contract by shepherds who undertake to collect, transport, graze, feed minerals and attend to health when the sheep are in the mountains, usually the Alps but sometimes the Massif Central. Transhumance gets the sheep off the land at a time when un-irrigated grazing is negligible or water is expensive, when heat is excessive and fly troubles might be expected. When at home the ewes are kept in a large dry barn and always have access to hay and purchased straw. The lambs are creep fed on compound pellets and crushed maize, and the whole flock (excluding new-born lambs) is grazed under the eye of the shepherd (who does other work) for 4 hours or so each day. The sheep manure is eagerly bought by vine growers and market gardeners. The main diseases are said to be pasteurellosis (associated with *Pasteurella oviseptica*). There are no signs of footrot. No vaccines are used. There are about 25% of twins and sales average about $2\frac{1}{2}$ lambs per ewe per 2 years. By U.K. standards, the flock is not of high reproductive or meat standards but the lambs meet a ready market. The manpower is the farmer and one man who also acts as shepherd. Fruit-picking, etc., is by casual labour.

An intensive tillage (rice) and tree fruit farm in the Rhône delta. South of the terraces of the Costières, the soil is firstly loamy and then, as one nears the sea, a heavy, structureless, calcareous clay soil which is slightly saline and on which rice-growing started in 1942. Although production is mechanised (e.g. rice combine harvesters are used), much of the transplanting of rice is still done by the seasonally imported labour from Spain but this supply is erratic and some of the larger farmers (rice is essentially a large-scale job) are now drilling direct. This large intensive tillage and tree fruit farm of 840 acres has 395 acres in continuous rice, 295 acres in vines, 56 acres in espalier or cordon apples and the rest (soon to be apples) in lucerne for sale as hay and in soft winter wheat. The soil is too heavy and too calcareous for peaches. The land is farmed in blocks. The rice, grown in 6-acre paddocks bordered by irrigation canals, is kept away from the vines (which are not irrigated), in case the salt taken into solution in the high water-table of the rice fields should penetrate through the soil to the vineyards (the farm lies only 5 ft above sea-level within a few miles of the sea and at $6\frac{1}{2}$ ft depth the water is quite saline, compare Israel, p. 317). Fortunately, however, the irrigation water is fresh and low in salts, so, whilst salinity has to be watched, it is not yet a serious problem on the rice which is relatively salt tolerant (p. 56). The soil pH is 8·0 to 8·5.

Rice is grown continuously because no break crop has yet been found which avoids the loss of a year's harvest. Between rice harvest (ends October) and the latest date for a spring crop, say, March, there is neither time nor right weather to make a reasonable seed bed for such crops as wheat or lucerne from the waterlogged, structureless, clay soil. So particular attention is paid to weeds and diseases. Initially, soaked rice was direct seeded in April by hand sowing on to flooded fields. The grain sank, germinated and grew (compare California, p. 161). Direct seeding of soaked grain is by centrifugal spinner on to water as soon as the temperature of the irrigation water reaches 54°F, as rice needs 55–57°F to germinate and grow. Before drilling 70 lb K_2O/acre and 120 lb P_2O_5/acre are applied in soluble form; 55 lb N are put on before drilling and 20 after. N is applied as the sulphate, as ammonium phosphate or as urea in granular form, in solution, or in the more concentrated water suspension. Some N is injected on contract as gas which is 'held' in the dry soil and, as clay is impermeable, is not leached when the land is flood irrigated. But direct seeding limits pre-drilling weed control operations, and grass weeds and broad-leaved species increased. Now seeding the rice in nurseries in early April and transplanting by hand in late May is increasingly used. Shortage of transplanting labour has, however, turned attention to chemical weed control despite its expense. The problem of finding the right chemicals to control both grasses and broad-leaved weeds is complicated by the danger of serious drift damage to vines, for example, in the *mistral*. Pre-emergence herbicides are being tried but have to be mixed into a dry soil which is not always easy to make on rice land.

The drilling rate for direct seeding is 180 lb/acre. In the nurseries for transplants, the rate is ten times higher but an acre of seedlings will be enough for 12 acres of main crop. The mean annual rainfall is 25 in but very unreliable and sometimes very intense (up to 6 in in 24 hours), hence there are dangerous and sudden floods. Irrigation averages about 90 acre-in* per season (April to October). Irrigation ditches are kept free of weeds by herbicides, burning or cutting. Fungal and insect diseases are controlled by spraying. Harvest is by two crawler rice-combines in October as soon as the seed is ripe. Harvest is urgent because the *mistral* may shake the grains out of the panicles and because, when the autumn rains come, unharvested grains may swell and split. The rice, harvested at about 21 % moisture, is dried and stored by a local co-operative (which also handles wheat) at 15%. Yields of 'long grain' Italian varieties average 4,500 lb/acre and of round grain 5,500 lb/acre (unhulled in both cases). Rice has been grown every year on the same land for 15 years without any signs of trouble. In addition to the combines and heavy levelling and other machines for flood irrigation, there are 12 tractors. Some of these can be given extra high clearance to crawl over vines when intertilling, some are crawlers, and some can be fitted with 'skeleton' wheels to give grip in rice fields under water. On the whole farm there are 30 regular men.

A farm at Bormes-les-Mimosas, Var. Bormes is a village, mostly of small farms, 60 miles east of Marseilles. The mean annual precipitation is 25 in and is unreliable; summer is very dry. The *mistral*, which here occurs chiefly in winter, is not so severe as it is further west. This farm (one of the two large ones) has a slightly acid sandy clay and 100 acres of vines, 37 acres of vegetables and also forest. The vines, of several acid-tolerant varieties, are norm-

* 1 acre-in = 22,600 Imperial gallons of 10 lb.

283

ally left for 30 to 40 years and then pulled up and burnt when the land is rested for 2 or 3 years and then replanted $7\frac{1}{2}$ ft between and 3 ft within rows. This spacing permits cross cultivation. Pruning goes on all through the winter, the prunings being normally burnt. In alternate years farmyard manure or fertiliser (N.P.K.) is applied. To control weeds the vines are cultivated with shallow ploughs which can, by an ingenious sideways spring, ride right up against the vine plant (no herbicides are used and little hand hoeing is done). They are dusted with a copper base powder 5 to 10 times (according to humidity) during the growing season to control mildew. The grapes are harvested by hand in September and carted to the farmer's own winery where the pressing is mechanical and the grape juice is pumped through pipes into large tanks for fermentation and storage. The young wine is sold wholesale from March onwards. The grape residue is sold for further fermentation and distillation of commercial alcohol and what then remains is dried, fortified with N, P and K and sold back to vineyards for application to the vines. (The grape seeds are, in some areas, crushed to get grape oil, which is highly unsaturated and therefore favoured for cholesterol-free diets.) The 37 acres of vegetables are in a rotation of first (and sometimes second) year cauliflower or early potatoes (of which two crops are taken in a year—planted in August and January and harvested in December and May). Then from the 2nd or 3rd to the 4th or 6th year globe artichokes are grown. They are propagated by dividing clones and transplanting. The land in vegetables receives N.P.K. and farmyard manure ploughed in. The latter (purchased from ships bringing livestock to Marseilles from North Africa or from the stock-raising areas of the Alps) is one of the main expenses. Sheep manure is preferred. The vegetables are sprinkler irrigated. The machinery includes 3 wheeled and one crawler tractors, special plough-cultivators for vineyards, a sprayer which can either apply dust or blow on low-volume liquid. There are no leys and no livestock except 2 work mules. The regular labour force of 10 is augmented by casual workers during the grape harvest.

A village in the High Alps. To the north and east lie the Hautes Alpes which is the highest agricultural region in Europe with two-thirds of the land over 3,000 ft and half over 4,500 ft. The climate is 'Mediterranean alpine' with an autumn precipitation peak and a July trough. The C.V. of annual precipitation, which averages 40 in, is about twice that of the Paris basin and the rains often come in very heavy falls. Much of the rain (or snow when it melts) is lost as run-off. Insolation is high and there is a summer and sometimes a spring drought. For climatic and relief reasons, grassland is sometimes the only practicable land use, and, as farms are small (many of them less than 30 acres), and as grass output per acre is low, the income level is poor. At Saint-Léger (4,000 ft) the population of 159 has 24 farms averaging 26 acres (of which 10 are in potatoes, 6 in cereals and the rest in grassland and forage crops), plus 15 cattle and 24 sheep. In the snow-melt, streams are diverted to trickle over the fields, but this is wasteful of water and aggravates loss of plant nutrients by leaching. However, communal water control and co-operative sprinkler irrigation projects based on rotating both water rights and sprinklers round farms may have a future in this region (Chauvet and Coulet 1965).

East to West Transect

MASSIF CENTRAL (*Sector (v)*)

This great outcrop of highland dominates the physical geography of France, and, though only of minor farming importance, has wide agricultural repercussions on adjoining regions (for example, on the supply of irrigation water). It is summarised in Table 2.6.4. Most of the farms are owner-occupied and rather surprisingly are small in area; gross sales per acre and per man are the lowest in France (Fig. 2.6.4).

Farming systems are either *extensive grazing* or *semi-intensive alternating*. More than half the farmed area is in permanent grass and a third of the tillage is in fodder crops. Cattle raising is the dominant enterprise. But on the Causses to the west and south, with calcareous, well-drained stony soils, the vegetation is tufty grass and shrubs, and sheep are important. Flocks, including flocks from the lowlands for summer agistment, are large and many ewes are milked to provide Roquefort cheese.

AQUITAINE (*Sector (vi)*)

This sector starts in the foothills of the Massif Central and crosses the Aquitaine basin which is bordered on the south by the mountainous Pyrenees. The climate may be classed as warm humid, though moisture deficits are frequent from July to September. In the lowlands, in the thermal growing season of 11 months, the P–E–T averages 27–28 in, the A–E–T 25 in, and the mean temperature about 56°F. Bordeaux has a January mean temperature of 41°F, a July mean of 69°F, 45 frosty and 170 rainy days and averages 33 in annual precipitation. Toulouse, inland and further south, has the same January and July mean temperatures but less rainfall and fewer frosty and rainy days. The relief is mainly gently rolling plains with some forest-topped hills, whilst the flat, poor sand-dunes of the pine-covered Landes occupy much of the western seaboard. Apart from the thin mountain soils, there are, firstly, the clayey, calcium-rich, hard-to-work soils of the plateaux, and secondly, the sandy, acid, easily worked, drought-prone soils of the terraces of the large river valleys (Lot, Dordogne and Garonne) which have, however, good alluvial soils at their valley bottoms. The terrace soils respond well to fertilisers and irrigation. The infra-structure is not as good as it is in northern France but industry is growing; more farm people are working part-time in newly established local factories. Farms are small and often fragmented, and though most of the area is owner-occupied, *métayage* is still significant. Considering the climate and soils, gross sales per acre and per man are low (see Table 2.6.2. Region *K*, and Fig. 2.6.4). This is, in fact, one of the more backward regions.

Table 2.6.4

France — East to West Transect

Transect (b) — Long. 3° 55′ E, Lat. 44° 30′ N ←→ Long. 1° 15′ E, Lat. 43° 40′ N

	(v) Massif Central	(vi) Aquitaine Basin
1. Sector	(v)	(vi)
2. Name	Massif Central	Aquitaine Basin
3. From	Villefort	Figeac
4. To	Figeac	Coast (Cap Ferret)
5. Miles	95	160
6. Climate Type(s)	Mountain (Mediterranean)	Warm Humid (Mediterranean)
Thermal Growing Season:		
7a.—months at or >42½°F	7	11
7b.—months at or >72½°F	0	0
8a.—P–E–T (in) in T–G–S	21–23	27–28
8b.—A–E–T (in) in T–G–S	21–22	25
9. Growth limited by	Cold	Drought
Precipitation:		
10.—mean in (annual)	20–40	40–60
11.—C.V. % of Precipitation (annual)	10–15%	10–15%
12. Special Climate Features		Hot, dry summer months
13. Weather Reliability	3	3
Soil:		
14.—Zone/Group	Podsolic	Podsolic
15.—pH		
16.—Special Features	Limestone soils. Poor, thin granite soils. Volcanic soils	Clays rich in calcium. Light sands poor in calcium
17. Vegetation	Mixed forest	Mixed forest
18. Relief	Mountains	Rolling plains, hills and mountains
19. Market access	3	4
20. Farming Systems	Extensive grazing with a little semi-intensive tillage or alternating	Intensive or semi-intensive tillage or alternating
21. Major Enterprises	Sheep, cheese (ewe), beef, cheese (cow), cereals, vineyards and orchards in sheltered valleys	Cereals, maize, beef, wines, tobacco, fruits and vegetables, milk

286

But, with a good P–E–T and with growing supplies of irrigation water, it has a great potential and some enterprising farmers, some of them settlers returned from North Africa, are exploiting it and showing their neighbours what can be done.

Farming systems. The choice of farming systems is dominated by the threat of summer drought and the questions 'when will it start, how long will it last?' This risk often rules out spring wheat or barley; it forces reliance on deep-rooted plants such as vines or plum trees, on drought-resistant legumes such as lucerne and sainfoin (*Onobrychis viciifolia*), and on crops which mature *after* the drought hazard such as grain sorghum (on acid soils), grain maize (where summer moisture is more adequate) and tobacco, which is an important cash crop. In some valleys the mild winter is exploited by growing early vegetables which are sold before the likely drought. Though arable (tillage and ley) acreage is twice that of permanent grass, about a third of the average farm is in fodder crops (including lucerne and other leys); the remainder is in cereals (winter wheat, maize, sorghum) and horticultural crops, including vineyards and tobacco. Traditionally, this sector followed a 2-year wheat or maize and tobacco rotation with vines as a risk-spreading enterprise which often produced a small but good quality yield of wine in a dry summer when cereal yields were low. Recently, however, temporary leys have lengthened the rotation and provided, with cereals, a better basis for cattle production, whilst pigs, poultry and geese are now also important as converters of cereals into cash.

Intensive alternating tobacco and livestock farms in Dordogne. In France 60,000 farmers grow, on average, about an acre of tobacco each on contract for the State Tobacco Monopoly, which also provides close technical control and advice. It is grown on slightly acid-free working soils in Alsace, in the lower Rhône valley and especially in the south-west, i.e. in this sector where the main production centre is Bergerac on alluvial soils in the Dordogne valley. In the Dordogne one farmer in three grows tobacco, and a typical tobacco farm of 30 acres would have 10% of its area in tobacco, 5% in vines (for wine), 10% in vegetables, 50% in winter wheat or maize (for grain) and 25% in short-term (one-year) ley, and livestock equivalent to 5 cows. Leys are lengthening, for example, to 2-year ley of Italian rye-grass and white clover where the summer is not very dry or a ley of cock's-foot and bird's-foot trefoil (*Lotus corniculatus*) where it is dry. (Normally P_2O_5, K_2O and a little N are applied). For wet soils meadow fescue (*Festuca pratensis*) or tall fescue (*F. arundinacea*) are recommended. Wheat is drilled in the autumn and harvested in late June. Maize is planted when the frost risk has ended (April–May), and the cobs are harvested in September–October; the stalks are usually not harvested but left standing through the winter. Tobacco is sown in a nursery in the second half of March and transplanted in May to 28 in rows with 15 in between plants. Up to 250 lb of N per acre is applied together with P_2O_5 and K_2O but the muriate (i.e. potassium chloride) is avoided as it adversely affects quality.

By August a leaf area index of 7 or 8 is achieved. Either the whole plant is cut by hand in late August or the individual leaves (starting with the lower ones) are plucked from July to September. The crop is then air dried in special barns for 4 to 8 weeks. Unirrigated yields run up to 8,000 lb of dry matter per acre which compares well with a dry-matter output of 9,000–12,000 lb from a well-managed pasture in England. Experimentally there have, in south-west France, been good responses to sprinkler irrigation, but the high capital equipment cost (though sometimes State subsidised) and the low mean acreage of tobacco per farm, makes its use uneconomic unless some irrigable crops with a high gross cash output per acre are also grown, which is unusual. Cash receipts from tobacco run up to £500 per acre, so it may be asked why on these farms the number of enterprises is high? It is partly because of the traditional 'polyculture' of the Dordogne, partly for disease control in the tobacco and partly because leaf yield varies substantially with weather so other enterprises (especially those not affected by the later summer drought) help to 'buffer' a poor tobacco yield in a dry year. But perhaps the main reason is that tobacco growing requires intelligence and detailed attention; so less exacting semi-intensive complementary enterprises (such as veal or store cattle production) are acceptable, particularly if their seasonal work load offsets that on tobacco.

A tobacco and veal farm near Massidan. One farmer near Massidan in the valley of the Isle has 65 acres and concentrates on tobacco, which brings in 30% of his gross sales, and on selling veal calves, and/or stores from his 26 cows which he hopes to raise to 40, i.e. one cow to $1\frac{1}{2}$ acres. Apart from his flood-liable permanent pastures, his rotation is tobacco; cereal (wheat or barley); then Italian rye-grass/white or red clover ley for 3 years. He applies N, P_2O_5, K_2O and, as necessary, lime. His land is heavy, so cattle are housed (tied-up in stalls) from November to March. Though grass growth starts several weeks earlier, he rarely starts grazing until May. No concentrates are fed in summer. In winter his cows get hay and a few fodder beet or cabbage; his young stock receive home-grown grain and minerals in addition. Calves are usually single suckled for 3 months and then if not fit for veal or, if not needed for breeding, are fed like the cows and sold as stores (feeder cattle). Protein supplements are not used. The farmer and his wife (who have seven children) do all the work with the aid of a tractor and simple implements; a contractor harvests the cereals.

An intensive tillage (maize-growing) farm near Bergerac in the Dordogne valley. This unusual farm is a good example of what is being done by progressive pioneers in the rather conservative south-west of France. It has 95 acres of flat, deep, alluvial loam in the Dordogne valley and is all in continuous maize. There is also a small park and about 10 acres of buildings (the farm was once a small country estate) some of which are let as dwellings or as a 'rest-home' for bulls from a local A.I. centre. The maize stubble is ploughed in in the autumn or early spring, worked down with nitrogen and a first crop of weeds chitted and then scarified before the maize is planted with an American hybrid seed (Funk 75) in the second half of May. Behind the 2-row planter runs a 2-band fertiliser, which applies a concentrated phosphate and potash mixture fortified with Aldrin (a chlorinated hydrocarbon) for insect and other control, and behind that a 2-band sprayer applying Simazine (as a pre-emergence weed control) over the rows. The weeds between the rows are controlled by frequent tractor hoeing. Eight or 9 years out of 10 there is a serious moisture deficit in July and/or August and 6–8 in of water is applied in 2 in doses. Watering is stopped in mid-August when the owner goes on holiday. Nitrogen is put on as top-dressing or in the (sprinkled) irrigation water. The total fertiliser applied per acre per annum is 140 lb N, 140 lb P_2O_5 and 110 lb K_2O, whilst the pH has been brought up, by liming from 4·5 to 6·5. Harvesting is done on contract by a mechanical corn-picker in October at a moisture content of about 30%, which is reduced by a dryer to 15%. With care there are few pest and disease problems, the main one being *Fusarium*. This system of continuous maize has been followed for 8 years and yields now average 6,250 lb/acre. The owner is aiming at 9,000 lb/acre. There are two tractors, the equipment already mentioned, a large bulk dryer with a cool-air bin followed by a hot-air bin and an American-style man-high tripod sprinkler irrigation system. The water mains are mostly fixed but the laterals consist of movable aluminium pipes or of plastic hoses mounted on wheeled drums which can be uncoiled and recoiled by a tractor. The whole is most impressive.

Efficiency and Stability

France is the largest agricultural producer in western Europe. Farming contributes nearly a tenth of her Gross National Product (G.N.P.) but occupies nearly one-fifth of her manpower. Productivity of farm manpower and capital is low for an advanced country (Van den Noort 1968, Sharp and Capstick 1966). 'Productivity per agricultural worker in 1962 was only half that of workers in the rest of the economy. . . . Farm mechanisation, though progressing rapidly is still insufficient. The number of tractors in 1962, expressed in H.P. per hectare of agricultural land, was 0·64 H.P. as compared with 1·47 in West Germany and 0·95 in the Netherlands' (U.S.D.A. 1966). The average farm is only about 30–40 acres and its inputs of fertiliser, technical chemicals and sophisticated feeding stuffs are not high by western European standards; although, as noted (pp. 224, 283, 288), some farmers are very advanced. The sources of this relative lack of *efficiency* are complex. They include the long period of high-tariff protection; the belief in the social value of a peasant agriculture; the ageing farm population; the small and often fragmented farms; the lack of specialisation in some areas (e.g. the Dordogne), and the excessive specialisation in others (e.g. the vineyards in Herault); the differential rural exodus (which has reduced farm labour in industrialised areas but left an excess in others, e.g. Aquitaine); the retention of surplus production capacity of low labour productivity; the difficulty of standardising inputs and marketing processes in a country rather obsessed with food and wine and which boasts of its many local wines and its 300 varieties of cheese; and the rather reactionary political powers of farming interests. But the Government is well aware of these problems and is tackling them vigorously, with measures such as land consolidation, irrigation schemes, better marketing and processing. Its modernisation policy is, in fact, well ahead of the farm population.

Instability due to weather factors, especially spring frosts and drought, is marked in the southern half of the country, though the drought problem is likely to be corrected by better knowledge of plant/soil moisture relationships and the irrigation schemes already noted. Other sources of instability spring from the above efficiency factors or lack of them. Thus, protection tends to encourage high cost production of products (e.g. rice, maize, butter-fat) for which profitable outlets cannot be found on world markets, although the E.E.C. price stabilisation and 'guarantee' schemes may help. Secondly, the relative immobility of resources 'trapped' in agriculture tends to generate surpluses and instability (p. 493); this lack of mobility seems more marked in France than in, for example, the U.S.A. or the United Kingdom. Too many French farmers are imprisoned by the tendrils of the vine or some other inertial factor.

Conclusions and Outlook

The northern plains are well farmed though their output could be further increased by more industrial inputs and larger farming units. The southern half with its higher P–E–T and insolation could, if it were able to secure adequate irrigation water and use it efficiently, and to remedy the defects listed above, become one of the great low-cost food-producing areas of a Europe which will increasingly demand fruit, vegetables and animal products. The possibilities of the latter are substantial in the south, where the climate, on the eastern side, is similar to that of (north) California and, on the western side, to say, Georgia or North Carolina. But the south-west, like the western Capes (Brittany) is far from the mass urban markets of the E.E.C., and exploitation of its potential must go hand-in-hand with the better marketing which the Government is promoting. In brief, French agriculture is a giant which, though it has some lively limbs, is rather lethargic. If and when it is fully roused it could play a vital part in a Europe which, by the end of the century, may be unable to rely on imports from other continents. The combined effect of intra-community competition within the E.E.C., and of the Government's modernisation schemes may prove to be spurs which awaken the giant to exploit its undoubted potential.

Acknowledgements. For research assistance to S. Manrakham, Kathleen Down; for information to many professional men and farmers in France; for the data for Fig. 2.6.4 to Professor M. Cépède; for comment and criticism to T. Beresford, D. R. Bergman, G. W. Ford, A. Marcellin and A. Salles.

References and Further Reading

Anon. Tableaux de l'agriculture française – *Paysans*, **10** (61) Aout.–Sept. 1966.

Arléry, R., Garnier, M. and Langlois, R. 1954. *La Météorologie*, Quatrième série, No. 36, Oct.–Dec.

Beresford, T. 1967. Personal Communication. Chilmark, Wiltshire, England.

*Birot, P. and Dresch, J. 1953. *La Méditerranee and le Moyen–Orient, Vol. I. La Méditerranee Occidentale*. Presses Universitaire de France. Paris.

*Blanc, A., Juillard, E., Ray V. and Rochefort, M. 1960. *France de demain séries. Les Régions de l'est*, pp. 51–91. P.U.F. Paris.

*Carrere, P. and Dugrand, P. 1960. *France de demain série. La Région Méditerranéenne*, pp. 50–98. P.U.F. Paris.

Cépède M. and Madec A. 1966. Personal Communication. Min. of Agric. Paris.

Chauvet, P. and Coulet, B. 1965. *Bulletin Technique d'Information des Ingénieurs des Services Agricoles*, **201**. Juillet–Août, pp. 663–72.

*Demangeon, A. 1947. *Géographie Universelle*. Tome vi, Pt. II, pp. 151, 216–35, 218, 236, 254, 255–85, 286, 301, 302–23, 324–47, 348–68.

*Finch, V. C., Trewartha, G. T., Robinson, A. H. and Hammond, E. H. 1957. *Elements of Geography: Physical and Cultural*. McGraw-Hill, New York and London. 4th edn.

*George, P. and Randet, R. 1959. *France de demain série. La région Parisienne*, pp. 107–14. P.U.F. Paris.

*Gervais, M., Servolin, C., Weil J. 1965. *Une France sans paysans*. (Ed. du Sevil). Paris.

Godard, M. and Nigond, J. 1954. *La Météorologie*, Quatrième série, **36**. 319–44. Oct.–Dec.

*Gory, G. and Gauthier, J. 1963. *Notions d'Agriculture*. Périgueux, 1, Rue M. Hardy. 7th edn.

*Labasse, J. and La Ferrere, M. 1960. *France de demain séries. La Région Lyonnaise*, pp. 40–78. P.U.F. Paris.

*Laurent, C. 1965 and 1966. 'Enquête au 1/10e sur les exploitations agricoles en 1963 (Enchantillon-maître)', *Statistique agricole suppl. série 'Etudes'* (5), Juin 1965, (7), Sept. 1965, (18) Nov. 1966.

*Ministry of Agriculture. 1964. *France and her Agriculture*. Paris.

*Nistri, R. and Precheur, C. 1959. *France de demain séries. La Region du Nord due Nord Est*. Chap. 11, pp. 34–59.

O.E.E.C. 1960. Agricultural Regions in the E.E.C. *Documentation in Food and Agriculture*. **27**. O.E.C.D. Paris.

Severac, G. 1961. *Economie Rurale*. **48**. 3–25.

Sharp, G. and Capstick, C. W. 1966. *J. Agric. Econ*. **17**. 1. 2–16.

U.S.D.A. 1966. 'The Western European Agricultural Situation', *Econ. Res. Service*. Foreign 149. U.S. Dept. of Agric., Washington, D.C.

Van den Noort, P.C. 1968. *J. Agric. Econ*. **19**. 1. 97–103.

Vernet, A. and Marger, T. 1946. *Ann. Agronomiques*. **3**. 3.

* Asterisked items are for further reading.

CHAPTER 2.7

The Eastern Mediterranean

Warm dry regions with hot summers and cool, wet winters occur round the Mediterranean basin, in California, U.S.A. (Chap. 2.3), along the coast of Chile (Chap. 2.8), in the south-west tip of South Africa and from the State of Victoria to Western Australia (Chap. 2.4). These 'Mediterranean' climates occur on the west of continents in mid-latitudes. To the east of such climates, and/or as latitudes decrease the climate becomes more arid and the rainfall less reliable. No distinction is made here between these two major types of warm dry climates. This is because, in these Mediterranean and semi-arid climates, problems are the same, viz. a lack of soil moisture in the season of greatest P–E–T and therefore of greatest potential N.A.R.; a shortage of animal feed, especially that of low-fibre content; and heat stress in livestock. However, wisely used irrigation can do much to solve these problems. Examples of such climatic areas are discussed in Chaps. 2.3 and 2.4, whilst Sector (iv) in France (Chap. 2.6) is taken as our example from the northern rim of the Mediterranean basin. (See also Chap. 1.7 on the socio-economics of this basin.) The southern rim from Morocco to Libya is less watered and much less developed than the northern rim; but with irrigation, education, capital injection and political stability, the ancient ravages of erosion and neglect could be made good and parts could become highly productive. At the eastern end of the basin, there are better-watered and/or more progressive countries of which the United Arab Republic (Egypt) and Israel are studied here. The latter is treated in more depth than Egypt because it offers a greater ecological range and is well documented. Both are the scene of important agrarian developments.

United Arab Republic (Egypt)

'Egypt is as big as France and Spain together, but its cultivated and inhabited area is about the size of Holland' (Mansfield 1965). Only about 6·6m acres (2·67%), of the land area is cultivated but, because perennial irrigation from the Nile has largely replaced the old seasonal 'basin' irrigation, two crops per year are usually taken, and the area of crops harvested each year is in fact nearly twice the farmed (i.e. cultivated) acreage. The Nile, however, is only a beautiful green gash in the face of a yellow red countryside, and over 90% of the country is desert. The cultivated and cropped areas and their yields will be increased when the Aswan High Dam and its related irrigation schemes are completed. But even then Egypt's basic problem will still remain: how to support a rapidly growing population, of which 70% is rural, in a

Table 2·7·1

Crop	ISRAEL Area (000's acres)	ISRAEL Yield (average for 1963–65) (lb per acre)	JAPAN Area (000's acres)	JAPAN Yield (average for 1963–65) (lb per acre)
Wheat	178	1,606	1,176	1,897
Oats	2·4	1,270	153	1,766
Barley	128	1,023	1,043	2,013
Rye	—	—	2·4	1,629
Maize	2·4	3,535	74	2,242
Sorghum	74	2,487	2·4	1,082
Millet	—	—	54	1,421
Rice (hulled, unpolished)	—	—	8,043	3,446
All cereals	383	1,512	10,625	3,098
Sugar-cane	—	—	30	55,363
Sugar-beet	20	40,408	148	23,429
Potatoes	12	19,683	526	15,818
Sweet-potatoes and yams	—	—	635	17,959
Vegetables (a)	22		580	
Dry peas and beans	5	650	554	874
Soyabeans	—	—	455	1,106
Other oil-seeds (b)	61	1,665	403	1,345
Tea	—	—	121	1,307
Tobacco	7	401	213	2,075
Cotton (lint)	42	1,052	—	—
Flax fibre	—	—	12	526
		Production 000's lb		Production 000's lb
Tree crops (fruit) (c)		2,313,706		7,418,502

Livestock	Numbers 000's	Yield (average for 1963–65) (lb per animal)	Numbers 000's	Yield (average for 1963–65) (lb per animal)
Total cattle	217	(d) 257	3,175	(d) 498
Dairy cows	87	(e) 10,449	859	(e) 8,756
Pigs	—		3,976	(d) 60
Sheep	192	(d) 66 (e) 413	207	(d) 48
Wool		2·3		
Goats	153	(e) 2,282	325	(d) 31
Poultry (chickens)	7,050		120,197	
Buffaloes	225		—	
Camels	11		—	

(a) Onions, tomatoes, cabbages, cauliflowers, green peas and green beans.
(b) Groundnuts, sesmae seed (cotton-seed and sunflower seed in Israel; linseed and rape-seed in Japan).
(c) Apples, pears, plums, prunes, peaches, oranges, tangerines, lemons, limes and grapes (cherries in Japan; apricots, grapefruit, dates, figs, bananas and olives in Israel) (grape yields—Israel 6,167 lb per acre; Japan, 9,227 lb per acre).
(d) Live-weight of animals slaughtered.
(e) Milk per milking cow, goat or sheep.

country in which the rural levels of housing, nutrition,* health and education are still low (Tables 1.5.1, 1.5.2 and 3.1.3) and where industrialisation has yet to make a substantial mark on farming, either in drawing off surplus man-power or in providing the industrial inputs, the economic incentives and the health, educational and social services needed to raise yields further.

* Over four-fifths of the dietary energy comes from cereals, supplemented by vegetables, some legume protein and a little animal protein.

However, the land tenure reforms of the last 100 years, and more particularly since 1952 (Saab 1967), have done much to replace the apathy engendered by a combination of oppressive landlordism, based largely on share-cropping on giant holdings in a few hands, and more than 2m small, almost miniature, fragmented farms.* In 1952 more than 2m farmers (or 72% of all proprietors) owned or farmed less than one feddan (just over 1 acre); but 'two feddans of land constitute an absolute minimum from which a *fellahin* family can make a living. . . .' (Mansfield 1965). *The fellahin* uses a primitive plough pulled by buffalo or ox, clears the weeds with a short-handled hoe, harvests cereals with a sickle and 'threshes' them by piling them up and driving a heavy wooden sledge over them, and winnows them in the wind. For irrigation, which is the main part of his work, he raises the water with a wooden water-wheel pulled by an ox—or more probably by hand with the Archimedean screw or *shadoof* —a horizontal bar on a forked stick with a weight at one end and a bucket on the other. For cotton, child labour is widely used. 'Children are the right height for harvesting the cotton and for the important work of searching the backs of leaves for the cotton worm. . . .' Mansfield (loc. cit.). On the larger farms, estates and co-operatives and State farms, however, mechanisation is proceeding rapidly.

Climate, soils and farming systems. The contrast between the infra-structure and the ecological background is striking. The P–E–T and A–E–T per annum range respectively from 43 in and 7 in at Alexandria in the north to 59 in and 1 in at Aswan in the south. The large moisture deficit (P–E–T minus A–E–T) is partly compensated by flood irrigation—though some sprinkler irrigation is practised on the higher lands just inside the desert proper. The cultivated soils are largely naturally fertile silts. Fertiliser usage is growing steadily. Until recently a typical 2-year rotation was:

November–May	Winter wheat or Berseem (*Trifolium alexandrinum*)
	(This latter is fed to cattle or water buffalo used for draught and, being leguminous, raises soil nitrogen)
June–July:	Fallow
August–November:	Summer cereals: maize, millet, rice
December–January:	Fallow or Berseem
February–November:	Cotton for lint and seed

This rotation produces one cash crop (cotton) mainly for export and two food crops in two years but, especially if a second berseem crop is not taken, reduces the nutrient status, particularly nitrogen, of the soil. But a less common 3-year rotation raises mean yields per annum by about 20%,

---

* In one village of 1,850 acres there were 1,585 properties divided into 3,500 plots; an average of about ½ acre per plot.

viz. (i) cotton, (ii) maize or rice, (iii) berseem, was introduced by progressive large farmers, then applied by the supervised co-operatives and finally, in 1963 (Mansfield 1965, p. 180) applied generally. (Yields are given in Table 2.7.1.)

The supervised co-operatives are proving very successful and are to be extended gradually over the whole farming area of Egypt. Their structure appears to be unique. Many of them work on a three-field system (with cotton, grain, berseem as the rotation) and each member has his share, each year, of the product of each of the three fields. These are cultivated, drilled, fertilised and sprayed by the common services of the co-operative, but apparently harvested by the individual member. They are perhaps more like a modernised co-operative version of the old three-field system of western Europe than the Soviet collective, the Chinese commune or the Israelis' Kibbutzim or Moshavim; but they are none the worse for that, though Saab 1967 is critical of some aspects of their progress. Their success, if maintained, may well provide a pattern for many less-developed parts of the world.

Israel*

Israel has a land area roughly equal to that of Belgium, Massachusetts or Northern Ireland with a population of $2\frac{1}{2}$m, as in Northern Ireland. But in variety of climate, topography, farming types and products it has a range as great as California which has nearly 20 times its land area; whilst its farming structure runs from private landlords, tenants and merchants to Kibbutzim, i.e. collective villages which are socially but not politically to the left of the collective farms of the communist world. Some of these Kibbutzim and private farms are technically very advanced in their farming and management. At the other end of the scale, amongst the Jews, some of the individual smallholdings in the private sector or in the Moshavim villages are neither technically nor economically efficient, whilst the farming and herding of some of the Druze and Arab hill villages and of the Bedouin arabs in the Negev desert has probably changed little since Biblical times—except that some of the Bedouins are no longer nomadic. Economically, however, the non-Jewish population is not important and 88·5 % of the people are Jews.

Climate, Relief, Soils, Vegetation

After a narrow coastal plain, the altitude increases very rapidly as one moves eastwards into the hills and mountains; it then falls abruptly into the Rift valley of the River Jordan (Fig. 2.7.1), the main permanent water source available to Israel. The prevailing winds in the northern half of the country are from the west and north-west, bringing moisture from the Mediterranean Sea, but about 10 % of the time there are dry, easterly winds, called *Hamsin*, from the deserts. These are hot and are one of the weather hazards in farming.

* This section was drafted before the Arab–Israel war of June 1967 and has not been amended to the present *de facto* situation.

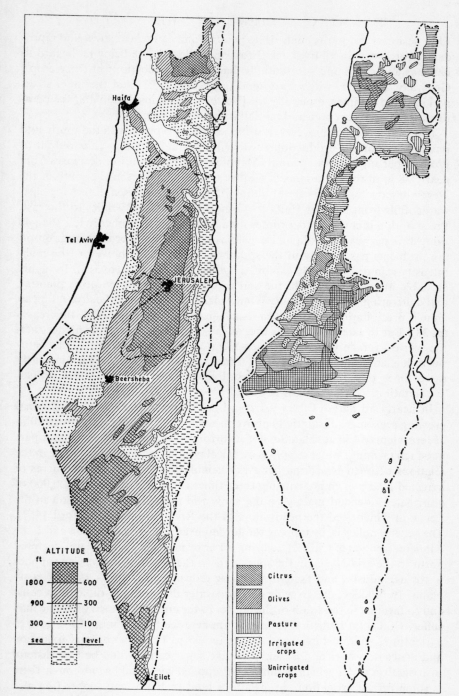

ALTITUDE

ft		m
1800		600
900		300
300		100
sea		level

Citrus

Olives

Pasture

Irrigated crops

Unirrigated crops

Fig. 2.7.1 Physical Features of Israel   Fig. 2.7.2 Farming Systems

297

They desiccate crops, may dehydrate poultry, and cause heat stress and reproductive problems in all livestock. Frost and cold are (see below) a hazard to
bananas and citrus and substantially determine the locations of these crops.
Generally, however, winters are mild (see Table 2.7.2) and the main stress
period is the summer, especially for livestock, when insolation and temperatures are high to very high and rainfall is extremely rare.

The main problem, however, is water shortage, especially in the south, and
the most important production coefficient in Israel farming is the output of
product per unit of water input. Mean annual precipitation decreases from
nearly 38 in in the mountains of Galilee to 1·2 in at Eilat in the south of the
Negev desert, i.e. at the rate of about 1·5 in per 10 miles, and moving eastwards, falls from 26 in at Haifa to 17 in at the frontier 35 miles to the east.
Precipitation is confined to 5 winter months and usually comes in less than 40
rainy days per year. So rainfall intensity is high and water erosion and run-off
are problems particularly in the southern desert where, however, the high-
intensity rain can be partly exploited by 'run-off' farming (p. 318). In highly
erodable areas in the north, the soil conservation service enforces planting
and cultivating on the contour. Wind erosion is a minor difficulty. As mean
annual precipitation decreases so its coefficient of variation (C.V.) increases
(p. 38). But in Israel the C.V.'s are, as in Australia, appreciably higher than
the world average C.V. for a given mean annual precipitation. Thus Jerusalem has an annual average of 22·6 in and a C.V. of 30% which is nearly
three times that found in south-east England for the same mean annual
precipitation.

In nearly all parts (Tables 2.7.2 and 2.7.3), mean P–E–T in the thermal
growing season is substantially above mean A–E–T which ranges from 20 in
at Jerusalem to 1 in at Eilat. So, not surprisingly, irrigation plays a major part
in Israeli farming; water supply is a critical political and military factor in the
relations between Israel and her Arab neighbours. 'Israel's 400,000 acres of
irrigated land (which are roughly three times as productive as the 600,000 of
unirrigated), depend mainly on the snow and rainfall of Mt Hermon in the
Lebanon which feed the headwaters of the River Jordan' (Beresford 1967).
The general policy is, by tapping the Jordan headwaters, the Yarkon and Lake
Kinneret (the Sea of Galilee), to bring water by pipe-line from the well-watered
north to the arid Negev in the south and on the way provide service supplies
to towns, industry and exchange-earning crops, such as citrus, in the coastal
plains. In practice, the growing *en route* water demands are difficult to resist
and so farming in the south or the farmer at the end of a branch pipe-line may
suffer. Of a total irrigation supply of 1m acre-feet* per annum 81% goes to
agriculture, 15% to towns and 4% to industry. There is no flood irrigation
and hence no wet rice in Israel. This, like sugar-cane, would be too wasteful
of irrigation water which is almost all applied by sprinkler, although there
is some furrow irrigation. Applications tend to be in acre–inches (see also

---

* 1¼m acre-feet by 1968.

Table 2.7.2

Climatic Data — Eastern Mediterranean

Station	Bright Sunshine hours/day (mean annual)	Thermal Growing Season (>42½°F) length in months	Thermal Growing Season Mean Temp °F	Mean Months >72½°F	Mean Annual Precipitation (inches)	Annual Precipitation Coeff. Var. (C.V.)	Evapo-transpiration in Thermal Growing Season Potential (mean inches)	Evapo-transpiration in Thermal Growing Season Actual (mean inches)	Moisture Index Im	Max. recorded rain/day inches (month of occurrence)
Beersheba		12	67	6	8·9	30%	39·3	8·9	−77 arid	—
Eilat		12	77	7	0·9	>35%	54·9	0·9	−84 arid	1·3 (April)
Haifa		12	71	6	26·2	25%	37·0	20·5	−29 semi-arid	10·7 (December)
Jerusalem		12	63	4	22·6	30%	34·1	18·1	−34 semi-arid	3·9 (January)
Luxor		12	77	7	<0·1	>40%	54·9	0·04	−99·8 arid	0·2 (October)
Aswan		12	80	9	<0·1	>40%	59·3	0·1	−99·8 arid	0·2 (May and October)
Cairo/Helwan		12	71	6	1·1	35%	46·1	1·0	−98 arid	1·5 (April)
Mersa Matruh		12	67	5	6·1	30%	37·5	6·1	−84 arid	3·9 (December)
Siwa Oasis		12	70	7	0·4	>40%	45·2	0·4	−91 arid	1·1 (December)
Amman		12	63	4	10·9	25%	35·0	12	−69 arid	3·1 (February and November)

299

Australia) in contrast to California where they are often in acre–feet; water
is usually only applied to citrus, top 'industrial' crops (such as cotton, groundnuts (peanuts) (*Arachis hypogaea*), sugar-beet), to vegetables and to the 'green'
fodder crops which supplement concentrates in the production of milk from
cows. Unfortunately, however, much of the irrigation water has a fairly high
salt content. Thus the exit water from Lake Kinneret has 270 parts of salt per
million (p.p.m.): this makes it unsuitable, without dilution by river water, for
intensive application to oranges (which have a safety limit for continuous
irrigation of 170 p.p.m. of sodium chloride in contrast to barley which has a
limit of 1,200 p.p.m.). Further, the river supply is highly seasonal and storage
dams are very expensive. (E. Monroe, *The Times* (London), 11 May 1965.)
Unlike California or the South Island of New Zealand, not enough trapped
snow melts and flows just when water demands of crops are starting to
rise.

Apart from the coastal sands (some of which tend to be saline), most of the
soils in the coastal and inland plains of the northern half of the country are
deep heavy clays of high pH and high in chalk ($CaCO_3$)—this latter *may* be
associated with the low incidence of bovine milk fever although this is held in
check by the addition of minerals to the concentrate ration (60–70% of total
ration) (see p. 310 below). In the Negev desert, particularly, the loessal soils
cap easily (which leads to run-off) so that the poor sandier soils near the coast
may yield better crops or carry more livestock because they can better 'receive'
intense rain.

Social and Economic Infra-structure

Socially, Israel is a unique compound of Socialism, Americanism and
Nationalism in which the Jewish religion still plays an important part even
though a minority of Israelis are practising orthodox Jews—much in the
same way that the Anglican Church in England still has an influence which
is substantially greater than the number of its communicant members. The
social ideals of Zionism (symbolised by the Kibbutz), a strong co-operative
movement, a dominant trade union, viz. General Federation of Labour (the
Histadrut), private enterprise capitalism, a large and very intelligent professional class, the powerful Jewish Agency financed by Jews in other countries,
poor Jewish immigrants from 40 different countries, well-established third
generation Jewish families—all these are somehow held together by the concept of a national home for the Jews, and by the defence needs of the delicate
strategic situation of a country whose existence is anathema to its neighbours.
Both Zionist ideals and strategic needs demand a rapid build-up of a dedicated population, especially to settle and farm the border areas or the unoccupied deserts. The high natural population increase (3%) is supplemented by
immigration of Jews (at the rate of 50,000 per year) to make an annual growth-
rate of nearly 4·5%. The defence budget is heavy for a small country and the
rate of capital investment in buildings, irrigation and strategic roads is high.

Inflation is a chronic problem. This, plus the desire to achieve economic independence, leads to taxes on many imports and a drive to earn foreign exchange, primarily from citrus exports, the re-export, after cutting, of diamonds, tourism and the export of manufactured goods. Despite these difficulties, the general standard of living is comfortable but not high (Table 1.5.1). Proportionately there are fewer rich people and fewer cars than in, say, the U.S.A. or the United Kingdom. There is some squalor and poverty even in thriving new purpose-built towns like Beersheba. Generally, however, the population is intelligent, vigorous, living simply but well and enjoying most of the advantages of a modern welfare-type state. Roads, transport, schools, medical services, trading facilities, etc. are good in both urban and rural areas.

Agrarian Structure

The agrarian structure has been, and is, determined more by social ideals and defence considerations than by free-enterprise economics. The result is interesting and, in many ways, instructive. In 1964, of 804 villages, about 250 were Kibbutzim and 347 were Moshavim; the rest consisted of 'private' farmers. One-third of the farm population are on private farms as owner-occupiers, tenants or employees; these farms tend to be rather larger than the average Moshav holdings. They are still small farms by U.K. or U.S. standards, though there are a few large private enterprises and public companies, especially in citrus or cotton on the coastal plains.

The Moshavim. About half the farm population are smallholders on Moshavim. 'The "Moshav" is a co-operative settlement in which there are 60–150 individual farmsteads forming one "village" co-operative through which they sell their produce and buy their supplies, and which is largely responsible for the management of the village as a whole' (Lowe 1963). The average smallholder on the older Moshavim works about 13 acres with the emphasis on dairy cattle, orchards (citrus, vineyards and deciduous fruits) and irrigated fodder crops. But each smallholder is free to farm as he likes; in 1958–59 net income ranged from less than IL 1,500 to more than IL 20,000 per farm.* So there is some scope for the enterprising go-ahead man who can, in some cases, rent additional land from his fellow members or elsewhere. In some Moshavim, unirrigated field crops, such as barley, wheat, sorghum, are worked for the members by the co-operative which uses its own or hired mechanised machinery. The co-operative owns and operates the central services (water, milk depot, incubators, heavy machinery, workshops, retail shops, kindergarten school, cultural activities, etc.). About one-fifth of the total man and woman power in a Moshav is employed in these 'common services', the remainder on the farms. (In Israel, a very high proportion of women, both in towns and farms, work, at least part-time, outside the home. On a farm the work unit is usually a man and his wife.)

* One U.S. dollar = 3 Israel pounds (IL); 1 pound sterling = 7.2 Israel pounds (IL) (1968). (Before U.K. devaluation of 1967.)

Table 2.7.3

Israel Transects
Sectors (a), (b), (c): Lat. 32°45′N. ←→ 32°30′N Sectors (d), (e): 31°40′N ←→ 31°20′N
Sectors (f), (g): 31°15′N

	(a)	(b)	(c)
1. Sector			
2. Name	*Coastal Plains* (Haifa)	*Carmel Hills*	*Inland Plains*
3. From	Megadim	Beit Oren	Kiriat
4. To	Beit Oren	Kiriat	Beit Sh'ean
5. Miles	10	15	35
6. Climate-type(s)	Dry Warm (Mediterranean)	Dry Warm (Mediterranean)	Dry Warm (semi-arid)
Thermal Growing Season: **7a.** — months at or $> 42\frac{1}{2}°F$	12	12	12
7b. — months at or $> 72\frac{1}{2}°F$	5	5	6
8a. — P–E–T (in) in T–G–S	45	30	30
8b. — A–E–T (in) in T–G–S	21	20	16
9. Growth limited by	Summer drought (winter cold)	Summer drought (winter cold)	Summer drought (winter cold)
Precipitation: **10.** — mean in (annual)	24	30	20–14
11. — C.V.% of Precipitation (annual)	25%	25%	27%
12. Special Climate Features	No frosts. High rainfall intensity	High rainfall intensity. Some frosts	High rainfall intens Some frosts
13. Weather Reliability	1	1	1
Soil: **14.** — Zone/Group	Chernozemic desertic	Chernozemic desertic	Chernozemic desert
15. — pH	7·2 to 7·5	7·2 to 7·3	7·3 to 7·6
16. — Special Features	10% $CaCO_3$. Alluvial Colluvial. Heavy clays. Stony	5% $CaCO_3$. Terra Rossa. Heavy, shallow, very stony	15% $CaCO_3$. Deep, heavy, up to 60% cl particles
17. Vegetation	Mediterranean woodland and scrub	Mediterranean woodland and scrub	Short grasslands
18. Relief	Plains	Hills	Hills with wide plai
19. Market Access	2	2	2
20. Farming Systems	Intensive tillage; Intensive tree fruits	Semi-intensive tillage; extensive grazing	Intensive tillage; extensive grazing
21. Major Enterprises	(Very intensive.) Irrigated citrus, bananas, cotton, sugar-beet, vegetables, forage. Cows' milk, poultry	(Not irrigated). Winter cereals, vineyards, some sheep and goat grazing	(Some irrigation.) Deciduous fruits, winter cereals, grai sorghum, cows', an sheeps' milk, beef, poultry

Table 2.7.3

Israel Transects
Sectors (a), (b), (c): Lat. 32°45′N. ⟷ 32° 30′N Sectors (d), (e): 31°40′N ⟷ 31°20′N
Sectors (f), (g): 31°15′N

(d)	(e)	(f)	(g)
estern Lachish	*Eastern Lachish*	*Northern Negev*	*Jordan Valley Oases*
kelon	Kiriat Gat	Gaza Strip	Dead Sea Coast
riat Gat	Lahav	Dead Sea	—
20	20	60	—
y Warm (Mediterranean	Dry Warm (semi-arid)	Desert	Desert
12	12	12	12
5	5	6	7
35	35	40	50
15	16	9	3
mmer drought	Summer drought	Summer and winter drought	Summer and winter drought
18–14	14	12–4	3
34%	28%	30%	35%
t dry winds. High nfall intensity	Hot dry winds. High rainfall intensity	Hot dry winds. High summer temperatures	No frosts. High summer temperatures
1	1	1	1
ernozemic desertic	Chernozemic desertic	Desertic	Desertic
7·5	7·6	7·8 to 8·0	7·5
% $CaCO_3$. Sands and dy clays	20% $CaCO_3$. Loessal; silty clay loams	20% $CaCO_3$. Loessal. Sandy to heavy	
diterranean woodland scrub	Mid-latitude grasslands	Desert scrub	Tropical evergreens
ns	Rolling plains and hills	Plains with mountains	Flat. Below sea-level
2	2	2	1
nsive tillage; intensive fruits	Semi-intensive tillage; extensive grazing	Extensive tillage; extensive grazing	Intensive tillage
ne irrigation.) Citrus, er cereals, vegetables, s for milk	(Not irrigated.) Winter cereals, grain sorghum, beef	(Some irrigation.) Shifting cultivation (winter cereals) and grazing, milk, sheep	(Irrigated by fresh-water springs.) Intensive out of season vegetables and flowers

The Kibbutzim. 'The Kibbutzim is a collective settlement, the members of which form a large-scale enterprise based upon common ownership of resources and upon the pooling of labour, income and expenditure' (Lowe 1963). Social life is communal, with feeding in common dining rooms; the children are, in part, brought up separately from their parents; no member may have private property apart from personal books, record-players, etc. The Kibbutzim, like the Moshavim were started by the Jewish Agency which acquired land from 1909 onwards from Arab landlords. 'To each according to his need, from each according to his ability' is carried to its logical conclusion. Everything is shared. The officials who are elected to provide the secretariat, to manage the farm or its several branches and to manage the industrial enterprises which are often parts of Kibbutzim, do not receive a greater share of the distributed goods and services, though they may, while they hold office, for say 3 years, be exempted from some common duties. At the end of his period of office, a farm manager, or a treasurer, may go back to tractor-driving—apparently quite happily, sometimes perhaps with relief, for the day-to-day administration of these democratic institutions is not always easy.

Kibbutzim, like the *Moshavim shitufim,** are relatively large-scale enterprises with from 200 to 600 adult members and their dependents. On them, about 63% of earning population is, on the average, engaged in farming, 20% in contract work for others in building or on installations on their farms and 17% in industrial enterprises (e.g. furniture making, printing, plastics, agricultural machinery or shoe manufacture, etc.). Though it may be doubted whether these industrial enterprises achieve adequate economies of scale, the farming is often technically and managerially at least as advanced as that of very large farms in the U.S.A. or the United Kingdom. The Inter-Kibbutz Union now maintains its own planning section using modern business mathematics (e.g. linear programming) to enhance efficiency of resource use. The Kibbutzim tend to concentrate on enterprises suited to their scale. It is not unusual for a Kibbutz or a Moshav shitufi to have up to 3,000 acres in mechanised grain production, up to 300 cows in milk, more than 30,000 laying hens or broilers and as much acreage of cotton, groundnuts and sugar-beet, i.e. industrial crops, as their irrigation water supply and production quotas (see below) allow. Their ability to command technological and other resources (including mobility of labour) has given them an effective monopoly of banana growing, of carp production in fish ponds (fertilised with inorganic and organic nutrients), of extensive beef raising on semi-ranching lines and of the use of large flocks of Awassi dairy sheep which are milked by machine. The Kibbutzim, and to some extent the Moshavim, jointly own and operate several 'wholesale' co-operatives for sugar-beet processing, packing citrus, storing grain, making cheese from sheep milk, artificial insemination and contract work by tractors, combine harvesters, cotton-picking machines, sugar-beet machinery, etc. They do not, however, own or operate any fertiliser or tech-

* These are large collective farms where work, but not living is communal.

304

nical chemical (crop-protection) factories. The theoretical advantages of such
vertical and horizontal organisation are obvious (Halperin 1963) but, it may
be doubted whether the factories are large enough to compete on a world
scale. But this scale problem affects much of the Israel economy as a whole.
There are, for instance, three beet factories for a total acreage of 12,000 (which
has, however, a very high yield per acre).

Israel is, in part, a planned economy in which social and strategic considerations may outweigh the economic factors which dominate a free-enterprise economy. This is well illustrated by the regional rural development
plans, such as that in the Lachish region which lies (Table 2.7.3) just to the
north of the Negev desert and was sparsely settled. Now water has been piped
in from the north and the whole area is being redeveloped. New villages are
being built to house immigrant settlers in Moshavim. For each 5 or 6 villages,
there is a rural centre with shops, schools, medical services, etc. and then, at
Kiryat Gat, a regional centre with hospitals, specialist services, processing
factories for farm produce (e.g. creameries, cotton-gins) and, it is planned,
partly independent manufacturing industries, e.g. for plastics. It is hoped that
the income of the farm family will equal that of the skilled urban worker. To
do this it is realised that, as farm productivity rises, more workers will have to
leave the land and work in local factories—even if they continue to live on the
farm. This, it is hoped will avoid the hardships and waste of fixed social resources (e.g. schools) which accompanies the drift from the land in an un-
planned economy (e.g. the Welsh hills in the U.K., the Ozarks in the U.S.A.).
But whether, in such a planned integration, one can achieve economically the
technological scale and capital investment needed by many modern industries,
remains to be seen. Agriculturally, a more serious problem in these and, indeed, in all Moshavim and smallholding areas, is whether, as productivity per
man rises, the size of the farm enterprises and/or the acreage per farm can be
increased rapidly enough. The Israeli authorities are well aware that many
land reforms have finished up as inefficient anachronisms (e.g. in Northern
Ireland). As one leader put it 'every settlement must be capable of continuous
re-development'. Finally, there is the strategic question. The military advantages of a larger and highly skilled population settled over as much land as
possible coincide with most of the social ideals. The manpower reserve on
Kibbutzim and Moshavim is significant. More significant is the location of
defensible Kibbutzim with young dedicated members, near the frontiers.
Military considerations have in earlier history helped to determine the location of farming systems (e.g. in Roman frontier colonies); they are still doing
so in Israel today.

Services. The majority of farm dwellings now have electricity, piped water,
reasonable sanitation, passable access roads, sound radios and access to a
telephone. Medical, social and school services seem to be good. The Extension
Service (run jointly by the Jewish Agency and the Ministry of Agriculture)
and the Government control of crop and livestock pest and diseases, genetic

improvement, water use, soil erosion are all well developed. Agricultural research is of a high standard. If the drive to develop the country quickly has
led to some short-cut 'look-see' field trials which cannot be classified as research, it has, equally, fostered, for instance, in agricultural meteorology, some
ingenious devices for the quick 'build-up' of climatic data. Thus, mobile 'jeep-
mounted' meteorological units search for frost pockets or collect records which
can be used, by extrapolation from long-record stations, to forecast the probable climate of areas intended for farming settlement. The input supply services are good if somewhat expensive. Marketing of output is highly, perhaps
over-organised. There are marketing boards, some with draconian powers, for
all major commodities. Thus, no farmer can grow potatoes unless another
farmer stops and he can take over the acreage. Maximum production quotas
are widely applied. If a farm or Kibbutz exceeds its sale quota, it receives no
subsidy for the excess. Prices for outputs are usually negotiated between the
Government and the agricultural side of the General Federation of Labour
(Histadrut), and the other two minor farmers' organisations. Internal prices,
except for citrus, tend to be above world levels and agricultural imports,
except for subsidised surpluses of U.S. cereals, are discouraged by high tariffs
or exchange control. With such a high rainfall variability in a dry country,
crop yields and hence animal costs vary greatly from year to year. There is no
general insurance against weather hazards, but in drought years the Government makes *ad hoc* payments to badly affected farmers and Kibbutzim who
farm north of the 8 in annual isohyet. (It was found that farms south of this
line were sowing crops in anticipation of crop failure and of drought payments; hence this limit.)

Crops and Livestock

As in some other warm dry Mediterranean climates (e.g. California, southern Spain) a very wide range of crops are grown (see statistics in Table 2.7.1).
The dominant tree fruit is citrus which, in quantity and value of output, exceeds that of all other tree fruits; these latter include bananas, grapes, olives
(of decreasing importance), and such deciduous fruits as apples, pears, plums
and peaches. Of growing importance are sub-tropical fruits such as avocados,
mangoes, etc. especially for export. Amongst the irrigated field crops, vegetables, potatoes and especially 'green' fodder crops, including fodder beet for
cattle, are most important, but 'industrial crops' such as sugar-beet, groundnuts, cotton (for lint and seed) are of growing significance. The main unirrigated (dry-farming) crops are winter wheat, winter barley, vetch hay and,
increasingly, sorghum for grain (grown as a summer crop). Dairy cattle
(mainly 'Israeli Friesian') and poultry for eggs and broilers are dominant in
livestock, especially on smaller farms, but beef cattle and mutton are of growing significance. Awassi sheep for milk are decreasing in numbers; they are
found especially on those easterly and southerly Kibbutzim which have rough
mountain or desert grazing or land otherwise unsuitable for tillage. Awassi

sheep are, together with shifting cultivation of unirrigated cereals, the main-
stay of the Bedouin Arabs in the northern Negev. Finally, carp can be and are
grown in fish ponds of rather saline water; the water authorities, however,
argue that the evaporation loss from fish ponds is excessive and that the water
could be better used elsewhere. Zohary 1962 lists the 'palatable grass and
leguminous species in the local flora which have been the source of pasture
for cattle and sheep from ancient times to the present days. . .' However, as
in parts of Australia 'the bulk of grazing during the rainy season is provided
by annual grasses including *Avena* spp, *Bromus* spp, *Cutandia memphitica*,
Hordeum murinum, *Koeleria phleoides*, *Aegilops* spp, *Brachypodium distach-
yum*, *Phalaris* spp *and Lolium gaudini*. Herbage legumes include *Trifolium* spp,
Medicago spp, *Lathyrus* spp, *Pisum fulvum*, *P. elatinus*, *Astragalus* spp and
Vicia spp, and grow mainly in the winter. They supply useful grazing on
natural pastures in February and March. Shrubs are also of some practical
and perhaps great potential importance as browse for livestock in the less
well-watered regions. Zohary 1962 mentions '*Atriplex halimus*, *Kochia indica*,
Colutea istria and others'.

The geographical distribution of crops and livestock is indicated in Table
2.7.3. However, the structural distribution is interesting. Kibbutzim, with
their large and specialised human, machine and technological resources
have a near monopoly of bananas, dates, cotton, fish ponds, milk sheep (other
than Arab holdings) and beef cattle. At least one Kibbutz has an experimental
herd of pigs—otherwise forbidden on religious grounds. (Christian Arab
farmers, however, are quite active in pig production.) The Moshavim tend to
specialise in cow milk production, poultry, sugar-beet and groundnuts—
enterprises which have lower technological and higher personal labour demands. The private sector farms are rather like the Moshavim but, almost
always being longer settled and on goodish land, grow citrus which is, however, also found on Moshavim and Kibbutzim in areas where frost is not a
hazard and water is available.

Crop Pests, and Diseases and Weeds

No crop is impossible to grow because of disease or pest problems. Crop
protection by integrated biological and chemical control is highly developed;
it is a 'must' for successful agriculture round the Mediterranean. The Israelis
dislike sole dependence on chemicals. They try to reduce an unwanted species
to sub-clinical levels; if they exterminate it they may find that they have got
rid of a useful predator. The main problems of *citrus* are the Mediterranean
fruit-fly (*Ceratitis capitata*) which is controlled by a malathion-loaded bait,
often sprayed from the air; the Egyptian black scale insect which is biologically controlled by the wasp (*Aphitis noloxanthus*) introduced from Hong Kong;
citrus moth (*Prays citri*) for which there is no real control; and mites which
are the biggest pest but can be controlled by spraying. Fungal diseases are not
serious whilst psorosis virus, which is transmitted by direct contamination, is

controlled by a bud-wood certification scheme. *Cotton* is protected by strict
plant quarantine. There is no black arm but spiny bollworm (*Earias insulana*)
and army worm (*Heliothis armigera*) are chemically controllable problems. As
elsewhere, sodium chlorate is used for defoliation before harvesting the boll.
There seem to be no virus diseases in *sugar-beet* but a fungal disease (*Cerco-spora apii*), which can be controlled by tin compounds, is a nuisance. *Nema-todes* cause some concern in bananas but most of the pests and diseases of
other tree crops are under control—at a price. In *potatoes*, late blight (*Phy-tophthora infestans*) is controlled by the usual copper compounds; virus
diseases are checked by using only certified seed from the United Kingdom.
Winter *cereals* present no great problems and the pests which attack summer-grown sorghum, can usually be controlled by husbandry means (e.g. selection
of the right drilling date, crop rotation, etc.). Weed control in cereals on
Kibbutzim is largely by spraying from the air and on small farms by ground
spraying. Sprays are also used in tree crops but inter-row cultivation and hand
hoeing are still widely relied upon. Zohary 1962 states 450 species of weeds are
found in cultivated crops in Israel and Jordan. Some came in with imported
crop seeds, others immigrated from near-by steppes and deserts, but the
majority are old native plants which still occupy 'primary habitats ecologic-ally similar to those that are now under the plough'. There are marked weed
communities which are associated more with soil type and climate than with
particular crops. Incidentally, in earlier times, the seed of fat-hen or goose-foot (*Chenopodium album*), a well-known tillage weed in several parts of the
world, is 'known to have been used in years of famine and wartime' for
making bread (Zohary 1962), and was also, it is said, used as food in pre-historic agriculture in Europe.

Animal Health and Reproduction

As in most advanced countries, veterinary interest is moving towards meta-bolic and fertility problems as classic pests and diseases come under control.
Israel is, however, geographically vulnerable to epizootics such as pleuro-pneumonia and some types of foot-and-mouth disease. Tuberculosis and con-tagious abortion have been, in effect, eradicated in cattle, where the major
problems seem to be tick fever and mastitis, apart from scours and other
diseases of infancy in calves (e.g. *Salmonella* infections) which are serious.
Ticks (*Ixodides*) are the vectors of duo-plasmosis, piro-plasmosis and *Tirenia
annulata;* although some protection can be obtained by vaccination (except
against piro-plasmosis), the main and costly control is spraying beef cattle
every 7–10 days during the tick season. Most dairy cattle spend their whole
life in open sheds. They rarely graze and internal parasites are no problem.
On Kibbutzim, mastitis has a high incidence in milking flocks of sheep which,
however, have few helminths. The latter are, however, found in large numbers
amongst Arab-owned sheep partly because they are often underfed and not
surprisingly have a higher incidence of pregnancy toxaemia than on Kibbut-

zim. The major poultry health problem is *Leucosis* (especially the neural and visceral forms). *Coccidiosis* is widespread (but not serious) in spite of routine use of coccidiostats and probably owing to undue reliance on them. The main metabolic disease seems to be milk fever in cows, but though yields are high in Israel, its incidence is lower than might be expected. Most Israeli soils are high in $CaCO_3$ and it may be more than coincidence that the blood of local dairy cows has higher than normal calcium content. When post-parturition trouble occurs, it is, however, sometimes associated with a low blood phosphate or a disturbed $Ca:P_2O_5$ ratio. Hypomagnesaemia seems to be almost unknown—possibly because of the high concentrate diet of cows. Infertility occurs in dairy herds (which are mostly artificially inseminated) towards the end of summer (Sept./Oct.); it is suspected that the cows rather than the bulls are to blame. The latter are, like the cows, given frequent showers but it is difficult to cool the scrotum and testes by this means. Economically this late summer fall in conception rates is not important, however, as most cows are in calf by June—the main calving season being November to March. Reproductive research is advanced. Thus in successful experiments in early calving, heifers of 14 months are producing calves and heifers of 23 months have calved and then gone on to produce 1,500 gallons of milk (15,450 lb) in 321 days (Beresford and Edwards 1966).

Sheep are kept mainly for milk. There is no problem of summer sterility in Awassi ewes and rams, possibly because their fat tails may insulate the scrotum of the latter. Semen remains of good quality through the hot weather and conception rates are high in midsummer. The usual high latitude photoperiodic control of oestrus does not apply; the breeding season is from June through to March followed by a period of an-oestrus in April–May when, however, oestrus can be hormone-induced. In practice, on Israeli farms, if artificial insemination is started or the rams are put to the ewes in, say, mid-June, then on some farms all ewes that have not come to heat are treated with P.M.S. (Pregnant Mare Serum) and progesterone to induce oestrus and so get a more concentrated lambing. A good lamb crop is 100% but selection for twinning is increasingly practised.

Heat tolerance in livestock. Apart from infertility problems associated with the climate, there are thermal difficulties due, not to heat loss by cold or wind or rain, but to insolation, heat and hot winds, each of which may reach clinical levels. In Israel, heat stress is not, apparently, a problem with Awassi sheep but it is with dairy cattle and poultry—both of which are mainly of breeds of cool temperate origin, e.g. Friesian cattle, Leghorn poultry (for eggs). Beef cattle are largely of Hereford or Hereford cross type—which, despite their geographical origin, have some capacity to withstand heat stress. Zebu-type cattle are, however, more heat tolerant and there are a few herds of Brahmin cattle for pure or crossbred beef production. In beef, and dairy sheep, one breeds the animal to fit its surroundings; but in dairy cattle and poultry, the Israelis have successfully adapted the environment to suit the

animal—so successfully indeed, that mean yields per cow in Israel are the
highest in the world (average of 12,500 lb/year compared with 8,270 in the
U.K.). (One of the highest yielding herds visited by Edwards 1967 was in the
Negev desert, where summer temperatures were around 100°F and rainfall
averaged little more than 1 or 2 in per annum.)

How has this been achieved? In part, by careful genetic selection,* by
milking three times daily and by personal attention to each animal. But
though important, these devices are used elsewhere without outstanding success. In Israel they are probably supplementary; here the keys to success are
meticulous attention, firstly, to reduction of the heat load the cow has to dis-
sipate and secondly, to easing the loss of heat by the cow. The first objective
is achieved by keeping the cows in covered yards so that there is no direct
body insolation and so that conversion of chemical energy into body heat by
exercise or during ingestion is minimal; and by feeding rations relatively high
in concentrates (which are largely imported from the U.S.A.) and relatively
low in fibrous feeds.† In high summer, the 'green fodder' may, in fact, be
fodder beet which has only a 10% fibre content compared to say 26–33% in
hay. This reduces the amount of heat produced in rumen fermentation, i.e.
reduces heat increment per unit of net energy available for tissue formation or
milk secretion. It has led to an interest in the possibility of feeding cows an all
concentrate diet which may raise, for instance, liver problems as in the U.K.
barley beef, and might presumably depress milk fat yields, but would, incidentally, reduce the water demand for irrigated green fodder crops. The
second objective is achieved by clipping the cows in summer to reduce the
thickness and hence the insulating capacity of their coats, and by showering
them with water.

Research on dairy cows is now being extended to the efficiency of housing.
It is already clear that the air movement is influenced substantially by shed
design and that the behaviour of the boss cow in a group may frustrate the
architects' concepts about efficient ventilation. Showering the cows in their
houses (as well as in the milking stalls or parlours) is well established. The
idea now is that, if the cows are kept on slats, the shower water that drips from
them washes the faeces through to the slurry pit, and is then, as in the United
Kingdom and elsewhere, pumped into the land through irrigation pipes. But
this work, and the readiness of dairy farmers (most of whom came to Israel
with no built-in prejudices) to accept advice based on research and on rapidly
gained practical experience, provides an excellent example of how science and
technology can adapt environment to make highly efficient milk production

---

* It appears, however, that the coefficient of heritability of heat tolerant characteristics
(e.g. sweating capacity) is too low to permit efficient genetic solutions to these problems
although the large U.S. type has higher heat loss capacity than the Dutch type of Friesian.

† High fibre diets are associated with a high production of acetates in the rumen which
are absorbed into the body but can only be metabolised into fat. Any acetate in excess of
requirement for fat formation is oxidised with the production of heat which may aggravate
heat stress. This, at least, is the theory.

possible under potentially adverse conditions. The application of such environmental control in other hot countries and to other types of livestock is likely to grow in importance.

Transects and Farming Systems

As noted, mean annual precipitation decreases rapidly eastwards and particularly southwards, and, at most latitudes, the coastal plain gives way, moving eastwards, to mountain or hill country, and then sinks down to the Jordan valley. Three latitudinal transects have been taken, therefore, one in the well-watered and long-settled north; one in the mid-south where new settlement is still in progress; and one further south, in the northern part of the southern desert (the Negev) where close settlement is still to come.

Northern Transect (*Table 2.7.3*)

COASTAL PLAINS (*Sector (a)*)

After the coastal dunes, this narrow strip has alluvial and colluvial soils of rather strong heavy clay. The annual precipitation is more than 24 in (C.V. 25%) with a marked winter peak and a thermal growing season of 12 months with temperatures averaging about 70°F. Except in occasional frost pockets, frosts are rare and almost all crops can be grown. The natural vegetation has long since been destroyed but is assumed to have been Mediterranean forest and scrub with some sub-tropical savanna.

Farming. There are few farms over 100 acres. Most of them run from 5 to 20 acres and, especially in the citrus groves, many are privately owned or rented. The dominant plantation crops are bananas (which can tolerate no frost) and particularly citrus, which is evergreen. Amongst field crops, sugar-beet, cotton, vegetables and green crops for milk cows are important. Dairying, and to a lesser extent poultry, are the main livestock enterprises and most farms, even citrus groves, have one or both of these as subsidiary enterprises. The young citrus plantations are often inter-cropped with for example, potatoes. But once the trees have almost met across the rows, cultivation is limited to weed control; though in many cases such control seems to be left to the competitive power of the established citrus trees and in particular to their ability to draw water from greater depths than ephemeral or perennial weeds. One grass, Bermuda grass (*Cynodon dactylon*) is, however, a major problem. Spraying for pest and disease control follows a fairly strict regime.

The choice of citrus species and cultivars and their location depends on (*a*) frost risk (lemons are very susceptible but most varieties of oranges can tolerate temperatures of 28°F, whilst grapefruit can withstand lower temperatures), (*b*) suitable soil—very heavy soils are avoided and a fairly high sand fraction is preferred, and (*c*) the availability of non-saline irrigation water. This latter is less critical in the north than further south. In general, bananas, citrus and green feed for milk get water priority. Other field crops

*

are irrigated as needed if water supply and equipment permit. Protection from
the wind, especially against the *hamsin*, is often needed, and most established
citrus groves are sheltered by belts of *Cupressus* spp which, as they mature,
replace the temporary shelter belts of elephant grass (*Pennisetum purpureum*).
Though spraying, fertiliser application (especially nitrogen) and cultivations
consume some time in spring and summer, the main work load on citrus is
picking; this extends through autumn and winter, starting with the early
varieties and finishing with Valencia oranges. This means that more time is
available in the summer for irrigating and for cutting and carting green feed
(which includes fodder beet) to the dairy cows.

CARMEL HILLS (*Sector (b)*)

Much of Israel (Fig. 2.7.1) is hilly limestone with shallow stony soils and,
before the advent of modern irrigated farming in the plains, was largely terraced for crop production. But centuries of erosion, wars, neglect and labour
shortage have led to the partial abandonment of much hill land. Some of it
has been afforested and much of the rest, except where there are strategic
Kibbutzim along the frontier, is farmed rather extensively from Arab villages.
Patches of winter cereals (wheat, barley), vegetables, vineyards and (decreasingly) olive groves scatter the hill-sides together with herded sheep and goats
on land not accessible for or suitable for cropping—there are few stock-proof
fences or hedges in Israel. On these hills the dry *hamsin* from the desert—
which blows almost 30 days a year and desiccates by excess turbulence, lower
humidity and higher transpiration—is more damaging and irritating (to
humans) than on the plains. Its effects on crops can be offset by irrigation but
on the hills there is generally no water to spare for this. The Carmel Hills,
occupied by the Druze people, are fairly typical of this type of farming. Precipitation at 30 in per annum (C.V. 25%) is higher than on the plains, and the
resultant leaching may account for the low (for northern Israel) $CaCO_3$ content of the soil which has a pH of 7·2 to 7·3. The 'natural' vegetation is Mediterranean forest and scrub. Farming on the Carmel Hills is unirrigated and
follows the usual pattern of Arab hill settlements, viz. *extensive grazing* and
semi-intensive tillage.

INLAND PLAINS (*Sector (c)*)

These plains run down from the east of the Carmel Hills for 25 to 35 miles
to the River Jordan. They are rich in history. Nazareth can be seen amongst
the hill towns which fringe them and Megiddo is in their midst. For at least
2,000 years they have had a reputation for high fertility; when the Romans
were occupying Palestine, wheat was sent from these plains to Rome. Today
they are still very productive and provide some of the most interesting and
exciting farming in Israel. The soils are described as deep, hydromorphic
brown grumosols. They contain up to 15% $CaCO_3$, have a pH of 7·3 to 7·6 and
up to 60% of clay particles. Some of the clays near the Jordan are hard to

work and require great skill in cultivation. Lying in the rainshadow of the Carmel Hills and being further inland, the annual precipitation averages from 20 in down to 14 in at their eastern edge. With P–E–T at about 30 in per annum, moisture deficits are common. The thermal growing season is 12 months but there are 6 months at or over $72\frac{1}{2}°F$, so heat stress in livestock and possibly crops is a problem. The climate is warm dry (semi-arid) and the vegetation changes as one moves eastwards from Mediterranean scrub to short grasslands.

Farming. The farms are larger than on the coast, and some of the Kibbutzim and collective Moshavi run up to several thousand acres. Frosts are too frequent or severe for citrus but the winter cold breaks the dormancy of apples, making deciduous fruit plantations possible. (In the hotter parts of Israel such dormancy may be broken by chemical spray.) Where irrigation water is available, it may be used for sugar-beet, cotton and green feed for the cows. But the main crops are unirrigated; they are winter wheat and winter barley and in summer, if moisture permits, sorghum for grain. Attempts at using Australian type rotations (see Sector (*e*)) do not, so far, seem to have been successful. On the hills and waste lands there is considerable grazing of 'natural' grasses, weeds and shrubs; this is used for rearing beef cattle—some of them of tropical type, for example, Brahmin—and for the production of milk from Awassi sheep, which are machine milked. The ewes are placed in batches on large turn-tables on which they are fed their concentrates. As each batch of 8 or 10 is milked, the turn-table is moved on, the milked ewes released and another batch let on to the turn-table. The milk, which is high in fat, is sold for cheese making. The seasonal lambing pattern has been outlined above. Cows for milk and also poultry are often major enterprises and the usual precautions are taken against heat stress in summer.

On one collective *moshav shitufi* (Moledet) of some 2,700 acres, plus 2,500 acres rough grazing, and 120 families, the soils are red-loam basaltic clays with pH 7·6 to 8·0, $CaCo_3$ content of 12–20% and 50–60% clay particles. The annual precipitation over the past 25 years has ranged from 3 in to 10 in and rainy days from 42 to 85. There have been 6 drought years with substantial failures of unirrigated crops—which can, however, then be used to provide some badly needed grazing. In a typical year there would be 750 acres of winter wheat, 200 winter barley, 400–450 sorghum for grain, 80 clover for seed, 400 acres for hay, 100 for silage, 150 for green manure, 75 unirrigated Sudan grass (*Sorghum sudanense*) (pastured) plus 50 acres irrigated green feed. (Purchased irrigation water is supplemented by the collection of spring and run-off water.) There are 25 acres in olives, 10 in grapes and a small acreage of vegetables for seed because there is not enough water to produce them for human consumption. The nearer fields may be irrigated and are rotated as follows: (1) legumes for silage (usually receive 1 or 2 irrigations per year); (2) winter wheat; (3) Sudan grass pastured or sorghum for grain; (4) clover and vetches for hay; (5) winter barley. On the outlying fields, dry farming is practised and the rotation is: (1) green manure (*Trigonella foenum graecum*); (2) winter wheat; (3) sorghum for grain; (4) vetches for hay; (5) winter barley; (6) summer sorghum for grain or clover for seed.

The livestock comprises over 200 dairy cattle (of which over 120 are milked), 250 beef cattle (some Brahmins and their crosses), 1,000 sheep (750 milking ewes), 9,000 poultry (layers and broilers) and 135 beehives mainly for pollination. Yields average about 13,600 lb per cow, about 630–700 lb per ewe, and 200 eggs per laying bird. Wheat runs at about 2,250 lb to the acre and grain sorghum somewhat less, but in drought years yields may only be 30% of these figures. The sources of the energy intake of the dairy herd are 55–60% concentrates, 11% hay, 8% silage, 10% unirrigated and 12% irrigated green feed. The cows are kept indoors all the year, but the sheep and the younger beef cattle graze. The latter are fattened indoors. The equipment is modern and includes 8 caterpillar and 9 wheeled tractors, 3 self-propelled combine harvesters, hay balers, etc. The work force of 120 families is by U.S.A. or even U.K. standards, very high. There are also grain stores, cattle and poultry sheds, workshops, etc. About one-third of the expenditure is for living costs, one-fifth for purchased concentrates, 10–15% for fuel and machinery, but outgoings on seed and fertiliser are low by U.K. standards at about 5%. Sales of grain, cows' milk and poultry products each contribute one-fifth to the total sales, sheep 10%, vegetable seeds and orchards 10%, outside contract work 10% and beef cattle 5%.

Southern Transect

WESTERN AND CENTRAL LACHISH (*Sector (d)*)

This sector, together with eastern Lachish, Sector (*e*), is interesting, partly because the farming systems are somewhat different from those in the previous sector as the climate is drier and the soils rather lighter; partly because it is the location of a large-scale project in new farm settlement within a 'composite rural structure'; partly because it is also the site of a comprehensive pilot scheme in watershed control. The climate is warm dry (Mediterranean) and so is the vegetation, the soils are sandy on the coast or sandy (terra rossa) clays with a pH over 7·5 and a CaCo₃ content higher than that in the north. The terrain is flat but gently rising to the east. Precipitation is relatively low at 14–18 in annual mean with a C.V. over 30%, i.e. decidedly unreliable. There are 3 or 4 heavy rains a year and water erosion and loss of run-off water are problems, even on gentle slopes. The area is helped, however, by some piped irrigation water and by attempts at ground-water control (see below). The use of water is strictly controlled; it is applied only to high-value or export crops such as citrus, industrial crops (sugar-beet, peanuts and cotton), vegetables and green feed for cows.

The individual farms are mostly very small, and the farmers are mainly members of Moshavim. The main enterprises are citrus, winter cereals, vegetables, poultry and cows for milk. The inputs of fertiliser (mainly nitrogen and phosphate) and, on stock farms, of purchased feeding stuffs, are high. Husbandry practices are broadly the same as those in the north. Though poaching of the soil is a problem in the winter rains, more dairy cows seem to be grazed (generally in the evenings or mornings) than in most parts of Israel where zero-grazing is the usual rule for milking cows. Some green manuring is practised in the hope that it will improve soil structure.

Settlement in a 'composite rural structure'. To overcome the problems of small-scale operation and a rigid framework planners here 'have attempted to combine some of the production advantages of the large scale administered farm with the socio-economic advantages of the family farm by adopting a new approach to rural structure and organisation' (Weitz 1963a). Briefly, 70 to 80 family farms make a co-operative (*moshav*) centred on a village with a grocery store, a kindergarten and a village hall. Four or five such villages are linked to a centre, not more than 5 miles away, where there is a school, a clinic, a youth club, a tractor station, a citrus or other packing station, etc. The centres are, in turn, linked to a town, in this case Kiryat Gat, which has shops, hospital, administrative offices, cold storage and factories which process the farmers' 'industrial' crops, and may have some strictly industrial enterprises (e.g. textiles, electronics). These latter, it is hoped, will in due course provide 'local' employment for those released from the land by technological or economic change.

Each moshav is, in theory, laid out in groups of fields in such a way that common services (e.g. irrigation pipe-lines, mechanised cereal production, aerial spraying), are practicable. But the farmer is not compelled to use these services. 'The farm family ... lives in its own house, tills its own fields and makes its own decisions' (Weitz 1963a). Farm size depends on rainfall, soil and crop potential. In the wetter, better soiled western section, 7 or 8 acres, all irrigated, are allocated for a citrus farm (with $2\frac{1}{2}$ acres in this crop) or a dairy farm (with at least 8 cows); in the centre $11\frac{1}{2}$ acres (half irrigated) is allowed for a tillage farm growing 'industrial' crops rotated with vegetables. In the hillier, drier, eastern Lachish (see below) the plan provides for 60–80 acres per farm (of which 2 would be irrigated) and the growing of winter cereals in rotation with a fallow year in which, however, grain sorghum may be grown as a summer crop. In brief, this is a planned settlement which by putting the accent on flexibility and on the need 'for constant re-reform' of agrarian structure, *may* avoid the fate of schemes of land reform which in other lands have only too often been radical at their birth but become anachronisms long before they reach old age. It is also an experiment in the vertical integration of farm and factory advocated by Halperin 1963.

EASTERN LACHISH (*Sector (e)*)

The altitude here is higher, the terrain rolling, and the annual precipitation at about 10 in averages rather less than in the Western Sector. It is very

unreliable and, in some parts, in 4 years out of 10 winter cereals are a complete
failure or so poor they can only be grazed. The soils are mostly fairly deep
loessal, brown, rather silty clay loams, with a pH of over 7·6 and more than
20% CaCo₃ but further into the hills, they become shallower and somewhat
stony. The natural vegetation is transitional to short grasslands (or Turko-
Iranian steppe).

Farming systems. Irrigation water is very limited so dry farming (*extensive
tillage*), mostly on large holdings, is practised. The major enterprises are
winter wheat, winter barley and summer grains (sorghums), though there are
a few deciduous orchards and some vegetables where water can be applied.
On one kibbutz at Lahav, about 20 miles north of Beersheba, the altitude is
1,800 ft and the very variable rainfall averages 11 in per annum. 70% of this
comes in 3 or 4 big rains in the winter. There are more than 5,000 acres in
tillage (on useful loessal heavy loam which can, however, become unworkable
after rain) supplemented by plenty of rough *extensive grazing* on the hills. A
little piped water is used for the 50 acres of deciduous fruit. But a dam to
collect run-off has also been built, the water being pumped on to near-by fields
to bring them up to field capacity often *before* the corn is drilled. With this
water 'stored' in the soil, the emptied dam is then ready for the run-off from
the next big rain. The farm has 80 adults and 45 children, is highly mechanised
and has a large grain silo (1,000 tons). There are beef cattle, mainly Herefords
or crosses thereof, as a subsidiary enterprise. They are grazed on the hills in
the winter and spring and then, after cereal harvest in May/June, on the
stubbles.

On the tillage, in the first year, barley and some wheat are drilled in October
or November. The second year is fallow through the winter but if, after the
winter rains, the soil is damp down to 3 ft or more, then grain sorghum is
drilled. Mould-board (not disc) ploughs are used when possible. Winter
corn has a single dressing of nitrogen to help it get well established in case of
drought. Superphosphate is applied to every other winter corn crop, i.e. once
in four years. Herbicide, e.g. 2.4.D., is sprayed from the air at up to 1,500
acres per day at a suitable date between January and March; this is expensive
but it gets the large acreage done at the right stage to kill weeds with the
minimum damage to grain yield. In an average year there will be 2,500 acres
in winter cereals and 1,000 acres in grain sorghum. In a good year barley
yields 2,800 lb/acre but after a dry winter it may be a total failure. In brief,
this is a good example of modern, dry farming (if one overlooks the high
labour force) though it can be criticised for its lack of erosion control and
absence of leys.

Watershed management. The Nahal Shikma pilot project is operated jointly
by the F.A.O. and the Israel Ministry of Agriculture in the southern Lachish
(Sectors (*d*) (*e*)) and parts of northern Negev (Sector (*f*)). The 3 or 4 heavy
winter rains rush down off the land, with a heavy burden of eroded soil, into
wadis (valleys), which only flow intermittently even in the wet season, and

thence into the sea. The resultant problems are being tackled in three ways. First, after a rain, a dam near the coast collects run-off water. This is allowed 3 days to drop its sediment and is then pumped out onto surrounding land to raise the ground-water level for the local citrus growers, to check inward seepage of sea-water and to empty the dam before the next rain. This system, however, involves moisture loss *en route* to the dam, and of course, ultimately the silting up of the reservoir. Secondly, farmers are encouraged to trap and use run-off water on their own land as on the kibbutz at Lahav above. Thirdly, there are large-scale trials in alternate husbandry somewhat on the successful Australian wheat-and-sheep belt model (p. 183) and designed to make better use of water and reduce both wind and water erosions.

In these field-trials, leys of annual self-seeding grasses and legumes are established with superphosphate and then for 3 or 4 years are grazed by Merino sheep or cut for hay. Next they are ploughed and cropped 2 or 3 times with winter cereals before re-seeding either by broadcasting or allowing seeds from the previous ley to re-establish themselves. Some of the grass and legume seeds are washed down into the erosion gullies and form sward-covered erosion-checking waterways. The most successful mixtures are either *Vicia dasycarpa* or *Medicago hispida* (Cyprus Bur clover) with soft brome grass (*Bromus mollis*) or with Wimmera rye-grass (*Lolium rigidum*). Subterranean clover has not been successful. Some useful hay yields (2,800 lb/acre) have been obtained in good rain years. In drought years, the cereal crops will be fed to the sheep and use made of the *Acacia* trees and *Atriplex* shrubs which have been established on rough land (instead of forests) to provide drought-reserve browse.

Desert Transect

Negev Sector (*f*)

The sector, at the latitude of Beersheba, has a fairly flat terrain and loessal soils varying from sandy to fairly heavy clay, with pH of 7·8 to 8·0 and more than 20% $CaCo_3$. The P–E–T is more than 4 times the A–E–T and the annual precipitation (C.V. over 30%) averages from 12 down to 4 in as one moves eastwards. In summer, high insolation, high temperatures and hot dry east winds from the true desert make soil and crop desiccation as well as heat stress of crops and animals major limiting factors. Intensive farming is impossible without added water so perhaps the two most significant potential features of this sector are the Negev Arid Zone Research Institute (where research on heat and water stress in crops, on watershed and on desert soil management as well as on desalting sea-water are in hand) and the near-by Atomic Energy Research Station. (The United States and Israeli Governments are considering a joint nuclear power and desalting project.)

West of Beersheba agriculture is based on irrigation and there are no Bedouin, but east of Beersheba the farming is that of the semi-nomadic Bedouin Arabs who, in suitable areas, practise extensive cereal growing and

herd their Awassi milking sheep and a few camels on the very limited 'natural'
vegetation (very extensive grazing). Recent research has suggested that the
ancient system of scratch ploughing (used by the contemporary Bedouin)
makes much better use of the limited natural rainfall than mould-board
ploughing, and the equally ancient practice of trying to keep water on the soil
surface and guiding the run-off from say 50 acres on to a patch of 3 to 5
cropped acres may be technically sound. In the southern Negev, where the
terrain is broken by mountains and fantastic wind-sculptured formations of
red rock reminiscent of Ayers Rock in Central Australia or parts of Arizona,
U.S.A., there is no farming at all.

JORDAN (RIFT) VALLEY OASES (*Sector (g)*)

Along the Dead Sea, which is 1,292 ft below sea-level and is much too
saline for irrigation use, there are some fresh-water springs on which oases
with tropical vegetation have become established. The soil is mostly light and
gravelly. There are no frosts. Here a few settlements (e.g. Ein Geddi), though
remote of access, successfully grow very early vegetables and flowers in winter
and spring.

Efficiency and Stability

By orthodox economic standards, Israeli agriculture can hardly be termed
efficient. The farms are too small, the labour too plentiful, the industry too
highly protected and the economies of scale are not readily achieved in the
ancillary industries. These factors are most apparent in those parts of the
economy where general political and defence considerations outweigh agricultural. The situation, however, is not static and, given a greater international
stability, Israeli farming could readily become more efficient. In fact, improvements are now being made. In Moshavim, where the small farm problem is
most acute, specialisation, more intensive cropping and the pooling of resources are helping to raise productivity. Further, imports are being liberal-
ised and mechanisation is going ahead. For instance, 5 years ago cotton was
hand-picked; today it is nearly all machine harvested. Indeed, some of the
farming, especially in well-established Moshavim and Kibbutz, is technically
very advanced and exciting as witness the high yields per cow and the high
yields/acre of sugar-beet. Technical standards are, however, lower in the
settlements which have been only recently established for untrained Jewish
immigrants. Overall, the growth-rate of agricultural productivity is amongst
the highest in the world (Beresford and Edwards 1966). However, given the
social, political and strategic assumptions on which agricultural policy and
settlement are based, the low output per man and the small scale of many of
the ancillary enterprises are less relevant. Thus, Israeli agriculture can be regarded, on the one hand, as an interesting and indeed exciting experiment in
integrated farm, industrial and social development in an unstable environ-

ment, and, on the other, as a not very efficient cheap 'unemployment insurance', as in other developing countries.

The sources of actual and potential instability (apart from military factors) are three, viz. the heavy dependence, especially of cows and poultry, on imports of subsidised U.S. feed grains; the difficulty of finding markets for citrus and other farm exports, especially in the European Economic Community; and finally the variability of the rainfall which, as noted, is substantially greater than the average variability (C.V.) for an area with the same mean annual rainfall as Israel.

Conclusions

In the long run no doubt the above problems will be overcome as the country becomes more fully developed and especially if the country's irrigation water supply can be secured, or if, as seems possible, economic ways of desalting saline water can be devised and increases made in the return per unit of water. For despite the interesting infra-structure, the high technical level of much farming and research, and the energy of the people, in the long run the future supply of non-saline water is the future of Israel agriculture.

Acknowledgements. For data, comments and criticisms to I. Arnon, T. R. Beresford, A. Berman, Mrs K. M. Down, G. P. Hirsch, Yehuda Lowe, Michael Morag, Mrs L. M. Roman and G. Stanhill.

References

Beresford, T. R. 1967. Personal Communication. Chilmark, Wiltshire, England.

*Beresford, T. and Edwards, J. 1966. *Agriculture in Israel.* Anglo-Israel Ass. London.

*Brichambaut, G. P. de and Wallén, C. C. 1963. *A Study of Agro-Climatology in Semi-Arid and Arid Zones of the Near East.* Tech. Note No. **56**. No. **141** T.P. 66. World Meteorological Organisation, Geneva.

Dan, J. and Koyumdisky, H. 1963. 'The Soils of Israel and their Distribution', *J. Soil Science.* **14**. 1. 12–20.

*Dash, J. and Etrat, E. 1964. *The Israel Physical Master Plan.* Ministry of the Interior, Jerusalem.

Edwards, J. 1967. Personal Communication. Milk Marketing Board, England.

*Halperin, H. 1963. *Agrindus: Integration of Agriculture and Industry.* London.

*Israel Government, 1965. *Facts about Israel.* Information Division, Ministry for Foreign Affairs.

Lowe, Y. 1963. *Kibbutz and Moshav in Israel: An economic study.* Ministry of Agriculture and the Jewish Agency, Israel. Mimeo.

Mansfield, P. 1965 *Nasser's Egypt.* Penguin African Library, Harmondsworth, England.

Farming *Saab, G. S. 1967. *The Egyptian Agrarian Reform.* London.
Systems Weitz, R. 1963a. *Family Farm versus Large Scale Farms in Rural Development.* Artha Vijnana. **5**. 3. 225–40. Poona, India.
*Weitz, R. 1963b. *Agriculture and Rural Development in Israel: Projection and Planning.* Bulletin No. **68**. National and Univ. Inst. of Agric. Rehovoth, Israel.
Zohary, M. 1962. *Plant Life of Palestine, Israel and Jordan.* New York.

* Asterisked items are for further reading.

CHAPTER 2.8

Temperate South America (Uruquay, Argentina, Chile)

(*Note:* This chapter is only a broad summary and this region is not examined in the same detail as other temperate countries in this part).

Introduction

These three rather lightly populated countries cover an area almost as large as India and Pakistan but with a population of only 32m. They jointly display a wide and sharply defined range of climates, natural vegetation and soils. But, although there are large areas between latitudes 30°S and 39°S where the ecological conditions are naturally favourable for farming and have a great potential for intensive tillage and meat production, agriculture is, by comparison with broadly homologous regions in North America or Australasia, relatively extensive and underdeveloped.

Why is this? Market access and input access are problems which have been aggravated by tariff, disease control and other import barriers (for example, in the U.S.A. and Europe); these have not borne so heavily on Australia and New Zealand which are further from the main industrial markets of the world but unlike these three countries are free of foot-and-mouth disease. The export earnings of Argentina and Uruguay are almost exclusively agricultural with the emphasis on livestock but, in Chile, come largely from copper (70% of Chile's export earnings) and from nitrate. The main reasons, however, are internal, both economic and social, much of the land having been in the hands of large landowners who have had little real financial incentive to adopt modern agricultural techniques. The stage of economic development and industrialisation in these three countries, which are short of mineral and power resources, lags behind that of most temperate countries. Neither consumption of steel and power per head, nor the standard of living are high (Table 1.5.1). Recently, growth of oil-fields and sometimes ill-advised industrial development, though in the long run no doubt valuable, have in the short term diverted resources which might have been better devoted to agriculture.

In contrast to tropical America (Chap. 2.14) further north, these three countries share a common historical background. All were pioneered by the Spaniards in the 16th century but they have all been independent nations for a century or more. Development, both social and agricultural, varies widely in Argentina and Chile from one part of the country to another, owing, to a great extent, to the enormous distances from north to south and to the great differences in climatic conditions that apply in the different districts. These

Table 2.8.1

Crop	ARGENTINA Area (000's acres)	ARGENTINA Yield (average for 1963–65) (lb per acre)	URUGUAY Area (000's acres)	URUGUAY Yield (average for 1963–65) (lb per acre)	CHILE Area (000's acres)	CHILE Yield (average for 1963–65) (lb per acre)
Wheat	10,413	1,397	941	889	2,098	1,355
Oats	1.040	1,148	284	737	279	1,017
Barley	949	1,192	89	624	178	1,674
Rye	818	714	—	—	32	987
Maize	7,566	1,525	474	517	180	2,340
Millet	284	942	—	—	—	—
Sorghum	2,026	1,255	7	598	—	—
Rice	168	3,232	69	2,744	74	2,435
All cereals	23,363	1,379	1,880	835	2,842	1,431
Sugar-cane	618	46,681	15	32,171	—	—
Sugar-beet	—	—	42	20,307	44	34,134
Potatoes	504	8,742	35	4,222	212	8,206
Sweet-potatoes and yams	89	8,741	40	4,400	—	—
Cassava	57	10,347	—	—	—	—
Vegetables (a)	133		15		63	
Dry peas and beans	140	1,089	15	574	255	690
Other pulses (b)	57	791	—	—	114	472
Soyabeans	40	942	—	—	—	—
Other oil-seeds (c)	7,262	605	573	467	313	970
Tea	47	9,960	—	—	—	—
Tobacco	118	1,050	—	—	5	2,673
Cotton (lint)	1,320	211	5	229	—	—
		Production 000's lb		Production 000's lb		Production 000's lb
Tree crops (fruit) (d)		8,975,452		526,554		2,011,401

Livestock	Numbers 000's	Yield (average for 1963–65) (lb per animal)	Numbers 000's	Yield (average for 1963–65) (lb per animal)	Numbers 000's	Yield (average for 1963–65) (lb per animal)
Total cattle	46,709	(e) 891 (1964)	8,142	(e) ?	3,116	(f) 546
Dairy cows	?	(g) 926	?	(g) 911	310	(g) 3,601
Pigs	3,500	(e) 249 (1964)	438	(e) ?	1,074	(f) 139
Sheep	46,100	(e) ?	21,874	(e) ?	6,577	(f) 41
Wool						7·8
Goats	5,098	(e) ?	17	(e) ?	1,464	(f) 29
Poultry (chicken)	34,200		7,500		?	

(a) Onions, tomatoes, cabbages, cauliflowers, green beans and green peas (Chile: additional 26,700 acres of melons and pumpkins).
(b) Chick peas and lentils.
(c) Groundnuts, cotton-seed, linseed and sunflower seed; rape-seed (Chile only) and castor beans (Argentine only).
(d) Apples, pears, plums, prunes, cherries, peaches, apricots, oranges, tangerines, grapefruit, lemons, limes, olives and grapes (grape yields—Argentine, 8,036 lb per acre; Uruguay, 5,454 lb per acre; Chile, 5,534 lb per acre). Argentine only: figs, bananas and pineapples.
(e) Live-weight of animals slaughtered.
(f) Carcass weight of animals slaughtered. Approximate conversion:
    Cattle and sheep: live-weight = carcass weight × 2
    Pigs: live-weight = carcass weight × $\frac{4}{3}$
(g) Milk per milking cow.

wide differences are not evident in Uruguay because of its comparatively small size and the similarity of the topography and climate throughout the country. One common factor in all these three countries is the presence of very large *Estancias* (Argentina and Uruguay) or *Fundos* (Chile) (up to 250,000 acres each), which have provided a good return to the owners on the basis of low farming, but which are, as in Spain, ill-suited to or unwilling to apply modern technology to farming. This is discouraged, in part, by high tariffs. In Argentina a £600 foreign tractor costs £3,000 landed. However, Agrarian Reform

in Chile is now well developed towards the break-up of the large *fundos* into
smallholdings. It remains to be seen if this social plan will be conducive to
greater efficiency in agriculture. There is the danger in redistributing land
that, unless the technical assistance to farmers who have newly acquired land
is really good, the agrarian system might change from inefficient large land-
owners to badly educated and incompetent small farmers.

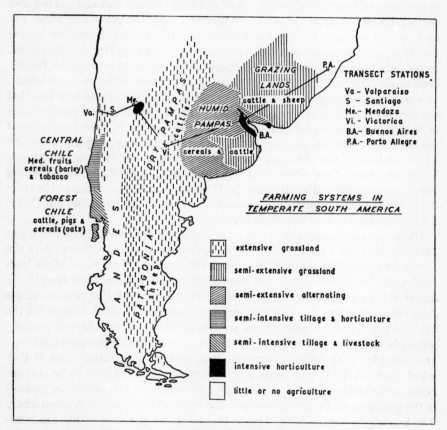

Fig. 2.8.1 Farming Systems in Temperate South America

Yields per acre and per animal (Table 2.8.1) are generally less than in other
temperate countries with homologous climates and soils, although cereal
yields in Argentina may be slightly higher. Though, on farms, labour inputs
may be high, industrial and scientific inputs (especially of fertilisers*) are low.
Such technological inputs, together with good climate, especially A–E–T are,
in these days, the key to high crop and livestock yields. Finally, there is the

* Fertiliser consumption/acre in Argentina is the lowest of the 57 countries included in
Fig. 1.1.9, although the acreage under cereals is the seventh largest.

problem of political stability which, though enjoyed by Uruguay, is not marked in the other two countries.

Ecological Background

Climate and soils. From the Tropic of Capricorn to Tierra del Fuego is some 2,500 miles (through 40° of latitude); this transect exhibits marked differences in climate and vegetation, which reflect not only latitude and altitude but also the combined effects of the Andes mountain range, the prevailing winds off, and coastal currents in, the Pacific and Atlantic oceans. Sweeping up the Pacific coast of South America from Antarctica is the cold Humboldt current; it not only lowers temperatures compared with the same latitude on the Atlantic coast, but radically reduces the moisture content of the on-shore winds. In northern Chile, the moisture content of the prevailing winds is so low that the area is completely rainless. In central Chile, the mean annual precipitation is 20 in at Valparaiso and 14 in at Santiago. Further south, from the region of the 'roaring forties' to Magellan land, precipitation increases southwards with 87 in annually on the Taito peninsula and 120 in at the western end of Magellan Straits. East of the Andes the Pacific influence is expended. The prevailing winds are easterly and north-easterly. Precipitation increases eastwards from 7·5 in at Mendoza at the foot of the Andes to 37 in at Buenos Aires, but (Table 2.8.2) is unreliable over the whole region.

High summer temperatures are recorded in the interior (Corrientes mean January temp. 82°F) and low winter temperatures on the southern tip of the continent (mean July temp. 32°F) and on the glacier-draped peaks of the high Andes. On the Pacific coast, the Humboldt current ensures that, even in the desert of the north, summer temperatures seldom reach 85°F in contrast to the Atlantic coast where January temperatures may reach 100°F. Except for the extreme south and the high Andes, winters are mild and the thermal growing season is 12 months. P–E–T also, of course, decreases with increasing latitude and altitude. With minor exceptions it is precipitation rather than P–E–T which limits A–E–T. Not surprisngly, therefore, the soils tend to be chernozemic or desertic over much of this region though there are, of course, podsols in the wetter parts. In detail, however, the soil map is too complex (even where it is known) to pursue here, though some data on the soils of the more important areas are given in Table 2.8.3.

*Vegetation.** The livestock industries are largely based on grazing of natural or sometimes introduced plant species, and an appreciation of the vegetation is essential to an understanding of this region's agriculture. In sharp contrast to the vast stretches of tropical forest and savanna to the north, the natural vegetation south of the Tropic of Capricorn is grassland and falls into three groups:

1. Good natural (tall) grasslands such as are found in the eastern, or humid pampa of Argentina and Uruguay with no extremes of climate. (The

* This section is largely based on Roseveare 1948.

northern half of Uruguay is officially tick-infested (*Boophilus annulatus microplus*) as is much of the provinces of Entre Rios, Corrientes and Missiones in Argentina).
2. Semi-arid (short) grasslands, mostly suitable for sheep, occurring in the western, dry pampa and over Chilean Patagonia.
3. Cool mountain grasslands, often used in summer, found in parts of the Andes.

The pampa of Argentina is a *completely* flat treeless plain unrelieved, except by three groups of low hills. It is virtually a 'sea' of grass only broken by the occasional groves of eucalyptus and poplars around *estancia* houses and by the isolated ombu tree (*Phytolacca dioica*) known as the 'lighthouse of the pampa'. There are few rivers, water being pumped by windmills from depths of 100–500 ft. Since the 16th century, the vegetation cover has changed from the indigenous tall, plumed bunch grasses on which the wild herds of cattle and horses first grazed.* From the late 19th century introduced grasses such as *Bromus* (*B. hordeaceus*), *Poa*, *Lolium*, *Briza minor*, *Cynodon*, *Koeleria phleoides* led to pasture improvement as did herbs such as *Erodium cicutarium* and *Echium plantagineum* and legumes such as *Trifolium repens* and *Medicago denticulata*. These introduced species provided 'tender' herbage, growing mainly in fertile moist soils or on tilled land. 'Strong' indigenous grasses, generally perennials (such as *Stipa neesiana*, *S. hyalina*, *Tridens brasiliensis*, *Briza subaristata*, *B. stricta*, *Piptochaetium* spp and *Paspalum quadrifarium*) being strongly rooted, are better adapted to the drier pampa but provide poor grazing.

The east of this region has a transition southwards from the semi-deciduous forests of the tropics to the treeless plains of the Argentine pampas. Between these two areas lie the undulating grasslands of Uruguay where abundant watercourses are marked by dense fringes of woodland and useful indigenous pastures. Trees occur as one goes *up* the Andes but the *paramos* (or Andine Meadow vegetation of low-growing plants, largely of the rosette and cushion type) are found over the great highland plain or *puna*. One species (*Alchemilla pinnata*), with a high content of P, K, magnesium and calcium, is particularly useful for sheep. In Chile, with its rugged terrain and diversity of climate, 'natural' grasslands are of less significance and the vegetation ranges, north to south, from dry xerophytic through the typical Mediterranean shrubs and grasses of central Chile and the dense primeval forests further south down to the bleak windswept plateaux of Patagonia. Here vegetation cover is determined by aspect, owing to the low angle of the sun's rays, by high winds (up

---

* In 1769 the wild artichoke (*Cynara cardunculus*) was brought into the La Plata area lodged in the coat of a donkey, and together with the giant thistle (*Sylibum marianum*) transformed that part of the pampa until they covered the land with a dense growth, sometimes up to 10 ft high, as far as the eye could see. Although the thistles presented a fire risk in December and January they provided fodder for cattle, horses, sheep and even pigs, although it caused tainting of cows' milk.

to 100 m.p.h.) and by cold from the Antarctic; the climax of vegetation is
stunted shrubs, many of them edible, but in more protected areas, particularly
in gullies, where deep rich alluvial soils have accumulated, good natural pastures exist.

In both Uruguay and Argentina the indigenous vegetation has been substantially changed by livestock and, in the former country, several centuries
of uncontrolled grazing have suppressed good indigenous herbage, particularly *Paspalum* spp, but left poorer grasses, especially 'hard' grasses of *Stipa*
spp. A Uruguyan commission emphasised the low-protein and high-cellulose
content of the herbage, the incidence of osteomalacia (in livestock) in phosphate-deficient areas, the excessive size of paddocks and the need for rotational grazing, the dangers of locusts, the grain aphis (*Toxoptera graminum*) (a
menace to cereal catch crops for winter grazing) and ticks. (In central Uruguay (Rio Negro district) more than half the area of otherwise good pasture
land is denied to livestock by inedible weeds. If the weed control law could be
enforced, the carrying capacity, as well as the condition of the cattle, would be
greatly improved.) Many of these problems occur, with the expected ecological variations, all over the pastoral parts of these 3 countries. Thus in Argentina, the advantages of the *eastern pampa*, viz. a good A–E–T, fertile soils and
relative freedom from ticks, have been reduced by indifferent management
and poor administration, for example, failure to eradicate or at least control
foot-and-mouth disease. Again, the introduction of lucerne (alfalfa) into
temperate South America, enabled high-grade European livestock to be bred
and fattened for the European meat markets and also stimulated cereal production. But excessive dependence on poorly managed lucerne in 3 to 5 year
leys after a succession of cereal crops has led to overgrazing and hence to invasion by unwanted herbage species and to wind erosion. However, heavy
use of cultivations, rather than chemicals, for weed control in the cereal phase
may be a greater source of such erosion. Simultaneously, the stock sold to the
frigorificos (meat packers) take with them mineral nutrients (especially P, Ca)
which are not replaced. Finally, the pumping of water for livestock has drastically lowered the water-table, in places up to 15 ft.

FARMING IN URUGUAY

Uruguay, with its population of 2·5m, mainly of Spanish extraction, is
trying to industrialise, probably at the short-term expense of farming 'for
which it is magnificently endowed, but which has been neglected' (Davies
1966, p. 437). The warm humid climate, a fairly reliable precipitation, a good
mean annual A–E–T (31–36 in) (Table 2.8.2) and predominantly dark brown
or black chernozemic potash-rich soils, make it suitable, in many parts, for
intensive grazing, alternating or even tillage systems. But the farming is, in
fact, mainly *semi-extensive grazing* producing beef cattle and sheep for the
frigorificos and for wool which is one of the chief exports (Table 2.8.3, Sector
(*a*), Fig. 2.8.1).

Table 2.8.2

Climatic Data — Temperate South America

Station and Country	Bright Sunshine hours/day (mean annual)	Thermal Growing Season (>42½°F) length in months	Thermal Growing Season Mean Temp °F	Mean Months >72½°F	Mean Annual Precipitation (inches)	Annual Precipitation Coeff.Var. (C.V.)	Evapo-transpiration in Thermal Growing Season		Moisture Index Im	Max. recorded rain/day inches (month of occurrence)
							Potential (mean inches)	Actual (mean inches)		
Porto Allegre (S. Brazil)		12	67	4	53	20–25%	37	36	43 humid	4·3 (May)
Buenos Aires (Arg.)	6·6	12	61·5	2	38	20–25%	32	31	19 moist sub-humid	7·0 (October)
Rosario (Arg.)		12	63·5	3	37	20–25%	34	33	8 moist sub-humid	10·0 (March)
Mendoza (Arg.)		12	61·5	3	7	30%	32	7	−78 arid	1·9 (January and February)
Bahia Blanca (Arg.)		12	61	1	21	25%	32	21	−34 semi-arid	4·6 (June)
Victoria (Arg.)	7·5	12	60·5	3	22	25%	32	22	−31 semi-arid	—
Santiago (Chile)		12	58·5	0	15	30%	29	15	−48 semi-arid	4·1 (June)
Valparaiso (Chile)		12	58·5	0	19	30%	28	19	−33 semi-arid	7·3 (June)
Valdivia (Chile)		12	53·5	0	105	15–20%	26	26	303 per humid	6·9 (August)
Antofagasta (Chile)		12	63	0	0·5	40%	31	0·5	−98 arid	2·2 (March)
Corrientes (Arg.)		12	71·5	5	48	15%	43	43	12 moist sub-humid	7·1 (March)

327

Table 2.8.3

Temperate South America Transect
Porto Allegre (S. Brazil) — Valparaiso (Chile)

	(a) South Brazil and Uruguay	(b) Argentine: Eastern Pampa	(c) Argentine: Western Pampa	(d) Andes Mountains	(e) Central Valley of Chile
1. Sector					
2. Name	South Brazil and Uruguay	Argentine: Eastern Pampa	Argentine: Western Pampa	Andes Mountains	Central Valley of Chile
3. From	Porto Allegre	Buenos Aires	Victorica	Mendoza	Santiago
4. To	Buenos Aires	Victorica	Mendoza	Santiago	Valparaiso
5. Miles	420	410	260	120	85
6. Climate-type(s)	Warm Humid	Warm Humid	Warm Dry	Mountain	Mediterranean
Thermal Growing Season:					
7a. — months at or > 42½°F	12	12	12	n.a.	12
7b. — months at or > 72½°F	3	3	3	n.a.	0
8a. — P-E-T (in) in T-G-S	32–37	32	31–32	n.a.	28–29
8b. — A-E-T (in) in T-G-S	31–36	22–31	7–22	n.a.	15–20
9. Growth limited by		Drought, frost	Drought, frost	n.a.	Summer drought
Precipitation:					
10. — mean in (annual)	35–50	25–35	10–25	n.a.	15–20
11. — C.V. % of Precipitation (annual)	20–25%	20–25%	25–35%	n.a.	30%
12. Special Climate Features				Mountains. Cause rain shadowing on eastern slopes	
13. Weather Reliability	3	3	3	n.a.	2
Soil:					
14. — Zone/Group	Dark Brown or Black Chernozemic	Black Chernozemic	Chernozemic Desertic	n.a.	
15. — pH	Low	Medium Low	High	n.a.	High
16. — Special Features	Compact soils overlying impermeable subsoil	Fertile black soils or deep beds of fine sand, clay, silt and wind-blown dust, unconsolidated	Brown and chestnut to grey most sandy, frequently saline or alkaline	n.a.	Mostly alluvial, deep rich loams
17. Vegetation	Grassland	Grassland	Transition from grassland in east to xerophilous woodland in west	n.a.	Mediterranean woodland and shrub
18. Relief	Flat plains in S. Brazil undulating in Uruguay	Dead flat plain	Flat plains rising to Andean Piedmont	High mountains with steep W. slope and gradual E. slope	Rugged
19. Market Access	2	2	2	0	2
20. Farming Systems	Semi-intensive grazing and some alternating or tillage	Semi-intensive grazing and alternating	Extensive grazing and some tillage or alternating	Andean grazing of paramos up to 5,000 ft Mendoza—irrigated oasis. Horticulture.	Semi-intensive alternating; tree crops
21. Major Enterprises	Beef and sheep. Milk and horticulture around Montevideo and along north shore of River Plate	Beef and cereals. Horticulture and milk near Buenos Aires	Cattle raising and sheep. Lucerne and cultivated leys.	Lucerne and barley	Mainly irrigated. Vines, beans, barley, maize and fruits (excluding citrus)

The *frigorificos* and refrigerated shipping, as in New Zealand and Argentina, traditionally play a key role in the economy; they obtain a large part of their livestock supplies from their own estates which have herds that, like the other livestock in this country and the Argentine, are genetically mainly of British origin (e.g. Herefords).

Tillage crops are grown on the deeper and better soils of the south-west, mainly for internal use, etc., but these only account for about 6% of the farmed area. The rest is devoted to grazing by sheep and beef cattle on 'native' pastures except for an area within 40 miles of Montevideo where dairying is concentrated. Although wool, milk and meat productivity per acre and per animal is low, there is good genetic potential in the livestock of the country and yields of crops and stock could be increased substantially by genotype selection, good management and the introduction of improved pastures. At the Agricultural Research Station at La Estanzuela, the yields of milk per acre and per Friesian cow have reached nearly three times the national average and at a great saving in concentrates. But, despite this potential, 'farming increasingly fails to export enough to pay for the raw materials needed by industry' (Davies 1966, p. 437) which employs one-quarter of the labour force as against 35–40% in farming.

FARMING IN ARGENTINA

Argentina has a population of less than 24m in an area equal to almost half that of Europe but, as in Australia, two-thirds of the people are in the large coastal cities and many parts of the interior are almost uninhabited. Industrialisation has proceeded further than in Uruguay; of the employed population 35–40% is in manufacturing industry, a fifth in farming and the remainder in government and service industries. Fat stock and cash crops are moved to the ports by the extensive but inefficient railway system or, increasingly, by motor truck. Half the exports are livestock products (chilled and frozen beef, some lamb, wool, hides and skin) and two-fifths are cash crops (maize and other cereals and oil and oil-seeds, especially linseed). Europe and the U.S.A. are the chief export markets.

The climate, in the *eastern pampa*, which embraces Buenos Aires in an arc of about 350 miles radius, is warm humid with a tendency for P–E–T to exceed precipitation as one moves inland. The mean A–E–T in the thermal growing season, which is of 12 months, is over 20 in (Tables 2.8.2, 2.8.3, Sector (*b*) and Fig. 2.8.1). But summer temperatures are high and the strong winds over the dead flat plains make moisture shortage and wind erosion* frequent problems. There are spring and autumn frosts. The soil here is dark brown or black chernozemic, has a high pH and is not markedly deficient in major nutrients.

The eastern pampa were formerly entirely devoted to extensive grazing, but

---

* It is said that, at overnight stops, the trains need to be tethered to keep them on the railway lines!

with the introduction of lucerne a novel alternating system developed on the better-watered lands. The established land-owners, whose holdings ran up to 250,000 acres each, let land for four or five years to immigrants. These share-cropping tenants took wheat crops for 3 years or so and were then required to put down lucerne (alfalfa) ley before they moved on to new land. The land-owner used the ley five or six times a year for 6 to 10 years and then perhaps let it again to a cereal grower. Out of this rotation of occupiers grew the current *alternating system* between 'camp' grazing and 'chicra' (extensive tillage crop-ping), which is now operated by the estate owners. The fields, fenced with smooth wire and termite-resistant wooden posts, are large, often 500 acres or so, and, in the tillage phase, grow maize, wheat and linseed for grain and some cereals for forage; little or no fertiliser is applied. The tillage acreage is, however, less than pre-war and semi-intensive grazing is the dominant land use. The cattle fattening zone runs from north La Pampa to Entre Rios; breed-ing and rearing are found both on the seaward and the western sides of this arc. At Buenos Aires are the *frigorificos* and grain-handling equipment, with some intensive milk production from cows and also intensive horticulture near by. Other crops include sugar, rice, cotton, tree fruits (apples and pears), grapes, tobacco, Indian tea and decreasingly yerba maté (a beverage).

Moving from the eastern to the *western pampa*, precipitation and A–E–T decline (and the coefficient of variation of the former increases). Nearing the Andes and also south of 38°S (Bahia Blanca) A–E–T is under 10 in in a ther-mal growing season of 12 months. In the better-watered parts, there is some *extensive* dry land *tillage* with sorghum for grain, wheat and other small grains but it is not easy to obtain good yields of lucerne (except with irrigation) and *extensive grazing* is the dominant system. Here, as in the eastern pampa, there is little surface or river water and drinking water for humans and livestock is obtained from deep bores. Along the transect line (Table 2.8.3 and Fig. 2.8.1) and to the north of it, both cattle and sheep are produced, but in the drier and colder grasslands south of the line, sheep become, in Patagonia, the main enterprise. This in turn gives way in the far south to the cold, humid extensive sheep-grazing system in Tierra del Fuego with one ewe to 15 or more acres.

The *Andes Mountains* (Sector (*d*), Table 2.8.3) divide Argentina from Chile and run up to 22,000 ft. As in most mountain systems, one cannot generalise about climate, soils or systems, but in the better districts, on gentle slopes, there is extensive grazing of sheep on the 'natural' vegetation as modified by the grazing animals. However, Mendoza is the centre of the wine industry. On the steeper western, Chilean slopes, there is little farming of any kind.

FARMING IN CHILE

Chile has a population of 9m, mainly located in 'Middle Chile' and con-centrated in and around the towns. It has a very high birth-rate, a high death-rate and a high incidence of malnutrition and under-nutrition, both in town and country. A third of the work force work is on farms, but they only produce one-third of the country's food; in the United Kingdom, one-twentieth of the work force produce more than half the food consumed. Although mean calorie intake is not unduly low (Table 1.5.1), protein and other deficiencies are serious; indeed 'Chile's most pressing economic problem is how to grow enough food for herself' (Davies 1966, p. 199).

This long, thin country has, as the epigram has it, 'a hot head and cold feet'. It runs 2,600 miles from the almost waterless deserts at the frontier with Peru, through forests, down to the cold, wet, windswept, extensively grazed sheep lands in Tierra del Fuego (p. 330 above) which is divided between Argentina and Chile. Agriculturally, the most important part is the Central Valley of Chile (Sector (e), Table 2.8.3) which lies south of Santiago and has a warm dry Mediterranean climate with moderately hot summers and cool, wet winters. As in California, however, the annual precipitation is unreliable (C.V. 30%). But fortunately there are plenty of rivers fed through the summer with snow-melt water from the Andes; this is distributed by gravity along simple irrigation canals over the gentle western slopes which run down from the Andean foothills (Pendle 1963, p. 39). The natural vegetation is Mediterranean woodland and scrub and the valley soils are mostly deep rich loam alluvium. Thus, in central Chile, poverty and malnutrition paradoxically stare at potential plenty which is not, or cannot be, exploited because the land tenure system is outdated and technology is poor. 'One percent of the owners control 63·7% of the land' (Davies 1966, p. 199) much of it more potentially productive than the poor land on which the growing number of small 'independent' farmers find themselves on the coastal ranges and in the Andean foothills. In theory, the developing mining and other industries, which account for 85% of the exports, could be increased to earn more foreign exchange to spend on imported food. But, apart from making Chile more dependent on the uncertain international copper and mineral markets, this course would fail to exploit the Central Valley's excellent and only partly marshalled farming resources. Thus, the irrigated area could probably be doubled and yields perhaps trebled, but, until the present system of large *fundos* and numerous small farms yields to modern technology and business management, the food prospect is poor and the farming potential of this ecologically and visually attractive area will remain barely half used.

The *patron* or owner of the large, often self-contained *funda* may or may not be resident. He employs tenant workers (*inquilinas*) who live in primitive conditions with a hut, 2 or 3 acres of land to produce food for their families or to earn a little cash, and drinking-water from the irrigation canals. They work, mostly without benefit of mechanisation, on the *patron's* land which produces irrigated tree fruits (excluding, surprisingly, citrus), grape vines, cereals (especially maize and barley), beans and Irish potatoes, as well as lucerne and forage cereals (especially oats) which are used for local livestock and also for stock brought down from the foothills and the coastal ranges when seasonal pasture growth is over. (Transhumance is, in fact, quite important (Burland

1960, p. 66); contrary to that practised in Alpine countries cattle are moved *up*
and out of the valleys where snow collects in winter, to sheltered ground and
down into valleys in summer.) Milk production, usually linking farmers with a
well-organised milk co-operative providing technical guidance and collecting
milk from 60,000 to 80,000 cows, is making progress which could usefully be
copied in other enterprises.

Stability and Efficiency

Apart from the variable political situation and the vulnerability of her export trade, *Argentine* farming suffers from a relatively unreliable rainfall (C.V.
20–35%), indifferent grass and crop hygiene, overstocked pastures, low fertiliser use, rampant inflation and currency uncertainty, all of which make for
instability of output and income. Such instability discourages inputs. Low
actual inputs and low input ratios often lead, in these days, to high unit costs,
as farmers in other lands have amply discovered. As in Uruguay, both the life
and the grazing management of the large, sometimes absentee, pastoral land-
owner on the *pampa* are often too easygoing. Thus, though attention is paid to
some aspects of animal health which, despite foot-and-mouth disease, is relatively advanced, fertiliser use, internal livestock parasites and crop hygiene
(pests, diseases, weeds) are too commonly neglected e.g. caterpillars, which
could easily be controlled by aerial insecticide spray, may devastate thousands
of acres of lucerne or sunflower. In other countries with the same climates,
cereal yields, for instance, have been rising steadily or rapidly but in Argentina
they are stagnant. Through the farm advisory service (N.T.A.), Government
action to promote efficiency, however, has had some success and, in recent
years, intensive tillage, often by immigrant farmers, has developed outside the
pampa in, for example, the irrigated areas.

In *Uruguay* the position is somewhat better. There appears to be less ecological and political instability but productive efficiency is not, considering
the better climate, markedly greater. The central technical problem is the
application of sound methods of animal husbandry and the introduction of
improved pastures to a significant proportion of the vast areas of natural
grasslands. In *Chile* reliable irrigation water offsets the unreliability of the
rainfall, at least in the important Central Valley which has a great food production potential. Its exploitation is, however, precluded by an antiquated
land tenure system or at least by an unwillingness to modernise, by an obses-
sion with mineral exports and the neglect of the countryside by the towns,
despite the prevalent malnutrition.

Conclusion and Prospects

Here are three warm temperate countries which lie between latitude 30°S
and 39°S and are, except in the Andes and the drier northern edge of the western pampa, ecologically suitable for modern technological farming with high
input ratios, and so for greatly increased production per acre, per animal and

per man. Admittedly economic nationalism, especially in the Northern Hemisphere, may have aggravated the problems of market access which they share, in part, with Australia and New Zealand but the main reasons for their agricultural lethargy are internal, socio-economic ones. Until these are remedied, and, unless the ecological criteria advanced in Chaps. 1.4, 1.7 and 3.1, are accepted, temperate South America is likely to remain one of the world's greatest *potential* food sources — a giant of great power that is only half-awake and is unwilling, perhaps, to drive itself into the mid-20th century.

Acknowledgements. For research assistance and a first draft to Kathleen Down, and for helpful comments and criticisms to J. M. A. Cross, J. H. Proctor, G. V. Short and C. W. Strutt.

References and Further Reading

*Burland, G. T. 1956. *Chile: An outline of its geography, economics and politics.* London.

Davies, H. (Ed.) 1966. *The South American Handbook.* London.

Finch, W. C., Trewartha, G. T., Robinson, A. H. and Hammond, E. H. 1957. *Elements of Geography: Physical and Cultural.* McGraw-Hill, New York and London. 4th edn.

*Hardy, F. 1942. *Soils of South America.* Chronica Botanica. Waltham, Massachusetts.

James, P. E. 1941. *Latin America.* London.

*Pendle, G. 1963. *South America.* London.

*Roseveare, G. M. 1948. *The Grasslands of Latin America.* Imperial Bureau of Pastures and Field Crops. Aberystwyth, Gt. Britain (now Commonwealth Agricultural Bureaux, Farnham, Bucks., England).

* Asterisked items are for further reading.

CHAPTER 2.9

Japan

Japan consists of four principal islands stretching in a north-east south-westerly arc for 1,500 miles. It has a land area of just under 143,000 sq miles which is about $1\frac{1}{2}$ times that of the United Kingdom. The population is just under 100 million, making it the sixth largest in the world. The population density is 686/sq mile (France 259, Netherlands 950). However, this figure is misleading as much of the country is mountainous (four-fifths lies at over 1,000 ft altitude); hence the density per sq mile of farmed land is well over 4,000 which is equivalent to 5 to 6 people per farmed acre. Despite the problems associated with overcrowding, Japan is a very atypical Asian country; it is highly industrialised and has many similarities with Western societies. Since the Second World War it has been a parliamentary democracy with a titular emperor. Statistics of living standards are given in Tables 1.5.1 and 1.5.2.

Climate, Relief, Soils, Vegetation

Japan spans the Temperate Zone. The northern tip has the same latitude as the Bay of Biscay and the southern tip that of Marrakesh in Morocco. Plentiful rainfall sufficiently distributed throughout the year ensures there are no real dry seasons. Mean annual precipitation increases from 40 in in the north to over 100 in in the south (Fig. 2.9.1), the latter receiving heavy precipitation in the main rainy season which is at its maximum in June and July. Between the end of August and October is the second, lesser rainy season. In winter, snowfalls are heavy, especially in mountainous central Japan. The prevailing winds are southerly in the summer and north-westerly in the winter. The north-westerly winds and the central chain of mountains cause heavy orographical precipitation, largely on the north-west facing mountain ranges in the winter. Especially in September, south-eastern Japan is subjected to typhoons of very high winds and heavy rain which sometimes cause 'mud flows', serious flooding, and hence serious disasters. Temperature ranges are roughly equal to those on the east coast of the U.S.A. from New England to Florida. The mean thermal growing season (T–G–S) increases from 6 to 12 months from north to south and latitude 37°N roughly separates those areas able to harvest two crops a year from those capable of only a single crop. Mean annual precipitation exceeds P–E–T in the growing season, so soils are leached. A–E–T, however, is high except in the north (Tables 2.9.2 and 2.9.3). Climates are, therefore, cool humid or warm humid.

Japan is situated on a line of weakness in the earth's crust along the western

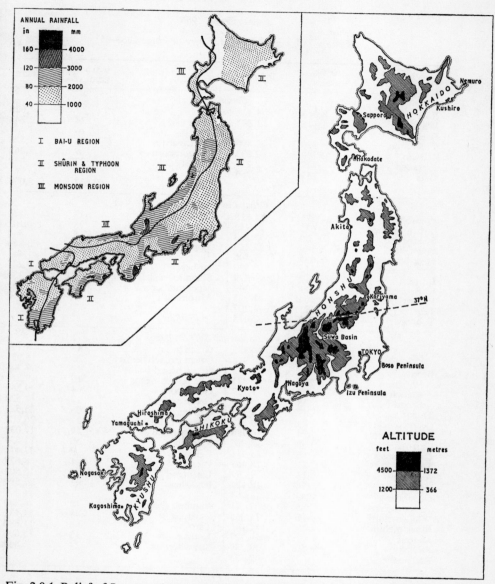

Fig. 2.9.1 Relief of Japan and Mean Annual Rainfall. Rainy season occurs in early summer in I Bai-u region, in autumn in II Shûrin and Typhoon region and in winter in III Monsoon region

Table 2.9.1
(Supplement to Table 2.7.1)

Japan: Crops, 1966

Area in 000's acres: Yield in lb/acre (average 1963–65)

	000's acres	lb/acre		000's acres	lb/acre
Cereals:					
Rice	8,043	3,446	Rye	2	1,629
Wheat	1,176	1,897	Maize	74	2,242
Barley:			Millet	16	1,468
'Rokujo'	326	2,490	Barnyard	28	1,526
'Nijo'	280	2,583	Broomcorn	9	1,062
Naked	437	2,227	Buckwheat	77	843
Oats	153	1,766			
			All cereals	10,623	3,098
Other Field Crops:					
Potatoes	526	15,818	Flax fibre	12	526
Sweetpotatoes	635	17,959	'Konnyaku'	38	6,102
Sugar-cane	30	55,363	Pyrethrum	5	569
Sugar-beet	148	23,429	Peppermint	6	3,966
Soya-beans	455	1,106	Hops	4	1,443
Sesame seed	15	590	Rushes	23	9,221
Linseed	12	279	'Shichito' rushes	3	6,133
Rape-seed	211	349	Paper mulberry	6	935
Peanuts	164	1,923	'Mitomata'	13	363
Tea	121	1,307			
Tobacco	213	2,075			
Vegetables and Soft Fruits:					
Peas	20	885	Green peas	47	4,198
Cow-peas	26	918	Green French beans	28	5,931
Beans:			Green broad-beans	27	3,345
Red bean	268	831	Green soya-beans	16	9,261
French bean	228	1,127	Green corn	64	9,287
Broad-bean	38	769	Green pepper	7	17,520
Tomatoes	47	24,345	Cabbages	106	23,523
Onions	83	20,683	Chinese cabbage	123	26,603
Welsh onions	73	16,984	Other cabbages	59	17,032
Carrots	59	14,706	Spinach	59	11,633
Cauliflowers	3	15,951	Celery	0·7	27,474
Turnips	23	18,467	Parsley	1	8,623
Asparagus	13	2,297	Lettuces	7	16,502
Egg plant	74	18,121	Radishes	243	27,590
Cucumbers	85	19,054			
Water melons	93	17,741	Bamboo shoots	18	6,782
White melons	10	14,265	Burdock	48	14,431
Musk melons	19	11,185	Taro	96	11,680
Melons	1·5	19,327	Lotus roots	13	15,696
Pumpkins	56	13,789	Strawberries	24	6,884
Red peppers	7	2,135			
Tree Fruits:					
Oranges:			Peaches	52	9,018
Mandarin	285	10,295	Plums	29	3,327
Summer	37	11,897	Cherries	4	3,948
Navel	2	9,303	Grapes	56	8,329
Others	20	7,634	Loquat	7	5,736
Apples	162	17,212	Chestnuts	66	1,190
Pears	5	5,973	Persimmon	95	9,366
Japanese pears	47	15,542			

Source: Embassy of Japan, London.

edge of the Pacific and is subject to the destructive forces of earthquakes. There are many volcanoes, 58 of which are active. Each island consists of mountainous ranges with small interior basins, narrow river valleys and coastal plains. River waters are used for hydro-electricity and irrigation of rice fields. The land in the plains tends to fall away from the river banks which is useful for irrigation but very damaging in times of flood (Wilman 1966). The soils are mainly of volcanic origin and are mostly acidic and of low fertility. This situation is aggravated on sloping ground (over 60% of the slopes are more than 15%) by shallow soil and the heavy rainfall which leaches out plant nutrients and causes erosion unless checked. The soils in the northern half are podsols or podsolic; in the south they are podsolic/latosolic. Most respond well to additions of farmyard manure, lime, phosphate and potash. The soils on the plains are replenished by river silt and tend to be loams rather than clays—an advantage for cultivation in Japan's wet climate (Wilman 1966). The 'natural' vegetation is coniferous forest (almost 70% of the land is classed as forest) of larch, spruce and fir in Hokkaido in the north, and mixed coniferous/deciduous forests with pine, cedar, maple, beech and chestnut in the central section of the country. In the south, the forest contains broad-leaved evergreen species including live oak (*Cyclobalanopsis* spp). Where the forests over shallow, infertile soils have been destroyed, they are replaced by scrub, bamboo (*Bambusa* spp) and invading poor pasture grasses including *Sasa* spp, *Miscanthus japonicus* and *Zoysia japonica*. There are no native useful grasses or legumes.

Social and Economic Background

'Major aims of Japan's agricultural policy are to assure food for the population, to equalize farm returns with urban incomes, and to keep prices of food in line with prices of other commodities' (U.S.D.A. 1967). As a result of the rapid growth of the national economy, in particular the industrial sector, Japan is becoming an urban society comparable with Western countries. The population engaged in primary industries (agriculture, forestry and fishing) is declining but is still about one-third of the total, which is high for an industrial country. Farmers and farm workers average about 1 per acre of farm land compared with 1 per 40–50 acres in the United Kingdom. The Japanese are very hardworking and their standard of living has risen greatly since the last war. Consumer goods are very widespread, even in the less affluent rural areas. For example, by 1961 72% of the urban and 29% of the rural population owned TV sets. Unemployment is low and the general shortage of labour and the search for increased efficiency is changing the traditional one-job-for-life situation into a more mobile labour scene. One factor that has contributed considerably to the spectacular growth of the economy has been the very low level of defence spending. With a higher standard of living has come a desire for a European type of diet. This has led to increased demands for meat, dairy products and fresh fruit and vegetables from Japan's farmlands and to

Table 2.9.2

Climatic Data — Japan

Station	Bright Sunshine hours/day (mean annual)	Thermal Growing Season >42½°F (length in months)	Thermal Growing Season Mean Temp °F	Mean Months >72½°F	Mean Annual Precipitation (inches)	Annual Precipitation Coeff.Var. (C.V.)	Evapo-transpiration in Thermal Growing Season		Moisture Index Im	Max. recorded rain/day inches (month of occurrence)
							Potential (mean inches)	Actual (mean inches)		
Akuna Kagoshima		12	62·5	3	78·7	10–15%	33·9	33·9	132 per humid	12 (June)
Nagasaki	5·6	12	61	3	75·9	10–15%	33·8	33·8	124 per humid	15·2 (June)
Hiroshima		10	62·5	3	60·0	10–15%	31·7	31·6	86 humid	13·4 (September)
Nr. Koyoto	5·3									
Tokyo	5·8	10	61·5	3	60·0	10–15%	31·5	31·3	88 humid	11·0 (June)
Akita		8	60	1	67·6	10–15%	26·8	26·8	153 per humid	7·0 (August)
Hakodate		6	60·5	0	46·2	10–15%	22·0	22·0	92 humid	6·9 (August)
Kushiro		6	55·5	0	42·4	10–15%	19·3	19·3	104 per humid	7·2 (September)

Source: See p. 107

considerable imports of meat, wheat, maize, sugar and soya-beans but Japan is still 80% self-sufficient in food.* These 'new' foods supplement protein sources which formerly consisted only of fresh or dried fish and such crops as soya-beans, in which however the protein is of fairly high quality. Improved nutrition has also been accelerated by the School Lunch Programme for primary school children.

Japan's demographic statistics are similar to Western rather than Asian countries. There is an annual population increase of less than 1%, and a long life expectancy. The country has had nearly 100% literacy since the end of the 19th century and there is a very high attendance at centres of higher education. Communications and transportation are efficient and modern. In many areas roads are narrow but unexpectedly paved, and the extensive and fast railways offer a good alternative form of both passenger and commercial transport. Besides a shortage of farming land, Japan lacks sufficient coal and most metals, has virtually no petroleum and must import all her requirements of cotton and raw wool. Besides agriculture, fishing and forestry are important. The fish catch is a major source of animal protein and is one of the biggest in the world. Forests cover about two-thirds of the total land area and provide the country's chief building material as well as paper pulp and charcoal for domestic fuel. These, plus mountains and urban needs, leave only 16% of the land area for arable (tillage and temporary grass), over half of which is irrigated.

Agrarian Structure

The average size of farms is $2\frac{1}{2}$ acres; about 40% are less than $1\frac{1}{2}$ acres and only 8% are 5 acres or more. Most of the larger farms are in the north (Hokkaido) where the T–G–S is short and inputs less intensive. Outside this area, inputs especially of manpower are high and every available piece of land is used, particularly in the south where quite steep hill-sides have been terraced and are irrigated. Drastic land reform was undertaken between 1947 and 1950 during the American occupation. Farm land, of which nearly half had been owned by absentee landlords, was purchased by the Government and sold in smallholdings to tenants on long-term, low-interest-rate loans. Now 90% of farm land is owner-occupied. The maximum size of holding was fixed at $7\frac{1}{2}$ acres (higher in the north). Land reform, however, did not deal with the chronic fragmentation problem. For example, a farm of $2\frac{1}{2}$ acres near

* 'Primary emphasis since the Second World War has been placed upon greater diversification of food consumption in an effort to upgrade the food intake. The consumption of foods and beverages *per capita* in 1965/66 was reported by the Japanese Government at 2,400 calories, compared with 2,100 calories pre-war. Despite efforts to increase consumption of protein foods, such as meats and dairy products, rice, wheat, and other cereals still account for 55% of total caloric consumption. Although the nutritional composition of Japanese diets has improved, the total caloric intake of foods is still low in comparison with Western countries. The caloric intake is low also in contrast to economic growth and the high standards of living which have developed in recent years.' (U.S.D.A. 1967.)

Kyoto is split into 13 scattered plots with 2 of them nearly 2 miles from the farmhouse (Yang 1962). It has been estimated that 15 fields is the average; this implies a mean plot size of less than one-fifth of an acre.

Despite an increased output per unit of labour, the rise in manpower productivity in farming has not kept pace with that in industry. This has led to an income discrepancy which, combined with the national labour shortage, has led to an enormous increase in part-time farming. By 1962 only 25% of farm households relied wholly on agriculture for their income (Wilman 1965, 1966). The average annual income of a farm household in 1962 was £525, of which £270 derived from the farm (Wilman 1965, 1966). Almost all farms are family farms with all adult members helping; hired labour is extremely rare. As the farm population declines, it is becoming older and dependence on women is increasing; they now represent over 50% of the farm work force. The falling farm work force, the rising population and growing demand for food, has led to rapid farm mechanisation. Because of the small size of the farms (and correspondingly small budgets) and of the tiny individual plots, the most useful machines have proved to be 3–7 h.p. hand tractors. By 1964 there were over 1¾ million such tractors on Japanese farms as against 7,000 in 1947. They are very adaptable machines, being used in flooded rice fields as well as for dry ploughing. Attachments such as sprayers can be fitted or they can be converted to provide power for threshing corn, for saws, etc. Their use speeds up field operations considerably, compared with draught animals, and thus lengthens the operational season. This in turn increases the area which can be double-cropped, i.e. where two crops per year can be taken. Thus, with hand tractors it is possible to increase both output per acre and per man. Some larger tractors are in use, especially in Hokkaido, and are expected to increase as Government plans to consolidate farms into larger units make progress. For it is now accepted that the present small farms are economically outdated and that having fewer but larger farms would raise productivity per farm worker.

Scientific research is well developed. Each of the 46 prefectures into which the country is divided, has experimental and research stations, and the widespread literacy greatly speeds the dissemination of up-to-date information. The Japanese are great newspaper readers (Table 1.5.1) and there are many farming publications. In 1960 nearly 6,000 students graduated from Agricultural Departments of Universities and nearly 60,000 from Agricultural High Schools. There were 12,000 extension advisers in 1962 (1 per 539 farms or 1 per 1,400 acres).

Crops and Livestock

Farm land in Japan is customarily divided into irrigated and dry field. The former covers slightly more than half of the total and can be kept flooded during the rice-growing season. More irrigated land is found in the south where it can be drained after the rice harvest and a second crop planted. Most farms have some of both classes of land. The total area of harvested crops in 1954–58 was 157% of the total cultivated area; thus more than half is double-cropped. Rice occupies over 60% of the total crop area. Some rice is grown without irrigation in 'dry fields', but in general these are devoted to other crops. Other important cereal crops include wheat, barley, etc., which are grown both on 'dry fields' and also, in the south, in drained rice fields after the rice has been harvested. Both sweet (*Ipomaea batatas*) and white (Irish) potatoes are important food crops. Soya-beans and pulses, e.g. peas and kidney-beans, are grown widely and, except for soya-beans, the acreage in legumes for human food has increased substantially since the Second World War at the expense partly of unirrigated starchy crops (e.g. barley). Other important crops include Chinese milk vetch (*Astragalus sinensis*), tea, tobacco, sugar-beet, mat rush (*Typha orientalis*) and paper bush (or mulberry) (*Broussonetia papyrifera*). Among important fruits are mandarins, apples, persimmons, summer oranges, grapes and pears. Vegetables include daikon (*Raphanus sativus*) (a cruciferous crop like white radish), taro (*Colocasia antiquorum*), onions, various kinds of cabbage, water-melons and cucumbers. Mulberry trees, the staple diet of silkworms, are grown in small patches on many farms in central and south-western Japan, and 10% of farm families spend part of their time raising silkworms. Thanks to active crop breeding and selection, modern methods and heavy fertiliser use, Japan has high crop yields, especially of the staple food, irrigated rice, which averaged 4,500 lb/acre for the 3 years 1963–65, and ranks after Australia and Spain in yield (see Tables 2.7.1 and 2.9.1* for yields and acreages).

Applications of plant nutrients (including organic manures) are very high; 98 lb N, 56 lb P_2O_5 and 67 lb K_2O per acre (a total of 221 elemental units) are applied on average to tillage land and orchards (Klatt in Yang 1962). In some cases, continuous high levels of fertiliser have had detrimental effects (Yang 1962, p. 25). and more leguminous and green manure crops are now grown. There is widespread use of chemical pesticides and herbicides and the chief rice pests, rice stem borer (*Chilo* spp) and paddy borer (*Schoenobius incertellus*) can now be controlled.

Before 1945, Japanese farmers kept few milch cows, livestock mainly consisting of draught cattle and horses. This was because very little milk and meat was consumed, and because the overall scarcity of land does not permit much grassland with its low output of human dietary energy per acre. Also, in summer, high temperature (3 months over $72\frac{1}{2}°F$) and high humidity over much of the country (Otsuki 1961) discourage animal production. But, since the war, the Government has encouraged cow milk production, and the rise in the standard of living has increased the demand for pork and beef. Poultry raising has also increased rapidly recently, and small numbers

---

* This table, unlike the other acreage and yield tables, is not from F.A.O. sources but by courtesy of the Japanese Embassy, London. Details of crops are given to show the wide range of crops grown in Japan.

of sheep and goats are kept on farms in the mountains and in Hokkaido.
In most areas livestock are kept in buildings and fed, as in Israel, largely on
concentrates and/or cut and carted green feed. Dairy cows are, however,
sometimes, tether-grazed on clover, etc.

North-West to South-East Transect (*Table 2.9.3*)

HOKKAIDO (*Sector (a)*)

Hokkaido is the most northerly of the four main islands. It is distinct cul-
turally and historically, having been settled later and still having few cities
and a much less dense population. It has long cold winters with considerable
falls of snow, especially in the west. The mean frost-free period varies from
120 days in the mountains inland to 145 days on the coast. The annual precipitation of 40–45 in is distributed through the year and the thermal growing
season is 6 months with a P–E–T and hence A–E–T of 19 to 22 in. The soils
are volcanic ashes of various ages, clays and peat. The vegetation of coniferous forests supports a thriving industry with paper and pulp mills. As round
all the coasts of Japan, fishing is important, and, as the climate only allows
one crop off the land a year, many coastal farmers fish in winter.

Farming systems. In Hokkaido the farmhouses are scattered over the plains
(unlike other parts of Japan, where they are usually grouped into villages)
and farms are larger than the national average of $2\frac{1}{2}$ acres. The systems are
tillage or *alternating* with some *extensive grazing*. On the north-eastern plains
it is not possible to grow rice, so the chief crops are oats, wheat, barley, white
(Irish) potatoes and sugar-beet. To the west of the central mountains specially selected strains of irrigated rice can be grown; these require only 90 days
to mature to harvest and are especially disease resistant. The practice of planting rice in sheltered seedbeds (nurseries) and then transplanting is widely used
for four main reasons. First, sowing can be earlier and, although transplanting the seedlings may delay maturation by a week, the net reduction in crop
duration is about a month. Secondly, less seed is required to ensure full stands
in the fields. Thirdly, weed control is more effective. Fourthly (a factor which
is only applicable in the southern half of Japan) seedbed nurseries only take
up a fraction of the whole area needed for the rice crop, so the land to be put
into rice can have another crop growing on it for about 1 month longer, thus
permitting double-cropping. The alternative of growing higher yielding rice
varieties with longer growing periods (crop durations) can also be used.

Much of the small acreage of pasture in Japan is in Hokkaido. Large-scale
Government projects are increasing this area; peat land is being drained and
volcanic ashes made more fertile and then seeded with pasture grasses and
legumes. These include common European and New Zealand species, which
are imported, but Japan has recently begun to produce some of its own seeds.
Livestock must be housed for at least 6 months of the year, so the emphasis
is on producing conserved forage, and, as rainfall is heavy at hay-making

342

Table 2.9.3

Japan (Lat. 46°N ⟷ 31°N)

	(a) Hokkaido	(b) Tohoku	(c) Central Honshu	(d) Southern Honshu	(e) Kyushu
1. Sector					
2. Name					
3. From	Nemuro	Hakodate	Kariyama	Nagoya	Yamaguchi
4. To	Hakodate	Kariyama	Nagoya	Yamaguchi	Kagoshima
5. Miles	300	340	250	350	170
6. Climate-type(s)	Cool Humid	Cool Humid	Warm Humid	Warm Humid	Warm Humid
Thermal Growing Season:					
7a. —months at or > 42½°F	6	6-8	10	10	12
7b. —months at or > 72½°F	0	0-1	3	3	3
8a. —P-E-T (in) in T-G-S	19-22	22-28	28-31	31-33	33-34
8b. —A-E-T (in) in T-G-S	19-22	22-28	28-31	31-33	33-34
9. Growth limited by	Winter cold	Winter cold	—	—	—
Precipitation:					
10. —mean in (annual)	40-46	45-70	40-70	60-70	70-80(-100)
11. —C.V. % of Precipitation (annual)	10-15%	10-15%	10-15%	10-15%	10-15%
12. Special Climate Features	Wetter on west coast		Drier inland. Typhoons on coast	Monsoonal and typhoons	Monsoonal and typhoons
13. Weather Reliability	3	3	3	3	3
Soil:					
14. —Zone/Group	Podsols	Podsolic	Podsolic/Latosolic	Podsolic/Latosolic	Podsolic/Latosolic
15. —pH	Low	Low	Low	Low	Low
16. —Special Features	—	—	—	—	—
17. —Vegetation	Coniferous forest	Coniferous forest	Mixed coniferous/deciduous forest	Mixed coniferous/deciduous forest	Broadleaved evergreen forest
18. Relief	Mountains and coastal plains	Mountains and coastal plains	Mountains and coastal plains	Mountains and coastal plains	Mountains and coastal plains
19. Market Access	3	3	3	3	3
20. Farming Systems	Intensive tillage and alternating; some extensive grazing	Intensive tillage and alternating; some extensive grazing	Intensive tillage with some livestock.	Intensive tillage with some livestock.	Intensive tillage.
21. Major Enterprises	Irrigated rice and other cereals, potatoes and sugar-beet. Store cattle	Irrigated rice and other cereals. Store cattle	Double-cropping Irrigated rice in summer. Fruit and vegetables. Winter cereals. Level (cows') milk	Double-cropping Irrigated rice in summer. Fruit and vegetables. Winter cereals. Level (cows') milk	Double-cropping Irrigated rice in summer. Winter cereals

343

*

time, silage is favoured. Farmers are encouraged to alternate grazing and
cutting to help maintain soil fertility and pasture productivity (F.A.O. 1964).
Beef animals for fattening and young and dry dairy cattle are best suited to
these reclaimed areas. Holstein (Friesian) calves are bought from the dairy
regions in the south, grown on and then fattened on grass or, in the winter,
on silage and grain. Sheep, generally Corriedale types, are kept on some farms,
but the long period of hand-feeding cuts down their economic viability.

TOHOKU (*Sector* (*b*))

See Summary in Table 2.9.3.

CENTRAL HONSHU (*Sector* (*c*))

The summers are warm and humid, and the winters are cool and relatively
dry. The rainfall may be intense, especially when accompanying typhoons.
There are 200–240 frost-free days near the coast, whilst, in the more continental inland areas there are 160–180. In the inland basins with hot summers
and cold winters, the mean annual rainfall is 40–50 in, but higher on the
mountain slopes. On the coastal plains the thermal growing season is 10
months, and the winters are mild enough for double-cropping. The P–E–T
and hence the A–E–T in the T–G–S averages 28 to 31 in.

Farming systems. The farms here and further south are very small, often
less than $2\frac{1}{2}$ acres. They have intensive *tillage systems* with the land almost
continuously in crops. They are usually dominated by rice growing, but some
more progressive farmers are branching out to include livestock (see the
sample farm in sector (*d*)). Rice is planted in seedbeds (nurseries) in April
(spring) and then transplanted by hand 6 to 8 weeks later into the paddies.
If a winter cereal crop has been grown, the ground will have been reflooded
and then ploughed. The traditional method of paddy management was to
leave the paddies flooded all the year round, which helped to reduce oxidation
of organic matter, and to control weeds, but nowadays they are generally
drained a month before the rice harvest.*

Certain areas (e.g. the Boso and Izu peninsulas) near large cities have developed dairying. Each tiny farm keeps only a few head of cattle, but by growing forage crops such as green-cut maize, soya-beans, turnips and Chinese
milk vetch (*Astragalus sinensis*) (which can be harvested after 3 months) fresh
green feed is available for most of the year (cf. Israel, p. 312). This reduces
the bills for bought concentrates such as barley, and makes very efficient use
of the land. The fresh forage is cut daily as needed, and silage is made to fill
the short winter gap. The cows (Holstein and Jersey) are kept in stables, and
average 9,700 lb per cow per year. Their manure is returned to the soil and,
with purchased fertilisers and leguminous crops, helps to maintain fertility.
Pigs and poultry are also kept on some farms, whilst others specialise in very

---

* In Taiwan (China) where systems are comparable, the next crop may be interplanted
in unflooded rice before the latter is harvested.

intensive vegetable growing. Cloches are often used for high-value early and
late season produce. In the uplands vegetables and fruit such as mandarin
oranges and grapes as well as tea, paperbush and mulberry are grown (Yang
1962). One of the intermontane basins, Suwa basin, is the centre of the silk
industry. The silkworms are kept on trays in the farmhouse and fed on fresh
mulberry leaves until the cocoons are spun, when they are sold. Mulberry
trees tolerate very poor, dry sloping soil. They are pruned in seasonal batches
so as to provide leaves for picking for each of the 3 silkworm hatchings per
year.

SOUTH HONSHU (*Sector (d)*)

This area (see Table 2.9.3) is very similar to the previous sector, but has
slightly higher summer temperatures. There are 3 months with a mean temperature above $72\frac{1}{2}°F$ but heat stress in cattle is minimised as they are kept in
buildings. Farming systems are the same as in the previous sector.

A small, intensive family farm near Kyoto. This farm of $2\frac{1}{2}$ acres has more
than 60 in of annual precipitation with A–E–T over 30 in, and is in a typical
irrigated rice district. Winters can be quite cold, and summers are usually hot
and wet as the monsoonal rains last from April to September, but the frost-
free periods of 7–8 months last until November. For the annual precipitation,
the mean annual sunshine, at 1,900 hours, is high. The farm, which has been
in the same family for 200 years, is split into 13 scattered plots, two of them
nearly 2 miles from the farmhouse. The work force is the farmer, his wife, his
two parents and a working cow. The equipment includes a motor-cycle, a
small kerosene engine, a rice thresher, 2 hand sprayers, a harrow, a half-share
in a cultivator and a mechanical crop-duster which is jointly owned by 10
families. Fig. 2.9.2 shows that the cropping system is complex and the land use
very intensive. Rice and other tropical or warm temperate crops such as sweet-
potato, are grown in the hot wet summer and are followed, through the winter,
by cool temperate food crops (wheat, barley and rape-seed), by fodder crops
which are grown for the single milking cow and her heifer, or by vegetables
(cabbage, etc.). So the land is cropped twice each year, and the underlining
in Fig. 2.9.2 shows how much of the year each field was occupied by crops.
There are also 200 poultry, which like the milk cow, are fed in part on purchased concentrated feeds.

The rice yields about 4,500 lb per acre per crop and the wheat and barley
about 2,700 lb per acre per crop, so that the land which is cropped summer
and winter with cereals yields over 7,000 lb per acre per annum. This is nearly
twice the annual yield in south-east England and about ten times the average
yield of grain in the Canadian Prairies. The cow yields about 9,000 lb of milk
per year, which is also good. These high yields, on what is admittedly an
above-average progressive farm, are obtained by skill, by genetically improved
crop varieties, by high labour input, by the careful husbanding and use of
poultry and cow manure supplemented by purchased inorganic fertiliser, and

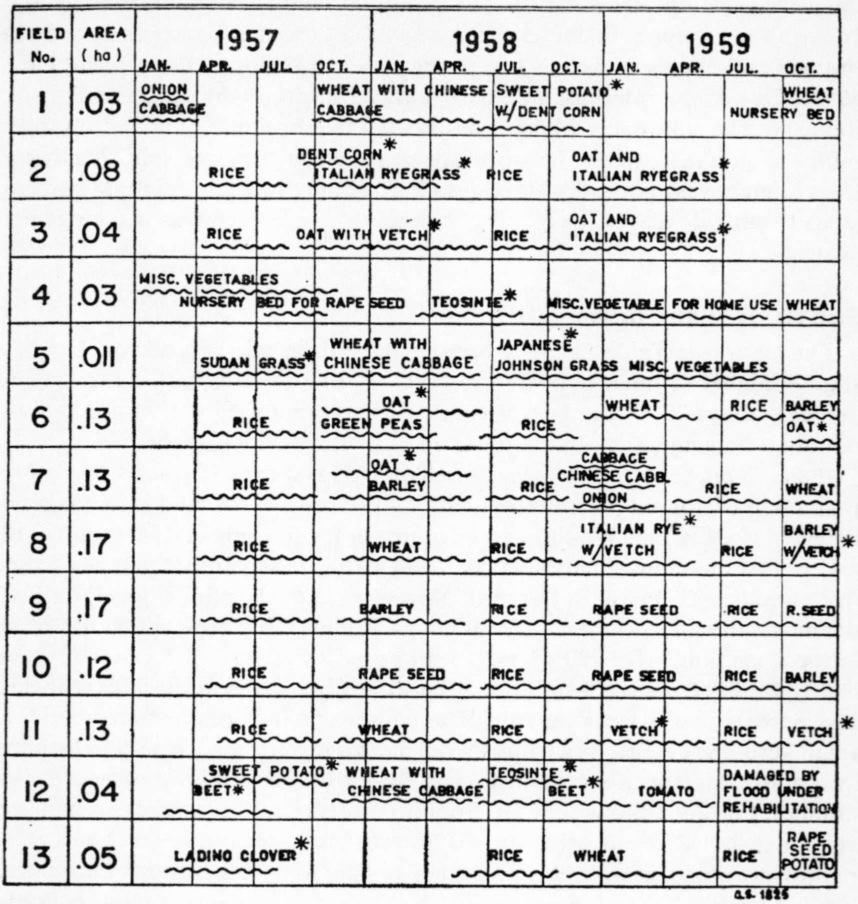

FIELD No.	AREA (ha)	1957 (JAN. APR. JUL. OCT.)	1958 (JAN. APR. JUL. OCT.)	1959 (JAN. APR. JUL. OCT.)
1	.03	ONION CABBAGE	WHEAT WITH CHINESE CABBAGE · SWEET POTATO* W/DENT CORN	WHEAT · NURSERY BED
2	.08	RICE · DENT CORN* ITALIAN RYEGRASS*	RICE · OAT AND ITALIAN RYEGRASS*	
3	.04	RICE · OAT WITH VETCH*	RICE · OAT AND ITALIAN RYEGRASS*	
4	.03	MISC. VEGETABLES NURSERY BED FOR RAPE SEED · TEOSINTE*	MISC. VEGETABLE FOR HOME USE · WHEAT	
5	.011	SUDAN GRASS* · WHEAT WITH CHINESE CABBAGE	JAPANESE* JOHNSON GRASS · MISC. VEGETABLES	
6	.13	RICE · OAT* GREEN PEAS	RICE · WHEAT	RICE · BARLEY OAT*
7	.13	RICE · OAT* BARLEY	RICE · CABBAGE CHINESE CABB. ONION	RICE · WHEAT
8	.17	RICE · WHEAT	RICE · ITALIAN RYE* W/VETCH	RICE · BARLEY W/VETCH*
9	.17	RICE · BARLEY	RICE · RAPE SEED	RICE · R. SEED
10	.12	RICE · RAPE SEED	RICE · RAPE SEED	RICE · BARLEY
11	.13	RICE · WHEAT	RICE · VETCH*	RICE · VETCH*
12	.04	SWEET POTATO* BEET* · WHEAT WITH CHINESE CABBAGE	TEOSINTE* BEET* · TOMATO	DAMAGED BY FLOOD UNDER REHABILITATION
13	.05	LADINO CLOVER*	RICE · WHEAT	RICE · RAPE SEED POTATO

* denotes fodder crops

Fig. 2.9.2 Crops growing, by field, on a Japanese farm in 1957, 1958 and 1959. *Source:* F.A.O. 1962

by the use of modern weed-killers and pesticides. An excellent example of successful modern technology *without* sophisticated mechanisation. (*Source:* Yang 1962.)

KYUSHU (*Sector (e)*)

This part of Japan is semi-tropical with a thermal growing season of 12 months (Table 2.9.3). The rains in the Bai-u season (June–July) and the Typhoon season (September–October) are very heavy, whilst typhoons also can cause considerable damage to the rice crop (e.g. by shedding, lodging). In fact, the use of seedbed covers has brought forward rice maturation dates so that many varieties can be harvested in August before the typhoons. An

August harvest also widens the scope for crops that can be planted after rice. **Farming**
Crops (F.A.O. 1964) for livestock can be grown all year round (3 or 4 crops **Systems**
a year, although it is not common practice), but otherwise the farming is
similar to that described above for sectors of southern Japan.

Stability and Efficiency

Although precipitation substantially exceeds P–E–T and the hydrologic
ratio (p. 462) is relatively low, the climate of Japan is good, especially in the
southern half. Though not so reliable as that of the North Sea basin and subject to typhoon hazards, the thermal growing season is long and rainfall generally adequate for the P–E–T at each season of the year. The wide range of
possible crops and the ability to double-crop permit a mixed year-round
tillage farming system which has a 'built-in' compensatory stability that is
enhanced by adequate pest and disease control. The extremely mixed cropping
of the farms, coupled with their small size and fragmented plots, precludes the
economies of scale and of specialisation, and severely limits output per man.
On this point, Allen (1965) states:

> The land reform solidified the traditional structure of small holdings,
> which is now anachronistic in modern Japan. This structure handicaps
> the further rationalisation of agriculture which, as in other industrialised
> countries, is highly protected. This is an embarrassment in an age of
> Trade Liberalisation. However, output per man in agriculture is relatively
> low in comparison with other industrialised countries, because of the
> relatively labour intensive pattern of production that still persists on the
> land.

On the other hand, the facts that, as a result of good husbandry, of high
industrial and scientific inputs, of mini-mechanisation and of double-cropping,
crop yields per acre per annum reach high averages, imply that Japanese
agriculture, like that of Taiwan, is ecologically efficient. It certainly provides
an object lesson in what can be done to meet population pressure in a good
reliable climate. But where are the less-developed countries with comparable
climates, comparable population pressures and comparable capacity or
opportunities to overcome social constraints?

In the future, food imports may assume greater importance and, unless net
output per person on farms can be raised further (for example, by larger,
specialised units), Japanese agriculture may feel the wind of competition.
Ecological efficiency may have to yield priority to economic efficiency.

Acknowledgements. For research assistance and first draft to Sarah E. Rowe
Jones, and for comments and criticisms to M. M. Yoshino, Fukuo Ueno,
Shozo Yamomoto, D. Wilman and Kathleen Down.

References and Further Reading

Allen, G. C. (Ed.) 1965. *Japan's Economic Expansion.* London.

*Assoc. of Japanese Geographers. *Japanese Geography, 1966. Its Recent Trend.*

F.A.O. 1964. *Pasture and Fodder Development.* Report to the Government of Japan. F.A.O. Report No. 1822. F.A.O. Rome.

*Hall, R. B. 1963. *Japan, Industrial Power of Asia.* New York.

*Hemmi, K. 1961. *Japan's Agriculture and Anticipated Trade Liberalisation.* Nat. Res. Inst. Agric. Tokyo. Bull, No. 6. Tokyo.

*I.G.U. (International Geographical Union). 1957. *Proc. Regional Conference in Tokyo.* Tokyo.

*(Japanese) Ministry of Foreign Affairs. 1962. *The Japan of Today.* Public Information and Cultural Affairs Bureau. Tokyo.

*Learmouth, A. T. A. and A. M. 1964. *The Eastern Lands.* London.

*Otsuki, M. 1961. *Japanese Agriculture and the Direction of its Development.* Nat. Res. Inst. Agric. Tokyo. Bull. No. 3. Tokyo.

*Ueno, F. *Japanese Agriculture.* Tokyo.

*Wilman, D. 1965. *Agric. Progress.* **40**. 103–106.

Wilman, D. 1966. Personal Communication. School of Agriculture, Cambridge.

*Yang, W. Y. 1962. *Farm Development in Japan.* F.A.O. Agric. Development Paper No. 76. F.A.O. Rome.

Yoshino, M. M. 1967. Personal Communication. Univ. of Tokyo.

* Asterisked items are for further reading.

East Africa

Introduction

For the purposes of this chapter, 'East Africa' will be taken to include three countries which are often grouped under this heading: Kenya, Tanzania (Tanganyika and Zanzibar) and Uganda (Fig. 2.10.1). These three countries have a common political heritage in that they were all under British administration before independence. The customs union, which dates from that administration and was continued after independence, also made for economic homogeneity between them. The countries have great similarities in their flora (savanna vegetation covers by far the greater part of the region) and fauna, and in the ethnic make-up of their populations. The great majority of the inhabitants are Bantu, with smaller numbers belonging to Hamitic and Nilotic tribes; and there are sizeable minorities, scattered through all three countries, of Indians, Arabs and Europeans with agriculture being practised by members of all these three races. Geographically, all three countries contain some high mountains (Kilimanjaro in Tanzania is at 19,340 ft the highest mountain in Africa) and more or less extensive tracts at high altitudes which are used for farming.

The agriculture of these three countries can, however, also serve as a representative sample for a very much wider area. Environmental conditions and agricultural practices found in northern Uganda stretch far into the southern Sudan, and the semi-desert conditions of northern Kenya are similar to those of Somalia and part of Ethiopia. To the south, the broad zone of tropical savanna extends from Tanzania with little change through several countries as far as the borders of the Republic of South Africa, and farming systems are basically similar throughout this vast region. Such differences as exist depend more on altitude than on any other factor, and these differences are as well exemplified within East Africa itself as anywhere else.

Ecological Background

As a generalisation, it may be said that in Kenya and Tanzania rainfall is a little too low and unreliable to be ideal for crop production and results in frequent food shortages, whilst Uganda is blessed with a better and more reliable rainfall of 40–60 in a year in most parts. At the latitude of the Equator (which passes through the middle of Kenya and Uganda), a minimum rainfall of about 30 in is required for general mixed cropping, and about 20 in to grow any crops with success. Rainfall probability maps of East Africa (East African Royal Commission 1955) show that the greater part of Kenya and about half

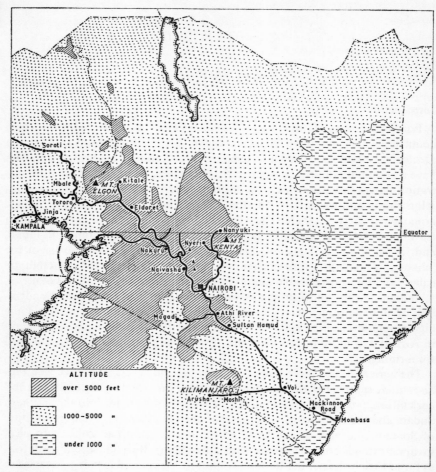

Fig. 2.10.1 Altitude and line of rail in East Africa

of Tanzania may fail to receive 30 in in about one year in three or more often, and a very large part of these areas may, just as often, fail to receive 20 in. Rainfall throughout most of the region follows the equatorial pattern of two main rainy periods which set in shortly after the equinoxes and alternate with two drier periods, though parts of Tanganyika have only a single wet season from about December to April.

Temperature varies with altitude, and in East Africa altitude is a very important determinant of farming systems. To East Africans, the 'highlands' (where temperatures are felt to be noticeably cooler than at lower altitudes) begin at about 6,000 ft above sea-level, and farming is carried on to an altitude of as much as 10,000 ft in a few places. The most extensive highland tracts are in western Kenya, where European settlement was mainly developed, but there are also considerable highland areas in northern and southern blocks in Tanganyika, and some small scattered areas in Uganda.

On the face of it, tropical highland areas would seem to combine the disadvantages for plant growth of both the tropical and temperate zones, since they have shorter hours of daylight than in the growing season of the temperate zone and lower temperatures than in the tropical lowlands. In the absence of exact experiments on plant growth, it is difficult to substantiate from agricultural experience how far this supposition is true. Certainly no spectacularly high crop yields have been recorded from the East African highlands. European farmers in the highlands obtain relatively high crop yields simply because their methods of husbandry are more advanced than those of the African peasants who farm the lowlands. Probably the reduction of evaporation at the lower temperatures of the highlands is important in a region where water is so often a limiting factor to crop growth; the highland soils may also suffer less from leaching and loss of organic matter because of the lower temperatures.

In spite of the dry climate of East Africa which would make irrigation so beneficial, there has been little development in this direction, mainly because of the paucity of large rivers and the reduction of their flow in the dry seasons when it would be most needed. The Chagga tribe on Mt. Kilimanjaro have long practised an elementary kind of irrigation from mountain streams, and some small modern schemes, as at Mwea in Kenya, have shown the high crop yields which can be obtained with irrigation. The installation of sprinkler irrigation on two sugar estates in Uganda has underlined the same point. A succession of draft projects for irrigation from the Tana river in Kenya has resulted in recent years in some cautious development which may show that a larger potential exists.

The soils of East Africa include some which are of good fertility, notably the best of the tropical red soils which occupy much of central Uganda and smaller areas in Kenya, as around Kiambu. There are also some excellent soils derived from volcanic rock around Mts. Elgon, Kenya, Kilimanjaro and in parts of the Rift Valley. The remaining East African soils are of more moderate fertility but generally easily workable with the exception of the tropical black clays, which occupy some depressions and poorly-drained areas. East Africa has many ridges and flat-topped hills whose soils are often too thin and stony for cropping. This is one of the parts of the world which are most seriously menaced by soil erosion. Amongst the reasons for this state of affairs are the frequently high intensity of the rainfall, the pronounced dry seasons which do not allow a continuous cover of vegetation to protect the soil, the abundance of steeply sloping land and the common overstocking of grazing areas. Soil conservation measures have been actively inculcated in all the countries, but there are still many farmers who do not practise them and the menace is a continuing one.

The predominant type of vegetation is the savanna, though how much of this is truly natural and how much is only maintained by man's activities in grazing and burning is a matter for argument. Much of the East African savanna is only lightly wooded and there are some extensive tracts of nearly pure grassland. Thus the extent of the potential grazing areas, coupled with the rather low rainfall for crops in many parts, make East Africa one of the parts of the tropics in which livestock are more important as compared with crop production. Of the other types of vegetation, Uganda has sufficient rainfall to carry dense forests, but these have been so much cleared for cultivation and grazing that only comparatively small relict areas remain. On all the high mountains there is montane forest, including large areas of bamboos, which has been better preserved and is now largely protected in forest reserves. The swamps which form in the valleys of sluggish rivers and are often filled with papyrus (*Cyperus papyrus*) are a characteristic feature of East African vegetation. At first sight they seem to offer great possibilities for drainage and agricultural development, but in practice their fertility, based on the accumulation of organic matter, is soon exhausted, leaving a residual generally sandy soil which is perhaps best utilised to provide dry-season grazing.

The larger fauna are abundant; East Africa is the world's big game region *par excellence*. These animals can do considerable damage to crops, of which trampling by elephants is the most spectacular form. Their consumption of grazing is, however, more important to the economy; there are parts of Kenya where improved paddocks have to be fenced more to keep out zebra and antelopes than to keep in cattle. It is often suggested that marginally dry lands would be more economically used by the controlled management of game for meat and skins than by domestic livestock; but there have not yet been sufficiently exact experiments to allow a critical judgement to be made. Tsetse-flies are the most serious insects, infesting considerable areas in Kenya and Uganda and some two-thirds of Tanzania; as vectors of the animal disease trypanosomiasis, they make the keeping of livestock other than poultry almost impossible in the infested areas. Progress made in recent decades in clearing considerable areas from tsetse has been partly counterbalanced by fresh encroachments elsewhere. The desert locust invades East Africa from the north sufficiently often to justify contributions by the three countries to maintenance of the Desert Locust Control Organisation.

Agrarian Structure

Although European and Asian immigrants have made notable contributions to the agriculture of East Africa, which have been especially important in the production of export crops, the overwhelming bulk of production has always come from African peasant farms. It is a characteristic of East African society that the African farmers do not live in villages but in isolated homesteads situated on the individual farms. Their land tenure in general follows the customary pattern of usufructuary rights only over the land, held by the occupier from the tribe although these tribal rights are theoretically embraced in a general governmental control over all unalienated land. The idea of private freehold ownership, which was first introduced on some African-occupied land in Buganda under the Uganda Agreement of 1900, has however been making steady inroads into the customary pattern and has been the basis of a very large programme of registration of land-holders in Kenya, particularly in the Kikuyu districts, in recent years.

Most African farms are worked by family labour in which the women play at least as large a part as the men. Traditionally the food crops are grown by the women and the cash crops by the men, but the rigidity of this distinction is nowadays tending to break down. The hours worked are often short, and have been estimated in some parts of Uganda at no more than an average of 3 hours per working day for men, though probably rather more by women on agricultural tasks. Fewer data on this subject are available for East than West Africa, but, in Malawi with a population of Bantu farmers in a very similar environment, a survey showed that men worked from 400 to 900 hours per year and women from 580 to 760 (Clark and Haswell 1964). The normal agricultural unit is the nuclear rather than the extended family, but particularly amongst the Nilotic tribes there may be co-operative cultivation and harvesting of crop areas by a group of neighbours. African employment of other Africans as hired farm labour is common in a few areas, such as the Kikuyu districts of Kenya and the Buganda kingdom of Uganda, where production is very dependent on the employment of immigrant labourers from Rwanda.

The size of African farms in East Africa is too variable to lend itself to easy generalisations. The most common amount of land actually cropped by a family is probably from 2 to 5 acres where cultivation is entirely by hand, and up to 10 acres where an ox-drawn plough is used. But some farms are much larger than this, as in some of the recent settlement schemes for African farmers in Kenya, or amongst coffee growers in Uganda where in 1960 the Department of Agriculture recorded that 15% of the 'estate' acreage of coffee was African-owned. Fragmentation of agricultural holdings is a very real evil in many parts. In some areas of Kenya, a single farm may consist of 15–20 small plots scattered within a radius of up to 5 or 10 miles; but this problem has been somewhat reduced by Government programmes for the consolidation of holdings.

Most cultivation, as well as weeding, is still done with the digging hoe but some use of ox-ploughs is made in almost all areas where the incidence of animal diseases is not too high, and there are some districts, as in northern Uganda, where this has now become the normal method of cultivation although weeding and harvesting are still done by hand. Tractors, in spite of many pioneer Government schemes of mechanisation, spread only slowly and the number in use by African farmers is still very small. The application of animal manure to the land has in the last thirty years become commonplace, though the supply is too small to have a very marked effect on production;

FARMING SYSTEMS OF THE WORLD

there is as yet no appreciable use of fertilisers by African farmers, though rock
phosphate and soda phosphate have long been produced and marketed from
local materials.

No crops are cultivated by the nomadic cattle-keeping tribes of East Africa,
who inhabit low-rainfall areas and subsist mainly on meat, milk and the
blood which they draw from their cattle. The most important of these numerically are the Masai, who straddle the border of Kenya and Tanzania. In
Uganda another Hamitic tribe, the Hima, have a similar economy. There are
smaller groups on the semi-desert northern fringes of Kenya who perforce
also live almost solely by their livestock (cattle, camels and goats). The place
of these peoples in the East African economy is almost exactly similar to that
of the Fulani in West Africa (Chap. 2.11).

The proportion of the European sector in East African agriculture is indicated by the percentage of the national land alienated or reserved for European occupation in each country at the peak of settlement (Hailey 1957): 7%
of Kenya, 0·9% of Tanganyika and less than 0·5% of Uganda. Nevertheless,
the importance of this sector has been much greater than these figures suggest
because of the high yields obtained from the land and the very large proportion of the agricultural exports of Kenya and Tanganyika derived from European farms. This European agricultural production has developed along two
main lines. First, there were farms of a size which one individual could afford
to develop and manage to run under his own supervision. Many such farms
would be of less than 1,000 acres, though some were larger especially in the
arid ranching areas. Many were mixed farms with both crops and livestock;
others were more or less specialised on the production of one crop such as
coffee, tea, wheat or pyrethrum, or on livestock such as dairy cattle, pigs or
woolled sheep. The other class of holding was the large single-crop estate or
group of estates run by a company with a paid managerial staff; such estates
have been particularly important in the production of sisal in Tanganyika
and of tea and some coffee in all the countries. Land for European agricultural occupation has been granted at different times both in freehold and
on leases of varying length up to 999 years.

Asians have taken a lesser part in East African agriculture, though Arab
clove-growing in Zanzibar has a long history and the Kenya Government did
at one time reserve blocks of land in the Coast and Nyanza provinces for
Asian settlement. This is because most Asian immigrants to East Africa were
traders, clerks or craftsmen rather than farmers. Asian enterprise has been
particularly important in sugar-cane production in all the countries, the larger
estates producing centrifugal sugar and the smaller, the cruder product
known as 'jaggery'. The scope of their enterprise is indicated by the fact that
two large Indian-owned sugar estates in Uganda have produced enough sugar
for the needs of the country and a surplus for export.

While the petty trade in foodstuffs and local exchange of agricultural produce has been left free of control except in wartime, the marketing of the

major cash crops, and especially those which are mainly exported, is largely
in the hands of statutory boards, mostly set up to handle a single commodity
within each country. Subject to overriding Government control, most of these
boards aim at setting a fixed price to the producer for a period of time, with
incentives to encourage quality production. Some of these boards possess
facilities for handling and processing the produce, and some operate auction
floors such as exist for cotton and coffee in Kampala and coffee in Nairobi.
The livestock trade has not been quite so rigidly controlled, but particularly
in Kenya there has been a long history of control of cattle buying by a board
or commission. The co-operative movement has played a considerable part
in some spheres of agricultural marketing. Some East African co-operatives
have a long and honourable record, such as the Kilimanjaro Native Co-
operative Union which was set up by Chagga growers in Tanganyika to
market their coffee, or the Kenya Farmers' Association and Kenya Co-operative Creameries originally founded by European farmers in Kenya. In recent
years the growth of co-operative marketing in Uganda has been spectacular,
and the societies now market a large proportion of the cotton and coffee crops
and possess their own cotton ginneries and coffee factories.

East African agriculture is fairly well served by research and extension services. Amongst the 'common services' supported by all three countries are the
East African Agriculture and Forestry Research Organisation (EAAFRO)
a similar veterinary organisation (EAVRO) and the East African Tsetse and
Trypanosomiasis Research Organisation. There are individual commodity
research stations concerned with tea, coffee, sisal and pyrethrum, and the
main research station of the Cotton Research Corporation in Uganda. Agricultural extension services are better than in many tropical countries, with
particularly intensive staffing in Kenya which dates from the period of active
agricultural development following the Mau Mau troubles of the 1950s.
Higher education in agriculture has been somewhat limited by the paucity of
candidates forthcoming, and the only university faculty of agriculture is at
Makerere in Uganda. East Africa has not lacked a variety of schemes of agricultural development since the Second World War, some of which have made
disappointingly little impact. The best known is the groundnut scheme which
was begun in Tanganyika in 1947 and failed in its main objective; the cleared
areas were later handed over to the Tanganyika Agricultural Corporation,
which has utilised them largely under a system of African tenant farming with
assistance in mechanical cultivation. In recent years, a major operation has
been mounted in Kenya in settling African farmers on purchased land formerly in European occupation. The first schemes involved high density, low
density and 'yeoman' settlement, with respectively increasing size of holding.
In 1965 an Agricultural Development Corporation was set up to organise and
operate farm units during transition from European to African ownership,
and was particularly important for the large-scale units which were being
preserved during the second stage of these transfers. The 'Swynnerton Plan'

such crops as coffee, tea and pyrethrum.

Crops and Livestock

Table 2.10.1 gives such figures as are available of the production of some
principal crops, and numbers of livestock, in the East African countries.
Maize is the predominant staple food of Kenya and much of Tanganyika.
The crop is mostly of white-grained dent types of mixed varietal origin, but
some good 'synthetic' varieties have lately been developed, and hybrid maize
is beginning to be introduced. In Zanzibar, rice replaces maize as the staple
cereal but the island does not grow enough for its needs and has to import
rice from Asia. In the more arid areas of northern Uganda, the Northern
Province of Kenya, and some parts of Tanganyika, the important cereals are
sorghum, finger millet (*Eleusine coracana*) and bulrush millet (*Pennisetum
typhoideum*). A good deal of the sorghum grown goes, however, into beer-
making, and finger millet ('wimbi' in Swahili) is the most important of the
three crops as a food; its excellent storage properties also make it useful as a
famine reserve. Wheat is grown in the highlands by European farmers with
full mechanisation. Originally produced to feed the European and Asian
population, a large proportion of the crop is now consumed by Africans, for

Table 2.10.1

Crop Production and Livestock Numbers in East Africa

Crop production in thousands of short tons per annum:			
	Kenya	Tanzania	Uganda
Wheat	144	34	—
Rice (paddy)	16	96	2
Maize	1,179	617	243
Millet and sorghum	353	1,214	739
Sweet-potatoes	496	285	1,543
Cassava	661	1,246	1,598
Groundnuts (in shell)	4	12	180
Coffee	57	49	246
Tea	22	7	9
Sisal	66	240	—
Cotton (lint)	4	74	87
Sugar-cane	496	937	1,764

Livestock numbers, in thousands:

	Kenya	Tanzania	Uganda
Cattle	7,206	8,837	3,627
Goats	6,300	4,462	1,998
Sheep	5,027	2,986	791
Pigs	36	21	20
Donkeys	n.a.	108	17
Camels	177	—	—

Source: F.A.O. Production Yearbook 1966.

bread-eating has increased as rapidly here as elsewhere in Africa. The main
problem has been caused by several types of rust disease (*Puccinia*) which have
necessitated the breeding in Kenya of a succession of resistant wheat varieties
suited to different altitudes. In central Uganda, the staple food and beer-
making crop is the plantain or cooking banana, and this is also true of some
parts of Tanganyika, such as the Chagga area and Bukoba. The two important
root crops, cassava and sweet-potatoes, are very widely grown in East Africa
but the largest production is in Uganda where the rainfall is most favourable.
Pulses are a substantial item in East African diets; groundnuts and dwarf
beans (*Phaseolus vulgaris*) are the most important, with cow-peas and pigeon
peas in a secondary position. Sesame (*Sesamum indicum*), known in East
Africa as 'sim-sim', is grown both as a food and cash crop; soya-beans are
almost entirely grown for sale. The Asian community are considerable pur-
chasers of pulses and vegetable oils for their own dietary needs.

Cash crops of importance vary more between the countries than do food
crops. Zanzibar is the world's leading producer of cloves with coconut products as her second export. In Tanganyika, sisal is overwhelmingly the chief
agricultural export although coffee and cotton are also important. Cotton for
long dominated the exports of Uganda, but since 1957 has been overtaken by
coffee. Kenya has a much more varied list of significant agricultural exports,
with no such exclusive importance attaching to any one of them. Coffee, sisal,
tea, pyrethrum and maize have fluctuated in their relative order of importance
to the economy largely according to current prices for these products. The
coffee exports of East Africa comprise two types: the higher-priced 'arabica'
which is only grown in highland areas and provides the bulk of the production
from Kenya and Tanzania, and the lower-valued 'robusta' which is produced
in far greater quantity in Uganda. Cotton is of the American Upland type but
has the advantage, as grown in Uganda, of a longer staple than 'American
middling'. The minor crop exports of East Africa are rather numerous and
include cashew-nuts, castor seed, tinned pineapples, wattle bark and extract,
and chillies. Tobacco is mainly grown for local manufacture.

The main problems confronting crop production in East Africa are the
twin dangers of soil erosion and soil exhaustion. Though soil conservation
measures such as contour strip cropping and the use of broad-base and
narrow-base terraces have been quite widely adopted in the last 30 years, the
fight against erosion is a continuing one in which any relaxation would be
disastrous. As population increases and it becomes less and less possible to
allow the land long resting periods under the system of shifting cultivation, the
spectre of soil exhaustion becomes more menacing in the most congested
districts. A typical example is the Nyanza province of Kenya, once known as
Kenya's granary, where the decline in maize yields began to cause considerable alarm in the 1950s. Perhaps more common is the case where the slow
decline in fertility has been about counterbalanced by the equally slow improvement in crop breeding and husbandry practices, so that yields have re-

mained stagnant in spite of all the effort put into research and extension work.

Livestock are of particular importance in East Africa, where Tanzania and Kenya have the highest cattle populations in the continent after Ethiopia and South Africa. The indigenous cattle, predominantly zebu but with an element of Hamitic Longhorns in Uganda and Tanzania, are for the most part not very productive. The most outstanding breed amongst them is the Boran from northern Kenya which is a useful beef producer but lacks resistance to the tick-borne diseases. Pure-bred European cattle and their crosses with the zebu are successfully kept in the highlands, especially where dipping and other measures of disease control have been carried out under European management. The Ayrshire, Friesian and Channel Island breeds are the most widely used for dairy purposes, and are beginning to be used for crossing also in more lowland areas as disease control can be achieved. More recently the Indian Sahiwal has been introduced to grade up local dairy stock. Government facilities for artificial insemination are helping these crossing programmes in all the countries. The large goat population of East Africa, kept solely for the production of meat and skins, consists of unimproved indigenous animals. The same is generally true of the rather less numerous small local sheep with hairy fleeces, though the Kenya highlands are one of the few areas in the tropics where woolled sheep are kept. This has been achieved by crossing Merino rams on local Masai ewes. Donkeys are kept in some number as pack-animals in some of the drier areas; camels feature in the economy only in the extreme north of Kenya. Pigs, kept only in small numbers by non-Moslem African farmers, are more important on European farms especially where they can utilise dairy by-products on farms which sell cream for butter making; Kenya is one of the few tropical countries making significant amounts of bacon. European breeds of poultry have been so extensively introduced by settlers as well as Government agencies that there is more exotic blood in the poultry of East Africa than in most parts of the tropics, though there are still many unimproved birds with a very low production.

Control over animal diseases, whose incidence in East Africa used to be particularly high, is gradually improving. Rinderpest has now been entirely eliminated from Tanzania. The tick-borne diseases include East Coast fever and Nairobi sheep disease which are peculiar to eastern Africa; but dipping and spraying are gradually extending control to larger areas and making the keeping of exotic or crossbred stock more widely possible. Trypanosomiasis carried by tsetse-flies still denies large areas to livestock, especially in Tanzania, and necessitates the maintenance of tsetse control operations to prevent even wider spread.

Hides and skins are the chief export of East Africa of animal origin. A butter industry developed naturally in the Kenya highlands where European settlers on remote farms found it easier to market cream than whole milk, and this has made East Africa self-supporting in butter with some export. Some ghee (clarified butter) is made, largely for consumption by the local Asian

community, and there have been small exports. Meat factories in Kenya and
Tanzania provide some export, particularly of tinned meat, although the consumption of meat in many parts of East Africa is less than would be nutritionally desirable. Some of the Kenya wool clip is exported and some is used
locally.

The greatest single problem of the East African livestock economy is probably local overstocking which often leads to gross soil erosion and denudation
of vegetation in pasture areas. The problem arises from the social valuation
put upon cattle by many East African tribes, according to which the number
of cattle owned, rather than their productivity, defines a man's status. This
attitude is reinforced by the customary use of cattle and goats to pay bride-
price, without which a wife cannot be obtained, and by the traditional investment of wealth in purchasing cattle as a form of savings which provides a
natural increase and does not lose its value by inflation. These factors have
produced extreme resistance to Government measures of compulsory de-
stocking, and although more rational ideas on livestock productivity are
slowly percolating, it will be long before all stocking rates on East African
grazing lands are logically adjusted. As against this, it must be said that there
are some progressive African farmers in all three countries, often belonging
to groups amongst which cattle-keeping has not been common, who are dis-
playing most encouraging qualities in the management particularly of small
dairy herds. The enclosure of grazing land by fencing is slowly spreading
amongst such progressive farmers and is the essential first step to the improvement of animal husbandry and disease control.

An East African Transect

Almost any transect drawn in a straight line through East Africa would be
bound to pass through large areas which remain agriculturally undeveloped,
not through lack of biological potential, but because of the absence of communications. It is, therefore, more convenient, in order to show the effect of
environment on agriculture, to take a transect along a line of rail or road
which has enabled agriculture to make its potential response to the environment. The transect selected is along the main railway line from Mombasa on
the Kenya coast to Kampala, the capital of Uganda. This has the advantage
of covering the whole range of altitude at which farming is carried out, and
thereby illustrating the effect of this dominant factor in East African agriculture.

THE COASTAL REGION

The coastline around Mombasa shows some of the agricultural features
typical of many tropical shores. Immediately above high-water mark, a thin
fringe of coconut palms on the still sandy soil fills an ecological niche which
no other crop could usefully occupy. Directly inland lies a strip of land underlain by coral rock where soil deep enough to justify the planting of crops
is only in pockets and even gardening is difficult. This strip is fortunately a

359

narrow one, and behind it lies a wide zone of red soil on which the real agriculture of the coastal region is carried out. With an average of over 40 in of annual rainfall, the choice of possible crops is a fairly wide one but has been influenced historically by the preferences of Arab land-owners in the past and the demands of Asian consumers in the present. These factors probably account for the local importance of the rice crop, and the prevalence of such fruits as citrus and mangoes. Maize, cassava and legumes are the other chief crops grown for food, in which the region is normally self-supporting. Cotton and cashew-nuts are important cash crops, and there is some estate production of sisal and sugar-cane. There is some infestation with tsetse-fly, and livestock are relatively unimportant. This agricultural region is, however, only a fairly narrow belt, for rainfall declines rapidly away from the coast, and the railway station of Mackinnon Road, about 50 miles inland from Mombasa, represents roughly the point at which annual rainfall falls below 20 in and crop production ceases to be worthwhile.

THE 'NYIKA'

From this point onwards for some 200 miles, low rainfall and scarcity of water supplies have combined to prevent any appreciable settlement or practice of agriculture. The vegetation of this tract, known as the 'nyika', which extends far towards the northern frontier of Kenya where it passes into more desert-like conditions, consists of dry bushland in which small and often thorny trees and bushes are scattered amongst drought-resistant grasses. There is a population of big game animals (particularly in the 8,000 square miles of Tsavo National Park), but it is generally sparse because of the distance between water-holes. The only agricultural oasis in the first part of this tract is the Teita Hills, some way to the south of the railway line, whose local climate provides a rainfall rising to 50 in annually in a small area. Here coffee and maize flourish and there has developed some production of vegetables and fruits which can be sent by railway to the towns. Because of the steep topography and high density of population, soil erosion has been a constant menace and this is one of the parts of Kenya where terracing and other soil conservation measures were first inculcated by the extension services.

Between the stations of Sultan Hamud and Athi River, there lies to the north the area inhabited by the Kamba tribe. This receives a rather better rainfall but is still marginal for crop production to the extent that this part of Kenya has a recurrent history of food shortage and famine due to drought. These conditions have been aggravated by high population density and much soil erosion. The area provides one of the classic instances in Africa of overstocking with cattle, and this factor also has led to much denudation of the vegetation. Throughout this long stretch the railway has been steadily gaining in altitude. Some distance before the station of Athi River (where the oldest meat factory in East Africa is situated) the 5,000 ft contour is passed, and from this station onwards we may be said to be in the highlands.

THE HIGHLANDS

Recent history has made it difficult to write about the conditions of farming in the Kenya highlands with contemporary accuracy, for great changes have been taking place and the situation is still fluid. On the lands farmed by Africans, extensive programmes of land registration, consolidation and enclosure have changed the face of the countryside so that in some areas it begins to resemble an English landscape with mixed farming practised in small hedged or fenced fields. Of the land formerly settled by Europeans, much has been bought by the Government for redistribution to Africans under various schemes of settlement, and the number of European farmers has correspondingly diminished. The future pattern of farming in these highlands, whose agricultural history in the past 70 years has been as chequered as that of any area in the world, remains as problematic as ever.

It is nevertheless possible to give some outline of the changing ecological conditions and the crops and stock which it is possible to raise, along our transect following the railway line as it runs westward from Nairobi. Leaving a little to the north the important coffee plantations established by European farmers around Kiambu, this line passes first through a part of the Kikuyu country, where African farmers grow mostly maize and other food crops on rich red soils which are amongst the best in East Africa. A prominent feature of the landscape is the black wattle trees, which are grown for the dual purpose of selling bark to produce a tanning extract and using the stems for building poles and firewood. Next along the railway line lies Limuru, an area of old-established European mixed farming including dairying, and where tea and coffee plantations are also important. A little further along the line is the large Uplands bacon factory.

The railway next descends the escarpment into the Rift Valley and an area of lower rainfall. Around Naivasha the growing of crops is for this reason only marginal, and cattle and sheep rearing must, therefore, be the main farming enterprises. To the west, the somewhat arid grassland ranges on which the nomadic Masai herd their cattle lie contiguous to pastoral areas in European hands producing both beef and dairy products; at Naivasha is located one of Kenya's co-operative creameries. A few miles north of the railway however the Kinangop plateau lying under the Aberdare Mountains has much better rainfall and was the scene of more intensive European settlement. On many of these farms pyrethrum, a crop which at this latitude needs a minimal altitude of about 7,000 ft to give its best production, was the most paying enterprise.

Continuing along the Rift Valley, the railway comes to Nakuru, one of the main market towns of the European settlers and lying on the edge of the most important area growing temperate cereals in Kenya. Centred on Njoro, a little to the south-west, this was developed by Europeans as a region of large farms specialising in the mechanised production of wheat, and of barley which is partly used in malting by local breweries whose product, like bread, is in

increasing demand by Africans. Near Njoro is the Egerton College of Agriculture, originally founded to train young European farmers but now preparing
students of all races to play their part in the future of Kenya farming. At
Nakuru is a factory which deals with much of Kenya's wool clip. Some of
Kenya's sisal estates are also located near Nakuru, this being a crop whose
distribution throughout the country is rather sporadic. It does not demand a
very high rainfall, but an ample and permanent water supply is essential for
factory decortication of the fibre.

Beyond Nakuru, the railway climbs again out of the Rift Valley and reaches
its greatest altitude of 9,150 ft above sea-level. Here, in the neighbourhood of
the railway stations of Equator and Timboroa, farming is carried on by Europeans up to an extreme altitude of nearly 10,000 ft. This is the only area along
our transect where frosts have to be considered as a factor in agriculture, and
consequently many tropical and sub-tropical crops cannot be grown. Wheat
and pyrethrum are staple crops; such temperate fruits as apples and pears can
be grown, and there is a commercial production of temperate vegetable seed,
partly for export; pure-bred European livestock thrive if they can be kept free
of the local diseases, and at this altitude European species of pasture plants
can be planted as well as the indigenous species such as Kikuyu grass (*Pennisetum clandestinum*) which are so important in the rest of the highlands. The
next place of importance along the railway is Eldoret, another market town
which was formerly the centre of a considerable settlement by Dutch South
African farmers. Here are situated flour mills and a wattle extract factory.
Maize is the chief crop of the locality, but to the north lies the important
Trans–Nzoia area of European settlement with Kitale as its centre and a more
varied agricultural production. Maize, wheat and cattle are all important
here, and the area includes the rich volcanic soils surrounding Mt Elgon
which have been extensively used for arabica coffee production.

WESTERN KENYA

About 40 miles westward of Eldoret, the railway sinks below 5,000 ft altitude and the highlands may be said to be left behind. For the remainder of its
passage through Kenya, the average annual rainfall is between 50 and 60 in,
as against 40 in at Eldoret. This area of western Kenya was part of the
Nyanza province until the reorganisation of regional boundaries in 1963, and
it is convenient for the present purpose to treat the old Nyanza province (now
the Western and Nyanza Regions) as a whole, since there is a great degree of
agricultural uniformity. With lower altitude and higher rainfall than the highlands, and no European settlement, farming in this part of western Kenya has
a somewhat different aspect. Conditions have favoured a very massive peasant
production of food crops, and the area has long been regarded as the granary
of Kenya, exporting in particular much maize to other parts of the country.
Sorghum and finger millet are also important cereals; rice, cassava and sweet-
potatoes are locally produced on some scale; legumes are widely grown and

there is often a surplus of groundnuts. The staple food crops are, therefore, in so far as they are in surplus, the main cash crops of the region. Cotton is the only other important cash crop.

The problems of the area arise from the density of the population (over 1,000 per square mile in some locations not far from the railway) and the intensity of the cultivation. The old 'shifting cultivation' mentality for long precluded the idea of manuring although in many areas there is hardly any resting land to be seen. Soil fertility is undoubtedly declining in many places and local overstocking of the decreasing area of grazing occurs. These troubles will take much time and effort to solve, though the increasing use of animal manure in recent years is a heartening sign.

EASTERN UGANDA

After passing the Uganda frontier, the first railway station is Tororo at 4,045 ft altitude. Deposits of phosphate rock in the neighbourhood provide the basis for a fertiliser production which was running at the rate of 30,000 tons per annum in 1966, though the manufacture of superphosphate has not yet proved practicable. Tororo lies only a few miles south of the foothills of Mt. Elgon, a massif which has an agricultural economy quite different to that of the Uganda plains. Cultivation is carried by the Gisu tribe up to about 7,000 ft; plantains (cooking bananas) and root crops are more important food crops than cereals. The sole important cash crop is arabica coffee, a peasant production which has long been sedulously fostered by the Government. Processing and marketing have been largely entrusted to producer's co-operatives which work under close governmental advice and own coffee pulperies and a processing factory at Mbale.

The main railway line, however, runs on through the lower country of the Eastern Province, a region which is somewhat transitional between the drier north of Uganda where finger millet is the main food crop, and the higher rainfall areas where plantains and root crops provide most of the food. Around Tororo the cereals finger millet, sorghum and maize predominate, but in Busoga district as Jinja is approached plantains occupy a greater acreage. Groundnuts, dwarf beans and sesame are grown throughout the area as subsidiary food crops and there is sometimes a marketable surplus. Cotton is, however, by far the most important cash crop, and the country is liberally dotted with ginneries for processing the crop. An Indian-owned sugar estate in Busoga provides half the country's sugar.

CENTRAL UGANDA

At Jinja, the railway line crosses the Nile into Buganda and runs on to Kampala through Mengo district, which is perhaps the most productive agricultural area in East Africa. The red soils are of excellent fertility and rainfall is adequate for growing perennial crops or two annual crops a year. Plantains are the main food (and beer-making) crop, with root crops (mainly sweet potatoes) and legumes (groundnuts and beans) subsidiary. Robusta coffee is

the main cash crop, this area producing the bulk of Uganda's very large exports. There are important coffee curing factories at Kampala, where coffee
auctions are also held. Cotton, which was formerly the chief cash crop, has
now sunk to a secondary position but there is still a considerable production
and many ginneries. Many farmers, however, derive a large part of their income from the sale of food crops, especially plantains and sweet potatoes, for
consumption in the towns of Kampala and Jinja. The production of hydro-
electric power from the Nile dam at Jinja powers industry in both towns, and
near Jinja is the largest cotton textile factory in East Africa which absorbs
annually some 20,000 bales of the Uganda crop.

The average production of farms in this area is higher than is usual in East
African peasant farming, for many Ganda farmers employ immigrant labourers especially from Rwanda, and a few large landowners may be said to operate on an 'estate' scale. There are a few European-owned plantations, growing
mostly coffee or tea, and another large Indian-owned sugar plantation. Livestock are of smaller importance in this area, where the people are not traditionally cattle-minded and there was a high incidence of animal disease in the
past; but the opportunities for selling milk in Kampala have attracted a few
enterprising African dairymen, some of whom are undertaking a most
promising development of fenced pastures and the use of grade animals.

This transect need not be taken beyond Kampala, for it has already provided illustrations of all the main farming systems which are practised in East
Africa. A résumé of the chief environmental data for points along the line of
the transect is provided in Table 2.10.2.

SAMPLE FARMS

(a) *Modern trends in Kenya.* Farming systems are changing so rapidly in
Kenya that it is perhaps as well to take an example from the modern trend
which is being established with the advice of the extension services. This advice often lays particular stress on mixed farming with both crops and livestock, as is well illustrated in the following example of an officially recommended farm plan for an $8\frac{1}{2}$-acre African holding in the 'high bracken zone'
of the Central Province (Clayton 1964). The acreage is divided between the
following uses:

	Acres
House and garden (growing yams and bananas)	1·0
Annual crops	1·0
Permanent pasture	4·0
Rotated ley	1·0
Tea	0·5
Pyrethrum	0·5
Shelter belts and fuel (incl. some wattle bark for sale)	0·5
	8·5

Table 2.10.2

An East African Transect

Place	Zone	Longitude	Altitude (ft above sea-level)	Rainfall (mean inches per year)	Relative Humidity (mean % in midday hours)	Temperature (mean daily maximum °F)	Soil Type	Vegetation	Chief Food Crops	Chief Cash Products
Mombasa	Coastal	39°39′E	52	47·3	70	84	Loamy sands	Savanna	Rice, maize, cassava, pulses	Cotton, cashew-nuts, sisal
Voi	Nyika	38°34′E	1,837	21·6	44	87	Brown loam with laterite horizon	Dry bushland	—	—
Nairobi	Highland	36°48′E	5,971	37·7	51	74	Red latosols	Savanna	Maize, pulses	Coffee, wattle, milk
Equator	Extreme altitude	35°33′E	9,062	45·9	53	65	Red latosols	Forest	—	Wheat, pyrethrum, dairy products
Eldoret	Highland	35°16′E	6,863	40·5	45	76	Dark brown clay	Savanna	Maize, potatoes	Maize, wheat, cattle
Tororo	Eastern Uganda	34°12′E	4,045	61·8	55	84	Dark brown clay	Savanna	Finger millet, sorghum, root crops	Cotton
Kampala	Central Uganda	32°36′E	4,304	46·2	64	80	Red friable clay with laterite horizon	Savanna	Plantains, sweet potatoes	Coffee, cotton, milk

In this district two annual crops per year can be taken, and the one acre available is divided between potatoes (*Solanum*) and peas in the first rainy season, and between potatoes, peas and wheat or millet in the second. The 5 acres of pasture are reckoned to be capable of supporting 3 cows and their followers, producing a total of 9,000 lb of milk a year. The tea is supplied as leaf to a factory which processes for a large number of small growers. The whole enterprise was computed to provide an income for the farmer of £142 in 1964.

(*b*) *A Uganda holding*. A very typical African farmer in the Masaka district of Buganda cultivates 7·3 acres of crops, and has nearly the same area of fallow land in reserve within the boundaries of his holding, for which he pays rent to the African landlord. This acreage is divided as follows:

	Acres
Robusta coffee	3·6
Cotton	1·4
Plantains	1·4
Sweet potatoes	0·2
Groundnuts	0·1
Mixed cereals and legumes (maize or sorghum with beans or groundnuts)	0·6
	7·3

This does not, however, represent 7·3 acres of land used, since some of the cotton follows cereals or pulses planted earlier in the year on the same land. To help him in his work, the farmer has the labour of his wife and of two hired labourers for at least part of the year. These are immigrants from Rwanda, who are not permanent since they either return to their own country or take up their own farms in Uganda, and hence have no opportunity to acquire special skills; they share a small hut on the farm. The only livestock kept are three small hens. The coffee berries are sun-dried on the farm and then sold, like the cotton, to Indian traders who have buying stores very close by, though an increasing number of Ganda farmers are now marketing both these products through their co-operative societies. All cultivation is done with the hoe, the woman being primarily responsible for the cultivation of the food crops. Plantains are the chief of these, but as much as one-quarter of the plantain acreage may be of beer-making varieties.

Efficiency and Stability

Under these headings it is best to consider each of the East African countries separately. In Uganda, generally good soils and a reliable rainfall provide a biologically stable environment and a solidly based peasantry has been little disturbed by historical vicissitudes. Crop failures and serious food shortage are rare, and the two great cash crops, cotton and coffee, have for long provided the basis for a modest prosperity. The very ease with which food and a moderate income can be obtained have perhaps tended to blunt the incentives for farmers to improve their methods, and only a few enterprising individuals are making any radical innovations in farming.

In Tanzania sisal, the chief agricultural export, is produced with efficiency under the plantation system. Although neither soils nor rainfall are as favourable as in Uganda for general agricultural development, the country has from its mere size an enormous potential for increased production. Development of this potential is held back largely by the lack of water supplies, lack of communications and the presence of tsetse-fly which together result in vast areas being almost uninhabited whilst most of the population is concentrated in a relatively small proportion of the country. Mitigation of these three problems is, therefore, amongst the prime needs for agricultural development.

Kenya has had the least stable agricultural conditions of all the three countries. The high density of population in areas of rather low and unreliable rainfall, coupled with serious soil erosion and overstocking have resulted in a history of chronic food shortage; political vicissitudes, locust invasions and epizootics of animal disease have added to the instability of farming. The presence of European farmers has not always increased efficiency to the extent that might have been expected because many of the early settlers were untrained and inexperienced in farming in this environment, and many were *rentiers* who did not need to maximise profits from their farms; soil erosion took many European farmers as much by surprise as it did Africans. Nevertheless, it is precisely because of these difficulties that both Government and farmers have been led to pay greater attention to agricultural improvement than in the neighbouring countries, and recent developments in African farming have been particularly impressive. Optimism and pessimism about the future of Kenya agriculture have waxed and waned with the years. In 1957 one author (Brown 1957) could write: 'When one looks back to 1946/47 and remembers the hopeless statements that were then made about how every African area in Kenya was grossly over-populated, one can feel with satisfaction that the present picture is a good deal more cheerful.' Nevertheless, the inevitable growth of population is bound to place agriculture in both Kenya and Uganda under severe strain within the next two generations to provide the food and income required by their populations.

References

Brown, L. H. 1957. 'Development and Farm Planning in the African Areas of Kenya', *E. African Agr. Jour.* **23**. 67.

Clark, C. and Haswell, M. R. 1964. *The Economics of Subsistence Agriculture.* London.

Clayton, E. 1964. *Agrarian Development in Peasant Economies.* London.

East African Royal Commission 1953–55 (1955). Report, Cmd. 9475. H.M.S.O. London.

Hailey, Lord. 1957. *An African Survey.* Revised 1956. London.

General Reading

Allan, W. 1965. *The African Husbandman.* Edinburgh.

Atlas of Kenya, 1962. Survey of Kenya, Nairobi. 2nd edn.

Atlas of Tanganyika, 1956. Dept. of Lands and Surveys, Dar-es-Salaam. 3rd edn.

Atlas of Uganda, 1962. Dept. of Lands and Surveys. Kampala.

Lock, G. W. 1962. *Sisal.* London.

Matheson, J. K. and Bovill, E. W. (Ed.) 1950. *East African Agriculture.* London.

Russell, E. W. (Ed.) 1962. *The Natural Resources of East Africa.* Nairobi.

Ruthenberg, H. 1964. *Agricultural Development in Tanganyika.* Berlin.

de Wilde, J. C. (Ed.) 1967. *Experiences with Agricultural Development in Tropical Africa.* 2 vols. Baltimore.

West Africa

The area to be considered in this chapter includes the belt of West African countries which lie between the Sahara and the sea along the Guinea coast, from Senegal in the west and extending as far as the (formerly Belgian) Congo in the east (Figs. 2.11.1 and 2.11.2). It thus includes the largest country in Africa (Congo) and the most populous one (Nigeria). This area, to a considerable extent, constitutes a natural region, since it includes practically the whole of the true rain forest of tropical Africa, though there are also vast stretches of savanna country and of desert scrub as the Sahara is approached. Politically, the area has a common history in that all the countries, except Liberia, have been, until recently, under colonial administration, although for climatic reasons there was no European agricultural settlement. This has left something of a common economic heritage, with western Europe still looked to as the main market for West African exports, and shipping lines and financial contacts still oriented in that direction. The great majority of the inhabitants are Negroes in the strict ethnic sense, although Bantu are preponderant in the Congo; in some areas there are minorities of Hamites, Arabs and pygmies. The still largely nomadic Fulani (known as Peul to French speakers) range across some frontiers with their cattle and extend through many of the countries. Islam, Christianity and paganism exist side by side in all the West African countries.

Food crops and export crops are determined mainly by latitudinal location and tend to be grown in east–west belts which extend across many national frontiers. The distribution of livestock is determined more by tsetse-free areas than by economic considerations. The general picture is one in which the pattern of farming has been formed far more by ecological influences than by political or economic ones.

Ecological Background

The Equator passes through Gabon, Congo (Brazzaville) and Congo (ex-Belgian), and in this belt there are two seasonal rainfall peaks in the year with no very marked dry season, so that crops may be planted in many months of the year. Mean annual rainfall is approximately 75 in. Along the Guinea coast from Cameroon as far west as Sierra Leone rainfall is high, with an annual average of over 150 in at several points. Exceptional areas are the Cameroon mountain with 400 in, and the dry plains around Accra in Ghana which lie in a rain-shadow caused by the configuration of the coast. However, rainfall decreases rapidly away from the coast, and a much broader belt of important farming country receives a rainfall of 60–80 in. By the time latitude 12°N is

369

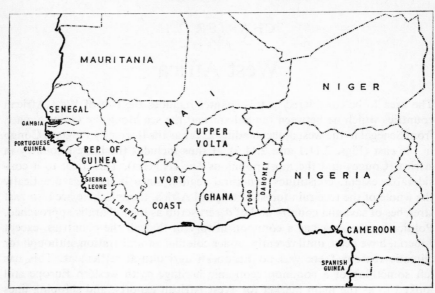

Fig. 2.11.1 International Boundaries in West Africa

reached, for example in Nigeria, rainfall is less than 30 in. At this latitude the two rainy seasons coalesce into one whose duration is only 4–5 months, the remainder of the year being too dry to permit planting of crops or even growth without irrigation. At Timbuktu in latitude 16°N rainfall is only 9 in and agriculture without irrigation has to be limited to stock raising. Extremely high temperatures are reached in the summer months on the fringes of the Sahara; a desiccating wind called the 'harmattan' blows from this area towards the south during the winter months and is very deleterious to crop growth.

The basic vegetation types of the region are only two—forest and savanna —but there are many grades within both. In the Congo (ex-Belgian) the central part of the country, usually known as the central Congo basin, supports dense rain forest, but there are zones of savanna to both north and south. Gabon is so covered with forest that the importance of forestry exceeds that of agriculture; in the Congo (Brazzaville) also, although it is less completely forested, forest products provide more than half the exports. Between the sea along the Guinea coast and the Sahara, it has become customary to describe a succession of vegetation zones running in broad belts east and west as follows (Keay 1959):

(a) Forest Regions

1. Mangrove Forest and Coastal Vegetation. This includes in some sectors large areas of fresh-water swamps which carry a different vegetation from that which fringes the salt-water creeks.

370

2. Rain Forest.
3. Mixed Deciduous Forest or Dry Forest. This is found under drier conditions than the rain forest, and the boundary between them has been roughly defined as the 64 in annual isohyet.

(b) Savanna Regions

4. Guinea Zone, comparatively well-wooded savanna.
5. Sudan Zone, with sparser trees.
6. Sahel Zone. This succeeds the Sudan Zone roughly north of the 20 in isohyet. Trees are still fewer, small and thorny.

Each of these regions has its own quite distinct flora and potentialities for crop production. Outside this zonal classification must be placed the montane flora (especially on Mt. Cameroon) and that of some plateau areas such as the Bamenda grasslands at an altitude of about 5,000 ft in Cameroon which are excellently adapted to cattle raising.

The fauna of West Africa no longer includes enough large game animals to cause any serious damage to agriculture. Attacks on rice and other grain crops by large flocks of *Quelea* finches which are difficult to control have been a serious hindrance to development in some places. The distribution of tsetse-flies is very important because their presence prevents the possibility of developing any important livestock industry. In the whole region, only the Saharan fringes as far south as a line roughly from the mouth of the Senegal river to Lake Chad are completely free of the fly, though small deviations from this line make parts of northern Nigeria and northern Cameroon suitable for cattle.

The soils of West Africa, although of varying geological origin, sometimes laterised and sometimes not, are in general characterised by a high sand fraction. There are considerable differences in fertility but West African soils are mostly poor. The most notable exception to this rule is the small area of rich volcanic soils surrounding the Cameroon mountain. In the high rainfall areas, excessive leaching has produced soils often very poor in plant nutrients, as in the central Congo basin or in the extremely permeable Benin sands of south central Nigeria. In the river estuaries, the poor flow of the rivers during the dry season means that salt-water penetrates far upstream and makes rice-growing along their banks somewhat problematic (see also France, p. 282). Trouble has also been experienced when empoldering low-lying lands for cultivation along some of these rivers by a development of sulphur compounds in the soil which are toxic to crops.

Although the rainfall in much of the West African hinterland is low, crop growth is often better than might be expected from the figures because all this rainfall is concentrated into one short season which is sufficient to grow a crop. Temporary waterlogging of the soil may actually occur, and this is probably the reason why in northern Nigeria the land is almost universally

ridged for crops during cultivation. A soil of light texture, supplied during the
growth period with ample water, is ideal for groundnuts, and many parts of
West Africa are particularly suited to this crop and produce some of the highest
average yields in the world.

An exceptional opportunity for irrigation is afforded by the 'inland delta'
of the Niger in the territory of Mali, and this was exploited under French
administration by the construction of a dam at Sansanding, begun in 1934
and completed during the Second World War. By 1953 675,000 acres were
planted to irrigated cotton and rice. At present rice is the most important
product of this scheme, and it would be possible to irrigate a much larger area,
but progress has been held up by the shortage of suitable settlers to farm the
land and the difficulty of producing and selling crops at economic prices from
this remote region.

Soil erosion (although numerous examples of it can be seen) is fortunately
not such a major menace in West Africa as in some other parts of the tropics.
Slopes are for the most part gentle, and in the high rainfall areas the soil is
naturally protected by a constant growth of vegetation. The layout of farms
to ensure soil conservation is not, therefore, a dominating factor in West
African agriculture.

Agrarian Structure

The agrarian structure of West Africa is based almost entirely on the small
peasant farm. Plantations are relatively few, though where they occur they
are large and mostly owned by companies with substantial capital resources.
In Liberia, the Firestone Corporation of the United States has rubber plantations on over 100,000 acres of land. The former British administrations in
West Africa did not look kindly upon plantations, though in West Cameroon
the Cameroons Development Corporation was set up after the Second
World War to manage a large block of ex-German estates producing bananas,
cocoa and other crops. The Belgian colonial government was more partial
to the plantation system and some hundreds of thousands of acres were developed in the Congo, mainly for oil-palms and coffee. In the ex-French territories, plantations were most extensively developed in the Ivory Coast,
Guinea and Cameroon and produced principally bananas and coffee.

No satisfactory data are available on the average size of peasant farms.
Many include some resting or forest land. The acreage actually under crops
is normally limited to the amount that a family can cultivate by its own labour,
which is seldom more than about 5 acres by manual methods or perhaps
two to three times this figure where an ox-plough is used. Farms may be
larger where perennial crops such as cocoa are grown, often with a minimum
of attention and the use of some hired labour. Systems of land tenure are
governed by local customary law. In general the farmer has the complete
right to the usufruct of the land he occupies, though ultimate ownership is
still often held to reside in the tribe, the chief or the village. Particularly in

cocoa-growing areas, many farmers are heavily in debt by pledging their crop **Farming** in advance to money-lenders, but actual deprivation of land through default **Systems** on debts is rare.

Amongst most of the West African peoples, women share the farm work with the men, and the older children also help on the farm though this is a declining source of labour as more and more of them go to school. However, in the Moslem Hausa communities of the north a fairly strict purdah of women is observed, though the women thus confined to the compound may still do economically productive work, especially in the spinning and weaving of cotton. Some of the poorer families find it a necessity for the women to do field work, but this is considered socially derogatory if a sufficiency can be achieved to avoid it. The use of hired labour is particularly common in cocoa farming. In some areas studied in Ghana (Hill 1956), from a quarter to a half of cocoa farmers were found to employ annual labourers in varying numbers, and many also employed more temporary labour. A survey amongst cocoa farmers in Nigeria showed that the average farmer worked for 6·3 hours a day on 229 days in the year (Galletti *et al*. 1956). Other figures (Clark and Haswell 1964) from different localities show: in Nigeria an average of 4 hours' agricultural work per day by men throughout the year, in Gambia 855 hours per year on 133 days by men and women on agricultural work, in Cameroon an average of 4 hours work per day by men and 5 by women for most of the year but 10 hours during a rush period of 3 months, in another Cameroon locality 194 days' work in the year by men and 189 by women. These figures, although partly explicable by climatic stress, poor health and seasonally slack periods, cannot lead to a high level of production.

By far the greatest proportion of cropped land in West Africa is still culti-vated and weeded with the digging hoe, which is imported in a great variety of shapes and weights. Ox-ploughs are slowly spreading in the tsetse-free areas, their use in the form of a ridging plough having been particularly successfully inculcated in northern Nigeria by the Government agricultural services; in Mali also their use has notably increased. Tractors are still very rare except on special Government schemes of agricultural development which have been particularly useful in producing rice and other crops on some of the 'fadamas' (bottom-lands) of northern Nigeria which often have such heavy soils that it is impracticable to cultivate them manually. Successful examples of minor 'mechanisation' are provided by the widespread use of knapsack (including motorised) sprayers (strapped on the back of the users) to control black pod of cocoa in Nigeria and mirid bugs on cocoa in Ghana.

Outside the rest of the agrarian structure lie the nomadic cattle-keepers, mostly of the Fulani tribe although some sections of this tribe have become settled and even grow crops. They herd their cattle to follow available grazing through the seasons, having no permanent home although most groups are limited to a fairly well-defined beat. To the conservative Fulani cattle are a way of life, and most West African governments have not interfered with their practices except for veterinary administration and for placing a limit on the number of stock that can be kept in some areas, such as the Bauchi plateau in Nigeria. In Sierra Leone attempts have been made to settle them on defined ranches in an ingeniously conceived scheme (Murray 1958). The nomadic Fulani occupy exactly the same position in the utilisation of West Africa's grassland resources as do the similarly nomadic cattle-keeping tribes of the Masai in Kenya and Tanzania and the Hima in Uganda (Chap. 2.10).

The social structure of West Africa is particularly characterised by the number of farmers who live in villages or even towns rather than in isolated homesteads situated on their holdings. Many farmers have to travel some miles on foot or by bicycle to reach the fields they cultivate. Such a form of rural organisation is, of course, only possible where larger farm livestock are, as over so much of West Africa, absent or unimportant.

The number of agricultural development schemes which have been attempted in West Africa is legion. Many, especially of those based on mechanisation have proved premature or abortive and have receded into the limbo of history. A number of others are still in their initial stages and it is too early to assess whether they will have any real impact. Some of the more successful ones must however be mentioned because of the large areas they cover or because of their influence as examples or as pointers to future development. The great French irrigation scheme on the middle Niger has already been mentioned. Other important schemes of French origin have involved the production of groundnuts by farmers assisted by mechanical cultivation in the Kaffrine and Casamance areas of Senegal. Also in Senegal is a large scheme for mechanised rice production at Richard–Toll which has involved a dam to keep out salty water in the Senegal river; originally developed by the Government, a private company was later brought in to assist in the enterprise. In Nigeria, the Regional Production Development Boards have established a number of large plantations of various crops which can serve as examples to smaller growers in the absence of privately-owned plantations. Most massive of all is the system of 'lotissements agricoles', generally known in English as 'corridor settlements', originally developed by the Belgian Government in the Congo and claimed to have included one-fifth of the country's farmers at the time of independence. Here crops were grown in defined rotations in strips cleared from the forest so that after four or five years' cultivation a natural reinvasion of forest could be allowed to take place to restore fertility.

The marketing of food crops in West Africa is for the most part by fairly simple methods, but in the case of export crops the free economy has been much modified, especially in the former British territories, by the creation of statutory Marketing Boards with a monopoly of the purchase and export of such products. These boards do not necessarily offer the producer a price corresponding to world market levels, since they may desire to accumulate their own reserve funds or to use these to subsidise low prices, and have sometimes been used as an instrument of Government policy in the creation of capital. The arguments for and against such systems have been much debated, but governments continue to favour their existence. Most countries of the region still retain commercial advantages from their previous or existing colonial connections, for example the former French colonies as associate members of the European Economic Community, and the former British territories by enjoying Commonwealth preference for their exports to the United Kingdom. The co-operative movement in the former British territories has been much concerned with marketing, especially in Ghana where cocoa farmers' co-operatives have a long history. In the ex-French territories the 'sociétés de prévoyance' have had somewhat wider aims and have done much to finance small rural and agricultural improvements. One limiting factor to efficient marketing almost throughout West Africa is still the lack of adequate communications, and much farming country could be more fully developed if good road systems were extended.

In agricultural research, West Africa is as well served as most tropical regions. The older well-known Government experiment stations such as those at Mbambey (Senegal), Ibadan (Nigeria) and Yangambi (ex-Belgian Congo) have been reinforced by more or less autonomous research institutes which include Adiopodoumé in the Ivory Coast, Tafo in Ghana (for cocoa) and the Oil Palm Research Institute near Benin in Nigeria. Universities have increasingly contributed to agricultural research in recent years, and in some cases have absorbed pre-existing experimental stations as at Samaru in Nigeria. Increasing opportunities for professional agricultural education have also been afforded by the spread of university institutions in West African countries, amongst which Nigeria is the leader with five universities. The supply of qualified recruits for the advisory services thus seems to be reasonably assured.

Crops and Livestock

Table 2.11.1 gives some figures (whose accuracy must not be overestimated) of crop production and livestock numbers in six representative countries of the region. The main differences to be noted are those between countries lying, like Senegal and Mali, wholly in the dry tropics and where therefore perennial crops are absent or unimportant, and countries which include regions of higher rainfall. In food crops of the coastal regions, there is a clear distinction between the western Guinea coast, from Portuguese Guinea to Liberia, where rice is the preferred staple food, and an eastern zone from the Ivory Coast to Cameroon where root crops are more important. There is, nevertheless, a tendency for rice acreages to increase slowly but steadily throughout West Africa. Maize is also an important, and in some localities the dominant, cereal in the wetter areas. The recent development of high-yielding maize types, from American material originally introduced to provide resistance against rust disease (*Puccinia polyspora*), appears likely to increase the attractiveness of the crop in future. In the drier north, the population has

FARMING SYSTEMS OF THE WORLD

Table 2.11.1

West African Crop Production and Livestock Numbers
(Selected Countries)

Crop production in thousands of short tons per annum:						
	Senegal	Ivory Coast	Ghana	Mali	Nigeria	Congo Dem. Rep.) (formerly French Congo)
Maize	80	197	198	110	1,251	261
Millet and sorghum	654	100	209	770	7,286	49
Rice (paddy)	121	265	45	165	386	62
Cassava	187	1,356	1,378	165	8,047	6,856
Sweet potatoes and yams	8	2,094	1,323	77	14,991	328
Groundnuts (in shell)	1,237	44	55	165	1,700	123
Cotton (lint)	—	3	—	22	49	6
Cocoa (beans)	—	133	459	—	203	4
Coffee	—	299	3	—	—	66
Palm kernels	4	14	13	—	449	134
Palm oil	—	31	47	—	568	230

Livestock numbers, in thousands:

	Senegal	Ivory Coast	Ghana	Mali	Nigeria	Congo
Cattle	1,920	350	505	4,640	7,465	1,200
Sheep	510	552	682	82	7,694	691
Goats	600	750	700	10,662	20,482	2,362
Pigs	45	115	250	19	674	388
Camels	6	—	—	209	12	—
Donkeys	84	1	9	468	1,240	—

Source: F.A.O. Production Yearbook 1966.

little alternative but to rely for its food on the more drought-resistant cereals which will grow there. The chief of these is sorghum, locally known as 'guinea corn', though as the Sahara is approached this has to be replaced by the even more drought-resistant bulrush millet (*Pennisetum typhoideum*).

The chief root crops of West Africa, grown only in the higher rainfall regions, are yams and cassava, followed by sweet potatoes and coco-yams (both *Colocasia* and *Xanthosoma* spp). Yams are a particularly characteristic staple food crop of south-eastern Nigeria; grated cassava meal, known as 'gari', is one of the commonest articles in the markets. Cow-peas and ground-nuts are the two chief leguminous crops of the West African farmer, but the latter is largely sold as an export crop. This list comprises the major food crops of West Africa, but a number of subsidiary ones are of considerable importance in some areas. These include the plantain (cooking banana) in regions of high rainfall; and the cereal *Digitaria exilis* and the Bambarra groundnut (*Voandzeia subterranea*) in some of the drier areas. Okra (*Hibiscus esculentus*) is a very widely grown vegetable which is to the West African taste.

Of the export crops, Ghana produces more than one-third of the world's
supply of cocoa, and West Africa as a whole more than one-half; Nigeria and
the Ivory Coast are the leading producers after Ghana. The main threat to
the crop in recent times has come from the 'swollen shoot' virus disease which
has affected large acreages in the main producing countries and can only be
contained by a drastic policy of cutting out infected trees and replacing them
with new plantings. Oil-palms (*Elaeis guineensis*) are native to the region, and
wild trees whose thick-shelled fruits are collected by peasant farmers provide
most of the production, though there is also an estate sector which is particularly important in the (formerly Belgian) Congo and which has the advantage
of using plantings of the higher-yielding thin-shelled variety. Palm-oil is an
important local food as well as an export product, and in some countries
such as Sierra Leone most of the oil is consumed locally and it is only the
kernels which are exported. Bananas are very important export products of
the Ivory Coast and Cameroon. Coffee, of 'Robusta' type, is mainly exported
from the Ivory Coast, followed by Congo (ex-Belgian) and Cameroon, with
Togo and the Central African Republic leading amongst the smaller producers.

Whilst these perennial crops are limited to the high rainfall areas, the drier
regions find their main export crops in groundnuts and cotton. Nigeria and
Senegal are the world's largest exporters of groundnuts; the crop provides
nearly all the export income of Gambia and is also an important export of
Mali; it is grown as a food crop in every country of the region. The cotton
grown for export is of the American Upland type; Nigeria and Chad are the
main producers, and the crop is also particularly important in Cameroon and
Mali. Sesame (often known in West Africa as 'benniseed') is widely grown for
food and substantial quantities are exported from Nigeria, where it is particularly important in the income of the Tiv people in the middle rainfall belt.
Amongst minor cash crops, kola-nuts which are valued for chewing by West
African peoples are widely grown in the high rainfall areas along the coast and
large quantities are sold into the drier areas where the tree cannot be grown.
The oil-rich fruits of the shea butter nut ('karité' in French) which grows wild
in the savannas are gathered both as a source of food and for export, especially when prices are high.

Livestock are comparatively unimportant in the high rainfall areas of West
Africa, as is to be expected both from the tsetse infestation of these areas and
because of the lack of grassland on which they could be pastured. The typical
cattle of these areas are the dwarf shorthorn (*Bos brachyceros*) or 'muturu',
very small animals which are casually kept in small numbers for beef but have
a certain survival value because they possess some degree of resistance to local
strains of trypanosomiasis, the disease transmitted by the tsetse-fly. Also to be
found here are the slightly larger 'Ndama' cattle, classified as longhorn
humpless, which originated in and around Guinea but have been widely
introduced elsewhere in West Africa because again of a certain resistance to

trypanosomiasis. The 'Borgu' or 'Keteku' cattle of Dahomey and Nigeria originated from crosses of muturu and Ndama with zebus.

The cattle of the savanna regions are predominantly of zebu derivation. Amongst the many types, some of them ill-defined, in the West African cattle population, a few have become known for their merit or peculiar characteristics, such as the White Fulani which is useful for draught and milk, and the large Kuri or Chad cattle with their remarkable bulbous horns. Historically, West African cattle have developed almost solely as a source of beef. Because cattle can hardly be kept in the wetter regions, there is a constant movement of live cattle from the savanna into these regions to supply their meat requirements which accounts for much of the income of the cattle-keepers. The demand for draught cattle is only a recent innovation. Production of milk is low, and although the Fulani obtain enough from their large herds for their own requirements, there is little milk available for sale in West Africa.

Goats are, after cattle, the most numerous livestock. The wetter regions are again characterised by dwarf types, but the dry hinterland produces much larger animals. Goats are kept to provide meat and skins. The export of goat-skins is financially important in many West African countries, and the famous 'red Sokoto' breed of north-western Nigeria produces some of the world's best skins. Sheep are more numerous in the dry parts of West Africa than in most tropical regions, but they are of small size and carry fleeces of hair rather than wool so that their productivity is limited. Pig products are not, of course, consumed by Moslems who form a large proportion of the West African population, but amongst Christians and pagans pigs are quite a popular form of livestock, particularly as these classes of the population inhabit mainly the wetter regions where other forms of meat are scarce. Camels are important in the driest countries of the region, as is shown by the figures for Mali in Table 2.11.1 and in their use as pack-animals they extend as far south as northern Nigeria. Donkeys are used in much larger numbers in most of the countries (Table 2.11.1). Local kinds of poultry are once again of particularly diminutive size in the wet regions, but introduced breeds of temperate origin tolerate the environment equally well and are rapidly replacing them. The failure of a large-scale British project for raising poultry in Gambia in the 1950s shows that the West African environment is not altogether an easy one in which to practise poultry husbandry.

The general absence of any integration between the raising of crops and livestock which might be described as mixed farming is particularly characteristic of West African agriculture. Efforts which have been made to induce this integration, as by the introduction of draught oxen on northern Nigerian farms, have as yet affected comparatively few farmers. Livestock (other than a few goats, pigs and poultry) are in general kept by different people and even different tribes from those who grow crops. Even farmers who keep pigs and poultry do not always realise the extent to which they need to grow crops to feed them. Amongst the growers of crops, most rely on a single cash crop to provide the bulk of their income, and this leaves them particularly vulnerable to fluctuations in price. In food crops there is more diversity amongst the crops grown on a single farm, since the majority of West African families expect to provide their food supply from their own land. But food crops are mostly selected from those possible in the environment to satisfy hunger rather than to produce a balanced diet. With little knowledge of food values, the choice falls predominantly on cereals and root crops and this, coupled with the paucity of livestock over so much of the region, leads to a widespread protein deficiency in the diet which is one of the major shortcomings of West African agricultural systems.

A Nigerian Transect

Since the main contrasts in West African agriculture are between the forest and savanna regions, with their intermediate or transitional phases, nearly all the main types of West African farming can be exemplified by any transect which runs across these divisions. A broad transect through Nigeria from south to north provides a particularly convenient illustration, because here the ecological zonation runs almost parallel to lines of latitude (Fig. 2.11.1). In general, a type of farming which occurs in Nigeria at one latitude is likely to be found elsewhere in West Africa at the same latitude or, if this is not always quite the case, at least in other parts of West Africa where the rainfall is similar.

COASTAL AREAS

The coast of Nigeria is deeply indented by creeks and lagoons. Much of it is fringed by mangrove swamps, at present almost useless but which might perhaps, if future regional food needs demand it, be reclaimed in some places for rice cultivation, as has been attempted in Sierra Leone. Where the soil behind the shore-line consists of marine sands, these are sometimes almost sterile for crop growth as in the immediate neighbourhood of the capital, Lagos; but elsewhere coconut palms can be planted to take advantage of this ecological niche which they tolerate better than other crops, and it is in this coastal strip that most of the coconuts of Nigeria and West Africa are grown, the chief centre of Nigerian production being around Badagri near the western frontier. For many of the population, fishing is a more important source of income than farming, which may consist therefore of only minimal plantings of food crops, mainly roots. The heavy consumption of fish does however ensure that there is less protein deficiency in the diet of the coastal community than amongst most Nigerians.

RAIN FOREST ZONE

This zone lies, with some deviations, between about 5° and 7°N. Rainfall varies for the most part between 65 and 100 in per year. The soils are classified as porous sands to sandy clays (Vine 1953). Some very dramatic erosion with

379

enormous gullies on hill lands in Onitsha and Owerri provinces is largely due
to the geological youth of these hills which are still being eroded by natural
processes but imposes the necessity for very careful agricultural practices in
these areas. Although the natural vegetation of this zone is high forest, from
which all land now used for agriculture has at some time been cleared, a great
deal of this zone is in fact now covered by secondary forest which has succeeded earlier clearings.

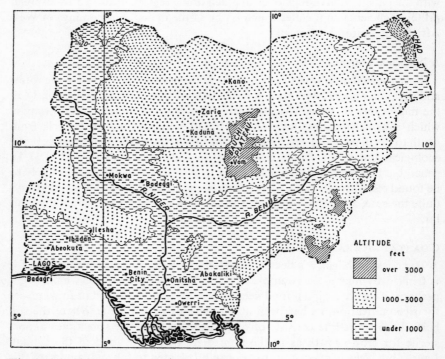

Fig. 2.11.2 Altitude and main rivers of Nigeria

In this high rainfall area, the main reliance for food is placed on root crops,
of which yams (especially in the eastern sector) and cassava are the most important, with sweet potatoes and coco-yams of secondary significance. The
yams are grown in hills, sometimes of enormous size, and some stems of small
trees and bushes are often left when clearing the forest to provide them with
support. Cassava suffers as severely as in any part of Africa from mosaic
disease which reduces average yields. Because of the reliance on root crops
in the diet, this ecological zone is the classic area in Africa for the incidence
of protein deficiency. This is especially seen in the form of 'kwashiorkor', a
disease typical in children at the post-weaning stage, which was first described
from an ecologically similar area in Ghana and is still responsible for much
child mortality. Maize is the only cereal of much importance, with some small

380

areas of swamp rice in the eastern part of this zone. Cow-peas, grown here, as elsewhere in Africa, in the dwarf erect form for the production of dried seeds, and groundnuts are the chief legumes grown for food.

By far the most important cash crop of this zone is the oil-palm, especially in the east. It is only in the secondary forest that the 'wild' oil-palms grow; often, and especially when encouraged by selective weeding, they form groves of almost pure stands on the village lands. Where fruit can easily be obtained from wild trees, many farmers are unwilling to undertake the labour of planting, although the advantages to be gained by planting better varieties of the palm at optimal spacings are undeniable, and successive governments have tried to encourage planting by many means, including a current scheme for the assisted settlement of trained young farmers on small plantations. Other Government-encouraged improvements which have been gradually affecting the industry include the use of hand oil-presses, mechanical crackers to produce kernels from nuts, and the erection of small 'Pioneer' oil mills to replace domestic extraction of the oil. Due in part to these improvements, and in part to extension work and price differentials, the quality of Nigerian palm-oil has greatly improved in recent years and it now has a much lower content of free fatty acid. The oil-palm is extraordinarily well suited to this environment because of its tolerance of soils of low fertility and high acidity. Another crop which has the same characteristics to some degree is rubber, and this is a less important cash crop in some parts of the zone. Rubber has long been grown by peasants in the Benin area on small acreages and with very poor husbandry involving over-close spacing, often over-tapping, and the smoking of sheets over fires in the huts. In recent years the area has been considerably extended and husbandry somewhat improved by extension work. A small part of the Nigerian cocoa acreage lies in this zone, on rather better soils in the western areas. In the same sector, kola is grown by many farmers as a cash crop on slightly poorer soils.

Livestock in this zone include some goats, pigs and poultry, all of which are more often kept in twos or threes per household than in larger numbers, and a few cattle mainly of the muturu breed. Natural grassland being non-existent, the roadside verges form one of the most important grazing grounds as elsewhere in the wet tropics. A Government dairy farm outside Lagos succeeds in maintaining milk production from guinea grass (*Panicum maximum*) pastures but has to maintain elaborate and costly precautions against disease.

The main agricultural problem of the zone is posed by the increasing pressure of population on land in the eastern areas, where population densities in some villages run at well over 1,000 per square mile. Yams, the staple food of the area, are a particularly exhausting crop and although it has been shown that they respond to fertilisers, particularly nitrogen and potash, the use of these has still hardly begun. A few Ibo villages have already, from force of circumstances, adopted the practice of returning human excreta to the fields.

THE DRY FOREST ZONE

This zone, lying roughly between $6\frac{1}{2}°$ and $8\frac{1}{2}°N$, has a rainfall generally of 50 to 64 in a year with a very pronounced dry season from November to February. The deciduous forest, by whose presence it is classified, is more symptomatic of this lower rainfall than in the rain forest zone. The soils are still similar to those in the last zone, but more of them are of slightly heavier texture and they tend to be more fertile, perhaps because they are less thoroughly leached. A notable feature is the presence, especially in western areas, of earth-worms in larger numbers than are usually found in tropical soils; they seem here to replace termites in a similar ecological niche, for termite mounds are rare in this area. The zone is one of the most populous and agriculturally important in Nigeria, and includes the important towns of Ibadan and Enugu.

The food crops are still the same as in the last zone, with the addition of an important and slowly increasing acreage of rice, which is grown in swamps in the eastern areas, particularly around Abakaliki, and as upland rice in some western parts. A characteristic minor crop of the area is the 'egusi' melon (*Cucumeropsis* spp) which is grown for its edible oil-rich seeds. Many crops are grown interplanted with each other rather than in pure stands. A straight-line transect near Ibadan through an area of typical farming land, which is primarily devoted to the production of food crops for the town, showed the following percentages of land in different uses:

Uncleared bush	44	Cassava and maize (interplanted)	8
Yams and maize (interplanted)	17	Cassava and yams (interplanted)	7
Yams alone	12	Cleared but not planted	3
Cassava alone	9		

As this transect suggests, the population in this zone is not yet so dense as to preclude the possibility of leaving a fair proportion of land resting under bush; soil fertility can therefore be maintained by a system of shifting cultivation for some while longer.

Oil-palms are of lesser importance in this zone which includes, no doubt because of slightly better soils than in the preceding zone, the main areas of cocoa production in Nigeria. These all lie in the western sector, along an arc extending west to east from Abeokuta through Ibadan and Ilesha to Ikare. Figures from an official survey showed the average cocoa-grower in Nigeria to have 3·5 acres under the crop (Galletti *et al.* 1956). Nearly all cocoa in West Africa is grown under shade, which is usually provided by leaving a few trees standing when clearing the forest for the crop. The spacing of cocoa trees is irregular and usually far from optimal. The low yields generally obtained could be enormously increased, as experimental work has shown, by reducing shade and using fertiliser. Higher yields could also be obtained by better spacing and weeding, the planting of superior varieties of cocoa, and better control of pests and diseases.

This ecological zone is better tolerated by cattle than the preceding one, though the Ndama and Keteku breeds are still valued here for their partial resistance to trypanosomiasis. Considerable experimental work has been done by Government and university staffs to develop grazing farms in the hope of showing that cattle can be profitably raised here for beef.

THE GUINEA ZONE

With this zone we move from forest to savanna, though there is usually a small transitional fringe between the two. This is the largest ecological zone in Nigeria, extending northward from the dry forest zone to about 11°N. Annual rainfall is for the most part between 35 and 50 in and is concentrated into the period from May to September. The soils are still mainly reddish porous sands to sandy clays, with an area of brown compact silty fine sands around Kaduna and Zaria; laterisation is shown over large areas by the presence of ironstone concretions in the profile. The southern part of this zone, often known as the 'Middle Belt' of Nigeria, is marked by a very low density of population, which is attributed to the depredations of slave raiders in past centuries. There is therefore no problem of population pressure and plenty of opportunity for development schemes.

In this drier zone, root crops are of comparatively slight importance in food supplies, and guinea corn (sorghum) becomes the staple food crop. Some maize is also very commonly grown. Rice, being less tied to rainfall than root crops provided that swampy land is available, is still important and the main rice research station of Nigeria is at Badeggi on the Niger in this zone. A transect through farmed land in the northern part of this zone showed the following percentages of land use:

Guinea corn	41·6
Land resting under grass cover	29·4
Maize and okra (interplanted)	14·5
Bulrush millet	11·5
House compounds*	3·8

* i.e. Farmstead.

The rainfall being too low for perennial cash crops, cotton is the main export crop of the zone. Cotton production has been very greatly increased since the Second World War, and the industry has been assisted by cotton breeding and experimental work carried out by the Cotton Research Corporation at Samaru research station. Sesame (benniseed) is particularly important as a cash crop amongst the Tiv people living south of the Benue river. Cattle-raising is still limited by the fact that much of the zone is infested with tsetse-fly.

An exceptional area in the middle of the zone is the extensive Bauchi or Jos plateau at an altitude of about 4,000 ft. This elevation is not by itself sufficient to create a markedly different ecological environment, but the farming is

differentiated by other factors. Slopes are often steep and the soil is of light
texture and often thin over the rocks, so that soil erosion is a greater danger
than in most parts of Nigeria and conservation practices are very desirable
in farming. The population is denser than the poor soils can easily support,
and considerable acreages of land have been put out of agricultural use by
tin-mining. The farmers belong mainly to rather primitive pagan tribes who
practise characteristic farming methods including the cultivation of the unusual cereal *Digitaria exilis*. This is good cattle country. The Nigerian Government veterinary services have their headquarters at Vom, where they have
pioneered the development of butter making in Nigeria and successfully
raised half-bred Friesian cattle.

The Guinea Zone as a whole has a particularly long history of attempted
schemes of agricultural development. One of the most elaborate but disap-
pointing schemes was the Niger Agricultural Project (Baldwin 1957) at
Mokwa for mechanised production of annual crops with settler farmers under
supervision. A more recent and successful venture has been the development
of the first sugar-cane plantation in Nigeria. Schemes for large-scale irrigation
from the Niger are being developed and offer considerable prospects for the
future.

THE SUDAN ZONE

This zone extends from about 11° to 13°N, with a rainfall in general of from
20 to 25 in, but concentrated into such a short season (June to September)
that crop possibilities are very limited. The predominant types of soil are
classified as orange-brown to red loose sands. The extensive 'fadamas' or
bottom-lands are characterised by black clay soils which are extremely difficult to work but have in some places been developed for rice-growing by
Government schemes of mechanisation with heavy tractors.

Human populations are denser than in the preceding zone, and there is a
particularly heavy concentration for some 40 miles round the city of Kano,
where resting land has disappeared from the farming system and increasing
use is made of animal manure, compost, slaughter-house wastes and fertilisers in the effort to maintain soil fertility. Katsina province is also very
heavily populated and will increasingly run into the same problem.

The chief food crop is guinea corn supplemented by bulrush millet. Because
the average rainfall is so low and critical for crop growth, comparatively slight
annual variations tend to produce bumper yields in some years and low yields
or even crop failures in others. There is a history of great fluctuations in grain
prices and frequent food shortages which have however stopped short of
disastrous famine. These troubles have been considerably alleviated in recent
times by more crop storage and the extension of transport facilities, both
roads and railways. The rice grown on wet lands in this zone is principally the
indigenous 'red' rice of West Africa (*Oryza glaberrima*). Because the grain

from each harvest has to last for at least the 12 months until the next harvest
can be taken, it is stored in small granaries made of dried earth or plant
materials, or in Bornu province sometimes in underground pits. Observations suggest that when unthreshed guinea corn is stored in this way, average
losses in this dry climate from insect damage do not exceed about 10% in a
year.

Groundnuts are the characteristic export crop and also important in the
diet. Acreage and production have been enormously expanded in the last 30
years, much assisted by road construction and the extension of the railway
into Bornu province. Nigerian exports of over 600,000 tons of groundnuts
annually come mainly from this zone. The use of phosphatic fertiliser on
groundnuts has been proved to be profitable and growers have begun to use
it although the practice still spreads only slowly.

With large areas free from tsetse-fly, this zone is the most important cattle
country of Nigeria, whilst goats and sheep are also numerous. Still bred
largely by nomads, cattle are primarily a source of beef and there is a very
large trade to the southern meat-deficient parts of Nigeria, some cattle also
being brought across the frontier from surrounding countries for this purpose.
Hides and skins are an important export product of this zone and also support
a small local tanning industry. The use of draught oxen for ploughing, though
still only practised by a small minority of farmers, has begun to make a significant contribution to productivity. The crops of these 'mixed farmers' often
stand out to the eye because of the effect of using animal manure on them.
Rinderpest, black-quarter and bovine pleuro-pneumonia are amongst the
animal diseases which present serious problems and necessitate painstaking
veterinary work to maintain the level of production.

THE SAHEL ZONE

This zone, which lies roughly north of latitude 13°N, has a rainfall of less
than 20 in. It only includes a very small part of Nigeria in the north of Bornu
province, but extends far to the north as a very wide belt of West Africa right
up to sub-desert conditions, and reaches the sea around the mouth of the
Senegal river. The soils in the Nigerian sector are similar to those of the
preceding zone except for areas of black clay soil which are poorly drained or
seasonally flooded near Lake Chad.

Food crops have to be of the most drought-resistant types, and bulrush
millet becomes the dominant cereal. Because of the ever-present risk of crop
failure, storage by the farmer of a carry-over of grain from good years to bad
is almost a condition of survival. There is no real cash crop in this zone of
Nigeria. Cattle and goats furnish much of the sustenance of the people
through meat and milk; most of the area is so remote that hides and skins
may be the only products it is possible to market.

Table 2.11.2 summarises the chief environmental and agricultural data of
this Nigerian transect.

Table 2.11.2

A Nigerian Transect

Place	Zone	Latitude	Altitude (ft above sea-level)	Rainfall (mean inches per year)	Relative Humidity (mean % in midday hours)	Temperature (mean daily maximum, °F)	Soil Type	Vegetation	Chief Food Crops	Chief Cash Products
Lagos	Coastal	6°27'N	10	72·3	74	86	Coastal alluvium	Mangroves	Roots, maize	Coconuts, fish
Benin	Rain forest	6°19'N	258	81·8	69	87	Acid sands	Rain forest	Yams, cassava	Oil palm, rubber
Ibadan	Dry forest	7°26'N	656	44·1	65	88	Reddish sandy to sandy clays	Dry forest	Yams, cassava, maize	Cocoa
Jos	Plateau	9°54'N	4,010	55·4	43	83	Light, stony	Grassland	Digitaria, bulrush millet	Tin
Kaduna	Guinea	10°35'N	2,113	50·1	41	89	Brown silty sands	Savanna	Guinea corn, maize	Cotton
Kano	Sudan	12°02'N	1,533	34·2	31	92	Orange-brown sands, some laterisation	Savanna	Guinea corn	Groundnuts, cattle, goats

SAMPLE FARMS

Owing to the general lack of numerical data, it is not easy to cite sample farms in West Africa for which precise figures, especially of acreage, can be given. However the following examples from forest and savanna areas known to the author have some backing of precision.

(a) *Forest zone.* At Turumbu, some 40 miles from Stanleyville in the (formerly Belgian) Congo, agriculture consists only of clearings in the primeval forest which otherwise covers the poor sandy soils. The farmers here were embodied into one of the 'corridor settlement' schemes of the Government; however, this has merely regularised rather than altered the basic rotation and choice of crops which they would have adopted outside the scheme. The rotation prescribed by the scheme consists of 4 years' cultivation followed by 16 years' forest fallow to restore fertility, during which the land is invaded by tree growth dominated by *Musanga cecropioides*. Within the cultivation period, the rotation is as follows:

1st year: Season A: maize.
             Season B: upland rice interplanted with cassava and bananas.
2nd year: cassava and bananas beginning to yield.
3rd year: cassava and bananas yielding, then uprooted and removed.
4th year: maize, beans and groundnuts on separate plots.

As one field is in each phase of the rotation at any given time, all these crops are available in each year. Groundnuts are the main cash crop, but surpluses of food crops are also sometimes sold. The diet is supplemented by some fishing. The area cultivated by each family is in this scheme adapted to the size of the family; in the 'extended family' this may be large, and cultivated acreages in these schemes range from about 5 to 18 acres. Livestock are scarcely kept in this environment, and may be represented at most by a few scraggy hens.

(b) *Savanna zone.* Some numerical data are provided by Grove (1952) for farmers on the Jos plateau of Nigeria, and the picture has not changed very much in succeeding years. A typical farmer with perhaps two wives and two children cultivates the following areas of field crops:

Digitaria exilis	5 acres	
Bulrush millet	2	,,
Finger millet	1·5	,,
Yams	1	,,
Cassava and *Coleus dazo*	0·5	,,

In addition, 2 acres around the house are used as an intensive 'garden', where manure and household waste are applied to yams, coco-yams, *Coleus*, sesame, cassava and guinea corn. There are finally about 4 acres under 'natural' grass fallow as part of the rotational system. These acreages are rather larger than would be manually cultivated by a family of this size in most parts of West Africa, but they are a necessity because crop yields are so low on these poor soils, and a possibility because the soils are so light as to be easily cultivable. The livestock owned by such a family would amount to a goat or two and a few poultry.

Efficiency and Stability

West African agriculture is perhaps only efficient in one respect: the region has always produced enough food to avoid catastrophic famines. In most other respects agriculture is, for a variety of reasons which have not been easily escapable, inefficient. Productivity could be much increased if farmers worked longer hours (which would imply better diets, better health services and less time wasted in marketing due to bad roads and distance of buying centres) and if more use were made of draught oxen where they can be kept (i.e. in tsetse-free areas) and land is abundant. The output of palm-oil could be enormously increased by the use of planted instead of wild palms. Cocoa yields could be greatly increased by methods which have been mentioned. The output of many crops could be improved by the judicious use of fertilisers, though it is by no means axiomatic that fertiliser use is profitable on all crops in West Africa, nor that the yield increases to be obtained are always startling (Watson 1964). Cattle could be bred and fed to attain the higher milk output which has been demonstrated at experiment stations. To achieve all these things is primarily the task of the agricultural extension services, though they may have to be supported by an injection of capital into peasant farming in the form of supervised credit. It is not to be denied that all these methods would be likely to increase gross production and force down prices for West African produce upon world markets, though there might be some corresponding increase in consumption through cheapness. As far as possible therefore increases in production should be aimed at raising local standards of living and increasing the exchange of goods and services between the farmer and the townsman.

The spectre of population increase outrunning natural resources is as yet a fairly distant one for most of West Africa, but already has to be faced in a few densely populated areas which must be used as training-grounds for the future in learning how to maintain and enhance soil fertility under local conditions.

A certain economic stability is inherent in any peasant farming society such as prevails in West Africa. Slumps and booms may come and go, but the peasant who uses only the unpaid labour of his family to produce their own food from their own land is immune from the worst effects of them. Nor can climatic instability be said to affect the high rainfall areas of West Africa; but in the dry areas it is a very real problem. Dry years not only deprive the people of food but distort the whole economy by the high prices which they cause. Where food crops are not stored because of either biological impossibility or improvidence, prices may fluctuate dangerously within the course of a single normal year. In Dahomey, the price of maize has fluctuated between 10 and 30 francs per kilo in the course of a single year, thereby tempting people to live on the less nutritious cassava which is in steadier supply; in a year of scarcity, the price rose to between 35 and 55 francs in different areas. The remedy for these conditions is Government storage of buffer stocks of grain, which is one of the prime improvements needed in many West African countries. In the longer term, the slowly increasing use of irrigation will also contribute to stability in production and prices.

References

Baldwin, K. D. S. 1957. *The Niger Agricultural Project.* Oxford.

Clark, C. and Haswell, M. R. 1964. *The Economics of Subsistence Agriculture.* London.

Galletti, R., Baldwin, K. D. S. and Dina, I. O. 1956. *Nigerian Cocoa Farmers.* London.

Grove, A. T. 1952. *Land Use and Soil Conservation on the Jos Plateau.* Geol. Survey of Nigeria, Bull. No. 22. Nigerian Govt.

Hill, Polly. 1956. *The Gold Coast Cocoa Farmer.* London.

Keay, R. W. J. 1959. *An Outline of Nigerian Vegetation.* Govt. printer, Lagos, 3rd edn.

Murray, A. K. 1958. 'The Fula Cattle Owners of Northern Sierra Leone, their Cattle and Methods of Management', *Trop. Agr. Trinidad.* **35.** 102.

Vine, H. 1953. *Notes on the Main Types of Nigerian Soils.* Lagos.

Watson, K. A. 1964. 'Fertilisers in Northern Nigeria: Current Utilisation and Recommendations for their Use', *African Soils.* **9.** 5.

General Reading

Coppock, J. T. 1966. 'Agricultural Development in Nigeria', *J. trop. Geog.* **23.** 1.

Doutressoule, G. 1947. *L'Elevage en Afrique Occidentale Française.* Paris.

F.A.O. 1966. *Agricultural Development in Nigeria 1965–1980.* Rome.

Gaudy, M. 1959. *Manuel d'Agriculture Tropicale.* Paris.

Irvine, F. R. 1953. *A Text-Book of West African Agriculture.* London. 2nd edn.

Johnston, B. F. 1958. *The Staple Food Economies of Western Tropical Africa.* Stanford, California.

Papadakis, J. 1966. *Crop Ecologic Survey in West Africa.* Vol. II. Atlas. F.A.O. Rome.

Phillips, T. A. 1956. *An Agricultural Note-Book (with special reference to Nigeria).* London.

Spitz, G. 1949. *Sansanding—Les Irrigations du Niger.* Paris.

Wills, J. B. (Ed.) 1962. *Agriculture and Land Use in Ghana.* London.

CHAPTER 2.12

India and Pakistan

Introduction

The area covered by this chapter consists of the national territories of India and Pakistan (Fig. 2.12.1). In so far as this area comprises a natural agricultural region, its uniformity is due to the dependence of agriculture, in both countries, on a common monsoon climate, and to their common political, economic and social history during the centuries when they were under Mogul and then British rule or influence. On the other hand, the agricultural diversities within the region are very striking. India stretches from within 8 degrees of the Equator to about 35°N latitude, with about half of its area within the tropics and the northern frontiers close to some of the highest mountains in the world. Pakistan, with only a small area within the tropics, has extremes of inundation in the eastern wing and aridity in the western, though both closely resemble adjacent parts of India. There is little ethnic or linguistic unity within India or between east and west Pakistan, the national consciousness of both countries being based rather on the respective dominance of their two great religions, Hinduism and Islam.

India is the second most populous country in the world, but is not so outstanding in size, in which it ranks seventh, whilst its population density of about 350 to the square mile is exceeded by a number of smaller countries. The most significant agricultural statistic is perhaps that India contains a greater population of bovines (cattle and buffaloes together) than any other country in the world. The agricultural economy of India is not closely paralleled by that of any other country in the world except parts of Pakistan. No great differences in agricultural practice are to be observed in passing the frontier from the one country to the other. The agriculture of parts of west Pakistan has, on the other hand, great affinities with that of the countries of the Middle East which lie to the west of it.

Ecological Background

The dominant climatic feature in the agriculture of the region is the occurrence of the rains brought by the south-west monsoon from approximately June to October. This period provides rain to the whole of the sub-continent, and to much of it almost its only useful rainfall. This rainy season sets in progressively from south to north. It starts at the southern tip of India at the end of May, reaches the neighbourhoods of Bombay and Calcutta in the second week of June, Delhi in the second half of June, and west Pakistan not until the first half of July. The amount of rain received in this period is however very

390

variable between areas; from 75 in upwards on the west coast of India and in Assam, but less than 5 in in parts of west Pakistan. Considerable areas of India and east Pakistan also receive rain at other periods. There is enough rain in the north in January–February to support the winter crops of wheat and barley. In March to June thunderstorms give heavy rain in Bengal and Assam, whilst Kerala in the south-west also gets useful rain at this time. In November–December the south-east eactually gets more rain than during the monsoon, and Kerala again benefits. The sub-continent presents the paradox of containing the station with the highest mean annual rainfall in the world —Cherrapunji in Assam with 428 in—and also dry semi-deserts in west Pakistan and the adjoining Indian state of Rajasthan. Variability of rainfall is a most important factor, especially where the mean total is low; in parts of west Pakistan and the Deccan, variability is more than 100% of the mean. Years of drought account for the only too abundant history of famine in India, whilst years of flood also cause very considerable loss of agricultural production.

Temperatures vary greatly, both geographically and seasonally. North and central India in January have temperatures comparable to Europe in July, though with a greater daily range, but in the pre-monsoon months daily temperatures of over 100°F are reached over large areas. The south is warmer than the north in the cold season, but cooler in the hot. Frosts can occur, even in the plains, as far south as a line drawn through Bihar and Madhya Pradesh, and may be heavy in Kashmir and the north of the Punjab.

Altitude is an important factor which has encouraged the growing of such plantation crops as tea and coffee in the Western Ghats, the Nilgiri Hills and some other regions in the south. In the north, agriculture is carried up the Himalayas to an extreme altitude of about 15,000 ft under increasingly temperate conditions. Kashmir, by reason of both altitude and latitude, is a region of purely temperate agriculture.

There are four main soil types in India (Randhawa 1958):

1. Alluvial soils. These cover huge areas, said to amount to 300,000 square miles—a quarter of the land area—in India alone. They are of the greatest importance to agriculture because of the immense concentration of population in the valleys of the Indus and Ganges systems in both India and Pakistan.
2. Black clays, locally known as 'regur' or 'black cotton soil', which occupy much of the western half of the Indian peninsula. Moderately good for agriculture, they are difficult to work because of their stickiness when wet and hardness when dry.
3. Red soils overlying metamorphic rocks. These have their greatest extent on the eastern side of the peninsula. They are generally poor in plant nutrients but respond well to fertilisers.
4. 'Laterites' in the sense in which this word originally received its classical

FARMING SYSTEMS OF THE WORLD

Desert soils occupy much of Rajasthan and west Pakistan. In general, the
soils of the sub-continent, although often of good original fertility, have been
much impoverished by cultivation and also by soil erosion which is still too
frequently allowed to occur in the plains whilst the attention of soil conserva-
tionists has been concentrated on the hill areas. India and Pakistan have,
during this century, lost to cultivation millions of acres of irrigated land which
have become unable to support cropping because of salinity and waterlogging
due to inadequate drainage, and which it is difficult and costly to reclaim (see
p. 74).

It would not be easy to give a concise account of the vegetation of the sub-
continent, which ranges from the tropical evergreen forest of the Malabar
coast to the desert scrub of Baluchistan or again to the montane coniferous
forests of the Himalayas. In fact, the natural vegetation has disappeared over
most of the areas which are used for agriculture and grazing; and those parts
of India and Pakistan which are not so exploited are mostly either topographically unsuitable or have too little rain. Forestry, however, remains a
most important industry in India, where approximately 275,000 square miles
or 23 % of the land area are under forest, including 20,000 square miles of the
valuable teak forests. *Casuarina* spp., a tree which has nitrogen-fixing root
nodules and makes good growth on poor soils, is very widely planted for
firewood production.

Amongst the wild fauna, we may note the population of monkeys which
take a regular toll of crops in many parts of India since most Indians are, for
religious reasons, unwilling to destroy them. India and Pakistan are still subject to periodic invasion by the desert locust.

Two facts stand out regarding the effect of environment on agriculture in
the sub-continent. One is the extreme heat and aridity of the dry season, which
places so much of the region, in spite of quite a high seasonal rainfall, in the
agricultural category of the 'seasonally dry tropics' (p. 42), typified by the
absence of perennial crops and by the importance of millets amongst the
cereals. The other is the part played by the rivers in agricultural production,
in two quite different ways. There are some areas, such as the plains of West
Bengal and more particularly of east Pakistan, where the rivers Ganges and
Brahmaputra debouch to the sea through such a network of channels that
during the monsoon season vast areas of country are under water by natural
flooding without deliberate irrigation; though dams or projected dams a little
higher up some of these rivers offer prospects of gradually increasing control
over the floods and are already providing useful hydro-electric power. In
contrast to this are areas such as in the west Punjab where rainfall is too low
for crop production without irrigation canals fed from the rivers, and, particularly in the basin of the Indus and its tributaries, farming is founded upon a

widespread network of such canals. India claims to be the most extensively irrigated country in the world. Out of 58 million acres, or 18% of the total sown area, irrigated in 1959, 36% was irrigated from Government canals, 5% from private canals, 20% from tanks (reservoirs), 29% from wells and 10% from other sources. West Pakistan consists essentially of a narrow strip of cultivated land along the River Indus, broadening out somewhat in the north-east between its tributaries, and flanked on both sides by lands which are fit only for grazing.

Agrarian Structure

It is difficult to generalise about agrarian structure because both India and Pakistan have constitutions of a federal type in which considerable powers reside in regional governments. In India, agriculture is a subject allotted to the state governments, which have thus been able to follow sometimes divergent policies. There are, however, many points of uniformity. In India since independence, the central government has nationalised the land held by land-owners under British rule, and the farmers have become tenants of the state with provision for them to purchase ownership. About 80% of the farmed land is however held by owner-occupiers. A ceiling has been placed by most state governments on the amount of land which can be owned by one person, which varies from state to state and allows larger holdings for plantation crops. The states have in varying degree undertaken programmes of consolidation of fragmented holdings. In east Pakistan, some 60% of farmers are owner-occupiers, the rest being share-croppers who pay half their rice crop to the landlord. In both countries, many farmers lease in or out small plots in order to get a better-adjusted holding or one which comprises land of different types.

The outstanding fact about land tenure in the sub-continent is the smallness of the holdings, the average farm size in most areas being lower than in most tropical countries. The average size of holding in India is 7·5 acres, with much smaller farms in the south (Kerala 2·4 acres, Madras 4·5) than in the north (Punjab 11·8 acres). Owing to the influence on the average of a few large farms, the common size of farm is however lower than these figures: in Kerala 0·5–1·0 acre, in much of Mysore 1·5 acres, in West Bengal about 2·0 acres. 69% of the national labour force was recorded as engaged in agriculture in 1961. In east Pakistan, the mode for farm size is 3·1 acres, though a few farmers have up to 30 or 50 acres. In west Pakistan, mean farm sizes vary from 0·9 acres in Rawalpindi district to 3·82 in Shahpur.

In 1952, India instituted a massive programme of community development in conjunction with the National Extension Service. This was organised through blocks of approximately 100 villages each, which have gradually been extended until they now cover the whole country. Each block is under the charge of a Block Development Officer, who is assisted on the agricultural side by an agricultural extension officer. Farmers are contacted primarily through the 'village level workers' who are multi-purpose extension agents, including agriculture amongst their responsibilities, and each covering from five to ten villages. In some blocks very good progress has been made in the use of fertilisers and better seed, control of plant and animal diseases, and the increase of wells and irrigation through loans; but the rate of improvement is very uneven. A more recent development is the Intensive Agricultural District Programme in which many simultaneous improvements are attempted in limited areas in what is popularly known as a 'package programme'.

In Pakistan, the Food and Agriculture Commission (1960) made recommendations, which have since been implemented, for the setting up in each of the two provinces of an Agricultural Development Corporation. Each corporation was to have a supply wing to produce or procure and distribute seeds, fertilisers, plant protection materials and farm implements, and a field wing to develop project areas. A major project, in which F.A.O. is assisting, is the Ganges–Kobadak project in east Pakistan, which involves the provision of controlled irrigation for 350,000 acres, of which 40,000 were enjoying perennial irrigation by 1965.

The main agricultural problem in both countries is to provide enough food for the requirements of the ever-growing populations. In India, grain production is not quite sufficient to meet these needs, and, although production has been rising slowly, the simultaneous increase of population means that there is still an annual deficiency, small as a percentage of the whole but nevertheless amounting to several million tons. This gap has been closed by imports, largely of wheat from the United States under special title. Famines, such as that in Bihar in 1967, have not involved catastrophic loss of life since the great Bengal famine of 1943, but it has been necessary in recent years to adopt consumer rationing of grain in all the major cities. In favourable years, consumers have been able to buy more of the rationed grains (rice and wheat) than the minimum guaranteed by the ration, and other foodstuffs can of course be purchased as available. The most deficient area is normally West Bengal including Calcutta, where sugar has also been rationed for some time, and it was estimated in 1965 that the average daily dietary intake was only about 1,800 kcal, of which the ration provided 1,200. East Pakistan is more or less self-supporting in food on a low dietary intake of about 2,000 kcal with much protein and Vitamin A deficiency. West Pakistan at the time of independence was exporting wheat to India and was thought to be a food surplus area, but it later became a deficit area which needed annual imports of food.

To meet these problems, India has laid great stress on increased grain production in the series of five-year plans which have been adopted by the Government. Many dams have been or are being constructed to extend irrigation, a notable example being the Damodar Valley project, where four dams had been built by 1958. Some improvement of rice production has been gained by the inculcation of the so-called 'Japanese method', the main features of which are a better and more regular spacing of plants and the use of fertiliser.

A very large investment has been made in the construction of fertiliser factories. By 1964, 15 factories were producing nitrogenous fertilisers and 27 were making phosphatic fertilisers, with another 15 and 11 factories in these respective categories projected (*Fertiliser Statistics*, 1963–64). Annual production in 1965/66, according to F.A.O. statistics, reached 270,400 short tons of nitrogen (as the element) and 131,000 tons of P_2O_5. Fertiliser sales to farmers are subsidised at varying rates in different states. In Pakistan there has also been construction of fertiliser factories, and production by 1965/66 had reached 93,400 tons of nitrogen and 1,200 tons of P_2O_5. Another notable feature of Pakistan's food drive has been the development of hybrid maize production in west Pakistan.

The common method of cultivation in both India and Pakistan is by the plough drawn by oxen or buffaloes. There is nevertheless a considerable acreage cultivated with the hand hoe by those who do not own and cannot borrow draught animals, The number of tractors, though still relatively small, increases steadily; in 1961 there were 34,000 agricultural tractors in India, almost four times the number of ten years earlier.

The importance of the plantation sector in the agriculture of the sub-continent lies mainly in the fact that tea is India's most valuable agricultural export. Most of the tea is grown in Assam, where the estates are still mainly in expatriate ownership; there is a smaller acreage in the hills of south India. A less important acreage of tea falls within the boundaries of east Pakistan. There is an important production of high-quality arabica coffee in the hills of western peninsular India. Rubber is mostly grown in Kerala; there are some large estates but many of the growers would come into the category of smallholders by the usual standards of this industry. Sugar-cane, which in many countries is a plantation crop, is in India almost entirely grown by smallholders, many of whom plant only 2–3 acres though some have up to ten.

In the sphere of marketing, the Indian Government has given much encouragement to co-operative marketing societies, of which there were 3,033 in 1960. The Government has also favoured the co-operative movement by entrusting part of the distribution of fertilisers to the societies, and by often selecting co-operative shops as 'fair price shops' in areas of food scarcity. With the prevailing scarcity of food, the prices of the main food grains have at times been rather rigidly controlled by the Government in recent years; the price levels chosen, and their effect upon production, have been the subject of much argument and there are frequent reports of black marketing. With the object of improving this situation, and also of ensuring better storage conditions, the Government set up the Food Corporation of India which began operations in 1965 in south India but is scheduled to extend gradually to the whole country. The Corporation is empowered to buy grain from farmers or from state governments at prices determined by the Agricultural Prices Commission. This grain may be stored or distributed at certain points to state governments, who in turn distribute it through selected wholesalers

or other channels. The amount of grain storage accommodation in both India
and Pakistan is very inadequate in proportion to their need for reserve stocks,
but is being steadily increased by the Governments of both countries.

Agricultural education in India is carried out at a very large number of
specialised institutions. In 1961, there were in the country 53 agricultural and
17 veterinary colleges, and under present policy a number of these colleges
are being progressively up-graded to full-scale agricultural universities. A
useful expansion is also taking place in training centres offering short courses
for young farmers. In Pakistan, there is an agricultural university in each
wing of the country, at Mymensingh for the east and Lyallpur for the west.
There are also in both countries some excellent research and experiment
stations which are not connected with undergraduate teaching. The Indian
Agricultural Research Institute at Delhi is a very large station with an international reputation, as has also the Sugar Cane Breeding Institute at Coimbatore. Specialised stations exist in India for research on rice, tea, coffee, coconuts and other crops. In Pakistan, an interesting station is the Academy for
Rural Development at Comilla which studies the general problems arising
in extension work.

Crops and Livestock

The most important part of farming in India and Pakistan is grain production. On this basis, the sub-continent may be divided into four agricultural
regions, as shown in Fig. 2.12.1. These are:

1. The rice region, extending from Assam through east Pakistan to include
   a very large part of north-eastern and south-eastern India, with another
   strip along the western coast.
2. The wheat region, occupying most of north-west India and the cultivated
   parts of west Pakistan.
3. The millet region, comprising mainly the dry Deccan plateau in the
   centre of the Indian peninsula.
4. The temperate Himalayan region of Kashmir and some adjoining areas.
   Here potatoes (*Solanum tuberosum*) are as important as the cereal crops,
   which are mainly maize and rice, and tree fruits form a large part of
   the agricultural production.

Table 2.12.1 shows the production of the main crops in the two countries,
and emphasises the overwhelming importance of rice. India produces about
one-fifth of the world's rice consumption, but still has not enough to meet her
internal demand. In east Pakistan, rice occupies something like 70% of the
cropped area. Only a very small part of the total area consists of upland rice,
most of it being swamp rice which is grown under conditions of natural
flooding; lack of control over the depth of water is a main cause of fluctuating
yields. Great efforts are being made to increase yields, both by using improved

varieties, of which some recently developed in the Philippines by the International Rice Research Institute are very promising, and by the use of fertilisers, mainly sulphate of ammonia. The increasing use of fertiliser demands, however, the breeding of stiffer-strawed varieties if lodging is to be avoided. In general an increase of about 50 % in rice yields is obtainable from an optimal use of fertilisers (Comhaire 1965). Wheat is grown in India and west

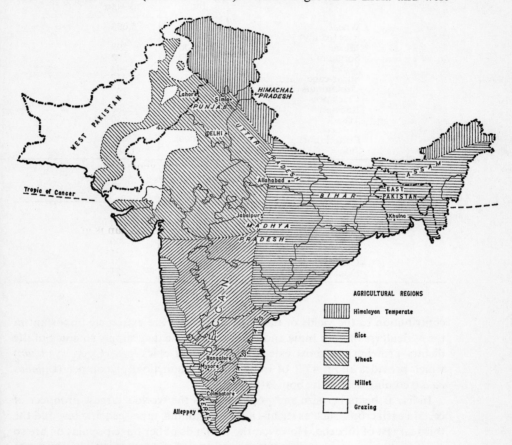

Fig. 2.12.1 Agricultural Regions of India and Pakistan

Pakistan as a winter crop, and is very important in the Punjab and around Delhi. Barley is grown to a lesser extent, mainly in Uttar Pradesh. In the dry millet region, sorghum is the most popular cereal where the rainfall is good enough, but in the driest parts there are very large acreages of both bulrush millet (*Pennisetum typhoideum*) and finger millet (*Eleusine coracana*). Maize, mainly of flint varieties, is widely grown as both a food and fodder crop. Plant breeding (the introduction of Mexican wheats, and the use of hybrid maize and synthetic varieties of maize and sorghum) is beginning to make a striking

Table 2.12.1

Crop Production and Livestock Numbers in India and Pakistan

Crop production in thousands of short tons per annum:		
	India	Pakistan
Wheat	13,547	5,098
Rice (paddy)	50,619	19,616
Maize	5,106	599
Sorghum	8,259	302
Millet	7,169	408
Sugar-cane	129,638	33,417
Groundnuts (in shell)	4,433	36
Chick peas	6,377	782
Rape-seed	1,616	338
Tea	402	30
Tobacco	408	121
Cotton (lint)	1,099	460
Jute	887	1,300

Livestock numbers, in thousands:		
Cattle	188,700	35,200
Buffaloes	55,460	8,600
Sheep	43,540	10,800
Goats	65,912	11,600
Pigs	5,451	94
Camels	1,003	601
Donkeys	1,170	925

Source: F.A.O. Production Yearbook 1966.

contribution to the yields of these cereals. Pulses are extremely important in the agriculture of both India and Pakistan because they supply so much of the dietary protein. The most extensively grown is chick pea (*Cicer arietinum*) which provides about 45% of India's pulse production, pigeon pea (*Cajanus cajan*) coming next with about 20%.

India, as befits her size and population, is the world's largest producer of certain cash crops, such as sugar-cane, groundnuts, rape-seed and tea, and the third largest of tobacco. However, the demands of her huge population are so great that she is not a major exporter of any of these except tea. Pakistan is the second largest producer of rape-seed; this crop and mustard seed are produced as oil-seeds grown in the autumn in the northern parts of the sub-continent to provide edible oil. In recent years, the Indian Government has encouraged exports of centrifugal sugar to earn much-needed foreign exchange even though sugar was at the same time being rationed in some Indian cities. Much of the sugar-cane production in both countries is utilised in the manufacture of the less refined *gur* sugar, which has been popular with producers because its price has not been controlled in the same way as centrifugal sugar. In tobacco, India is an important producer of flue-cured as well as other leaf although both here and in Pakistan most of the crop is for home consumption rather than export. The main area of tea production is in Assam, where many British-owned companies still have estates; the seasonality of temperature and rainfall in this area outside the tropics produces a seasonal flush of leaf rather than the more even production found nearer the Equator. Exports of jute, a crop which is grown on seasonally flooded land, are vital to the economy of east Pakistan, and almost equal quantities are grown in West Bengal in India, where most of the processing mills were situated before partition. There is a very wide range of minor cash crops. Some, like the areca palm and the betel vine, are grown to meet purely internal demands. India is one of the most important producers of certain spice crops such as pepper, cardamoms, chillies and ginger. Cotton is grown to supply the large Indian textile industry, and is particularly important in the economy of west Pakistan: kenaf (*Hibiscus cannabinus*) is also an important fibre in India. Temperate and sub-tropical fruits occupy large acreages in the northern part of west Pakistan and Kashmir, where there is production in the fresh or dried form of apples, almonds, cherries, apricots, peaches, pears and plums.

Taking cattle and buffaloes together, India contains nearly a quarter of the world's bovine population. These animals are not kept for beef, but for milk production and use as draught animals for cultivation and still to a large extent to pull bullock carts. Because the Hindu religion forbids taking the life of cattle, vast amounts of grazing and fodder are used to support senile animals which produce little or nothing for human consumption. In many parts of India, more buffaloes than cows are kept as dairy animals. The buffalo generally gives a higher milk yield with a greater fat content but as it is a larger animal its individual maintenance requirement is greater, and the balance of economic advantage between the two species is not clear. The Indian experiment stations have bred some of the best zebu dairy cattle in the world, such as the famous Sahiwal which has been widely exported to tropical countries. The Tharparkar is, under good conditions probably the next best milk yielder; but at the level of management of the ordinary Indian farmer, the Red Sindhi which originates in Pakistan often out-yields it. There are also famous breeds of large and strong draught cattle, such as the Ongole of south India. However the conformation and productivity of the average village cow in India is often deplorable, and at this level there is often little differentiation into recognisable breeds. Temperate breeds of cattle, especially Friesians and Jerseys, have been used pure-bred and in crosses especially at higher altitudes and where the disease risk can be limited. There are several good dairy breeds of buffaloes, amongst which the Murrah is perhaps the best known. The well-known dairy colony at Aarey which supplies milk to the city of Bombay uses entirely buffaloes, 16,000 animals being milked daily, Even on this scale, it is impossible to make the saving in labour that might be expected from mechanisation, because the teats of zebu cows are generally too small and those of buffaloes too uneven for machine milking. Both cows' and buffaloes' milk is utilised in the production of *ghee* or clarified butter, which is now by reason of its price mainly a rich man's food, the poorer classes using vegetable oils for cooking.

Sheep rearing in India is in the hands of a pastoral caste; wool is an important minor export. Goats are more numerous and are kept both in herds by nomadic shepherds and in small numbers by individual farmers. The goat is the major supplier of meat in Indian diets. A few types of goat in northern India, such as the Jamunapari, yield useful amounts of milk; the Himalayan goat yields the valuable 'cashmere' fleece; the diminutive black goats of Bengal have a propensity for twinning which is outstanding amongst tropical goats. In west Pakistan, sheep and goats are particularly important as a means of production from the arid lands which are fit only for a certain amount of grazing. Sheep breeds include the Karakul which is indigenous to this region of Asia and produces the valuable 'astrakhan' type of wool. A feature of the west Pakistan scene used to be the nomadic graziers or *powindahs* who annually brought in some 275,000 sheep and goats and 50,000 camels from Afghanistan for winter grazing, but the border was closed against them in 1966. The numbers of donkeys and camels in both countries are large enough to make considerable demands on grazing and fodder.

Pigs are unimportant in Pakistan, with its mainly Moslem population. In India there are more considerable numbers. The local type of Indian pig, often seen as a scavenger in the villages or herded in droves in woodlands, is one of the smallest and most poorly developed for meat conformation in the world, but improved breeds have been widely introduced. Poultry have not had a very significant place in Indian farming in the past, and this seemed natural enough since they compete with human beings in the consumption of food grains, which were already inadequate in supply. However in recent years it has been felt that the extreme shortage of animal protein in Indian diets justified an extension of egg production, particularly since eggs are one of the animal foods which are most acceptable to Indian consumers. In consequence, a considerable expansion of egg production has been encouraged both by the national extension service and other development agencies, and large numbers of pure-bred birds of such breeds as White Leghorn and Rhode Island Red have been distributed. About 12 eggs per person per annum were available in India in 1965, which was twice the amount of 1961. Amongst minor animal products, silkworm rearing is of some importance particularly in Mysore and Madras states, 1,473 tons of silk being produced in 1960. Lac, another important insect product, is produced mainly in Bihar and Madhya Pradesh.

The animal disease position is dominated by the fact that rinderpest is still enzootic in India and Pakistan, though kept within bounds by protective inoculation where necessary. A great deal of effort is also put into encouraging the inoculation of poultry against Ranikhet (Newcastle) disease.

An Indian Transect (*Table 2.12.3*)

The agriculture of the sub-continent presents such an enormous diversity due to differences of latitude, altitude and rainfall that it is impossible to include examples of every type of farming within a single transect. However a transect starting at Alleppey on the Kerala coast in the extreme south of India and running almost due north through the centre of the peninsula to pass ultimately through Delhi and end in the Himalayan region will exhibit most of the more important systems of agriculture which are practised.

KERALA

The state of Kerala in the extreme south-west of India, lying on both sides of latitude 10°N and with a general rainfall of about 100 in annually and only three rather dry months in December–February, exhibits the most extreme type of 'tropical' agriculture found in the sub-continent.

In a coastal strip up to some 10 miles wide, the land is flat and much intersected by both sea creeks and artificial canals, so that agricultural produce is often transported by boat. The soils are mostly sandy and the chief crop, indeed almost the sole one over large areas, is the coconut palm. Rice is grown wherever patches of less permeable soil occur, and this includes considerable polders which have been reclaimed from the creeks but which present special problems of salinity and seasonal inundation, the water being pumped out of them by electrically driven pumps provided by the Government and paid for by a cess on farmers. Inland of this coastal strip, the ground rises and the soils are red latosols. On this higher ground, rice is more widely grown in the valleys, where two crops a year can be taken, and root crops include tapioca, elephant-foot yam (*Amorphophallus* spp.) and *Colocasia* spp. There is a larger range of tropical cash crops including rubber, which is grown both on a plantation and smallholding scale, areca palms, cashew-nuts, pineapples, sesame and pepper.

Farm sizes throughout Kerala are amongst the smallest in India, holdings of between a half and one acre being very common. Because of this small size of farm, fewer farmers keep cattle than in other parts of India; much cultivation therefore has to be done with the hand hoe, and hand-carts are often used for transport of farm produce rather than the more usual bullock-cart. The farmhouses are not grouped together in villages to the extent usual in India, but are largely situated singly on the owners' holdings. There are experimental stations in the state for research on coconut and areca palms.

WESTERN MADRAS STATE

Our transect must now leap the mountain chain of the western Ghats, whose upper slopes are too steep for cultivation and are left under forest, to emerge in the western part of Madras state in the region of Coimbatore. Here the altitude is still about 1,400 ft but on this side of the mountains the rainfall is much lower, amounting to 27 in annually at Coimbatore. Soils are largely colluvial. Under these conditions, crops to be grown must be those which can tolerate this rather moderate rainfall, except in so far as irrigation can be provided from tube-wells; and for this purpose much of the water available, with a total solids content of about 130 parts per million, is regarded as more saline than is ideal. Sorghum is well suited to the conditions, and is the main food crop of the countryside, though in the towns people prefer to eat rice which is brought in from other areas.

Amongst cash crops, some cotton is grown though this is not one of the most important areas of India for this crop. Programmes of frequent spraying with rocker pumps which can be moved about the fields are recommended to control cotton pests and diseases. Current experiments with Sea Island cotton at Coimbatore are very promising, and it may be possible to extend the area of this crop. A certain amount of sugar-cane is grown, the main disease being red rot (*Colletotrichum*); mosaic disease is common but not serious. A feature of the landscape, though more common in the eastern than western part of Madras state, is the presence on the farms of singly standing Borassus palms; these are tapped, formerly for toddy manufacture, but since the enforcement of prohibition, mainly for the production of a crude sugar. Rice and ground-nuts become more important as one moves from the western to the eastern part of the state. The region carries the usual heavy Indian population of cattle for milk and draught purposes, the local Kangyam breed being specialised for the latter.

Coimbatore itself is notable for the complex of agricultural research institutions which have been set up here by the Government and the University of Madras. Besides an agricultural college, experimental farm, and botanic gardens there are the Sugar Cane Breeding Institute, an important head-quarters of the Botanical Survey of India, and a large Home Science College which is much concerned with food utilisation. The Sugar Cane Breeding Institute has produced varieties which are used in many parts of the world as well as all over India; but since its own work results in the constant supersession of one variety by a better one, there is little point in detailing those which are currently most planted.

THE NILGIRI HILLS

Only about 30 miles north of Coimbatore lie the Nilgiri Hills, which provide one of the most extensive areas of high-altitude environment for farming in south India. Agricultural development is chiefly centred round the towns of Coonoor and Ootacamund at altitudes of between 6,000 and 7,000 ft. The rainfall is about 65 in a year. The topography is rugged and the steepest slopes and summits have to be left under forest; though tea is often planted on steep slopes, there is constant risk of soil erosion, to combat which a certain amount of terracing has been done.

The tea and coffee estates, which are the main feature of this area, were originally pioneered by British companies or individuals, but many are now owned by Indians, and expatriate companies often employ Indian managerial staff. Tea is plucked at all times of year, but the heaviest yields are in the spring and autumn; it is pruned every 5 to 7 years according to altitude. Cuttings are now largely used for propagation. The main shade trees used are dadap (*Erythrina lithosperma*), *Acacia* and *Grevillea*. The coffee is *arabica*, producing a high-quality bean, and strains have been found which are resistant to the four physiological races of *Hemileia* rust which occur in India. Pruning consists of successive topping at $2\frac{1}{2}$ and $4\frac{1}{2}$ ft. The same shade trees are used as for tea, with the addition of *Gliricidia*. Wattle and cinchona are also grown in these hills. Amongst temperate fruits, a local experiment station has found that plums (Japanese varieties) and pears (varieties from north India and Israel) do best; apples can also be grown. Potatoes (*Solanum*) are the most important food crop. Livestock are not of much significance, but dairy cattle of European breeds or crosses do well. At Coonoor is the headquarters of the United Planters' Association of South India, an important private organisation which conducts both research and extension work.

MYSORE AND THE DECCAN

Some 100 miles further to the north, our transect passes close to Bangalore in latitude 12°58′N. This brings us to the Deccan, an area through which the transect runs for about another 700 miles to the northward; as our transect happens to enter it in Mysore state, a general description of the agriculture of that state will apply very largely to the whole of the Deccan region. The Deccan is one of the largest fairly homogeneous ecological areas of India. It is essentially the central plateau of the peninsular part of India, separated from the sea on both sides by ranges of mountains or strips of lowland. The altitude of the plateau itself varies, being for example 1,778 ft at Hyderabad but rising to 3,021 ft at Bangalore, which does not however prevent the temperature rising to over 100°F in the pre-monsoon heats. These high temperatures, coupled with the rather low rainfall (29 to 34 in over large areas) give the region a distinctive character of aridity; in agricultural terms, it is the great millet-growing zone of India. The soils are mostly red, 'laterised' to a greater or less degree; but there are also areas of black clay derived from the Deccan Trap.

All these factors are very well exemplified in the part of Mysore state lying around Bangalore. Here finger millet (*Eleusine coracana*) is the most widely grown food crop. Superior varieties are grown by registered growers for seed, which is distributed through the extension service. The average district yield of unfertilised millet is only 650 lb per acre, though many farmers get about 900 lb. These yields can be doubled on the red soils by optimal use of nitrogenous and phosphatic fertilisers, but few farmers use such dressings as yet. In the general scarcity of animal fodder millet straw is readily saleable and may fetch as much as a quarter of the value of the grain. Swamp rice is grown

403

on all the valley land which is capable of being seasonally flooded. In some of
the development blocks, all the rice now receives some fertiliser. Since rice
fetches a higher price than millet, most of the farmers sell their rice and eat
millet. Other crops grown in the area include potatoes (*Solanum*), sorghum,
groundnuts; mulberries to feed silkworms; chillies, onions, garlic, gourds and
okra (*Hibiscus esculentus*). Grapes, which command a ready sale in the towns,
are increasingly grown and in some areas are reckoned the most profitable
cash crop. Many of these horticultural crops are grown in small plots surrounded by earth bunds (ridges) which are cultivated by hand and irrigated
from open wells about 30 ft deep. Some of these wells have been constructed
by farmers with Government loans which are available for the purpose. Much
of the water is still raised with the 'picottah', a transverse beam with counter-
poise which lifts a bucket as two or three men walk on it with a see-saw effect;
but some of the richer farmers use electric pumps.

There are perhaps almost equal numbers of bullocks for draught purposes
and of buffalo cows for milk, the male buffalo calves being sold off. Some pigs
are seen as scavengers in the villages. Many farmers keep two or three unimproved hens. Small plantations of *Casuarina* are widely grown for fuel, and
this is not an area where cow-dung is generally used for the purpose. Although
bunds are seen on some of the cultivated land, too much soil erosion is allowed
to occur and land is uselessly wasted in systems of gullies, whilst sheet erosion
undoubtedly impoverishes the soil. The sale of crops takes place partly at
local markets, which are held weekly in the larger villages, but grapes are sold
on commission by agents in the towns.

Mysore is the Indian state with the largest acreage of coffee, a crop which is
mainly grown in the more mountainous areas on the western edge of the true
Deccan. More than half the acreage consists of *arabica* coffee producing a
very high quality bean; the economic life of the crop here is reckoned to be
some 50 years, which is longer than in some tropical countries. But the acreage
of *robusta* coffee tends to increase and now constitutes some two-fifths of the
whole in this state. At Balehonnur is the main Indian coffee research station.
Another large research establishment of the Deccan is the National Food
Research Institute near Mysore City, one of the most important institutions
in the tropics for research on the properties, processing and storage of foodstuffs.

MADHYA PRADESH AND UTTAR PRADESH

The Deccan is often said to end at the Vindhya range of mountains in the
state of Madhya Pradesh. In part of this state, therefore, and in Uttar Pradesh
to the north, we enter a country of higher rainfall with typical annual means
of 55 in at Jabalpur (Madhya Pradesh) and 42 in at Allahabad (Uttar Pradesh).
The Tropic of Cancer also passes through Madhya Pradesh, so that the agricultural contrast between the two states is partly that between a tropical and
sub-tropical area, and can be brought out best by considering some figures for

Table 2.12.2

Crop Production Percentages in Madhya Pradesh and Uttar Pradesh

Crop	Production, as % of all-India production	
	Madhya Pradesh	Uttar Pradesh
Rice	10·3	8·9
Sorghum	14·3	6·6
Maize	11·3	19·9
Wheat	20·2	33·2
Barley	5·6	55·8
Pulses	15·3	27·1

Source: Indian Agriculture in Brief, 6th edition. Ministry of Food and Agriculture, New Delhi, 1963.

the two states in conjunction, as is done in Table 2.12.2. From this table it will be seen that, in passing from the tropical state of Madhya Pradesh to the sub-tropical one of Uttar Pradesh there is a decline in the importance of rice and sorghum, and an increase in the importance of maize, wheat, barley and pulses (mainly chick pea). But there are many other contrasts between the two states. Most of Madhya Pradesh lies at between 600 and 1,600 ft altitude, whereas Uttar Pradesh is one of the states of the low-lying Ganges valley and lies mostly below 500 ft altitude. Madhya Pradesh has a population density of 189 people to the square mile and a mean farm size of 13·9 acres; Uttar Pradesh has the very high population density of 649 per square mile and a mean farm size of only 5·3 acres. Madhya Pradesh has mixed red and black soils, Uttar Pradesh has the alluvial soil of the Ganges valley.

For these reasons, Uttar Pradesh has the greater agricultural importance and therefore deserves consideration in more detail. It constitutes a type of the very densely populated and agriculturally productive Gangetic plain which is so important to India. The alluvial soils vary considerably in texture from clays to sands. This is one of the areas in which double-cropping (i.e. the taking of two annual crops off the same land in a year, a practice which is carried out on about 13% of all India's cultivated land) is most important. It is easy enough to grow a summer crop on the monsoon rains; the amount of rainfall available for a winter crop is more marginal, and double-cropping is largely carried out on land to which supplementary irrigation can be applied from canals or wells. Common combinations in double-cropping are: in the east of the state, a summer crop of rice and a winter crop of barley, and in the drier western part, a summer crop of sorghum and winter crop of wheat. There are also many other crops of importance. Bulrush millet is found to be a useful cereal on poor sandy soils in the drier western part of the state. Potatoes (Solanum) are a minor food crop. Much sugar-cane is grown on the better soils where it is often the chief cash crop; Uttar Pradesh provides 43% of India's total output of this crop. Other cash crops include linseed, rape and mustard seed (all grown as autumn or winter crops), mangoes, tobacco, coriander, and cotton at the dry western end of the state.

In animal husbandry, whilst the small size of farms severely limits the numbers of cattle that can be kept by individual farmers, fodder supplies are easier than in the Deccan because of the abundance of cereal straw and sugar-cane tops available and the greater ease of growing fodder crops. The state is the home of the Ponwar breed of cattle, excellent draught animals for speed and stamina but poor milkers, and the Kherigarh breed which provides fast bullocks for light draught.

DELHI AND THE PUNJAB

Our transect will pass nearly through Delhi, which lies on the boundary of Uttar Pradesh and the Punjab. From here a quite distinctive agricultural region stretches to the north and west through the Punjab (with which we may include for agricultural purposes the state of Hariana, which was separated from it in 1967). The most important crop, winter wheat, can be grown without irrigation around Delhi, where the rainfall averages 25 in annually. But rainfall declines sharply to the westward through the Punjab; at Lahore, which is just over the border in Pakistan, it is only 20 in, and in the western or Pakistan Punjab generally crop production is only possible with irrigation, which is provided by canals from the tributary rivers of the Indus and by tube-wells. In the Indian Punjab, the mean size of holdings is 11·8 acres, which is higher than in most of the country. Many of the people seem to be of more robust physique than further south, though whether this is due to racial differences, greater distance from the Equator or a better diet may be arguable. The lower temperatures in winter and longer hours of daylight in summer encourage the spending of more hours in agricultural work, which is made more possible by the larger size of holding.

Wheat is the main food crop. These north Indian wheats are classed as hard wheat of good quality, next after Canadian, though the grain is of a golden rather than red colour. Wheat provides the staple diet, in contradistinction to the rice of southern India, and is consumed mainly in the form of unleavened 'chapattis' and similar preparations. Barley, pulses (mainly chick pea and pigeon pea), and maize are the other main food crops. Cotton is the chief cash crop; others are sugar-cane, which is being grown on an increasing scale, and tobacco. Cereals are often interplanted with pulses; double-cropping may take the form of following wheat with sorghum, maize or cluster bean (*Cyamopsis psoralioides*).

The dry climate of this region seems to suit cattle, and relatively abundant supplies of fodder are available. A very well-known local breed is the Hariana which serves excellently the dual purpose of draught and milk. This area is also the home of the Murrah breed of buffaloes. Donkeys and horses are particularly numerous in this region, and some camels are used for draught purposes. Because of the general shortage of trees, cow-dung is very extensively used as a domestic fuel, and is sold in the markets in disc-shaped cakes for this purpose.

This region, so near the national capital of Delhi, is well served with agricultural research and extension services. The Indian Agricultural Research
Institute, which was transferred from Pusa to New Delhi in 1936, is one of the
best-equipped in Asia, with 1,250 acres of land available for experimental
work and a notable post-graduate research school of university status. The
development of improved wheat varieties with both rust and smut resistance
has been one of the most useful lines of work carried out.

THE HIMALAYAN REGION

Northwards from the Punjab, the last Indian state through which our
transect will pass is Himachal Pradesh in the Himalayan region. Altitudes
increase sharply as the great mountain range is approached, and a little farming and grazing is carried up to an extreme altitude of about 15,000 ft. Precipitation (as rain and snow) is much higher than in the plains. If we take Simla
as a representative climatic station, it stands at an altitude of 7,224 ft in latitude 31°N and has a mean annual precipitation of 62 in and mean daily
maximum and minimum temperatures for the year of 61°F and 50°F respectively. Agriculture in this region is more of a temperate than tropical character. The conditions prevailing in Himachal Pradesh are also common to parts
of Kashmir, and to a lesser extent to the eastern Indian Himalayas where,
however, rainfall is much higher. The main food crops are potatoes, maize and
rice. Cash crops consist very largely of tree fruits. Goats and sheep replace
cattle as the principal domestic animals, and there is a production of good
quality wool. Bee-keeping is quite extensively carried on.

Table 2.12.3 provides meteorological and other data for some stations which
are close to the line of this transect.

SAMPLE VILLAGES

The very small average size of farm in India and Pakistan, with the usual
concomitant of a high coefficient of variation in both the size and the nature
of farms, make it more appropriate for this region to take whole villages
rather than single farms as fair samples of farming practice. Two villages have
been selected for the purpose: one in the rather dry area of southern Mysore
state in India, and one in the seasonally very wet area of east Pakistan. These
illustrate respectively what we have described as the millet and the rice regions
of the sub-continent.

(a) *The Mysore millet region.* The hamlet to be described lies in the south-
east of Mysore state, near enough to Bangalore for that city to be its main
marketing centre. It has a population of between 150 and 200, and in this area
from two to four such hamlets make up a 'village' in the local government
sense. The main dry land crop is finger millet, but rice is grown in the valley
land, and most farmers have a plot of each type of land. The average farm
size is 1·5 acres, but this conceals great diversity in size between holdings. The
richest man is the headman of the village who has a farm of 10 acres on which

*   407

Table 2.12.3

An Indian Transect

Place	Region	Latitude	Altitude (ft above sea-level)	Rainfall (mean inches per year)	Relative Humidity (mean % in afternoon hours)	Temperature (mean daily maximum, °F)	Soil Type	Vegetation	Chief Food Crops	Chief Cash Products
Cochin	Kerala	9°58'N	10	115·3	77	88	Coastal alluvium	Tropical evergreen forest	Rice	Coconuts
Ootacamund	Nilgiri Hills	11°24'N	7,364	54·9	73	66	Red latosols	Evergreen hill forest	Potatoes (solanum)	Tea
Bangalore	Deccan	12°57'N	3,021	34·2	49	85	Red latosols	Dry thorny scrub	Finger millet	Rice, grapes
Jabalpur	Madhya Pradesh	23°10'N	1,327	55·4	44	88	Red latosols and yellow soils	Dry forest	Wheat, sorghum	Cotton
Allahabad	Uttar Pradesh	25°17'N	322	41·8	43	90	Alluvial	Dry forest	Wheat, barley	Sugar-cane
Delhi	Punjab	28°35'N	714	25·2	38	89	Alluvial	Xerophytic forest	Wheat, pulses	Cotton
Simla	Himalayan	31°06'N	7,224	62·1	58	61	Mountain soils	Oak and conifer forest	Potatoes, maize	Tree fruits

he employs some hired labour. He has a well-built homestead of tiled buildings round a courtyard, with the principal rooms lit by electric light. In his cowshed he has always at least two bullocks for ploughing and carting, and two buffalo cows in milk. He also has an electric pump for irrigating his dry land from a well, a pit for compost manure and a small flock of hens. Below this status are many social gradations, but most farmers are much poorer in land and amenities. Only some of them own cattle, and these are in half-starved condition near the end of the dry season when grazing has practically disappeared and they are eking out life on a little (rice and millet) straw. At the bottom end of the social scale are several landless labourers who work on the farms of others and live in one-roomed thatched houses without even a garden. The plight of these people is pitiable in old age, when they are usually dependent on gifts or remittances from grown-up sons. In recent years the grape crop has become the most important source of income to many of the farmers, as it sells well in the town and provides a high return per unit of area. The vines are trained up a trellis on plots of a small fraction of an acre, and are irrigated from wells with water raised by 'picottah' or bucket.

(b) *The east Pakistan rice region.* The sample village is near Khulna, just inside the tropic. It consists of about 15 households. The average size of farm is 3 acres, but here again there is considerable variation though no farmer has as much as 10 acres. The houses are closely grouped on a small piece of rising land where they are surrounded by small gardens in which are planted bananas, coconut palms, a few date palms and various vegetables and spices. The whole of the rest of the village lands form part of a plain which is completely inundated during the monsoon season. Rice is grown at this season and is almost the sole crop, though a little jute is also grown in the district. This is quite typical of east Pakistan, in which, as a whole, rice occupies about 70% of the cropped area, jute 8%, pulses 6%, and no other crop as much as 2%. During the dry part of the year, the land lies parched and uncultivated and no irrigation is practised although the area is intersected with small canals and backwaters fed from a neighbouring large river. However the next village, whose land abuts on this river, is able to take a second rice crop on the wetter land along its bank.

Purdah is still fairly strictly observed in the Moslem households of east Pakistan, and most of the village women do no field work. Livestock are comparatively rare, and little milk is drunk. Only one farmer in this village owns a small herd of cattle, which are raised for draught work and beef, and there are no buffaloes. The diet is monotonously based on rice; the average daily intake is estimated at 2,000 kcal, only a little better than in West Bengal across the neighbouring Indian border, and there is much protein and Vitamin A deficiency. However, some protein is derived from the many small fish which are caught in the canals. A sample of evening meals in this village showed that only one family was eating rice alone; the others all had either some fish or pulse with their rice, whilst in the one rich household which owned

409

the cattle meat was being eaten. Even such a minimal degree of prosperity may
be due to the fact that this village is near a town where many members of the
village households had whole-time or part-time work.

Efficiency and Stability

The agriculture of India and Pakistan can hardly be called efficient when
both countries fail to produce enough food to feed their populations in spite
of the employment of, in India, some 70% of the national labour force in
agriculture. Yields of crops are low by any standards; to take two examples,
India has 42% of the world's groundnut acreage but supplies only 33% of
world output of this product, whilst tobacco yields per acre are only about
half of those in the leading countries.

These failings are, however, as much due to natural as to human difficulties.
The primary problem is that the enormous supply of solar energy falling on
the region cannot be fully used for plant growth because of shortage of water
during so much of the year. The classic answer to this problem is irrigation, in
which both countries have made immense efforts and are still steadily expanding. But the land lost from cultivation through salinity and waterlogging provides a warning of the difficulties that may be incurred; and how far the continued increase of well irrigation can be pushed without seriously lowering the
water-table is an unknown factor. Subsidiary, though very valuable, contributions to increased yield can be made by raising the output of rain-fed crops
through plant breeding, greater use of fertilisers and improved husbandry.
Here, too, the correct policies are being adopted by both governments, but
progress in the field is pitifully slow.

Instability of output arises mainly through the variability of the monsoon
rains, leading to drought in some years and flooding in others. A subsidiary
factor is the incidence of cyclones which can also cause disastrous flooding.
The most practicable alleviation of these problems is again by better water
management, involving both the control of rivers by dams and more comprehensive drainage systems. These measures could have their greatest effect in
the Bengal region, involving both India and Pakistan, where two mighty rivers
spread uncontrolled inundation at certain seasons, and where also the frequency of cyclones is greatest. A comparatively small percentage increase in
crop yields would solve the food problems of both countries; but because of
the enormous human populations involved, the scale of the problem is greater
than in any other tropical region. India has reached a state of chronic food
scarcity in which many thoughtful Indians see birth-control as the only possible remedy; a policy of establishment of family planning clinics has been
officially adopted by the Indian Government.

References

Comhaire, M. 1965. *Rice Manuring*. Brussels.
Fertiliser Statistics. 1963–64. Fertiliser Association of India, New Delhi.
Food and Agriculture Commission, Report of, 1960. Govt. of Pakistan, Karachi.
Randhawa, M. S. 1958. *Agriculture and Animal Husbandry in India*. New Delhi.

General Reading

Chatterjee, S. P. 1949. *Bengal in Maps*. Calcutta.
Ghose, R. L. M., Ghatge, M. B. and Subrahmanyan, V. 1960. *Rice in India*. Revised edn. New Delhi.
Haswell, M. R. 1967. *Economics of Development in Village India*. London.
Indian Agriculture in Brief 1963. Delhi. 6th edn.
Indian Council of Agricultural Research, 1958. *Soils of India*. New Delhi.
Indian Council of Agricultural Research, 1966. *Handbook of Agriculture*. New Delhi.
Nafis Ahmad, 1958. *An Economic Geography of East Pakistan*. London.
Tayyeb. A. 1966. *Pakistan: A Political Geography*. London.
Thorner, D. 1964. *Agricultural Co-operatives in India: A Field Report*. London.
Whyte, R. O. 1967. *Milk Production in Developing Countries*. London.
Wilber, D. N. 1964. *Pakistan: Its People, Its Society, Its Culture*. New Haven, Conn.

South-East Asia

Introduction

In this chapter the four southernmost countries of Asia—Thailand (Siam), Malaysia, Indonesia and the Philippines —will be taken as providing a representative sample of agricultural conditions in south-east Asia, with particular examples taken from Malaya to illustrate general practices in more detail. Of these four countries, Indonesia and the Philippines are archipelagoes, Thailand is part of the Asian mainland and Malaysia is a federation of the geographically separated Malaya and the less-developed Sarawak and Sabah (formerly British North Borneo). There is nevertheless a remarkable similarity between the agriculture of all four countries. All have a typically wet tropical climate and grow much the same crops, amongst which rice is overwhelmingly the most important. Livestock production is relatively unimportant in all of them.

The peoples of Indonesia and Malaysia are predominantly of Malay race, and the Filipinos are ethnically related to them; the Thai people, though of different racial stock, are of similar stature and physique. All four countries have more or less numerous Chinese communities, who also take an important part in agriculture and trade; and all have more primitive peoples in the mountainous hinterlands who practise rather simple forms of farming. History has also imposed similarities. Three of the four countries have experienced the rule of European colonial powers. All four were occupied by the Japanese during the Second World War, with much damage to the agricultural economy. All have experienced political change and insecurity during the present century. It is at least partly due to these levelling factors that the economic infra-structure and standards of living are today very similar throughout the region. Nor is there any great disparity between the four countries we have chosen for description and the rest of south-east Asia. Agricultural conditions in Cambodia, Laos, Vietnam and Burma are in many ways very similar to those which will be described in this chapter.

Ecological Background

South-east Asia is a region of high rainfall, and in fact provides the largest area on the globe with a rainfall of over 100 in annually. Mean annual rainfall in the lowlands is nearly everywhere between 80 and 120 in; in mountainous areas it may be higher and sometimes exceeds 150 in. The only exception to these conditions is in some interior districts of Thailand which lie in a rain-shadow and have in places less than 40 in of rain. In some parts of the region

412

the south-west monsoon during the months of the northern summer, or the
north-east monsoon during the winter, produce peaks of rainfall, especially
where these winds coming from the sea meet mountains near the coast. But
there is so much orographic and instability rainfall that these effects are often
masked, and in fact no part of the region (except for the parts of Thailand
mentioned) ever normally experiences a dry season sufficient to provide a
serious check to plant growth. The rain commonly falls in short periods at
high intensities. Mean sunshine records throughout the region are notably
high in relation to the rainfall, providing exceptionally favourable conditions
for plant growth. The Philippines lie in a zone which is visited by occasional
typhoons in the months of the northern summer and autumn, capable of causing great destruction of buildings and crops.

Temperatures throughout the year are mostly extremely stable. Midday
temperatures of about 88°F to 90°F near the coast, and some five degrees
higher inland, are characteristic of the region. Lying across the Equator, and
with so little fluctuation in temperature or rainfall, large parts of the region
are almost seasonless and crops may be planted or harvested in any month of
the year.

There are hilly districts or mountains of moderate altitude in all the countries concerned; but, although the highest peaks exceed 16,000 ft in Indonesia
and 13,000 ft in Sabah, there is no important farming at such high altitudes
that agro-climatic conditions become drastically different. The soils of the
region are much more variable and on the whole less favourable to agriculture
than the climate. It is impossible to particularise all the soils of such a vast
region, but certain types are of widespread importance and support a great
deal of the agriculture. The best of these are the rich volcanic soils which
occupy large areas in Indonesia and the Philippines, and on which many of
the cash crops to be mentioned later are grown. The rice-growing valleys and
plains consist mostly of alluvial soils, the majority of heavy texture which provides the necessary imperviousness to water, but not outstandingly fertile.
There are also large areas, mostly of hilly or undulating land, covered with
rather poor soils overlying granite or sometimes other rocks; of these poor
soils Malaya has large areas. Red tropical soils or latosols can be found in all
the countries, and there are occasional areas of limestone soils which can be
very productive. A notable feature of the region is the occurrence of lowland
peats to an extent unusual in the tropics, especially in Indonesia and Malaysia.
These peats present special problems for agricultural development; they need
drainage, but when drained they shrink and tree crops may lose their roothold, and they contain buried tree trunks which are an obstacle to cultivation.

Soil erosion is particularly serious in the Philippines, where much of the
topography is geologically young and slopes are steep. It is also a special
problem in the cultivation practices of some of the more primitive peoples in
the hilly hinterlands, such as the Sakai in Malaya or the Iban in Sarawak.
These people clear forest on steep slopes to plant upland rice or tapioca

(cassava), causing such erosion that they have to move on after a year or two
to devastate further areas, leaving the first site with so little soil that the forest
may be unable to regenerate. The solution to this problem which is generally
being attempted is to persuade them to settle in the valleys and adopt swamp
rice cultivation. Other measures against soil erosion include the marvellous
systems of terraced rice fields in the Philippines and parts of Indonesia which
were laboriously created by manual effort in the past. In Malaya, considerable areas of valley soils have been disturbed by tin mining and need laborious reclamation before they can again be made fit for productive agriculture.

The natural vegetation of the whole region is tropical rain forest, with the
exception of the dry districts of Thailand where it is replaced by thin scrub
on the Korat plateau and dry scrub forest in the south-east. The rain forest of
south-east Asia is especially characterised by the prevalence of dipterocarp
trees, of which some genera, such as *Shorea* and *Dipterocarpus*, yield valuable timber; in Thailand it also includes natural teak forest. The forest still
occupies vast areas, since except in Java population densities are low throughout the region. Even in Indonesia, 23% of the land is under forest. In the
Philippines, less than 25% of the land surface is farmed, and in Thailand only
about one-third of the country is put to any effective use. Besides the virgin
forest, there are also large areas which have been cleared in the past and reverted to secondary jungle known in Malay as 'belukar'; this is often a more
tangled and impenetrable growth than the original forest. The complete dominance of forest and absence of grassland naturally make this region particularly ill-suited to any large-scale raising of livestock. Even if clearings are
made for grazing, it is laborious work for the peasant with his limited tools
to prevent re-invasion by woody growth. For this reason, any small areas of
grass which occur fortuitously, such as road verges, are avidly sought after
and frequently used for the tethering of cattle and goats.

One weed species of the native flora, the grass *Imperata cylindrica* which is
known as 'lalang' in Malay and 'cogon' in the Philippines, is of great agricultural significance. This grass, which also occurs in other parts of the tropics,
is under the conditions of south-east Asia an aggressive invader of cultivated
land and farming operations have always to be designed so as to keep it under
control. It is particularly feared in tree crop plantations such as rubber from
which, once established, it is difficult to eradicate; it is often treated with
sodium arsenite, which raises a danger of grazing animals being poisoned, and
more modern herbicides have not yet completely solved the problem. A more
beneficial group in the flora are the blue-green algae which grow in the water
of the rice fields and have been proved to make a useful contribution to fertility by their fixation of atmospheric nitrogen (Singh 1961).

The larger wild fauna of south-east Asia are not sufficiently numerous to
have much effect on agriculture. The scarcity of mammals is sometimes attributed to the enormous population of leeches in the perennially wet vegetation
of this region. These leeches are a real agricultural nuisance, both as ecto-

parasites of livestock and because they abound in rice fields and other irrigated land and attach themselves so frequently to the legs of farm workers. Other small animal pests include crabs (both sea and land) and rats, which are inimical to rice cultivation both by destroying plants and by damaging the bunds (earth ridges) on which maintenance of the water-level depends.

Agrarian Structure

Throughout south-east Asia, the peasant who owns the land he cultivates is the most common type of farmer. In all the countries there are also some tenant farmers who rent land either from a private landowner or from a national government or regional authority. In the Philippines, about 50% of farmers are owner-occupiers, 11% own part of their farm and 39% are tenants; in Thailand, 83% of farmers are owner-occupiers; in the northern states of Malaya, about half the farmers on rice land are tenants. In the Philippines most of the tenants are share-croppers paying half their crops to the landlord; agricultural reform legislation has marginally reduced their numbers and bettered their condition, but its impact as yet has been small. In Thailand the tenants pay rent mainly in kind but at a fixed rate which usually amounts to more than half the rice crop. Fragmentation of holdings is a handicap to farmers especially where, as in Malaysia and Indonesia, there are large populations who follow the Moslem laws of inheritance with their insistence on subdivision.

The average size of farm varies considerably between the countries. In Indonesia it is only 2·1 acres; in the Philippines 8·6 acres; in Thailand, about 13 acres in the rice-growing central plain but only 3·6 acres in the north. The peasant relies almost entirely on his own and his family's labour. A survey in Indonesia showed that the average smallholder worked for 4·75 hours per day, but here there are considerable local and racial differences. The Malays of Indonesia are probably harder workers in farming than those of Malaya; but this is largely because the latter traditionally regard themselves as sailors and fishermen and many relegate agriculture to the level of a subsidiary interest. In contrast, the Chinese who have settled as farmers in all the countries work hard and often for very long hours. Their farming is often an intensive form of horticulture, in which by quick work and interplanting one crop with another they can take several crops of vegetables in a year. By the construction of fish ponds (stocked with carp or *Tilapia* spp.), the keeping of pigs, and the return of their own excreta to the land, they use every by-product of the farming system and achieve one of the most intensive and fertility-maintaining forms of farming in the world.

The plantation sector is currently most important in Malaya, where the estates have a little less than half the rubber acreage but provide considerably more than half the production. The important palm-oil production (from the species *Elaeis guineensis*) comes wholly from estates. Estates in both these categories are mainly in expatriate ownership. In Indonesia, where plantations were formerly of equal importance, history has taken a very different turn. The plantations suffered severely from damage and neglect during the Japanese occupation in the Second World War and subsequently again during the fighting between Indonesian and Dutch forces. They had not recovered from this when the 45% of estates which were Dutch-owned were placed by the Government under the management of the Government Estates Central Bureau. Further restrictions on estate ownership or management have followed, and the result has been a catastrophic fall in the agricultural exports of Indonesia, which before the Second World War was perhaps the world's largest exporter of tropical produce. By 1959, Indonesian exports of certain produce had fallen to the following percentages of what she exported in 1938: copra 38, tea 46, tobacco 33, palm-oil and kernels 51, sugar 3, coffee 57, spices 24, kapok and sisal both under 25. By 1957, half a million peasant squatters had invaded land formerly used by estates.

Many estates in south-east Asia, and particularly in Malaysia, are run under a system of management which is peculiar to the Asian tropics and subtropics, the managing agency system. Under this arrangement, many of the smaller estates, or groups of estates owned by all but the biggest companies, are taken under the wing of agency firms operating from cities like Singapore and with head offices in London. These firms provide the estates with central secretarial, accountancy, and supply services and with representation in London, as well as exporting their produce. The firms also supply supervision and advisory services in the management of the estates, largely through the medium of well-qualified visiting agents who make frequent inspections. The system has stood the test of time and proved practical and economical (Macfadyen 1954).

Cultivations for rice, the major annual crop, are commonly carried out by draught animals, both cattle and buffaloes being used. There are also however many holdings in the region worked by hand labour, especially where these are only of horticultural size. Tree crops are established and maintained by hand labour. Mechanisation has as yet made little impact in the area, though a great deal of experimental work has been carried out especially in Malaya to try to find economic methods of mechanising smallholder rice production, and motor rotary cultivators are used to a small but increasing extent. Fertilisers are used in quantity on rubber plantations but as yet only by an enlightened minority of small farmers. 'Prawn dust' is an unusual local manure provided by the shells of prawns which are a popular food.

The countries of the region are to a varying extent self-sufficient in food supplies, the most important of which to them is rice. The Philippines are approximately self-sufficient in the staple foods. Indonesia in recent years has been a frequent importer of variable quantities of rice to make up her needs; efforts to increase production have achieved considerable success but are constantly overtaken by the growth of population. In this country the Food Supply Board imports or controls the import of rice, maintains storage, and procures a percentage of the internal crop for distribution. Malaya has long been a rice-deficit country, finding it more profitable to devote land and labour to export commodities such as rubber or tin and to buy rice with the proceeds. However, the proportion of rice consumption which is home-grown has been raised from the former level of one-third to over a half. Sarawak and Sabah also have to import rice to feed their populations. In contrast, Thailand is one of the three great rice-exporting countries of the world (the others being the U.S.A. and Burma) and supplies, amongst others, the deficit areas which have been mentioned.

Primary buying of most crops throughout the region is conducted largely by petty traders, amongst whom the Chinese are conspicuous. The co-operative movement has been encouraged by all the governments of the region, and there are numerous co-operative marketing societies. This is particularly true in Indonesia, where the development of co-operatives has been one of the main planks in a Government policy of socialism. In Malaya, great stress has been laid on the formation of co-operative groups for the processing of small-holders' rubber, as a means both of improving and of standardising quality.

The distribution of population in south-east Asia must be considered an important part of the background to agriculture. In Thailand, Malaysia and the Philippines, although some districts are heavily settled, the overall density of population is low and there are still vast reserves of unused country so that there is no real pressure of population on resources. Indonesia presents an extraordinary picture. The island of Java is one of the most heavily populated areas in the world. with density varying from 900 to 1,400 people per square mile in different districts. But the remaining islands of the archipelago have a population density of only 44 per square mile. Put in another way, 74% of the population live in Java on only 9% of the national land. The obvious remedy for this situation is to promote emigration from Java to the outer islands, and this the Government has long been trying to do. But the Javanese are very unwilling to emigrate, and such movement as has taken place has not succeeded in reducing the population of Java because of the high rate of natural increase. Since the internal food supplies of Indonesia are not now adequate to provide a satisfactory diet for the people of Java, the problem is perhaps the major one which emerges in an agricultural consideration of the region.

In Indonesia, the proportion of the population employed in or directly dependent on agriculture reaches the very high figure of about 80%. In the Philippines and Thailand, lower figures of the order of a little under 70% have been estimated, which is about the same as for India.

The agricultural research institutions developed in south-east Asia in colonial days were perhaps the best in the tropics, Dutch scientists being particularly numerous and active. Particular mention may be made of the Rubber Research Institute of Malaya, and of the College of Agriculture in the Philippines which became under American rule and still is one of the largest

training centres for tropical agriculture. The botanic gardens established by
the Dutch at Buitenzorg (now Bogor) in Java were amongst the most famous
in the world, and were also a great centre of research, but it has not been possible to maintain the same scientific impetus since independence. A very important modern creation is the International Rice Research Institute which
was established in 1962, with financial aid from American foundations, at Los
Baños in the Philippines; it aims to undertake research for the benefit of all
Asian rice-growing countries.

Agricultural extension services in south-east Asia have not generally been
as good as the research services, and although the plantation industry (to
whose needs a great deal of the research was directed) has been quick to take
advantage of new techniques, science has as yet made little impact on the
practice of the peasant farmer. A notable exception is the rubber replanting
scheme undertaken by the Government of Malaya since the Second World
War, which has been described as one of the biggest operations ever conducted
in tropical agriculture; by 1962, over half a million acres of smallholders'
rubber had been replanted with high-yielding material, as well as a much
larger area on estates. It is historically unfortunate that the emphasis on plantation agriculture, which under Dutch rule made Indonesia one of the richest
countries in the tropics, conduced so largely to the later decline of the economy
in that country where under independence the plantation system was not
found acceptable and peasant agriculture was ill-equipped to take its place.

Crops and Livestock

Table 2.13.1 shows the production of the main crops in the four countries
concerned. Rice is overwhelmingly preponderant amongst annual crops to an
extent which is hardly paralleled in any other agricultural system. In some
extensive plains areas, and most notably in central Thailand, it is practically
a monoculture. In Thailand as a whole, 78% of the cultivated area is under
rice; in the Philippines, rice and maize together occupy 55% of the farmed
land. Rice is everywhere the staple foodstuff. A small proportion of the total
rice acreage consists of 'upland' rice grown on dry land; in Indonesia, where
this proportion is relatively high, it amounts to 15%. The remainder is swamp
rice, but not all of it is irrigated; in many areas reliance is placed simply on
impounding rainfall by making bunds (ridges) round small fields in flat areas
with an impermeable soil. In both the Philippines and Indonesia, the proportion is about 60% of rice growing on impounded rainfall to 40% irrigated.
Yields are lower on the impounding system because there is no means of adding water when rain is short or of draining the surplus when rain is excessive.
All the governments are therefore trying to provide more canals for irrigation
and drainage, and the irrigated area tends slowly to increase.

Within the swamp rice areas rice growing is continuous only in an irregular
way. Land which is suitable for rice growing by reason of heavy rainfall, impermeable soil and flatness, and which has usually been levelled and bunded

long ago, is regarded as rice land and is too valuable for this purpose to use for **Farming** any other. But a farmer does not necessarily plant the whole of his rice land **Systems** every year, though most of it is so used; the exact amount he plants will depend on such variables as the number, strength and health of the family and their draught animals. Over most of the rice areas, only one crop of rice is taken in the year, and the fields are allowed to dry out during the fallow period. This is partly because weed control depends on a rotation in which dry-land weeds are killed by drowning in the wet phase and aquatic weeds are killed by

Table 2.13.1

Crop Production and Livestock Numbers in Selected Countries of
South-East Asia

Crop production in thousands of metric tons per annum:

	Thailand	Malaysia	Indonesia	Philippines
Rice (paddy)	10,569	1,217	14,600	4,490
Maize	1,102	10	2,517	1,521
Cassava	2,579	359	11,324	677
Copra	23	186	496	1,574
Rubber	238	1,046	790	7
Sugar-cane	3,858	—	10,584	12,897
Palm kernels	—	78	40	—
Palm oil	—	344	180	—
Tobacco	55	3	133	51

Livestock numbers, in thousands:

	Thailand	Malaysia	Indonesia	Philippines
Cattle	5,236	335	6,364	1,560
Buffaloes	6,878	361	2,990	3,346
Goats	30	348	5,052	606
Sheep	14	44	2,360	14
Pigs	4,291	827	2,802	6,939

Source: F.A.O. *Production Yearbook 1966.*

drying in the dry phase; and partly because the rice stubble and weeds in the dry phase provide almost indispensable grazing for the draught animals in an environment where other pasturage is so scarce. Sometimes other crops, such as pulses, are grown in the dry phase in rotation with rice, but this is rare. Especially where the impounding of rain is relied on, there is not usually sufficient water even in this region of high rainfall to allow more than a single crop of rice to be grown at the wettest time of the year. In fact, double-cropping either of rice or other annuals, is surprisingly rare in the region in view of the high rainfall. In the Philippines, only 15% of the cultivated land is double-cropped; in Thailand only 12% of farms are reported to practise double-cropping. In Malaya, there is only one extensive area (Province Wellesley) where two crops of rice a year can be practicably taken, and even here it involves

extremely quick work in harvesting and preparing the land for the second
crop which is difficult to achieve without mechanisation.

Rice cultivations consist in ploughing followed by a series of harrowing
or 'puddling' operations with locally-made harrows or rakes, carried out on
the wet soil largely with the object of sealing any pores in it so that it will retain water. The rice is almost universally transplanted from nurseries. The
varieties grown are *indica* types, though much experimental work has been
and is being done to try to improve the yield by crossing with *japonica* varieties. The preference, except for very small acreages, is for non-glutinous rice;
long-grained varieties, which are generally regarded as the highest quality,
form only a small proportion of the crop which is mostly medium-grained.
Yields per acre are poor throughout the region by world standards, though
Malaya has a slight lead over the other countries due to careful plant breeding
and the extension of irrigation in the past.

Maize is the only other cereal of significance, and is chiefly important in
Indonesia and the Philippines, where it is the principal crop in the island of
Cebu. Adlay (*Coix lachryma-jobi*) is a curious rather than important cereal
which is grown on a small scale in the Philippines. Tapioca (*Manihot utilissima*), known elsewhere in the tropics as 'cassava' or 'manioc', is the third
most important food crop, and heavy yields are easily obtained in this region
of high rainfall. It is used largely by the Chinese as a food for pigs as well as
humans, and there are also tapioca factories which prepare starch and pearl
tapioca for export. Pulses are important in providing protein in the diet; cow-
peas (*Vigna sinensis*) and groundnuts are widely grown, and soya-beans
especially by Chinese farmers. Amongst minor food crops, the sago palm
(*Metroxylon sagu*) grows both wild and cultivated, and sago flour is exported
from Indonesia and Sarawak. The areca-nut palm is planted on quite large
acreages, and palms of *Arenga* and *Nipa* spp. are a source of sugar. Two tree
fruits which are almost peculiar to this part of the world, the durian (*Durio
zibethinus*) and rambutan (*Nephelium lappaceum*), are particularly popular with
the Malay races, and the total acreage of them in Indonesia, Malaysia and
southern Thailand must be considerable although no statistics are available.

As would be expected in a region of such high rainfall, a high proportion
of the cash crops are tree crops. Rubber (*Hevea brasiliensis*) is the chief of
these, and Malaysia and Indonesia are the two chief producers of natural
rubber in the world. It is notable that although production of estate crops has
declined so much in Indonesia, the output of rubber in 1961 was nearly twice
that of 1938 (mainly by increase of smallholding production), although it has
since declined. Coconut palms are another extremely important cash crop,
the Philippines being the world's largest producer of copra and Indonesia the
second largest. Oil-palms are a considerable estate crop in Indonesia and
Malaysia, with the former still having a slightly higher production of oil in
spite of a decline from earlier figures. Amongst fibre crops, the Philippines
supply most of the world demand for manila hemp (*Musa textilis*), which is

grown on a much smaller scale in Sabah; Indonesian production of sisal (*Agave sisalana*) has fallen to a small figure.

Sugar-cane in this region is now of the greatest importance in the Philippines, where much of the crop is grown by smallholders although there are also numerous 'estates' of 200 acres upwards. Sugar for export is extracted in 'central' factories, but there is an even larger production of non-centrifugal muscovado sugar for local consumption. Thailand also has a large sugar-cane industry; for Indonesia, which was formerly one of the world's largest sugar exporters, figures of current production are not available. Tea is grown in all the countries, largely for local consumption; the formerly large export from Indonesia has greatly declined. Indonesia, the Philippines and Thailand are all large producers of tobacco in that order, but Indonesia no longer exports valuable amounts of the best cigar leaf. Indonesia is a substantial exporter of *robusta* coffee produced by smallholders; in Malaya, *liberica* coffee is grown for the internal market. Spices are a continuing export from Indonesia, to which cloves and nutmegs are indigenous, and pepper is exported from Sarawak and Sabah. Considerable efforts have been made in Malaya to foster the pineapple industry, which provides useful exports. Minor cash crops of the region include the opium poppy in the hills of northern Thailand, from which it is difficult to market bulkier products.

Livestock are perhaps less important in the agriculture of south-east Asia than in any other major region of the world. The need for animal protein in the diet is largely supplied by fish, which are in ready supply in countries with such long coastlines. Bovines are kept mainly for draught purposes. Buffaloes are particularly well-adapted to this work in wet rice fields, and Thailand and the Philippines are amongst the few countries in the world where buffaloes outnumber cattle. Little milk is produced or consumed; in Malaya dairy cows are kept mainly by the relatively large Indian community. Both cattle and buffaloes are generally of small size and nondescript type. The cattle are almost all zebus; there is one other domesticated indigenous species, the Bali cattle (*Bos banteng*), which is of small importance but has been under observation at experiment stations as a beef animal.

Goats are a subsidiary source of meat, especially in Indonesia, which is also the only one of the four countries with any appreciable number of sheep. The keeping of pigs depends on race and religion, since to Moslem populations this animal is unclean, whilst by contrast the Chinese are large consumers of pork. Pig-keeping is common amongst Chinese farmers, and the Chinese types of pig are the best of those developed in the tropics. The Chinese slaughter for pork at a heavier weight than is customary in Europe, and they like a soft flesh which is produced by feeding the pigs largely on root crops and green fodders. In these hot climates they believe in washing the pigs once or more daily to keep them cool; a practice which is not found in other parts of the tropics. Poultry in the region are fairly numerous but not outstanding in quality, though temperate breeds can be introduced with success and are steadily expanding; some Chinese farmers keep Cantonese fowls, which have less advantage over the indigenous types. There is a good demand for duck eggs from the Chinese population, and ducks are kept in considerable numbers in the rice-growing areas, where they can forage in the fields. In Thailand, the proportion of ducks to chickens was as high as 7 million to 23 million in 1962.

The incidence of animal disease in the region is less than in some other parts of the world which are equally hot and humid. Rinderpest, which formerly occurred in Thailand and Malaya, has been eliminated from both. Buffalo calves are always rather delicate and tend to suffer particularly from worm infestation.

A Malaysian Transect

In south-east Asia the conditions of climate, altitude, terrain and agricultural practice are in general more uniform than in any of the other tropical regions with which we have dealt. It is therefore possible to illustrate them from a shorter transect (Table 2.13.2). The transect chosen is in northern Malaya and runs approximately along latitude 5° 70′N from the west coast in the central part of the Kedah coastline, to pass through part of the state of Perak and reach the east coast somewhat south of Kota Bharu in the state of Kelantan. Although only about 150 miles long, this transect includes samples of the main types of south-east Asian agriculture. It illustrates the contrast between a mangrove-fringed and an open coast, it includes rice-growing plains and land occupied by rubber estates, and part of its length lies in jungle-clad hills occupied only by a few groups of primitive farmers.

THE WESTERN COASTAL FRINGE

The coast of Kedah, like most of the western coast of Malaya, is typical of those tropical coasts which are fringed by a dense growth of mangrove swamp precluding any access to the sea by land. The only communication is by boat through the river mouths, and it is at these mouths that most of the few villages in the coastal fringe have developed. Because there is no dry land on which to build, these villages are constructed of a row of houses built on stilts and touching each other, so that a continuous planked walk built at the same level along their frontages provides a village 'street' on which shops as well as houses are situated. The population consists entirely of fishermen, who carry out no farming unless they happen to own plots of land further up river; this maritime existence is of the essence of the local Malay tradition, and explains the lack of interest shown by so many of the population in farming. These people are not however entirely without useful plants. They cut the super-abundant mangroves and sell the poles as building timber. The mangrove bark is carefully stripped off and sun-dried in their villages before sale; used as a tanning material, it is an export of considerable importance to Malaya. The brackish water of the estuaries also supports the Nipa palm (*Nipa fruticans*) which grows prolifically in the shallow water near the banks. Like many other **Farming** palms, these can be tapped (from the spathe of the fruit) to obtain the sap, **Systems** which in many parts of Malaysia and Indonesia is evaporated down to yield a crude sugar. The leaves are also used like those of the coconut palm to make 'attap' thatching for houses. Rather higher up the rivers, since it is not quite so tolerant of salt, is found the sago palm growing on the banks or in marshy places. To obtain sago, the stems are cut and the pith rasped to provide a material from which the starch is washed out.

THE WESTERN RICE PLAIN

Immediately inland of the mangrove fringe, the transect enters a broad tract of completely flat country which is used for growing rice. The central rice-growing plain of Kedah, extending from about Sungei Patani in the south to somewhat north of Alor Star, the state capital, is one of the chief 'rice-bowls' of Malaya. Rice is here practically a monoculture. The landscape of this or any other of the great rice-growing plains of south-east Asia is something quite unique in world agriculture. If traversed at the season when the young rice is only just showing above the water, the whole countryside appears to be under water from horizon to horizon (except where, in Kedah, the distant hills are sometimes visible to the east). The plain is dead level, the scudding clouds are reflected in the water; the embankment which carries road or railway appears to be the only firm land in sight, and the traveller has the impression of being at sea rather than on land.

This concentrated cultivation of rice is made possible by various favouring factors in the environment. The rainfall at Alor Star, in the heart of the rice-growing area, averages 96 in a year of which 81% falls between March and October, and this normally provides plenty of water for one rice crop in a year. However, there are years when the rainfall is inadequate or alternatively may be excessive and drown the crop. To remedy this situation, the Government's Irrigation Department has been active in constructing canals which collect the run-off of the foot-hills and bring it to the rice fields, acting also at need as drains to remove surplus water; and these serve some parts of the Kedah rice area. A rice experiment station is maintained by the Government near Alor Star. The soils of the area have been defined as low humic gley soils of very variable fertility (Panton 1964); but the most important point is that, especially with puddling operations, they are sufficiently impermeable to hold the impounded rain-water for the period of the rice crop.

Rice in Malaya is grown by both Malay and Chinese farmers, but in Kedah it is the Malays who predominate. In this area very little is grown of any crop except rice. The 'kampongs' or settlements, which may consist of a single homestead or of several grouped together, appear as islands amongst the rice fields though very often the land on which they stand is hardly higher than the latter, and the rain only drains off the house compounds because they are not bunded. In the kampongs, the houses are usually surrounded by a few coconut, areca or sago palms, but there is often no garden; on these dry spots also are stalled the buffaloes or cattle which are used to cultivate the rice fields.

The rice-growing area of Kedah extends in places almost to the border with Thailand, beyond which an almost similar agriculture is practised, the people of this southern part of Thailand being largely Malay by race. The chief differences to be noted are the presence in Thailand of Borassus palms dotted amongst the rice fields (as in southern India, Chap. 2.12) and of rather more numerous sago palms.

THE RUBBER ZONE

To the east of the Kedah rice plain, the land rises and becomes gently undulating even before the foot-hills are reached, so that it is not suitable to use for rice growing. In such situations in western Malaya, rubber generally replaces rice. But it would not be true to say that a rubber belt lies everywhere to the east of a rice belt; in some places the higher land comes nearer to the sea, and even in Kedah the rice plain gives place to higher land in the south where much rubber is grown. To the north of Kedah, the small state of Perlis has more of this undulating land than of rice land, but there is little rubber and peasant farmers grow more of the dry land crops such as maize, tapioca, groundnuts, soya-beans and tobacco. Rainfall rises as the hills which run down the spine of the Malayan peninsula are approached, and at Kulim in the rubber-growing area of Kedah it is 124 in a year. The soils of this area are lateritic and are described as of average to below-average fertility.

Rubber holdings in Malaya are classified as estates if of over 100 acres, and as smallholdings if below that figure. Malay rubber growers usually only have very small acreages, and any smallholding of more than 20 acres is likely to be owned by a Chinese. The size of individual estates is very variable and is not of any great importance, since small estates are generally part of a group owned by a company and share common managerial services. Malays have never taken any great part in work on rubber estates, and the need for tappers has historically been one of the main reasons for the immigration of foreign labourers into Malaya; Indian workers constitute a large part of the tapping force.

Most of the land now planted with rubber in Malaya has been cleared by the planters from virgin forest; the steeper land is often terraced for rubber planting to minimise erosion. The plantation rubber industry in Malaya is one of the most efficient sectors of tropical agriculture, readily adopting new techniques proved by research and following closely the advice of the Rubber Research Institute which it supports. This scientific approach is shown in such aspects as the planting of high-yielding clonal material, the use of leguminous cover crops often with inoculation to ensure good nodulation, the adoption of carefully tested tapping systems including the use of plant hormones to stimulate yield, and the application of fertiliser mixtures recommended for the particular soil type. The result is that average yields of plantation rubber have very greatly increased in recent decades, and costs of production have been **Farming** held at a level where natural rubber is so far still able to compete with the ever- **Systems** increasing output of synthetic rubber.

While some estates and smallholders still produce smoked sheet rubber as their end-product, some estates manufacture crepe rubber which is produced from suitable clones by rolling the sheet at higher pressures and drying without smoking. There is also a substantial export of liquid latex which is concentrated by centrifuging, and increasingly of comminuted crumb rubber. Rubber is by far the most important agricultural export of Malaya and, with the export of tin, has mainly contributed to giving this country one of the highest national incomes per head and standards of living in Asia. In the rubber-growing areas, hardly any land is devoted to other crops, so that rubber in this zone is as nearly a monoculture as was rice in the preceding one.

THE CENTRAL HILL ZONE

The centre of the Malayan peninsula is occupied by a spinal range of hills which runs down its length from north to south. Although the highest peaks in these hills only reach just over 7,000 ft, the whole of the topography is exceedingly steep and broken and the entire surface is covered by heavy forest except where a few small clearings have been made. The soils are generally poor, being lithosols or shallow latosols. Throughout most of the hill area there are no lines of communication whatever, only two roads passing across the peninsula from east to west. Under these circumstances it is not surprising that there has been no agricultural development in these central hills; most of the area is in fact uninhabited. However, as in other south-east Asian countries, there are primitive peoples in the hills who have little contact with the more sophisticated communities of the coastlands. Along the line of our transect, which runs through the hills in parts of Kedah, northern Perak and Kelantan states, this primitive way of life is represented by groups of the Sakai people. These have no fixed home and few material goods. They live by clearing small patches of forest to grow upland rice and some tapioca and maize, moving on to other areas as soon as the inevitable soil erosion and decline of fertility set in. They also hunt the limited fauna of these forests, and preserve the use of the blow-pipe for this purpose.

There is only one area in these hills where any extensive agricultural development has taken place, and that is at Cameron Highlands about 80 miles south of the line of our transect. Here a small area in the hills has been opened up by a road system as a multi-purpose development. At about 5,000 ft, the area performs some of the functions of a 'hill station' to provide some relief from the heat of the plains, and its hotels attract some tourist trade. The water supplies of the hills can be tapped (the rainfall here averages 111 in a year) and dam construction enables the development of some hydro-electric power. On the agricultural side, several villages have been created by Chinese settlers who obtain most of their income by growing temperate vegetables which are in demand in the lowlands but cannot well be grown there. The mean annual temperature here is 65°F as against 81°F at Kuala Lumpur in the lowlands. Amongst the chief vegetables grown are English and Chinese cabbage, leeks, tomatoes, sweet pepper, beetroot, carrots, sugar peas and French beans. The seed used is largely imported from Australia.

There has also been some estate development of tea in this area. The slopes on which it is grown are often excessively steep, but here as in India it has been found that tea, once it is mature, is one of the best crops for giving protection against soil erosion. The highland tea is of finer quality but lower yield than the lowland teas of Malaya; most of it is consumed within the country. A tea research station is maintained here by the Government. These developments at Cameron Highlands provide at least a pointer to the possible agricultural development of the large hill areas of Malaya if and when this is necessitated by increasing population.

THE NORTH-EASTERN LOWLANDS

The eastern side of the Malayan peninsula has been much less developed than the western; it is still mainly covered with forest which often extends almost to the sea, with development confined to towns or villages at the river-mouths. In the north-east however a large plain area which extends south of the important town of Kota Bharu along the Kelantan river has provided the possibility for greater agricultural development, and it is through this area that our transect next runs. The soils here are similar to those of the Kedah rice plain. The average annual rainfall, 124 in at Kota Bharu, is as high as any along our transect. However, this comes mostly during the north-east (winter) monsoon, 54% of the rain falling between November and February, and in consequence the area presents for part of the year a slightly more arid appearance than is usual in Malaya.

The state of Kelantan contains the largest area of rice in Malaya after that of Kedah, but the crop here is not such a monoculture as on the Kedah plain, and the methods of growing it are more various. Kelantan has the largest area of upland rice in the country, and rice is also grown in the Kelantan valley under the 'tenggala' system, in which varieties are used which flourish either under dry conditions or when the land is flooded or waterlogged for uncertain periods. Crops other than rice are also grown by the Malay farmers who predominate in this area. Besides maize, legumes are particularly important amongst these and include cow-peas, soya-beans, groundnuts, green gram (*Phaseolus aureus*), and to a lesser extent French beans. It is noteworthy that throughout south-east Asia, cow-peas and French beans are grown in their climbing forms, usually supported on sticks, and cooked and eaten as whole young pods; this is in contrast to the use of dwarf-erect varieties, and the eating only of the dried seeds, which is usual in the African and American tropics. For this purpose, varieties with long pods are desirable; some cow-pea varieties, with very long pods indeed, are known as 'yard-long' beans.

It is a curious and unexplained fact that many Chinese farmers in Malaya prefer to use seed of this group of crops imported from China rather than to save their own seed.

Cattle and buffaloes are again important in Kelantan as plough animals, and this is one of the few areas in Malaya where sheep (kept for meat) are much in evidence; the local breed are of small size, the hairy fleece is usually white and the males are horned.

THE EASTERN COASTAL FRINGE

The east coast of Malaya is entirely different from the western in that there is no mangrove fringe but hundreds of miles of open beaches which are a considerable tourist attraction. We therefore again find the coastal fringe of coconut palms growing just inland of the beach which is characteristic of so many tropical shores. Around Kota Bharu the coconuts extend for some distance inland, and copra sales are an important source of farm income. Here, as elsewhere in the tropics, it is increasingly difficult to find pickers willing to undertake the arduous and dangerous work of climbing the palms. A local solution in Malaya and Thailand is the use of trained monkeys to climb the palms and throw down the nuts. Control is kept over the monkey by a long rope attached to a collar, and contractors who own these monkeys will harvest other farmers' trees at a set price per nut. Another interesting feature of this coast is the prolific growth of *Casuarina* trees along the shore just above the tide mark, apparently providing another instance of the ability of this species, with its nitrogen-fixing root nodules, to thrive on poor soils. The trees provide useful firewood and poles. Immediately inland of the coast, the soils are coarse-textured grey-brown podsols of poor fertility, which are little used for crops; most of the scanty coastal population earn their living mainly as fishermen.

Full meteorological data are not available for stations immediately along the line of this transect, but Table 2.13.2 gives some figures referring to the different zones through which the transect passes although some of the stations included lie rather further to the south.

SAMPLE FARMS

It is possible to give an excellent description of some sample farms in Malaya by drawing on the series 'Master Farmers of Malaya' which has been published in the *Malayan Agricultural Journal*. From these we have selected two farms: one small (Lim 1957) and one large (Hussein 1957), one worked by a Chinese and one by a Malay. However, it should be remembered that these farmers have been publicised as outstandingly good and to that extent are not typical of the whole farming population.

(*a*) *A rice farm in Province Wellesley.* In this area on the mainland opposite Penang, Kum Yeng, a Chinese farmer, has a holding of only 2 acres of rice land, but by hard work (probably 7–9 hours a day) and intensive use he makes a surprisingly good living out of it. In this area two crops of rice a year can

Table 2.13.2

A Malayan Transect

Place	Zone	Longitude	Altitude (ft above sea-level)	Rainfall (mean inches per year)	Relative Humidity (mean % in midday hours)	Temperature (mean daily maximum °F)	Soil Type	Vegetation	Chief Food Crops	Chief Cash Products
Penang	West Coast	100°19'E	17	107·7	67	90	Latosols, etc.	Forest	Rice	Rubber, coconuts, etc.
Alor Star	Rice Zone	100°23'E	—	95·7	—	—	Low humic gley soils	Forest	Rice	Rice
Kuala Lumpur	Rubber Zone	101°42'E	127	96·1	63	90	Latosols, lithosols	Forest	Rice, tapioca	Rubber
Cameron Highlands	Highland	101°23'E	4,750	104·1	77	72	Lithosols and shallow latosols	Forest	—	Vegetables, tea
Kuala Trengganu	East Coast	103°08'E	105	115·0	72	87	Coarse textured podsols	Forest	Rice	Fish, coconuts

be taken, and this is what Kum does regularly, securing a yield of not less than 3,200 lb of paddy (unhusked rice) per acre from each crop. One-third of the crop is required to feed his family, and two-thirds is available for sale. His other source of income is from the annual sale of 1,000 to 2,000 ducklings. He rears these mainly on purchased fish meal, of which he uses about 130 lb a day which he fetches from the town on his bicycle. When harvesting of the rice begins, the ducklings are turned loose in the fields to forage for insects and spilt grain and to trample the rice straw into the soil. The ducklings are sold at 70–80 days old. He applies duck droppings to the rice land at the rate of over half a ton per acre per season, and this obviates any need of fertilisers for the rice. The value of sales from the holding in 1957 was £112 from rice sales and about £87 from ducklings, no poor return by the standards of tropical peasant farming and from a holding of such small size.

(b) *A mixed farm in Kelantan.* Yaacob bin Sulong, a Malay, farms in the Kota Bharu district on the line of the transect we have described above. His family labour force consists of himself, his son, and both their wives. He has 4 acres of rubber, 2·5 acres of rice land, 4 acres of fruit trees (coconuts, rambutans and bananas), and a small plot of vegetables and tobacco. This acreage was partly inherited and partly purchased as he was able to save money. The rubber was replanted with high-yielding trees with the aid of a Government grant in 1955. The rice is manured with ammonium phosphate in the nursery, and in the field with cattle manure and a recommended fertiliser mixture at 100 lb per acre; the average yield is 2,560 lb per acre. Like the previous farmer, Yaacob reckons that one-third of the crop is needed to feed the family. The whole of the fruit area is manured, mainly with cattle manure. The rambutan trees are of clonal origin. *Stylosanthes gracilis*, a fodder legume, is planted under the young rubber and fruit trees to feed the cattle. These consist of two draught animals, which are bedded with rice straw to make plenty of farmyard manure; for this purpose the farmer hires a baling machine from the Department of Agriculture. There is a poultry flock of 50 local hens and a Rhode Island Red cock for grading-up; young birds are inoculated against Ranikhet disease (p. 400) at six weeks old.

Efficiency and Stability

Some sectors in south-east Asian agriculture are undoubtedly efficient. This category includes the plantation rubber industry, which by promoting research and taking every advantage of scientific findings has achieved dramatic increases in yield and productivity during this century. It also includes many Chinese farmers in all countries of the region, who by hard work and intensive methods have often achieved a very high output per acre, especially in vegetable production. But their systems are often very costly in labour, which could be reduced by the adoption of more modern techniques such as a degree of mechanisation or even of simpler methods such as mulching to increase soil moisture and save some of the laborious hand-watering which they so often apply to crops. For the rest of peasant agriculture, the degree of inefficiency can be judged by the generally low level of rice yields. Average national yields of paddy rice are only 1,103 lb per acre in the Philippines, 1,057 in Thailand and 1,627 in Indonesia; though the higher figure of 2,089 for Malaya and the yields quoted for the exceptional farmers mentioned above show what can be done with a few improvements. The result of generally low yields is seen in the fact that, even in an environment so favourable to plant growth, three of the four countries fail or barely succeed in providing enough food for themselves although some 65–80% of their population is employed in agriculture.

The region ranks higher for agricultural stability than for efficiency. Climatically this stability is very marked. No failure of the rains has ever caused a serious famine in the area, and natural disasters such as typhoons or volcanic eruptions (which devastated much agricultural land in the Indonesian island of Bali in 1963) are rarities. Economically, the agriculture of most of the region is solidly based on peasant farmers, the great majority of whom own the land they farm. The strength of this system is seen in the fact that the Indonesian population has been able to survive, without large-scale privation, recent years of galloping inflation and economic chaos. The long-term future of the plantation industry may be more uncertain, but it survives chiefly as rubber plantations in Malaya, where the proportion of smallholder acreage tends in any case to increase. Whether the rubber export industry so vital to Malaya and Indonesia can indefinitely face the competition of synthetic rubber is a question of first-class importance to the region to which no one can foresee the answer. The other chief problems concern not only the general state of the shaky Indonesian economy, but the particular question as to how much longer the teeming millions of Java can be fed without bringing into play the agricultural potential of the outer islands. This remains perhaps at present the most urgent agricultural problem of the region.

References

Hussein bin Isa. 1957. 'Master Farmers of Malaya III. Yaacob bin Sulong', *Malayan agr. J.* **40**. 284.

Lim, T. T. 1957. 'Master Farmers of Malaya', *Malayan agr. J.* **40**. 217.

Macfadyen, Sir E. 1954. 'Managing Agents in the Eastern Plantation Industry', *Trop. Agr. Trinidad.* **31**. 267.

Panton, W. P. 1964. 'The 1962 Soil Map of Malaya', *Jour. trop. Geography.* **18**. 118.

Singh, R. N. 1961. *Blue-Green Algae in the Nitrogen Economy of Indian Agriculture.* Indian Council of Agr. Research. New Delhi.

General Reading

Conklin, H. C. 1957. *Hanunoo Agriculture in the Philippines.* F.A.O. Forestry Development Paper No. 12. Rome.

Freeman, J. D. 1955. *Iban Agriculture*. H.M.S.O. London.

Grist, D. H. 1950. *An Outline of Malayan Agriculture*. London.

Jacoby, E. H. 1961. *Agrarian Unrest in South-east Asia*. London.

Levy, S. 1957. 'Agriculture and Economic Development in Indonesia', *Economic Botany*. **11**. 3.

Roe, F. W. 1952. *The Natural Resources of Sarawak*. Kuching.

P

The American Tropics

The American tropics constitute an enormous land area which is in some respects rather homogeneous and in others extremely varied. The homogeneity lies mostly in the ecological environment, for nearly the whole of the area falls clearly within the definition of the wet tropics, where perennial crops can be grown in an environment whose natural vegetational climax is forest. The exceptions to this are small arid areas on the coasts of Venezuela, Colombia and Ecuador, and actual deserts in a narrow coastal zone of Peru and Chile; the deserts of northern Mexico lie mostly outside the tropics.

The differences between parts of the region are mainly political, economic, social and racial. A very obvious difference is in the size of national units. At one extreme we have Brazil, the country with the largest land area in the tropics and almost three times the size of India. At the other are the numerous small republics of Central America and some even smaller island countries in the West Indies. The racial composition of the populations cannot be ignored in its effect on agricultural practice. In the Spanish and Portuguese-speaking countries of the mainland, the chief elements in the population are the same and consist of people of European or American Indian descent or the 'mestizo' element in which these bloods are mixed. The proportions of these groups however vary enormously from, for example, Bolivia with an Indian majority in the population to Costa Rica with a very high proportion of pure Spanish descent. Most of these countries have Negro or mulatto minorities, which in most of the West Indian islands become majorities; Haiti has an almost pure Negro population. Still other racial groups have a local importance; in Trinidad and Guyana (formerly British Guiana) Asian Indians form a very large part of the population, and in Brazil Japanese settlers have made a significant contribution to agriculture.

History and economics have combined to produce effects on agricultural trade which differ between the South and Central American countries which threw off the colonial yoke more than a century ago and those others, mostly West Indian, which remained colonies until very recently and whose agricultural exports are still largely oriented (in some cases by preferential agreements) to the markets of the mother-country. Cuba is in the special category of a Communist state where agriculture has been reformed on Marxist lines and trade is largely directed to other members of the Communist bloc.

Amongst this diversity, it is not possible in this chapter to give an adequate description of the agriculture of the whole of the American tropics; nor is it as easy as for other regions to describe agricultural transects or sample farms which give a fair representation of the region as a whole. The plan of this

432

chapter will therefore be first to sketch the agricultural background conditions
for the American tropics as a whole, and then to give a more detailed description of the agricultural geography of the four northernmost countries of South
America—Colombia, Venezuela, Ecuador and Guyana—taken as a sample
illustrative of the greater region. These four countries provide in fact examples
of most of the factors, natural and economic, which affect farming systems in
the American tropics. Three of them have long been independent, but one
(Guyana) has only recently emerged from colonial status; three of them depend mainly on agriculture for their national income, whilst one (Venezuela)
has developed an important alternative resource in oil; between them, they
include all the altitudinal variations of agriculture from lowland to high
Andean, they represent most of the racial traditions of agriculture found in
the American tropics, and within their borders are produced all the important
kinds of crops and livestock known to the region.

Ecological Background

Whilst most of the American tropics, with the exceptions already mentioned,
possess a 'wet tropical' climate, some parts only marginally do so, and there
are very considerable differences between the rainfall of different areas. In
South America, the typical annual rainfall of the lowlands within the tropics
is 60–80 in. Wetter areas include part of the Pacific coast of Colombia, where
mean annual rainfall reaches 280 in at Buenaventura; the upper Amazon
valley, with 80 to 100 in; and coastal areas from the mouth of the Amazon to
the Guianas, where Cayenne has 126 in of annual rainfall. Drier areas include
part of the Venezuelan and adjoining Colombian coast, where some stations
have less than 20 in of rain a year; an area of north-eastern Brazil, of which
the state of Ceara is typical, where again rainfall is in places under 20 in and
there is a history of intermittent crop failures due to drought; and many parts
of the Andean range where, to name only some national capitals, Bogotá has a
mean annual rainfall of 42 in, Quito 44 in and La Paz only 23 in.

In Central America, typical annual rainfall figures are 74 in for Belize and
70 in for San Salvador. But Mexico City has only 29 in, and this declines to the
northward into desert conditions; the Yucatan peninsula is also a drier area,
with rainfall declining to 20 in on its northern coast. In the West Indies, the
islands of Curaçao and Aruba share the aridity of the neighbouring coast of
Venezuela. The remaining islands have generally between 50 and 80 in annual
rainfall, but some are drier than others and in Antigua the mean of several
stations is only 44 in. Even in quite small islands there are often considerable
differences between windward and leeward coasts or due to the presence of
mountains. Irrigation has been found profitable to agriculturists in some of
the southern parishes of Jamaica (where Kingston has only 31 in of annual
rainfall) and in part of Barbados.

Within these annual totals, the distribution of rainfall through the year
mostly follows the usual pattern according to latitude. Within a few degrees

of the Equator, there is no marked dry season but there are maxima of rainfall following the two equinoxes. As higher tropical latitudes are reached,
there is a marked differentiation into one long wet season and one dry one.
Thus in Trinidad, the wet season lasts from June to December, and the dry
from January to May; but the 'dry season' here, and in the American tropics
generally, does not imply such conditions of aridity as would be conveyed by
the same phrase in Africa or Asia or Australia, for in Trinidad some 15 in of
rain fall during these five drier months. In the same latitude, the Venezuelan
'llanos' or grasslands experience heavy rain from April to October when vast
areas are flooded, and very dry weather for the remainder of the year.

Temperatures in the continental lowlands are high, with daily maxima
often around 90°F. In many cases there is very little seasonal variation; at
Georgetown, the capital of Guyana, for instance there is an annual range of
only 2°F in mean temperature, and at Quito in Ecuador this range is less than
1°F. Altitude is however a factor of the greatest importance in influencing
temperature owing to the immense extent of the Andean ranges. It is customary in South and Central America to classify lands according to their altitude, and a typical classification is given in the following table, though the
exact altitudinal definition of the different land types may vary somewhat
with latitude:

Altitude	Land Classification	Mean Annual Temperature	Land Use
Below 3,000 ft	Tierra caliente	75–83°F	Tropical agriculture
3,000–6,000 ft	Tierra templada	65–75°F	Sub-tropical agriculture
6,000–10,000 ft	Tierra fria	54–65°F	Temperate agriculture
10,000–13,000 ft	Paramos	43–54°F	Grazing
Above 14,000 ft	Perpetual snow		None

One other climatic factor to be noticed is the incidence of hurricanes, the
word used in the American tropics to describe circular tropical windstorms.
South America is outside their range, but the West Indies, the eastern parts of
Central America from Nicaragua northward, and the south-east coasts of the
United States are all liable to visitation. The months when they may be expected are usually from July to September. The damage to agriculture may be
locally catastrophic. Bananas, a crop which is always particularly vulnerable
to wind damage, are very widely grown in the region and often suffer the most;
but all tree crops may be damaged in any degree up to total destruction, and
sugar-cane too can suffer severely. The risk, which has led to the introduction
of crop insurance schemes, some of them supported by governments, in
different parts of the West Indies, is a factor which has to be taken into serious
consideration in the agricultural development of the region.

It is difficult to generalise about the soils of the American tropics because of the vast areas where no kind of soil survey has yet been attempted. In a few small areas, excellent soil maps have been prepared in considerable detail; a good example of this is the work carried out in the British West Indies, British Honduras and Guyana (Regional Research Centre 1958–63, Romney 1959). Soil maps on a much smaller scale have been published for a few other areas, for example Venezuela (Marrero 1964) and the description of Cuban soils by Bennett and Allison (1928) is classic. For the rest, recourse has to be had to generalised soil maps such as that prepared by Hardy (1942), which only purport to supply indications of prevailing soil types.

The West Indian islands have many soil types which rank high amongst tropical soils for fertility. This is particularly true of those islands which are of volcanic origin, but there are also some excellent soils derived from limestone and some agriculturally useful latosols which are particularly extensive in Cuba. On the mainland, the general picture of fertility is poorer. Most of tropical South America is covered by four main classes of soils:

1. Podsols or highly-leached soils of low to very low fertility. This designation actually covers a rather wide variety of soils from highly acid grey or pale brown infertile sands (often with a high quartz content) and silts to highly-leached latosols. Such soils occupy a larger area than any other single class, including most of the Amazon and Orinoco basins and the Guiana highlands.
2. Red earths or latosols of good fertility. These also occupy large areas, especially in southern Brazil where they are the typical soils on which coffee is grown, and in part of the interior of Venezuela.
3. Alluvial soils, which extend over considerable areas unmapped in detail along the major river valleys, especially in the Amazon basin, and in many smaller Andean valleys. Like most alluvia, they are of varying fertility and in some areas would need considerable drainage for agricultural development.
4. Lithosols; skeletal stony soils of generally poor fertility which occupy most of the sloping land in the Andean ranges.

It is easier to define these soil classes than to map them, for much of the territory is still agriculturally unexplored, as is indicated by Hardy's remark that his soil class of 'ground-water laterite podsol' is 'presumed to occupy a large part of the Amazon basin'. All observers are however agreed on the basic poverty of at least a large proportion of the soils of the interior of South America. These interior regions, and especially the enormous basin of the Amazon, which is still largely unpopulated, represent the largest remaining habitable region of the world which still lacks agricultural exploitation and fails to contribute a quota to the world's need for food supplies. Although factors such as absence of communications and distance from markets have

contributed to this lack of development, it is basically the poverty of the soils
which has discouraged agricultural enterprise and which should warn us
against setting too high a potential on this region.

Soil erosion is locally severe in many hilly or mountainous parts of the
American tropics of which Venezuela presents an outstanding example because so much of the country's cultivation is in its Andean region where it is
estimated that 70% of the crops are grown on slopes steeper than 35°. The
discoloration of the Atlantic by the silt carried down by the Orinoco river can
be discerned for as much as 100 miles from the river's mouth. The problem of
erosion is all the more threatening because, except in a few isolated instances
such as the island of St Vincent in the West Indies, so few steps have been
taken either by governments to inculcate or by farmers to carry out any
measures of soil conservation. Irrigation, although locally practised, is not
such a pressing agricultural need in this relatively wet region as in the more
arid parts of the Old World tropics, and there are no large irrigation schemes
on the scale on which they are to be found in Asia and Africa.

The natural climax vegetation of by far the greatest part of the American
tropics is forest; centred on the Amazon basin is the largest continuous block
of tropical rain forest in the world. Forestry is therefore an important industry throughout the area. Some areas have speciality timbers which have been
exploited for export, such as balsa wood (*Ochroma lagopus*) in Ecuador, or
wallaba (*Eperua falcata*) and greenheart (*Nectandra rodiaei*) in Guyana. Many
products of wild forest trees have figured in exports from tropical America.
Some of these products have now lost most of their significance, such as wild
rubber and cinchona bark which are produced commercially on a plantation
scale elsewhere, or logwood which has been replaced by synthetic dyes. Others
are still a useful source of revenue, such as chicle gum (from *Achras sapota*)
in Central America, brazil nuts (*Bertholletia excelsa*) in Venezuela and Brazil,
and to a lesser extent balata (*Mimusops globosa*) in Guyana and Venezuela.

Such zones of savanna as do occur in South America are chiefly located, as
in Africa, north and south of the main belt of equatorial forest, but they are
not so dry as in Africa. The absence of trees in these areas may sometimes be
due to soil conditions, sometimes to frequent flooding, and more widely to
the recurrent grass fires which, as also in some of the African savannas, prevent a reversion of savanna to forest. The northern belt of savanna is known
as the 'llano' and extends through much of Venezuela and into Colombia. It
consists largely of almost treeless grassland, with occasional groups of trees
or stands of palms in the wetter spots and fringing forest along the rivers. In
the south, a similar belt is known as the 'campo' in southern Brazil, extending
into Bolivia and Paraguay. Here the grassland is more interspersed with
small trees and more closely resembles the African savannas; in some
areas, the 'campos' shade off into the xerophilous woodland known as the
'chaco'.

The Andes reach their highest point in the peak of Aconcagua (22,834 ft)

in Argentina; in the northern sector, the highest peak is Chimborazo (20,500
ft) in Ecuador. The natural vegetation is montane forest, which consists in the
tropical sector of hardwoods, and continues up to the tree-line at about 10,000
ft. Above this line are the mountain grasslands known as 'paramos' which can
be used for grazing at all seasons of the year. In these Andean regions cultivation is mainly concentrated on the valleys, plateaux and the so-called 'intermont basins'. Occasionally volcanic activity is deleterious to agriculture, the
most conspicuous example in recent years being the continuous eruption of
Mt Irazu in Costa Rica since 1963, which has killed several hundred cattle
and all the crops on the slopes of the mountain.

The New World is fortunate in escaping many of the agricultural pests and
diseases which affect the Old World. Several of the most serious crop diseases
of the tropics, such as coffee leaf rust, cassava mosaic and rosette disease of
groundnuts, are absent; though there are a few contrasting cases, such as South
American leaf blight of rubber (caused by *Dothidella ulei*) which does not
occur in the Old World and is the historical reason for the plantation rubber
industry developing there rather than in the American tropics. Carnivorous
wild animals do not present a serious threat to the ranching of livestock. There
are no tsetse-flies, and such locusts as occur in South America do relatively
little agricultural damage. Animal diseases are fewer than in other parts of the
tropics, probably because of the absence of all the major kinds of livestock
until the 16th century; there is no rinderpest, and trypanosomiasis is practically limited to equines. A local hazard is the vampire bats which occasionally attack livestock.

Agrarian Structure

Perhaps the most pressing problem in the development of agriculture in
Latin America is the question of land reform (see also Chap. 2.8). To under-
stand the nature of the problem, it is necessary to go back a little in history.
The early Spanish conquerors and settlers, having an understandable desire to
appropriate large estates in the territories they had won, were themselves
neither numerous enough nor willing to provide the labour for cultivation.
The Spanish Government solved this problem by the system of the 'encomi-
enda'—a grant of a large area of land together with the free labour of those
Indians who happened to reside on it. It is this system which, with comparatively slight modifications, has so widely survived to form the basis of tenures
today. Many very large holdings are still owned by the descendants, white or
mestizo, of early settlers. Much of the land of these estates, from lack of enterprise or capital, is still undeveloped or only partially used in present ownership.
Many owners are absentees who live in the towns and entrust the oversight of
their estates to managers of little skill or education. Labour tenancies, which
derive directly from the feudal theory of medieval Spain (but have had an
exact parallel in the 'squatter' tenants with labour obligations on European
farms in Kenya), still exist today; Dumont (1965) describes an instance of a

tenant on an estate in Colombia holding 16 acres on condition of giving two
days' labour a week (of 10 hours each) without wages to the land-owner.
Elsewhere, such tenancies have been transmuted into share-cropping, or into
cash rents with the land-owner providing some paid work for which, at the
general low level of agricultural production, wages cannot be high.

In contrast to the large and generally only partially developed estates which
are often described as 'latifundia', are a class of excessively small holdings at
the other end of the scale. Many of these are worked by Indians who, left to
themselves in some of the areas which Spanish settlers found unattractive,
have become overcrowded in some regions. These very small holdings are
spoken of, according to their size, as 'parvifundia', 'minifundia' or even
'minimifundia'.

In areas, such as the West Indies, where the Indians were few in number or
died out quickly, Negro slaves were introduced for agricultural work. After
the abolition of slavery, although many West Indian estates have continued
to be employers of paid labour up to the present day, many Negroes took up
holdings either as tenants or owners. It is one of the agricultural tragedies of
the region that these people had lost any tradition of self-reliant peasant-
farming, and were not given any assistance or instruction until the development of modern and still inadequate extension services. This is largely the
reason that accounts for the poor standards in many areas of West Indian
peasant farming in an environment that is unusually favourable for tropical
agriculture.

Different Latin American countries have tackled these defective systems of
land tenure in different ways. In Mexico, following the revolution of 1911, an
ancient Indian system of land tenure was revived as the 'ejido'. Most of the
ejidos consist of land which is communally owned but individually worked by
the cultivators; but a small proportion, mainly on irrigated land, are genuine
collective farms. Most of the land involved has been found by the expropria-
tion of large estates. In 1960 there were 20,000 ejidos comprising about half
of the arable land of the country; the standard of cultivation is not necessarily
higher than that of other peasants but there is considerable variation. In
Cuba, the revolution of 1959 and adoption of Communist philosophy led to
the breaking-up of estates and their redistribution to cultivators to be organised as co-operative farms, of which there were over 700 large ones by 1965.
Most of the remaining Latin American states have programmes of land reform and many have passed legislation giving wide powers to carry it out, but
the actual degree of implementation of these programmes is very variable and
in many cases extremely slight. The true state of affairs that still exists is shown
by the estimate of Johnson and Kristjanson (1964) that for Latin America as
a whole, two-thirds of the agricultural areas comprise farms of over 2,500
acres which only represent 2% of all holdings.

Another great weakness of the agricultural economy is the lack of capital
and difficulty of obtaining credit for development. Even in the ejidos there

have been many members who failed to take up their land owing to lack of **Farming**
implements and capital. Nevertheless, Mexico's 'Banco Nacional de Crédito **Systems**
Ejidal' is one of the older credit institutions with a useful record of achievement. The need for better credit facilities for farmers is widely realised and
has attracted some help from outside the region. The American International
Association for Economic and Social Development, a private philanthropic
foundation in the United States which has done some excellent work in agricultural development in Latin America, has pioneered supervised credit and
been instrumental in the formation in Brazil of ABCAR (Associação Brasi-
leira de Crédito e Assistência Rural) and in Venezuela of CBR (Consejo de
Bienestar Rural). In Mexico, UNESCO and the Organisation of American
States sponsored with others a credit scheme which since 1951 has provided
an effective stimulus to the poultry industry (F.A.O. 1964).

The means of cultivation in the American tropics are rather varied. Ox-
ploughs, hand-hoeing on the smallest holdings, and tractors especially on the
large estates all play a significant role but it is difficult to estimate their relative
importance precisely. Brazil has by far the largest number of tractors in the
region (63,000 in 1962) and has had its own tractor manufacturing industry
since 1957. In 1962, Colombia had 23,500 tractors and Venezuela 10,000; but
by contrast Ecuador had only 1,620 in 1963 and Bolivia only 20 in 1957.
Mules are more important here than in any other region of the tropics, but
their use is for transport rather than cultivation. Some manufacture of fertilisers is more widely carried out in the American tropics than in most other
tropical regions. Apart from the long-established guano industry, chiefly in
Peru, manufacture of nitrogenous fertilisers is recorded in five countries, and
of phosphatic fertilisers in four. Except however in Mexico, output of these
materials is still only on a modest scale.

The marketing of farm products generally in the region is free between
farmers and traders, but there are also many instances of special arrangements.
In the British West Indies under colonial administration, associations of
growers were formed in some islands for the marketing of crops such as
bananas and citrus, whilst in other cases co-operative societies handled food
crops and vegetables. In Cuba after the revolution of 1959 the Government
took control of trade and directed most of the country's agricultural exports
to other Communist countries, largely under barter arrangements. In Brazil,
the Government's efforts to maintain coffee prices through so-called 'valorisa-
tion' programmes have led at different times to prohibitions on coffee planting, the burning of surplus stocks and large-scale storage to withhold the crop
from the market. Some plantations in many countries are owned by large
companies based in the United States and the United Kingdom which have
their own transport and marketing facilities. In some small countries it has
happened that a single company operates on such a scale as to have a determinant effect on the national economy; such situations inevitably tend to
concentrate public attention on the policy of the company and do not always

make for easy relations between it and the Government. This type of position
is exemplified by the very large production of bananas by the United Fruit
Company (based in the United States) in Honduras, and of sugar by the
Booker group (based in Britain) in Guyana.

One of the most depressing features of the agricultural scene in the American tropics is the extent of food imports into a region which should be inherently so productive. It is understandable that the West Indian islands should, as they have done for three centuries, find it necessary to import food both because of their crowded populations and because, with land at a premium, it may pay better to grow specialised export crops and buy food with the proceeds than to use the land for food crops. But it seems less logical that Brazil, with its huge area, should be a large food importer, that Venezuela should need to import about 10,000 tons of rice annually, or that Panama should import 60 % of its food supplies.

The poor standards of cultivation which are widespread are in part perhaps due to the lack of agricultural research, education and extension work, which have been later in coming to Latin America than to most parts of the tropics. It is noteworthy that two of the best-known regional centres for research and education have been provided by stimuli from the outside. These are the Imperial College of Tropical Agriculture (now part of the University of the West Indies) in Trinidad, which was established by the British Colonial Office in 1924 and used to train agricultural staff for all the British colonial territories; and the Inter-American Institute of Agricultural Sciences at Turrialba in Costa Rica, which was largely financed and staffed from the United States. Since the Second World War there has however been a great increase in the number and effectiveness of research stations, some of which have already made notable contributions. The breeding of higher-yielding wheat varieties in Mexico, with international help, has achieved a success which has attracted students from many other tropical and sub-tropical countries. The Campinas Institute of Agronomy in Brazil has acquired an international reputation in the genetics of coffee. Facilities for agricultural education at university level have also been increasing rapidly. By 1962 there were 54 university faculties of agriculture in Latin America. These however only produced 1,113 graduates, and there are still small countries with no such facilities, whose students must go to other countries in the region if graduate agriculturists are to be obtained for the national services. Agricultural extension work is still only slowly becoming effective; most countries in the region now have at least a nominal extension service, but in some, real results from its work have yet to be seen.

Crops and Livestock

The basic food crops of Latin America are still those which are native to the region and were already grown in pre-Columbian times. The most important of these are maize, beans (*Phaseolus vulgaris*) and cassava (*Manihot utilissima*) —also known in this region as 'manioc' and 'yucca'. Maize meal is here

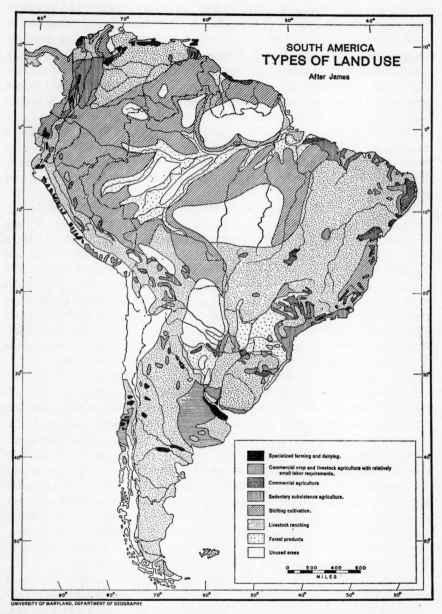

SOUTH AMERICA
TYPES OF LAND USE
After James

Legend:
- Specialized farming and dairying.
- Commercial crop and livestock agriculture with relatively small labor requirements.
- Commercial agriculture
- Sedentary subsistence agriculture.
- Shifting cultivation.
- Livestock ranching
- Forest products
- Unused areas

0   200   400   600
MILES

UNIVERSITY OF MARYLAND, DEPARTMENT OF GEOGRAPHY

Fig. 2.14.1 Types of Land Use in South America *Source:* Van Royen, 1954

441

cooked, not as a gruel as in Africa, but in pancakes (*tortillas*). In Venezuela,
the acreage of maize grown is more than that of all commercial crops combined. In Brazil, maize production increased by 50% in the decade 1955–65,
showing no sign of a decline in popularity. Other native American crops, such
as sweet potatoes and groundnuts, are common but not so important as those
already named. In the West Indies, with their Negro population, Old World
crops such as yams and pigeon peas become more important. The high altitude regions of the Andes, which gave the potato (*Solanum tuberosum*) to the
world, have many special and generally little-known crop plants, some of
which might perhaps prove on further study to have possibilities for development. These include the two non-gramineous grain crops quinoa and kañahua
(both species of *Chenopodium*) and the tubers ulluco (*Ullucus tuberosus*) and
apio (*Arracacha esculenta.*)

Of the food crops introduced from the Old World, rice has become much
the most important. Brazil is by far the largest producer, with an annual crop
of the order of 4 million tons; Colombia ranks second with only about a tenth
of this amount and is followed in descending order by Mexico, Guyana and
the other countries of the region, all of which grow some of the crop. The
acreage of this crop has for many years expanded steadily and is mainly devoted to meeting the domestic food demand, though Guyana has deliberately
developed rice cultivation for export and has now become the leading exporter
in the region. Wheat can of course be grown in the Andean highlands but few
of the countries have achieved self-sufficiency in this cereal though great
efforts are being made to get nearer this target in Colombia and elsewhere.

Sugar-cane is the oldest cash crop of the American tropics, having been
introduced by Columbus himself, and the economy of many countries is still
heavily dependent on it. Cuba is one of the world's largest producers and
exporters of sugar; there was some decline in production following the ex-
propriation of estates in the years after 1959, but this was later made good,
and exports are now directed largely to Communist countries. Brazil produces
almost as much sugar as Cuba but exports far less. Mexico and Peru are the
next biggest producers, but on a much smaller scale. Several West Indian
islands, although smaller producers because of their size, are extremely dependent on sugar exports for their income; these islands include notably
Puerto Rico, the Dominican Republic and Barbados. Besides refined sugar,
there is widespread production, as in Asia and Africa, of cruder non-centrifugal sugars which are here known as 'panela', for local consumption.

Coffee, though introduced much later than sugar-cane, is the other most
important export crop, about 60% of world supplies being grown in the
American tropics. This is practically all *arabica* coffee, though a little *robusta*
is grown in Trinidad and Jamaica. Brazil's share of the world crop, which
was once over half, has declined to some 40%, but she is still by far the largest
producer. Her product consists of 'hard' (dry-processed) coffee of a low grade
for *arabica*, and much of the crop is grown at a poor standard of cultivation:

local practices include the absence of shade trees, which are claimed not to be
beneficial in this climate, and the planting of several coffee trees together at
one planting point; there has been much soil erosion. Colombia produces a
much smaller crop but of higher quality. Mexico, Venezuela and other countries produce crops which are smaller again. Coffee is nevertheless very important in the economy of some small countries; in El Salvador it accounts for
some 75% of exports, although efforts are now being made to diversify the
economy, and the coffee of Costa Rica has a particularly high reputation for
quality. Cocoa (often and more correctly spelled 'cacao' in the literature of
this region) is the other most important beverage crop. The American tropics,
to which this tree is native, have long been outstripped in its production by
West Africa, but Brazil is still a large producer, and the crop is economically
important in Ecuador, Mexico, Colombia and Venezuela, and on a small
scale elsewhere. Another beverage crop, 'yerba de maté' (*Ilex paraguensis*) is
produced in some quantity in Brazil and Paraguay and in temperate south
America (p. 330) to meet the large local demand, but has as yet found little
place on world markets.

Bananas are the most important export fruit crop. The American tropics
have a monopoly of supplies of this fruit to temperate North and South
America, and also send large quantities to Europe. Production comes both
from large company-owned plantations, especially those of the United Fruit
Company in several of the Central American republics, and from middle-
sized and small holdings. Ecuador is the leading exporter, providing in recent
years about a quarter of total world exports of this crop. Her exports are
followed in order by those of Honduras, Costa Rica, Panama, Brazil and
Colombia. There is hardly a country in the region which does not have some
export of bananas. In the West Indies production is on a smaller scale than on
the mainland; the leading exporters are the French West Indies, the Dominican Republic and Jamaica. Citrus fruits are the other most significant export
fruits; there are important exports of oranges from Brazil and, on a smaller
scale, of grapefruit from Trinidad. Coconuts are a widely distributed crop,
especially in coastal regions, and copra is produced both for local industry and
as a minor export; Mexico is much the biggest producer. Amongst other oil
crops, Brazil is the world's largest producer of castor seed and exporter of the
oil. Groundnuts hardly figure amongst exports except on a small scale from
Brazil. Oil-palms have been grown experimentally in a number of places, but
have as yet achieved a small commercial status only in Costa Rica.

Amongst fibres, cotton tends to be grown in the drier areas. The leading
producers are Brazil and Mexico, followed by Peru and Colombia on a smaller
scale. Brazil has increased sisal production till it now ranks second only to
Tanzania; Haiti and Venezuela have small crops. Mexico produces both
henequen (*Agave fourcroydes*—a fibre resembling sisal but selling for a slightly
lower price), mainly in the Yucatan peninsula where a lowish rainfall and
calcareous soils suit the crop, and istle (*Agave heteracantha*). Brazil has

Farming successfully achieved, alone amongst non-Asiatic countries, a modest output of
Systems jute. Amongst export crops of less widespread importance must be reckoned
tobacco in Brazil, the Dominican Republic, Cuba and Colombia, and some
rubber in Brazil. There are also some interesting speciality crops in particular
areas. Thus the island of St Vincent in the West Indies produces almost the
whole world supply of arrowroot (*Maranta arundinacea*); Jamaica is almost
the sole producer of allspice (*Pimenta officinalis*); Sea Island cotton is practically confined to St Vincent and the Leeward Islands; Grenada specialises in
the production of nutmegs and mace. The coca shrub (*Erythroxylon coca*)
from which the drug cocaine is obtained is indigenous to Peru and Bolivia,
and these are the world's main suppliers.

Amongst plants grown for use by livestock, special mention must be made
of Pangola grass (*Digitaria decumbens*), a pasture grass quite recently introduced into the American tropics. It has proved so well suited to the environment that it has been very widely planted in the West Indies, Central America
and northern South America and has done much to improve the stocking
capacity of grazings.

Table 2.14.1 shows the production of the principal crops, and the numbers
of livestock, in the four countries selected for special description.

The only livestock indigenous to the Americas are llamas, turkeys and

Table 2.14.1

Crop Production and Livestock Numbers in Selected Countries of
Tropical South America

Crop production in thousands of metric tons per annum:

	Ecuador	Colombia	Venezuela	Guyana
Wheat	73	117	1	—
Maize	211	1,071	574	1
Rice (paddy)	173	742	220	308
Sugar-cane	8,914	16,535	4,409	3,571
Cassava	280	2,439	332	11
Beans (dry)	34	47	46	—
Bananas	3,638	1,064	1,356	6
Coffee	73	529	60	1
Cocoa (beans)	40	19	24	n.a.

Livestock numbers, in thousands:

	Ecuador	Colombia	Venezuela	Guyana
Cattle	2,000	14,116	6,636	332
Horses	227	983	401	3
Mules	106	380	70	2
Donkeys	167	366	453	3
Pigs	1,246	2,326	1,893	66
Sheep	1,718	1,223	79	87
Goats	165	374	1,241	32

Source: F.A.O. Production Yearbook 1966.

Table 2.14.2

Transcontinental Transect

Country	Guyana	Guyana	Venezuela	Colombia	Colombia	Colombia
Region	Coastal strip	Rupununi savannas	Llanos	Magdalena Valley	Andean	Pacific Coast
Altitude (ft)	6	300–650	0–600	0–600	3,000–9,000	0–300
Rainfall (approx) (annual inches)	89	50	30–40	40–80	42 at Bogota	280
Soil type	Clay	Light, poor	Bleached earth	Lithosols	Mountain soils	Forest
Vegetation	Swamp forest	Savanna	Grass, some trees	Savanna	Mountain flora	Forest
Farming systems	Plantations and peasants	Ranching	Ranching	Farms of varying size	Farms of varying size	Food crops only
Chief products	Sugar, rice	Cattle	Cattle	Bananas, sugar, tobacco, maize, etc.	Coffee, potatoes (Solanum), wheat	Sweet potatoes, maize

445

guinea-pigs. None of these are of any great agricultural importance today, though the llama is still used as a pack-animal principally in Peru and provides also wool, meat and hides. Better qualities of wool are obtained from its relatives the alpaca and vicuña; all three animals are mainly found in very high-altitude tablelands above 10,000 ft.

The cattle of the American mainland tropics were until the 19th century almost exclusively derived from Spanish and Portuguese introductions of types from their homelands, of which the long-horned black Andalusian breed was probably the most important. In the West Indies, there were also introductions by the colonists of English, French, Dutch and Danish cattle. This mixture of animals, which came to be known as 'criollo' or 'creole' cattle, gradually became acclimatised but are by modern standards poor producers of either milk or beef. They still provide the bulk of the cattle population but during the last century there have been frequent introductions both of modern European breeds and of zebu cattle. Friesians (more often known here as Holsteins) have been widely introduced as dairy animals, particularly at the higher altitudes although they can be rather more widely used than in the Old World tropics owing to the lower incidence of animal diseases. Brown Swiss and Channel Island breeds have also been conspicuous amongst introductions and in Jamaica a stabilised cross between the Jersey and zebu with some Holstein blood added later, known as the 'Jamaica Hope', has been widely used for dairy purposes. The zebu (often, following United States usage, called Brahman) has chiefly been used as a beef animal, for example on some of the ranches of the Venezuelan llanos.

The population of sheep and goats is again mainly derived from early introductions from Europe, and with poor genetic quality and generally poor management their productivity is low. There are more sheep than goats in most countries, but in Venezuela the reverse is the case. Pigs are equally or more numerous than either of these animals but their history and performance are much the same. Poultry, usually of the same mixed derivation, are fairly widely kept, and in Colombia for instance they are estimated to provide 8·5% of the total meat consumption. Improved types of all these kinds of animals have of course been introduced by enlightened land-owners and by experiment stations, but it will take time for them to affect significantly the mass of the animal population. Horses have been an indispensable element in the herding of cattle on fenceless ranches, and are still kept in considerable numbers. Donkeys and mules are widely used as pack-animals, the former being generally more numerous; without their donkeys, many small farmers would find it difficult to get their crops to market.

The Agriculture of Four Selected Countries

In describing the agriculture of the four selected countries—Ecuador, Colombia, Venezuela and Guyana—we shall proceed in that order from west to east, in an arc around the northern rim of South America. The descriptions will thus in a sense provide a curving transect across the continent from the Pacific to the Atlantic shores.

ECUADOR

Ecuador, as its name implies, lies astride the Equator which nearly passes through its capital, Quito; this city stands however in the Andes at an altitude of 9,446 ft and its climate, with a mean annual temperature of 55°F showing extraordinarily little variation, has been described as one of perpetual spring. Apart from a few arid patches on the coast, the rainfall is everywhere adequate for the growth of perennial crops. The agriculturally developed parts of the country fall into two regions. The first is the coastal region and lower river valleys, where tropical crops are grown. The second consists of the hills, foothills, mountain valleys and intermont basins where grazing and dairying and the production of cereals and potatoes (*Solanum*) prevail. Outside these developed areas, the remaining and greatest part of the country is covered by forest, but could not all be considered as cultivable owing to the mountainous topography and extent of land at very high altitudes. It is from the first of the two agricultural regions that most of the exports come. The second is more concerned with the production of foodstuffs for local consumption. Besides what it can provide, some wheat has to be imported into the country. The production of sugar-cane has been increasing and it is becoming a crop of major importance.

Bananas, cocoa and coffee constitute over 80% of the country's exports. It is only as recently as 1953 that Ecuador become the world's leading exporter of bananas. Ecuador actually held a similar position as the chief exporter of cocoa at the beginning of this century, and cocoa was still her main export item in the 1930s. The decline in the importance of cocoa was due at least partly to attacks of disease (caused by the fungi *Monilia* and *Marasmius*). The rapid growth of the banana industry has been ascribed by some observers to the satisfactory land tenure system in Ecuador, where there is a higher proportion of middle-sized holdings than in many Latin American countries. This is not to say, however, that there are not also too many very large estates and too many very small holdings. Statistics show that 2% of the country's land-owners with estates of over 250 acres own 64% of the land; and an Agrarian Reform and Colonisation Law was passed in 1964 to try to improve the situation. In the banana industry itself, the mean size of holding is 116 acres (Preston 1965), but many producers have more than one holding; this compares with a mean size of 43 acres for all farms in the country.

The banana industry also owes its rapid growth partly to the construction of roads which opened up new areas for production. Low wages in the country also contribute to making its production competitive. Perhaps warned by the earlier effects of disease on the cocoa industry, the Government provides through the Direccion Nacional del Banano disease control services for banana growers including oil spraying from the air against leaf-spot. Corporation-owned plantations play a minor part in the industry, and corporations func- tion more importantly as buyers and exporters. One of the problems of the industry is the high proportion of fruit which has to be rejected as unsuitable for export. About half the exports go to the United States; other leading cus- tomers in recent years have been Germany, Belgium and Chile.

Pyrethrum is a recent addition to the export crops to which the higher country seems well suited, and if world demand for this insecticide is sustained it may perhaps have an important future.

The Ecuadorean population consists roughly of 40% mestizos, 30% Indians, 10% whites and smaller numbers of Negroes and mulattoes. The annual rate of population increase from 1953 to 1960 was 3·2%, a high figure typical of Latin America where the 'population explosion' in recent years is one of the most violent of any region of the world. With a national population density of only 20 to the square mile, it would seem that any danger of over-population is still far off, but the limited proportion of land in the Andean region which is really worth cultivating has to be borne in mind.

COLOMBIA

With the exception of the Guajira peninsula in the north-east which has an arid climate, the rainfall is everywhere favourable to agriculture. Nevertheless, only a very small proportion of the country's land is actually cultivated. From 50 to 70% has been estimated to be under forest, and the south-eastern third of the country is almost undeveloped. Farming is mainly concentrated in the northern (Atlantic) coastal region and in the Andes where the two river valleys of the Cauca and Magdalena are especially important; the capital, Bogotá, is in this Andean region at an altitude of 8,730 ft. On the Pacific coast, a sparse Indian population practises a shifting cultivation of food crops according to their ancient methods.

The main determinant in agricultural production is altitude. The northern coastal plain and lower river valleys are suited to the production of tropical crops. Here sugar-cane is especially important, occupying nearly a million acres. Besides centrifugal sugar, 'panela' (a cruder product) has a large place in local diets. Rice is also an increasingly important crop for the internal market. The middle altitude zone from about 3,000 to 6,500 ft is essentially that in which coffee is grown. The sub-Andean belt also provides wild cin- chona trees for exploitation, whilst the lower forests contain wild rubber (a few rubber plantations have also been started) and copaiba trees (*Copaifera* spp) which are tapped for their resin. Going higher up the Andean valleys, maize is grown up to an altitude of about 8,800 ft, wheat up to 9,800 ft and barley and potatoes up to 10,500 ft. A peculiarity of local land use is that the flatter land of the valley bottoms has often been used for grazing, whilst crop production was carried out on steeper slopes where the soil was poorer and much erosion took place; the extension services are now attempting to correct this pattern.

Coffee is by far the most important export crop of Colombia, which is the **Farming** world's second producer after Brazil and the biggest producer of 'mild' coffees, **Systems** for its *arabica* production is of a higher quality than that of Brazil. Most of the coffee is grown by smallholders on family-sized farms; there are some 150,000 growers. The climate enables the coffee harvest to be spread throughout the year, thus avoiding market congestion and making transport problems easier. An interesting and less familiar cash crop is 'fique' (*Furcraea macrophylla* and to a less extent other species of *Furcraea*), a fibre crop of which some 12,000 tons are produced annually which is enough to manufacture the country's requirements of sacks and cordage. Local experiment stations are beginning to make their effect felt in crop improvement; the station at Palmira in the Cauca valley has improved the strains of beans, and the experimental vineyard at Bolivar has demonstrated that vines are a promising crop.

Cattle, kept mostly for beef, are widely distributed on the plains. There is a large export of hides, and tanning is an important industry.

The Colombian population is divided approximately into 47% mestizos, 30% white, 15% mulatto, 4% Negro, 2·5% Indian and small numbers of others (Arango Cano 1964). This is considerably different from that of Ecuador, which has more Indians and smaller white and Negro elements—an important agricultural point, since each race still practises, to a large extent, its own particular system of farming and choice of crops. The rate of population increase, at 2·2% annually in 1953–60, was lower than Ecuador's but the overall population density is greater at 34 per square mile. This, although one of the higher population densities for South America, is still very low compared with Asian and many African countries.

VENEZUELA

There are four main ecological zones in Venezuela. The coastal strip in the north, which at its western end adjoins the Guajira peninsula of Colombia, continues the same arid climate with an annual rainfall of less than 20 in in places and is too dry to offer much attraction for agriculture. Southward of this lie the Andean ranges, which in this northernmost part of South America curve round so as to run ultimately from west to east, and it is in the mountain valleys and basins that most of the country's crops are grown. The annual rainfall in most of the zone is between 45 and 64 in, though the capital Caracas, which stands at 3,418 ft but rather near to the coast to be in touch with its port of La Guaira, has only 33 in. The third zone, lying between the Andes and the Orinoco river, consists of the grassy plains of the llanos with their marked seasonal alternation between aridity and widespread flooding, which are mainly used for extensive ranching. Finally, about half the country lies south of the Orinoco and has hitherto been almost totally undeveloped. This area, known as the Guiana Highlands, is largely covered with forest but has also much interspersed savanna; the distribution of forest and savanna in this region, as in some other parts of South America, has proved something of a puzzle to ecologists as it is difficult to correlate it either with edaphic or climatic variations. In total, only 18% of the country is used for agriculture.

The background to the agricultural economy of Venezuela is rather different from that of the neighbouring countries because of her massive oil production, whilst there are also large known reserves of iron ore and bauxite which provide the base for an industrial complex now arising at Santo Tomé de Guayana in the east. Petroleum products have in recent years contributed more than four times as much as agriculture to the gross national product. This has had the effect of making more funds available for agricultural improvement than in neighbouring countries, whilst at the same time reducing the incentive to the development of export crops. It has also made Venezuela a high-wage country with high food prices sustained by tariffs on imported foods, which, nevertheless, the country is wealthy enough to afford in considerable quantity.

The system of land tenure has been, at least until very recently, that which is typical of South America. In 1956, 1·7% of land-owners owned 75% of the farmed land. Of the 400,000 farm families, it was estimated that half were hardly more than subsistence farmers, mostly with from 2½ to 5 acres under crops and cultivating entirely by hand. The remaining half, who sold appreciable amounts of produce and could therefore be classed as commercial farmers, ranged from holders of large estates through middle-sized farms to small but specialist farmers (I.B.R.D., 1961). Amongst these last, animal-drawn ploughs are common only in a few areas in the Andes, but tractors are steadily growing in importance. However, a more considerable dent in the old system has been made in Venezuela than in the other countries we have considered, probably because more money has been available for land purchase and resettlement operations. Land reform here stems from the Agrarian Reform Law of 1960, which empowers the Government to settle farmers both on vacant land in its possession and on land expropriated from undeveloped estates. By 1966, 85,000 families had been settled on more than 5 million acres of land, of which approximately one-third was private land purchased for redistribution.

In spite of the employment given by mineral resources, it was reported in 1959 that 51% of the population gainfully employed were in agricultural work. Of the total population, 65% are recorded as mestizo and 20% white; the 7% Indian element live mostly in the forests, and the 8% of Negroes along the coast. The annual rate of population increase in the period 1953–60 reached the very high figure of 4·3%, which is explained partly by immigration attracted by the oil industry and high Venezuelan wages. Spaniards were the most numerous national group amongst the immigrants. Local overpopulation has already become apparent in certain Andean valleys but has so far corrected itself by natural emigration to other parts of the country which are less crowded. With an overall population density of only 23 to the square mile, there cannot be said to be any real pressure of population on resources.

The food crops are the same as in adjacent countries. Maize, beans, cassava and rice are staples of the diet. A number of irrigation schemes carried out or projected by the Government are aimed largely at increased production; the one on which most has been spent is the Guarico project in the llano region, where so far about 25,000 acres have been developed. Sugar-cane is less important than in Colombia or Ecuador. Sesame is a locally important crop, especially in the state of Portuguesa. Wheat grain and flour are amongst the most important food imports. Coffee and cocoa are the only significant agricultural exports, but the amounts are not large on a world scale; Venezuela produces less coffee than several much smaller Central American countries, and less cocoa than Colombia or Ecuador though it is of a fine quality. There are reported to be 63,000 coffee growers, indicating a high proportion of small producers. Amongst minor sources of income, there are a large number of wild forest products: wild rubber, cinchona and tonka beans (all of which have to compete with the plantation-produced product), copaiba, divi-divi (*Caesalpinia coriaria*—a tanning material) and balata.

The cattle of the llanos are raised for beef, and are mainly of criollo type though there has been some introduction of zebus. Further north there is dairying, especially to supply the fresh milk market of Caracas, but much of this milk supply comes from the Maracaibo region in the north-west and is derived from criollo cattle kept on a ranching system and milked in bails (Fr. abri à traite) on the pastures; such milk is transported three or four hundred miles in refrigerated tankers to the capital. The industry includes the manufacture of milk powder, though this could not compete on a free market against the imported article which is one of the major food imports of Venezuela.

GUYANA

This country shares with the others already described the common feature of agricultural underdevelopment; some 85% of the land is covered with forest. The agricultural area is, however, far more concentrated than in the other countries, being practically limited to a coastal zone of not much more than 10 miles in width. These 'frontlands', which lie partly below sea-level, are intersected with a network of canals which serve the triple purpose of drainage, irrigation, and transport especially of sugar-cane, and were originally constructed under Dutch rule.

Rainfall, which averages 88 in annually at Georgetown, the capital, is everywhere adequate for crop growth. The soils of the frontlands are fertile and consist mainly of alluvial clays with some sand ridges and, on the landward side, areas of peaty soil which are known as 'pegasse'. The soils of the interior are mainly sandy and, with the exception of a few patches, of low fertility; this factor, with the almost entire absence of communications, is mainly responsible for the lack of development of an area about the same size as Great Britain.

The population, amounting only to some 600,000, comprises two main
groups: people of African descent are now outnumbered by 'East Indians',
the descendants of indentured labourers who came from India to work on the
sugar-cane plantations. Minority groups include a few thousand indigenous
Indians, here known as Amerindians to distinguish them from the East Indians, in the interior and small numbers of Europeans and Chinese.

As regards food crops, the Amerindians of the interior still practise their
traditional production of maize, cassava and beans. For the rest of the population, rice is by far the most important food crop, although its production is
mainly carried on by the East Indian population, who derive principally from
rice-growing regions of India. Subsidiary food crops include plantains, cassava, maize, cow-peas and pigeon peas. Guyana has more abundant food
supplies and lower food imports than most countries in the Caribbean.

Agriculture provides about 60% of the exports of the country, timber and
bauxite providing other contributions. Amongst the crop exports, sugar is
the most important. The history of Guyana is typical of a sugar-producing
colony, and until far into the 20th century the sugar plantations were the overwhelming basis of the economy. With favourable conditions of soil and water
supply, sugar yields in Guyana have always been high and steady. The majority
of the plantations are in the ownership of a single company which therefore
derives advantages from economy of scale in working; the technical level of
production is highly efficient, with heavy use of fertilisers and of machinery in
cultivation, and a considerable amount of research carried on by the company
itself. The other most important cash crop is rice, production of which was
originally undertaken to supply local food needs but has expanded until now
it is an export crop second only to sugar in importance. Rice is the main field
crop of the smaller farmers in Guyana. The Government has taken great
pains to foster this industry in order to reduce the dependence of the economy
on the single product of sugar, by research and extension services, by special
development areas to demonstrate the use of mechanical cultivation, and by
settlement schemes to provide holdings for rice producers. Marketing and
export are a Government monopoly entrusted to the Rice Marketing Board,
and exports go largely to neighbouring countries, especially the West Indies.

Other cash crops are of very minor importance compared with sugar-cane
and rice. There is a small export of cocoa, and of low-priced coffee of the
liberica species although about half the production is consumed locally.
Coconuts are grown and copra processed locally for the production of oil.

Cattle are the chief form of livestock. The country is self-sufficient in meat,
and there is a moderate production of milk by small farmers in the frontlands,
derived from Creole cattle or their crosses with Holsteins which have been
quite widely introduced. The most significant development in the interior has
been the attempt to develop ranching of beef cattle on the Rupununi savannas
which provide wide areas of natural grazing. Communications between the
coast and this distant area are so poor that the carcases are flown out by air,

which imposes heavy marketing costs. The land itself is of low fertility and
capable of carrying only 10–20 cattle to the square mile; research work in
which the Government has taken a hand has not so far discovered any practicable means of improving these pastures by the use of fertilisers or other methods. The limited and moderate success achieved by even a well-managed project of this nature illustrates the extreme difficulty of any agricultural development of these interior lands.

As a final comment on the agriculture of northern South America, it has been said that, starting from the 'back canal' which defines the frontlands of Guyana some ten miles from the Atlantic, one could proceed in a straight line for 1,300 miles across the continent without encountering any human settlement or farming until one arrived within ten miles of the Pacific on the Colombian coast—and it is probably possible to select a line of which this would be true. The fact illustrates the emptiness of South America and the smallness of the attractions which the forested interior has offered for agricultural development.

Efficiency and Stability

Real agricultural efficiency in the region is confined mainly to the plantation industry and the estates of a few enlightened land-owners in South and Central America. The plantation sugar industry in particular is, like the plantation rubber industry in south-east Asia, one of the very few forms of tropical agricultural production which are efficient by world standards. An example like that of Barbados, a small island which carries a population of 1,470 to the square mile without any significant industrial development and mainly supported by sugar production, is one of the triumphs of tropical agriculture. But on the great mass of farms in the American tropics, the level of production is, considering the general favourability of the environment to plant growth, pitiable. This favourable nature of the environment, the enormous scope for research and extension work which in many areas has scarcely yet been begun, and the relatively good level of education of at least some sections of the farming population as compared with tropical Africa and Asia, suggest that this is one of the parts of the tropics which offers the greatest opportunities to an energetic effort to improve farming techniques and standards of living.

Natural elements of instability in farming are few, the main example being the incidence of hurricanes in part of the Caribbean region; in particular, the region is much less subject to droughts than many parts of tropical Africa and Asia. The political instability of many countries in South and Central America is notorious; but changes of regime do not in general much affect the day-to-day operations of the farmer unless they either cause monetary difficulties or go to extremes as in Cuba. The population explosion in the American tropics is sometimes held to be a cause for alarm, since rates of population increase are currently even higher than in most other parts of the tropics. If the region could be taken as a whole, there would be no pressing danger of over-population since, as we have seen, so much of South America is agriculturally empty. But the maldistribution of the population undoubtedly poses a problem of rural congestion which already exists in certain areas, particularly some of the Andean valleys and some West Indian islands. It is difficult to see how any possible increase in agricultural productivity can provide an adequate livelihood for the increased populations which, on current statistics, are forecast for islands such as Puerto Rico and Barbados in the next decades. For such cases the only conceivable remedy seems to lie in either birth-control or emigration, and of these the latter is not always either easy or popular.

Meanwhile the vast undeveloped forest areas of South America stare at us from the map as the greatest remaining habitable but unfarmed region in a hungry world. We must beware of expecting too much from this untapped resource. The greatest obstacle to development is the poverty of the soils which seems to deny prospects of any really intensive use of the land. The second great obstacle, absence of communications and the natural difficulties of the terrain for constructing them, is made even more difficult to remedy because a population using the land only extensively would hardly support the expense of construction. Pressure of population will no doubt gradually enforce a slow nibbling into these undeveloped areas, and some patches may be found which are more fertile than others; but dramatic developments on a large scale hardly seem an immediate likelihood.

References

Arango Cano, J. 1964. *Geografia Fisica y Economica de Colombia*. Bogota. 5th edn.

Bennett, H. H. and Allison, R. V. 1928. *The Soils of Cuba*. Washington.

Dumont, R. 1965. *Lands Alive*. Merlin Press, London.

F.A.O. 1964. *New Approach to Agricultural Credit*. Rome.

Hardy, F. 1942. 'The Soils of South America', *Chronica Botanica*. 7. 213.

International Bank for Reconstruction and Development. 1961. *The Economic Development of Venezuela*. Baltimore.

Johnson, V. W. and Kristjanson, B. H. 1964. 'Programming for Land Reform in the Developing Agricultural Countries of Latin America', *Land Economics*. 40. 353.

Marrero, L. 1964. *Venezuela y Sus Recursos*. Caracas.

Preston, D. A. 1965. 'Changes in the Economic Geography of Banana Production in Ecuador', *Inst. British Geographers Trans. and Papers*. Pubn. No. 37.

Regional Research Centre of the British Caribbean. 1958–63. *Soil and Land-Use Surveys*. Nos. 1–14. Trinidad.

Romney, D. H. (Ed.) 1959. *Land in British Honduras: Report of the British Honduras Land Use Survey Team*. H.M.S.O. London.

General Reading

Agriculture in the West Indies. 1942. H.M.S.O. London.

Atlas Agricola de Venezuela. 1960. Caracas.

Cole, J. P. 1965. *Latin America: An Economic and Social Geography*. London.

James, P. E. 1959. *Latin America*. London, 3rd edn.

Ordish, G. 1964. *Man, Crops and Pests in Central America*. London.

Powelson, J. P. 1964. *Latin America: Today's Economic and Social Revolution*. New York.

Roseveare, G. M. 1948. *The Grasslands of Latin America*. Imperial Bureau of Pastures and Field Crops. Aberystwyth, Gt. Britain (now Commonwealth Agricultural Bureaux, Farnham, Bucks., England).

Smith, T. Lynn. 1963. *Brazil: People and Institutions*. Baton Rouge.

Part 3

CONCLUSIONS

The Location, Intensity, Stability and Efficiency of Farming Systems

Chap. 1.7 completed, with a catena (Table 1.7.1), the analysis of the factors influencing the location and intensity of farming systems and postulated an empirical process of synthesis. The practical effects of influents were examined, on a comparative basis, for a number of countries or regions where it was possible either (*a*) to hold ecological conditions relatively constant and study the effects of variations on socio-economic factors (p. 98), or (*b*) to hold socio-economic factors relatively constant and study the effect of changes in ecological factors (Part 2). Chap. 2.1 gave a broad working classification of farming systems (Table 2.1.1).

Based on the above evidence, and in part on Duckham 1963, the present chapter outlines *first*, for temperate areas, a mainly qualitative synthesis of the location and intensity of the classes (p. 105) of farming systems; *second*, it suggests some models of syntheses which could be, at least in part, quantified and possibly have predictive use; *third*, briefly discusses factors affecting stability and efficiency, and *fourth*, it considers the application of such syntheses and principles in the tropics.

The Influence of Ecological and Operational Factors

THE SPECTRUM OF TEMPERATE FARMING SYSTEMS

In advanced temperate regions with a Beckerman Index (B) (p. 11) greater than say (40) and where, even though population density (D_s) is high, resources per head are ample, there is, for a country or countries such as North America (U.S.A. and Canada), north-west Europe, Australia and New Zealand, in which socio-economic factors can be regarded as constant, a well-marked spectrum of farming systems.* The spectrum (Table 3.1.1 and Fig. 3.1.1) ranges from extensive grazing systems in warm, dry areas (Bands I and II) and then, through tillage (Band III), alternating (IV) and cultivated grass-land systems (V), back to extensive grazing systems in cold or cool wet mountain areas or on cold, often dry, scrub and tundra (Bands VI and VII).

The spectrum (and its apparent influents and resultants) is not intended to be precise; is necessarily somewhat arbitrary; and is painted in with a broad

* For interesting eco-climatic gradients in Africa and India, see Whyte, R.O., pp. 308–15, in Hills 1966.

459

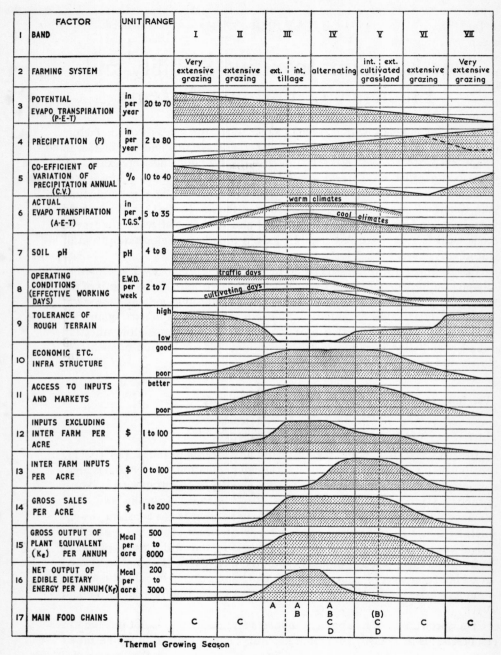

Fig. 3.1.1 Spectrum of Temperate Farming Systems

460

brush. But it does suggest that, in the centre there are Bands (III Int., IV, V Int.) where *either* mean annual precipitation (P) is not greatly in excess of mean potential evapo-transpiration (P–E–T) or hereafter T in the T–G–S *or* where T does not greatly exceed P. These Bands may be called hydrologically neutral; in them, the hydrologic ratio, i.e. h in models 3.1.1, 3.1.2 and 3.1.3 below is, say, greater than 0·8. Mean actual estimated transpiration A–E–T (or A in the models below) is, as noted in Chap. 1.2, a very useful climatic parameter. In these Bands it is usually over 20 in in the thermal growing season (T–G–S) (or, in warm semi-arid areas, in the hydrologic growing season) and is relatively stable. Such Bands can usually support intensive systems with high actual (other than inter-farm) inputs, or high input ratios, even though such land is not always intensively farmed (e.g. Uruguay).

Within the hydro-neutral zone there is flexibility in farming systems as well as in input intensity. Thus, the 'hydro-neutral' Canterbury Plains of New Zealand, now in intensive grassland and intensive alternating, *could* be in tillage systems; but limited access to distant markets and high transport costs make the conversion of primary plant production, in this case grass, into animal products the better land use. In Argentina, the limited, tall grass 'hydro-neutral' area (Band III Int.) could probably be in intensive tillage, as in the U.S. Corn Belt, but is in extensive grassland and extensive tillage. Or again, within the hydro-neutral zone, the proportion of tillage may vary in time with changes in economic or external pressures or with technology; for example, the great increase in tillage area in the United Kingdom in the First and Second World Wars and the recent development of 'continuous' cereal growing.

Within this zone relatively small differences in soil type (e.g. in clay content) or in soil moisture status may have great influence on systems and enterprises. Visser, W. C. in Rutter and Whitehead 1963 at pp. 356–65 brings this out very well. He shows that, in the Netherlands, high water-tables are more frequently in permanent grass and that low water-tables are more frequently in tillage crops. Neglecting such local influences, normally, however, *within* the hydro-neutral zone, towards the drier bands, i.e. as $T > P$ and h in model 3.1.1 below decreases, comparative advantage lies with tillage crops. These have, in general, a lower leaf area duration and hence have lower transpiration water needs than pasture swards, and also require more cultivating and traffic days (Duckham 1963, p. 429) than grassland systems. Towards the wetter and cooler bands, i.e. as $P > T$ and h in model 3.1.1 decreases, the advantage lies with grassland systems. These have greater moisture needs but less exacting operational requirements than tillage; they start and end the growing season with leaf area index (L.A.I.) of 1·0 or more; if well managed they can (in contra-distinction to annual tillage crops which often use only half the T–G–S (Duckham 1963, p. 477)), grow throughout a T–G–S which may be too short for the seeding, growing and harvesting of tillage crops as in parts of Sweden (p. 233).

A.E.T. and the hydrologic ratio

To the left of the hydro-neutral zone, in Band III (Ext.) and Band II, the choice of systems, enterprises and species is increasingly limited. A–E–T (i.e. A) falls, excess of T over P increases, i.e. the $\frac{P}{T}$ ratio and h decrease, pH and salinity rise, precipitation (P) becomes less reliable (i.e. has a higher C.V.). Therefore, high actual inputs or high input ratios (p. 6) are not justified; hence systems become more extensive, enterprise choice becomes limited and crop yields, stocking rates and cash outputs become lower. This is well shown in the U.S.A. (Table 2.3.5, Figs. 2.3.1 and 3.1.12).

To the right of the hydro-neutral zone, i.e. in Band V (Ext.) and Band VI, A falls, excess of precipitation P over T increases (i.e. the $\frac{T}{P}$ ratio and h decrease), leaching increases, effective working days decrease, the mean thermal growing season (T–G–S) declines and its effective duration becomes less predictable. Therefore, high actual inputs or high input ratios cease to be economic; choice of systems is restricted and extensive grazing becomes the dominant farming system with forestry and tourism as competing land uses.

Thus, at the extremes (viz. Bands I, II, VI, VII), tillage, even if operationally practicable, is uneconomic. The best agricultural way to use the land is to employ livestock to harvest and concentrate, as muscle tissue or wool, what little vegetational growth there is. The productive potential (model 3.1.2 below) is poor, often partly because the relief is difficult. Actual inputs and outputs per acre are low (Tables 2.3.4 and 2.3.5 and Fig. 3.1.12) and human population is usually sparse. Not surprisingly, even within rich countries such as the U.S.A., Sweden, the United Kingdom or New Zealand, the infra-structure tends to be poorer in these Bands (e.g. roads difficult, schools may be a long way away, the doctor a day's journey by car). Partly as a result, market access is locally poorer than in areas with higher productive potential, such as the U.S. Corn Belt or East Anglia in the United Kingdom or the northern plains of France. (The preceding paragraphs are summarised in diagrammatical form in Fig. 3.1.11 whilst the supporting climatic data is shown on the maps in Figs 3.1.2 to 3.1.10 and in Table 3.1.3.)

Effect of Irrigation

However, in Bands II and III (Ext.), and occasionally in Band I (e.g. irrigated valleys in the arid south-west mountains of the U.S.A.), especially in warm climates, any increase of effective moisture by irrigation raises both A and its reliability. Such areas can thus become effectively hydro-neutral, and naturally enjoy more effective working days than, say, East Anglia or the Netherlands. Enterprise and system choice is widened. These factors justify high actual inputs and high input ratios into intensive systems, especially where relief is no problem; this often results in very high crop or livestock

462

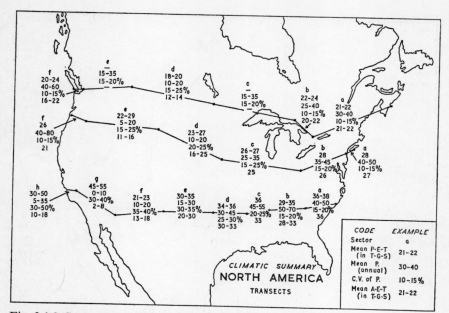

Fig. 3.1.2 Summary of **Climate** by transect sectors: U.S.A. and Canada

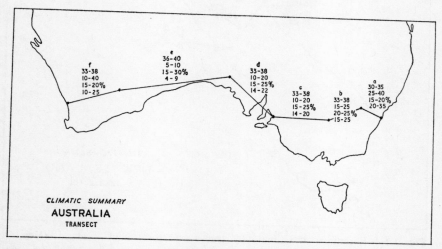

Fig. 3.1.3 Summary of Climate by transect sectors: Australia. (Key on Fig. 3.1.2)

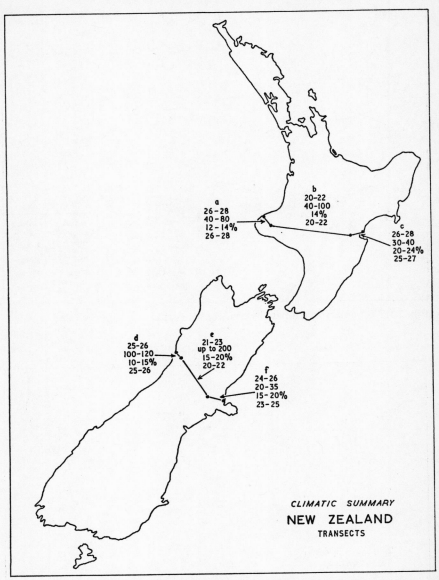

Fig. 3.1.4 Summary of Climate by transect sectors: New Zealand. (Key on Fig. 3.1.2)

yields if salinity is controlled (e.g. San Joaquin valley in California (Fig. 3.1.12. Col. IX), Murray River area in Australia, valleys in southern Spain, parts of Israel). Finally, in Band III if the terrain and soil are suited to mechanisation, extensive unirrigated tillage (e.g. the plains of western Texas, parts of

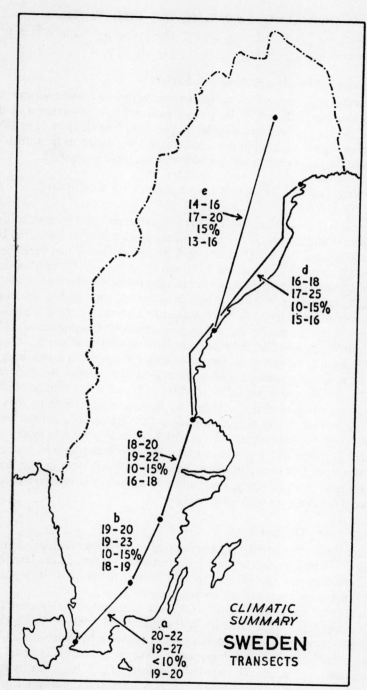

e
14 - 16
17 - 20
15%
13 - 16

d
16 - 18
17 - 25
10 - 15%
15 - 16

c
18 - 20
19 - 22
10 - 15%
16 - 18

b
19 - 20
19 - 23
10 - 15%
18 - 19

a
20 - 22
19 - 27
< 10 %
19 - 20

CLIMATIC SUMMARY

SWEDEN
TRANSECTS

Fig. 3.1.5 Summary of Climate by transect sectors: Sweden (Key on Fig. 3.1.2)

Australia, Israel) is found even though the hydrologic ratio h (see p. 488
below) is below 0·8.

The Influence of Social and Economic Factors

How is this spectrum, which so far assumes advanced economies and access
to mass markets, affected (i) by poorer national infra-structure and lower
national output and consumption levels per head (see Chap. 1.5, p. 76), (ii)
by lower or higher population densities, (iii) by long distances from such mass
markets and by other major market access problems (e.g. tariffs)?

EFFECT OF INFRA-STRUCTURE AND LEVEL OF ECONOMIC DEVELOPMENT

The positive correlation between (a) infra-structure and level of economic
development and (b) potential and/or actual productivity as measured by the
level of available industrial inputs and by cereal yields, has been stressed
(Chaps. 1.1 and 1.5 and models 3.1.2 and 3.1.3 below). In 'rich' countries,
areas of good productive potential are usually intensively farmed, whilst their
advanced technology extends ecological control, thus giving flexibility in
the choice and intensity of farming systems in a given climate. Such countries
have high input ratios (R_1, R_2, R_3), high output of plant equivalent (K_e) per
head and obtain more of their calories from animal products (p. 12); these
latter, depending on ecological factors or national consumer tastes, may come
either from tillage systems, grassland or alternating systems. But, where both
market access is limited and infra-structure and 'real' living standards (Chaps.
1.1 and 1.5) are less well developed (as in Argentina, Uruguay and Chile
which are, in fact, nearer rich mass markets than Australasia but are 'poorer')
then the tendency (as suggested by model 3.1.1) is to grassland systems, even
in potentially alternating or tillage areas. Further, actual inputs, input ratios
and output per unit area and/or per man are all lower than in, for example,
New Zealand. Paradoxically, this results in as high a percentage of animal
calories in the diet as in 'rich' countries.

POPULATION DENSITY

Where total population (D_s) per unit of farmed area is high, farming is
usually as intensive, in relation to productive potential, as the level of economic development (B) permits (p. 12, 79 and below p. 473). Input ratios
(R_1, R_2, R_3), p. 10 are positively correlated with B.

Where both population density (D_s) and economic development (B) are
high and the farm input intensity is absolutely and relatively high, then the
animal products which are prominent in the diet may, in hydro-neutral zones,
as just noted, come from tillage, alternating or grassland systems (U.K.,
Netherlands). In Japan, the population density (D_s) is high, the living standards (B) are rapidly rising but still not very high, and input ratios are fairly
high. Here, despite great excess of P over T and the fact that h is only about

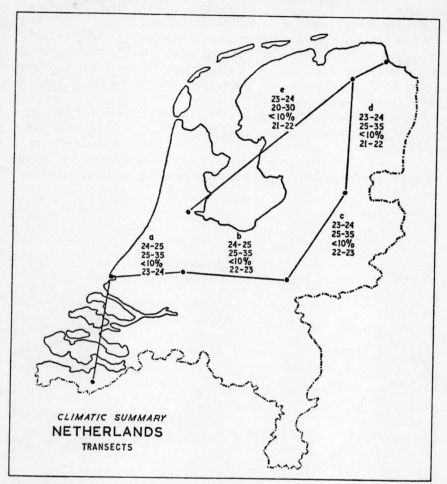

Fig. 3.1.6 Summary of Climate by transect sectors: the Netherlands. (Key on Fig. 3.1.2)

0·5, the emphasis is still on tillage crops for humans, though imports of animal products are increasing.

Where population (D_s) density per unit farm land area is high but economic development (B) is low (e.g. Egypt) and the area is hydro-neutral (or is made so by irrigation) then input ratios tend to be low and the imperative need for calories, or for foreign exchange to buy calories, places the emphasis on tillage systems for food and export crops, e.g. cotton.

Where population density (D_s) is low and market access is limited, then systems tend to be more extensive and more in grassland (Uruguay and eastern Argentine) despite *h* being > 0·8 and irrespective to some extent of stage of development (*B*) and of productive potential (model 3.1.2).

Where population density (*D*$_s$), economic development (*B*) and productive potentials are all low, systems are extensive (Morocco). Table 3.1.2 attempts to illustrate these generalisations by a few examples. But this is a complex subject which cannot be pursued further here; however, model 3.1.3 may be useful, although subject to many exceptions.

Distance from market (*L*$_m$ in the models). Almost irrespective of productive potential, poor market access forces farmers into crop enterprises which either have high-value outputs per unit weight (e.g. tea, cotton or dried raisins) or into livestock which can concentrate land and climatic resources by converting grass into transportable high-value meat or milk products or wool which

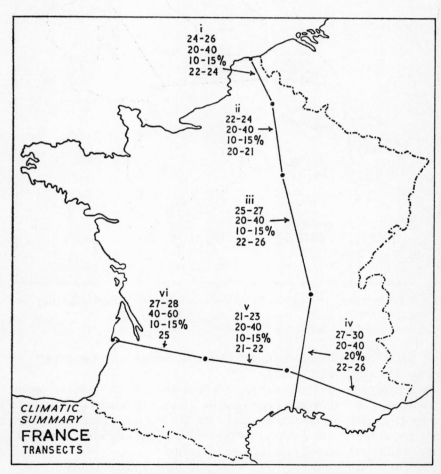

Fig. 3.1.7 Summary of Climate by transect sectors: France. (Key on Fig. 3.1.2)

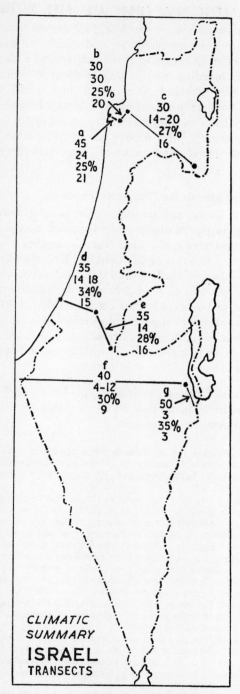

b
30
30
25%
20

c
30
14–20
27%
16

a
45
24
25%
21

d
35
14 18
34%
15

e
35
14
28%
16

f
40
4–12
30%
9

g
50
3
35%
3

CLIMATIC
SUMMARY
ISRAEL
TRANSECTS

Fig. 3.1.8 Summary of Climate by transect
sectors: Israel. (Key on Fig. 3.1.2)

Conclusions can bear heavy transport costs.* Areas of poor market access (i.e. high L_m), which *also* have low and unreliable A, and which are outside the hydro-neutral zone, are often in extensive grassland systems with the emphasis on wool or livestock breeding and raising without fattening (Table 2.4.3: Sectors (d), (f); New Zealand: Sectors (b), (c)). Tariff and other trade or disease-control barriers in export markets have a broadly similar effect in reducing the proportion of tillage area at home. Finally (as noted p. 466), remote flat semi-arid areas (with $h < 0.8$) are often suitable for tillage, espe-ically transportable cereals and make an important exception to model 3.1.1.†

Tentative Predictive Models for Temperate Zones

The next step is to attempt to make potentially quantifiable predictive models which summarise the above generalisations. Because the interactions cannot easily be illustrated multi-dimensionally, algebraic symbols are used, not to imply accuracy but to simplify thinking. Unfortunately, limitations of time and available data preclude formal statistical analysis and the presenta-tion and testing of mathematically substantiated predictive models.‡

The evidence in the transects in Part 2 and in Tables 3.1.1, 3.1.2, 3.1.3 and Fig. 3.1.1, justifies an attempt at potentially quantifiable models. Very broadly, the class (viz. plantation, tillage, alternating, grassland, p. 105) of tem-perate farming systems in a given site or district may be expressed and theo-retically predicted in terms of five main variables:

   (i) $\S A$ = Actual Evapo-transpiration (p. 40) (A–E–T) in the thermal growing season (T–G–S) or, in warm dry climates, in the hy-drologic growing season.

---

* Chisholm 1962 discusses the relations between product, perishability, transport to market and enterprise location. He quotes the following product weights, in metric tons, per million litres of liquid milk: butter, 39; cream, 82; cheese, 99; full cream dried milk, 124; full cream condensed milk, 372.

† Fig. 3.1.1, Tables 3.1.1, 3.1.3 and some of the models below, assume causal chains which may oversimplify, for it is often hard to disentangle cause from effect or influents from resultants. Thus, are inputs and outputs per acre high in the U.S. Corn Belt because the ecological and operational potentials encouraged, or at least contributed to, the develop-ment of a good infra-structure, to high level technology and to a good level of demand? Or did economic development and the growth of technology and of strong dietary demand, permit or encourage the exploitation of the ecological and operational potential? In practice, no doubt, both have contributed to the productivity of the Corn Belt. But this 'hen-and-egg' argument can be very important in less-developed countries where the problem is how to get the economy 'air-borne'. Future 'growth' models of agricultural systems must allow for the considerable, perhaps, synergistic 'two-way' interactions and 'feed-backs' between ecologi-cal, operational and socio-economic factors.

‡ Note that, although the constants (or correction factors) in these tentative models cannot yet be quantified, the main variables (viz. A, T, P, h, B, D_s) can be and that L_o and L_m might be partially quantifiable, at least in economic terms, for example local differences in land values, and differences between prices at farm gate and in main consumption centres. Pending this the inverse of Visher's rating (p. 2.1.000) is used for L_m.

§ T, P and A are all taken from Thornthwaite Associates 1962–64 which are based on

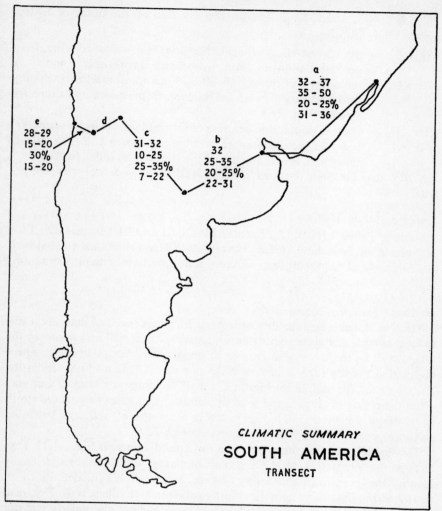

Fig. 3.1.9 Summary of Climate by transect sectors: South America. (Key on Fig. 3.1.2)

(ii) h = is the ratio of mean annual precipitation (P) to potential evapo-transpiration (T) in the thermal growing season (T–G–S) (p. 34) with the *smaller* quantity used as the numerator thus:

h = $\frac{T}{P}$ where $P > T$, i.e. on the wet side of the point of hydro neutrality, *or*

Thornthwaite and Mather 1955. The limitations of the Thornthwaite method are noted on p. 22.

$h = \dfrac{P}{T}$ where $T > P$, i.e. on the dry side of the point of hydro-neutrality.*

(iii) L_o = the adverse effect of local difficulties (e.g. awkward relief, frost pockets, liability to flood, unworkable clay soils, etc.) and

L_m = the adverse effect of difficulties of access to market (including transport facilities, actual mileage, tariff, quota and exchange rate barriers, etc.)

(iv) D_s = Human population per acre of farmed land (*superficie agricole*)

(v) B = Index of socio-economic development, when U.S.A. = 100 as measured, for simplicity, by the Beckerman Index (p. 11).

N.B. *Unless the context otherwise requires all these models are per unit per year.*

(Model 3.1.1)

System Location Model

Study of Tables 3.1.1 and 3.1.3 and of Figs. 1.1.1 and 3.1.1 suggest that if the proportion of farm land surface devoted to tillage (as defined p. 106) is S_t and to grassland or grazing (e.g. of scrub) is S_g then, as a first approximation:

$$\frac{S_t}{S_g} = g.A.h - i(L_o + L_m) + j.D_s - k.B$$

where g, i, j and k are constants.

Thus, as A declines, as aridity or wetness increases (i.e. as h decreases) and as local and market access constraints increase so there will be a tendency for tillage systems to give way to grazing or grassland systems. However, where population density (D_s) is high and development (B) is medium (India) or low, the need for 'edible calories', i.e. for tillage crops for human use, may offset the effects of climate and local difficulties which elsewhere would result in grassland. Further, as noted on p. 488 in areas where $T > P$, e.g. Australia, sector (b), extensive tillage may be found where $h < 0.6$.

This model, which is illustrated in a notional diagram in Fig. 3.1.11, neglects plantation systems (e.g. permanent orchards) and does not specifically identify the location of alternating systems. But, applying the definitions on p. 107, it implies that alternating systems (viz. where tillage is $\leqslant 75\%$ and $\geqslant 25\%$ of the farmed area *and* tillage and grassland are alternated), will not be found where $\dfrac{S_t}{S_g} \geqslant 3$ or $\dfrac{S_t}{S_g} \leqslant 0.25$ and are likely, but not certain, within areas where $\dfrac{S_t}{S_g} \leqslant 3$ and $\geqslant 0.25$. (The existence of alternating systems cannot,

---

* Complete hydro-neutrality occurs, in theory, where weekly (T) equals weekly (P) and where if income per acre (or £100 invested) from tillage or grassland is equal, it is indifferent to the farmer whether he has tillage or grassland or both. This logical possibility is very unlikely because neither weather nor costs and market prices behave like this; further, absolute hydro-neutrality is not operationally suitable for tillage or indeed grass conservation. A realistic point of hydro-neutrality might be slightly on the dry side of $h = 1.0$, say, $h = 0.9$. But to make a workable model $\dfrac{T}{P} = \dfrac{P}{T} = 1.0$ is used here.

incidentally, be shown by analysing census acreage statistics, because, within the range where alternating systems are likely, one farm (p. 19) may, indeed, practice three systems.)

Productive Potential Model   (Model
                                                                    3.1.2)

Productive potential (Rows A to Q in Table 3.1.1) may be said, when input ratios are optimal, i.e. at unity (p. 6), to be a function of A, multiplied by h and less a factor for local difficulties. Productive potential, therefore, is:

$$b\,(A.h - c.L_o)$$ where b and c are constants.

In practice, of course, input ratios rarely approach unity as the great difference between experimental and commercial results (Holliday 1966, Monteith 1966, Bawden 1967, De Vries *et al.* 1967, U.S.D.A. 1969) demonstrate.

Input Ratio Intensity Model   (Model
                                                                    3.1.3)

The input ratios, R_1, R_2, R_3 (p. 5) can be summed or averaged as R_m and are, broadly, positively correlated with development level (B) and probably with $A . h$ (see Fig. 3.1.12* and p. 487). So input ratios R_m will, as a first approximation, be

$$R_m = \mathrm{m}.A.h. - n.(L_o + L_m) + \mathrm{q}.B^f$$

where m, n, q are constants and B^f is development level (Beckerman Index) corrected for the proportion of total consumer income spent on food. Income elasticity of demand for food is higher in developing countries with a low B than it is in advanced countries (see Simantor and Tracy 1966 for some useful data on income elasticity).

Food Output Model   (Model
                                                                   3.1.4)

Food output will be, per unit farmed area (see p. 5):

$$R_1.A.G. - (R_2.W_p + R_3.W_a) = K_p + K_a = K_f$$

where K_p = edible tillage kcal, K_a = edible animal kcal, K_f = total edible human food energy (p. 5).

Food Supply and Food Resource Consumption Model   (Models
                                                                   3.1.5 and
So plant equivalent food supply, which is $K_e = K_f + 5.5\,K_a$ = plant equi-   3.1.6)
valent, (p. 10) and, in a self-contained community, food resource consumption per head will be:

$$\frac{K_e \text{ per unit farmed area}}{D_s} = K_e \text{ per head of population}$$

whilst dietary kcal will be:

$$\frac{K_f \text{ per unit farmed area}}{D_s} = K_f \text{ per head of population}$$

---

* Fig. 3.1.12 shows the input and output (in dollars) per inch of A for 8 types of U.S. farming and relates them to A and h.

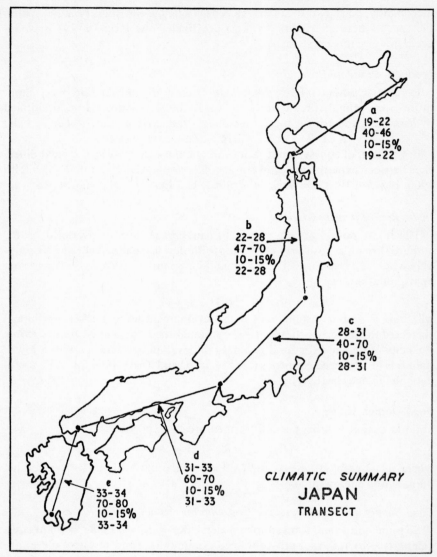

Fig. 3.1.10 Summary of Climate by transect sectors: Japan. (Key on Fig. 3.1.2)

Summary

In brief, input ratios (p. 6) are positively correlated with development (B) (model 3.1.1) and probably with $(A.h)$, local problems and market access; food resource consumption (K_e) per head is closely positively correlated with input ratios and, in turn, with development level (B). Thus, both input intensity per unit of $(A.G)$ and of productive potential (model 3.1.2) and plant equivalent supply (or food resource consumption per head) (K_e) tend to rise as

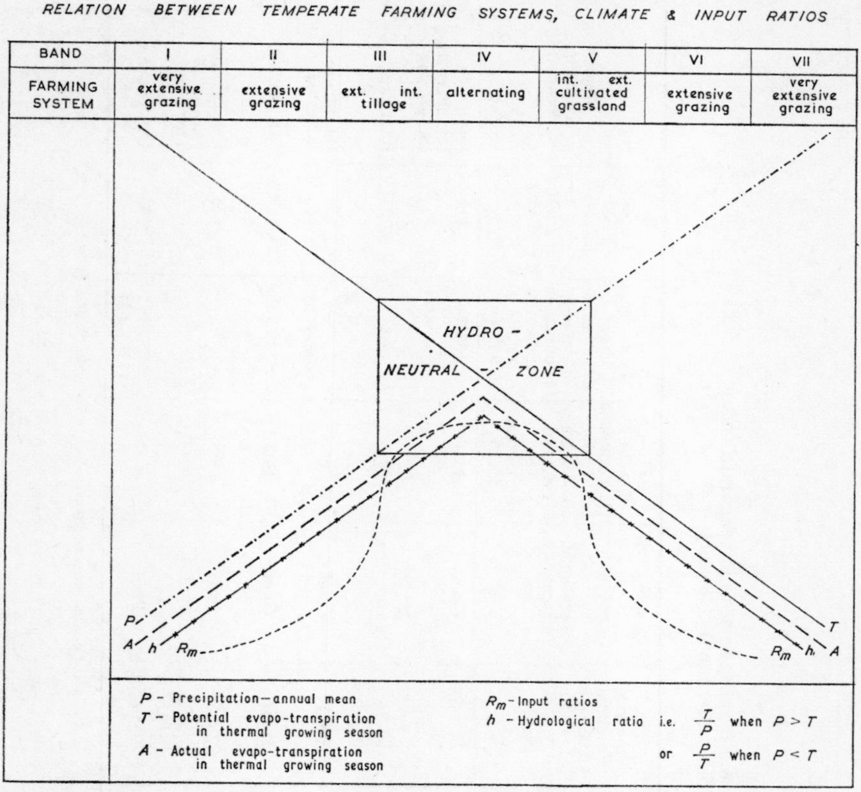

RELATION BETWEEN TEMPERATE FARMING SYSTEMS, CLIMATE & INPUT RATIOS

BAND	I	II	III	IV	V	VI	VII
FARMING SYSTEM	very extensive grazing	extensive grazing	ext. int. tillage	alternating	int. ext. cultivated grassland	extensive grazing	very extensive grazing

HYDRO –

NEUTRAL   ZONE

P – Precipitation—annual mean

T – Potential evapo-transpiration in thermal growing season

A – Actual evapo-transpiration in thermal growing season

R_m – Input ratios

h – Hydrological ratio i.e. $\dfrac{T}{P}$ when $P > T$

or $\dfrac{P}{T}$ when $P < T$

Fig. 3.1.11 Relation between Temperate Farming Systems, Climate and Input Ratios

level of development (B or B^f) increases. But low ($A.h$) as well as major local difficulties (L_o) and poor market access (L_m) *may* depress input ratios (i.e. actual input intensity per unit of productive potential).

Further, the spending or feeding habits of a particular population may depress or raise input ratios, for example a preference for bread rather than meat will reduce plant equivalent consumption (K_e) per head. Thus, even if development (B) is high, input ratios (R_m) need not be so high as they would have to be if the preference was reversed. In any case, as noted, at higher income levels the proportion of consumer income spent on food tends to be less than at lower incomes, which means that input ratios do not necessarily rise *pari passu*, with development levels (B). But in advanced countries, they often appear to do so (Chap. 499), thus creating surpluses.

The above models relate to an hypothetical area which neither imports or exports foods. But, they could, without difficulty, be modified to cater for imports and exports. More serious is their failure to take account of non-food enterprises (e.g. tea, cotton, sisal, wool) which may be grown and exported,

Table 3.1.2

Population Density and Farming Systems in Temperate Zones
(1965 or nearest available year)

(1) Country	(2) National Population per acre of farmland (D_s) (Number) (1965)	(3) Dietary Energy (kcal.K_f and K_f) per head per day (1964)	(4) Plant Equivalent Consumption ($K_e = K_p + K_a . 5·5$) kcal per head per day (1964)	(5) Ecological Potential (A–E–T in T–G–S) (A) inches	(6) Operational Potential	(7) Level of Development 'Real' consumption per head (B) (Beckerman 1960)	(8) Fertiliser Usage lb per head of population (as indicator of industrial inputs) (1965–66)	(9) Cereal Yield lb/acre per annum 1963–65	Major Farming Systems
Netherlands	2·06	2,890 (840)*	6,670	20·5–24	Good	45·0	100·8	3,428	Int. tillage Int. alternating Int. grazing
Japan	5·72	2,320 (255)	3,468	19–34	Medium/Poor	29·7	43·5	3,797†	Int. tillage
Egypt	4·56	2,930 (176)	3,722	0–7‡	Good	6·4	22·7 (1964–65)	2,956	Int. tillage (irrigated)
Uruguay	0·07	2,970 (1,306)	8,847	31–36	Good	16·2	26·5	867	Ext. grazing
Morocco	0·15	2,480 (n.a.)	—	2–25	Poor	8·1	7·3	676	Ext. grazing Ext. tillage Tree crops
Libya	0·06	1,910 (153)	2,598	0–19	Poor	n.a.	7·1 (1961–62)	274	Ext. grazing

* Figures in brackets are animal meal.　　† Including double-cropped areas.　　‡ Excluding irrigation.

as Egypt exports cotton, to earn foreign exchange to buy other foods (e.g. wheat), consumer goods or capital goods and services for development. But despite these and other defects, such as the apparent anomalies in Tables 3.1.2, 3.1.3 and Fig. 3.1.12), the models may serve, firstly to clarify the complex of factors determining the location and intensity of farming systems and their relation to economic development, population density and food consumption standards; and secondly to encourage others to test the hypothesis and to produce better models. Fig. 3.1.11 summarises the climatic aspects graphically but notionally.

The world situation. Finally, we may tentatively apply these models to the world situation (see Chap. 3.4) in which food and feeding stuff exports and imports balance. As, from models 3.1.4* and 3.1.5:

$$R_1.A.G - [R_2.W_p + R_3.W_a] = K_p + K_a = K_f$$

and $K_p + C.K_a = K_e$

where C = energy conversion ratio of plant product energy to animal product energy here assumed arbitrarily to be 5·5 (p. 10). Then for a mixed diet, the world model is:

$$\sum_1^n [S_p.R_1.A.G - R_2.W_p] + \sum_1^n S_a.R_1.A.G - (R_2.W_p + R_3.W_a) = K_p + K_a + K_f$$

(Model 3.1.7)

where S_p and S_a are the farmed areas used respectively for crops for human food and for crops and grass, etc. used in livestock production for human food, where animal work, wool, etc. are neglected and $\sum_1^n$, is the sum of such areas in the countries in the world.

But, in energetic terms we can say that:

$$\frac{S_d(K_p + CK_a)}{D_s.365} = K_e \text{ per head per day}$$

(Model 3.1.8)

where S_d = farmed area per 1,000 of the world's population, and:

$$\frac{S_d(K_p + K_a)}{D_s.365} = K_f \text{ per head per day}$$

(Model 3.1.9)

Assuming a temperate population needs 3,000 kcal/day of K_f of which $\geqslant$ 10% must be K_a and assuming that $C = 5\cdot5$, then, for satisfactory nutrition, model 3.1.8 must provide not less than 4,350 K_e/head/day. In some countries K_e exceeds twice this (Fig. 1.17).

Assuming S_d and A are relatively inelastic, that is unless large new areas can be opened up, for example, in South America p. 332 or Africa, or

---

* Since the text of this book went to press, practical tests with real data have revealed the need for some changes in Models 1.1.1, 3.1.4, 3.1.7, 3.1.8 and 3.1.9 whilst Morris, T. R. (University of Reading, personal communication) has pointed out two logical inconsistencies. Amendments to these models will accordingly be made to a chapter by one of us in Dent, J. B. (Ed.) University of New England, Australia and University of Reading, *Systems Analysis in Agricultural Management* (in preparation).

unless irrigation can be greatly increased in areas where $T > P$ and $h \leqslant 0\cdot8$ then the main potentially controllable variables are:

$R_1, G, R_2, C, S_p, S_a, D_s$

(For definitions of symbols see p. 5 and footnote* below.)

Theoretically, the number of people that could be fed at $K_e = 4{,}360$ kcal/day could be greatly increased if developed countries kept S_a constant, reduced their K_e and hence S_a and increased S_p for export to developing countries. But apart from moral, economic and political issues this would raise insuperable transport problems.

Reduction of D_s (or at least of rate of increase of D_s which is about 3%/annum in some developing countries) by population control methods may help in the future but is not an immediate remedy (p. 523). This leaves improved input ratios R_1, R_2 and improved C (mainly by better R_3) and also G. There is ample scope for raising G by better crop plant spacing, by better operational timing and by other husbandry devices, by better genotypes and by higher input ratios and to reduce wastages (W_p and W_a). But higher input ratios are positively correlated with economic development (B). Low B characterises developing countries with inadequate K_f and K_e. The question is how far can K_f or K_e be raised without full economic development? Which aspects of B are essential to increased food production? Further, B is often positively correlated with low increase rates in D_s. If this link is causal, how much and what kind of B is needed to reduce rate of increase in D_s without positive population control programmes? These questions are considered again and in more realistic terms, in Chaps. 3.3, 3.4 and, briefly, in 3.5.

RELIABILITY, PERSISTENCE AND EFFICIENCY

The requirements of a biologically efficient food-producing system, as set out on p. 3, include the capacity to adjust to changes in population, in technology and in demand or supply factors (Chaps. 3.2 and 3.3); they also include reliability and persistence.

Reliability and persistence are needed, firstly, because any system which fails to meet the day-to-day food needs of man throughout the year, i.e. any system which has chronic short-term ecological, economic or political instability, cannot be nutritionally or socially efficient for a species like man which is

* The algebraic symbols used in Chap. 1.1 and in this Chapter are defined on p. 5 and pp. 471–473. In addition,
1. C = gross energy conversion of plant energy to energy in animal products.
2. K_p, K_a and K_f = respectively, the edible tillage kcal, edible animal kcal, and edible total kcal available for human diet.
3. K_e = the plant equivalent or food resource use equivalent of K_f.
4. R_1, R_2, R_3 = input ratios (p. 5).
5. R_m = the mean or sum of input ratios.
6. S_a = farmed area used for the production of livestock for human food.
7. S_d = farmed area per 1000 population.
8. S_g = proportion of grassland in farmed area.
9. S_p = farmed area used for the production of crops for human food.
10. S_t = proportion of tillage in farmed area.

almost completely aseasonal (Duckham 1963, p. 45) and requires two or three meals a day. Secondly, because current production or consumption at the expense of long-term food supply (for example by a food-producing system which generates either irreversible long-term instability (p. 72) or uncompensated loss of food production capacity), cannot be biologically efficient from the point of view of man as an increasing species and as the last link in each of the four food chains (p. 9).

Increased *short-term* ecological and economic *stability* is a marked technological and institutional feature of the current agricultural revolution in advanced temperate economies (pp. 72, 135, 165). But it is not, unfortunately, noted in developing countries, whether temperate or tropical (Chap. 3.3), where famine is still a hazard calling for governmental action (Masefield 1967). Nor, in developed countries, have measures to improve *long-term stability* (p. 500), proved very effective. However, several industrial economies, e.g. U.S.A., U.K., Netherlands, are attempting to assess, prevent or remedy the probable long-term adverse effects, on farming, of urban development, of agricultural chemicals and of industrial and consumer pollution of land, water and ecosystems (p. 504). The effects of long-term instabilities on farming systems, and their intensity and efficiency cannot yet be effectively quantified. However, except for such disasters as nuclear war, the evidence suggests, so far, that, if they have the will to do so, technologically advanced societies can now, and will in future, be able to master (i) their own resource use, (ii) their pollution, and possibly (iii) their population problems.

Less-developed countries have not, so far, mastered the first and third (Chaps. 3.3 and 3.4) nor seriously had to face the second of these problems, viz. pollution. In tropical subsistence farming, long centuries of experience of the instability of the environment for agriculture have imbued the people with psychological attitudes which still affect their farm planning and are difficult to eradicate even when the instability has been reduced. For example, the heavy mortality due to cattle diseases in the past led stock-owners to insure themselves against it by retaining more cattle on the land than they really needed, and as noted in Chap. 1.4, is still a cause of overstocking even where disease has been greatly reduced by veterinary improvements. (In exactly the same way, more children are bred than are needed, based on former experience of heavy mortality, few would survive to care for the parents in their old age, and these contribute to the modern 'population explosion'.) Folk memories of past famine sometimes lead to a lack of confidence in the availability of food supplies which, even though unjustified, may induce hoarding and thus itself create food shortages. Long experience of insecurity of tenure of land and of political upsets may deter even those who might emerge as competent farmers from investing in farm improvements.

Efficiency may be measured in economic (e.g. return on capital), social (e.g. number of people adequately fed per man or woman on the land) or ecological terms. No good comparative data on economic efficiency are available. It is, however, clear that, although international comparisons are often misleading, the number of adequately fed people per man on the farm is highest in technically advanced industrial countries. If 'real' total national consumption standards such as the Beckerman Index (p. 5) can be accepted as indicators of economic development and also as economic parameters of overall national economic efficiency, then, both farm people and farming in developed countries are more efficient than in developing countries even although, or indeed because, farming in the former countries is increasingly dependent on inputs of urban origin.

Ecological comparisons are probably best made in energetic terms (Duckham 1963, p. 297). Only about 1% of solar radiation reaching the earth's surface is used for photosynthesis and only 0·3% to 0·002% become available as edible calories (K_f) (Table 1.1.1). This often quoted 1% statistic is misleading, partly because only part of solar radiation (i.e. that of wavelength 0·4 — 0·7 μ) is available for photosynthesis; but mainly because the solar radiation used for transpiration and evaporation is many times that used in photosynthesis. Hence (p. 40 and p. 65) P–E–T (T) and A–E–T (A) are perhaps more useful energetic indicators of the productive potential of an area than is total solar radiation. The energetic efficiency of a farming system or area may thus be defined as the proportion of productive potential (model 3.1.2 above) (or where irrigation is available and T exceeds A then the proportion of T), used to form organic plant dry matter *minus* the various wastages between photosynthesis and final consumption by man as set out in Figs. 1.1.2, 1.1.3, 1.1.4 and 1.1.5 and Table 1.1.1.

Application of the models to the temperate farming spectrum

It will be noted that, in the model on p. 5, the controllable, or partly controllable, terms are the wastages (W_p and W_a), the crop plant genotype (excluding weeds*) (G) and the input ratios (R_1, R_2, R_3, etc.). Excluding irrigation, water conservation, land drainage and flood prevention, A–E–T (A) and precipitation (P) are largely uncontrollable, although some control can be achieved by cultivations, by influencing plant population per acre, by use of crop genotypes adapted to dry or wet soil conditions, and of animal genotypes adapted to local climate. Realising that only parts of the models are currently controllable, let us apply the models on p. 5 and in this chapter to the Temperate Farming Spectrum (Table 3.1.1, Fig. 3.1.1):

(*a*) In Bands I, II and III (arid, semi-arid and short grassland) energetic efficiency (as measured by output per unit of productive potential) is usually low. The mean annual P and hence A in the T–G–S are low and unreliable; therefore yield variability (i.e. the C.V. of K_p or K_a) is high, partly because there are few buffers in extensive tillage or extensive grazing systems; though

---

* Weeds are normally wastages which compete for A and inputs and thus reduce K_f.

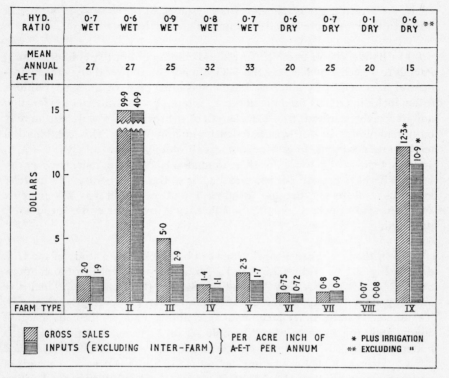

GROSS SALES AND INPUTS PER ACRE INCH OF A-E-T FOR NINE U.S. FARM TYPES

Fig. 3.1.12 Gross Sales and Inputs per acre inch of A–E–T for Nine U.S. Farm Types. (Key to columns on p. 143). The high inputs and sales from irrigated farming in column IX are per inch of *natural* A-E-T but include the cost of purchased irrigation water. *Sources:* Fig. 1.2.5 and Tables 2.3.2, 2.3.3, 2.3.4

the latter perhaps have more buffers (e.g. compensatory growth, capacity to seek food) than the former. In view of the high risks, farmers, therefore, tend to keep actual inputs and input ratios low (Fig. 3.1.12). This is particularly so in extensive grassland systems where *either* low herbage growth (i.e. low R_1. $A.G$) per unit area increases maintenance and exercise wastages (W_{a3} and W_{a4} in Fig. 1.1.3) *or* excessive heat (see Israel, p. 309) constrains feed intake, and so reduces the metabolisable energy available for tissue or milk formation. On the other hand in some advanced areas, wastage control input ratios (R_2 and R_3) for pest and disease reduction are high in crops and livestock (see Australia, U.S.A., Canada).

(*b*) In Bands IIIb, IV and Va (i.e. in the hydro-neutral zone) both *P* and *A* are higher *and* more reliable. So better plant genotypes (*G*), which respond

481

well to fertilisers, are practicable, and weed and plant and animal pest and disease control more sophisticated. Not only are inputs actually higher but input ratios (R_1, R_2 and R_3) are usually higher, i.e. nearer unity, in advanced areas. As a result, the efficiency of the systems, in terms of food yield, are higher not only per acre but also per unit of productive potential.

(c) In Bands Vb, VI and VII (forest and cold humid or dry cold grassland) P tends to exceed T, whilst A in the T–G–S tends to be lower than in the hydro-neutral Bands (IIIb, IV and Va). Leaching is often high, operating conditions (often including relief) are difficult (i.e. L_o is high) both for alternating farming and intensive grassland, whilst the length of the thermal growing season may be unreliable (as in northern Sweden or upland Wales). This combination tends to reduce inputs in grazing systems absolutely and relatively (i.e. R_1, R_2 and R_3 tend to be low). Further, animal wastages are increased, partly because herbage growth per unit area is less so the energy wasted in exercise increases and partly because wind-chill and rain-chill (p. 91) increase energy losses or depress appetite. So efficiency is lower per unit of productive potential.

These rather speculative conclusions can be checked by a study of the U.S. data in Fig. 3.1.12, Tables 2.3.4 and 2.3.5, and by reference to the chapters on the U.S.A., Canada, New Zealand and Sweden (Chaps. 2.2 to 2.9 inclusive and Appendix to Chap. 3.1). It does seem, however, that biological efficiency, if evaluated in terms of (i) flexibility, (ii) reliability, and (iii) the conversion of energy and moisture receipts into edible K_p or K_a (plant or animal kcal) is better in the hydro-neutral belt or where deficiency of precipitation is made good by irrigation. Thus, whether the test is food output per acre per unit of productive potential (model 3.1.2.) or per acre-inch of A or other limiting climatic parameter, the highest and most reliable energetic efficiency tends to be found in well-favoured areas even though man's ingenuity, skill and endurance in exploiting less-favoured regions excites our admiration.*

Farming Systems in the Tropics

In the tropics, it is less easy to quantify farming systems both because of the paucity of statistics and because in many areas there has been less stability due to the frequency of political changes. Where land use consists of nomadic pastoralism, it is impossible to state a 'farm size' or a consistent stocking rate; and at the other extreme in parts of overcrowded Asia, the farming system may be identical on a farm of a half acre and on one of two acres, although one is four times the size of the other. It is also less easy to relate agricultural productivity to any simple climatic statistic. This can be illustrated by a

---

* The above energy models can be applied, with modifications, to all nutrients in the human diet. (Duckham in Dent.) See footnote on p. 477.

perusal of the following figures which give inches of actual annual evapo-
transpiration (A–E–T or A) for certain stations in Africa (Thornthwaite 1964):

	in
Freetown	52·1
Kinshasa (formerly Leopoldville)	51·8
Kano	40·8
Kampala	40·0
Nairobi	31·9
Johannesburg	28·8
Salisbury	28·8
Cape Town	21·5
Tunis	16·2
Khartoum	7·1
Cairo	1·0

On the basis of these figures, it might naturally be supposed that Tunisia
was more agriculturally productive than Egypt; but actually the reverse is the
case, because Egypt has the Nile water to draw on for irrigation and Tunisia
has no such source. Again, Kano in Nigera and Kampala in Uganda have
practically the same annual actual evapo-transpiration (A–E–T); but this
conceals the fact that (as we have seen in earlier chapters) the former has an
annual rainfall of 34 in within a short rainy season which only allows annual
crops to be grown, whilst the latter has 46 in of rain with two main wet
seasons and the most important crops are two perennials, plantains and
coffee. Nairobi, with an even lower actual evapo-transpiration, is again in the
centre of an area where coffee and tea are successfully grown. In fact, with
Thornthwaite's formula, a higher figure for potential evapo-transpiration
(P–E–T or T) in one place than another is often due merely to a higher temperature which may simply mean that crops mature more quickly rather
than that they yield more.

Such instances exemplify the fact that, in the tropics, the theoretical environmental possibilities for plant growth often cannot be fully exploited
because a perfect 'fit' of cropping plan to climate cannot be obtained for
either natural or economic reasons. Thus it is impossible to plant all crops
exactly at the beginning of the rainy season when they would derive most
benefit from it, both because the soil may have been too hard to cultivate in the
preceding dry season and precious weeks have to be wasted in this operation
after the soil has been wetted, and because so much simultaneous planting
would present an impossible demand on labour resources. Again, with a
single wet season of say five months, and a choice of crops each of which
matures in three or four months, the planting of any of these crops will 'waste'
one or two months of useful rain, which is however too short a period in
which to mature a second crop. Another difficulty in applying meteorological

data to agriculture in the tropics is that, with high rainfall variability, statistics of mean annual rainfall provide little indication of the risk to the farmer in growing crops with a certain rainfall requirement. A statistic which is therefore now coming widely into use is the probability of a certain minimum rainfall being attained in a year. Two interesting maps on this basis have been published by the East African Royal Commission (1955) showing the percentage probability of a minimum rainfall of 20 and 30 in respectively being attained in any one year in different parts of the region.

The intensity of farming systems in the tropics is often determined not by the natural environment but by the density of the human population. In a subsistence economy, denser human populations must tend to produce smaller individual farms and at the same time a greater abundance of farm labour, thus inducing more intensive production from the land whether or not it is best suited for such use. The extreme case is a Chinese farmer with less than an acre of land, whose farming becomes indistinguishable from horticulture and who may take five or six crops of vegetables a year, fertilising them largely with human excreta and watering them manually, irrespective of the soil type or rainfall. Population density also affects technical methods in farming. Thus transplanting of rice, which generally gives a higher yield but needs more labour than sowing direct into the field, is commonly used in the rice-growing regions of Asia which are densely populated but is rarer in the more sparsely populated African continent. Stall-feeding of cattle, as opposed to grazing, is another adaptation to dense human population and is practised in some heavily populated islands in the West Indies, Bermuda and Mauritius, and in certain densely populated parts of India. This very important factor of population density is itself, in the tropics, as often as not unrelated to particularly favourable factors of the environment but simply due to historical accident. Thus we have seen in Part 2, Chapter 11, that the depopulation of the 'Middle Belt' of Nigeria was due to slave-raiding in earlier centuries; and a low population density in parts of the East African highlands is similarly traceable to the raiding propensities of the Masai in the past. Again there is no difference in the natural environment to explain the great disparity in population density between Java and the other main islands of the Indonesian archipelago (Part 2, Chapter 13), which is due rather to the local history of migration, administration and development. The sparse population and extensive farming of South America compared to tropical Asia are not dictated by the environment but are a result of the late migration of man into this continent and, at a more recent period, the number of willing Spanish and Portuguese emigrants which was limited by social and economic rather than environmental factors.

Acknowledgements to H. J. Critchfield, W. E. Russell, L. P. Smith, J. A. Taylor, R. W. Willey, for comments and criticism, and for research assistance to H. Farazdaghi. Thanks to T. R. Morris for valuable help with the models.

References

Bawden, F. C. 1967. *Intl. J. Agrarian Affairs.* **5.** 2. 115–29.

Chisholm, M. 1962. *Rural Settlement and Land Use: An essay in location.* London.

Dent, J. B. (Ed.). *Systems Analysis in Agricultural Management.* (In preparation.)

De Vries, C. A., Ferwerda, J. D. and Flach, M. 1967. *Netherlands J. Agric. Sci.* **15.** 4. 241–48.

Duckham, A. N. 1963. *Agricultural Synthesis: The Farming Year.* London.

East African Royal Commission 1953–55 (1955). Report, Cmd. 9475. H.M.S.O. London.

Hills, E. S. (Ed.) 1966. *Arid Lands: a Geographical Appraisal.* London.

Holliday, R. 1966. *Agric. Progress.* **41.** 24–34.

Masefield, G. B. 1967. F.A.O. Nutritional Studies No. 21. *Food and Nutrition Procedures in Times of Disaster.* F.A.O. Rome.

Monteith, J. L. 1966. *Agric. Progress.* **41.** 9–23.

Rutter, H. J. and Whitehead, S. H. (Eds.) 1963. *The Water Relations of Plants.* Oxford.

Simantov, A. and Tracy, M. 1966. *O.E.C.D. Observer,* **22.** 30. O.E.C.D. Paris.

Thornthwaite, C. W. Associates. 1962–65. *Publications in Climatology.* **15.** 2; **16.** 1; **16.** 3; **17.** 1; **17.** 3; **17.** 3; **18.** 2. Laboratory of Climatology, Centerton, N.J., U.S.A.

Thornthwaite, C. W. and Mather, J. R. 1955. *Publications in Climatology.* **8.** 1. Laboratory of Climatology, Centerton, N.J., U.S.A.

U.S.D.A. 1969. *An Analysis of Agricultural Research in Relation to the Demand for Agricultural Products.* Production Research Dept. 104. Agric. Res. Service. U.S.D.A. Washington, D.C.

Appendix to Chapter 3.1

Table 3.1.3 below attempts to test the validity of some of the quantifiable location variables suggested on p. 470 to p. 472. The values for T, P and A are arbitrarily taken as the median of the range of each parameter given in each sector of the transect tables in Part 2, the climatic data of which are summarised in Figs. 3.1.2 to 3.1.10. This method obviously introduces errors in T, P and A and hence in h and $(A.h.)$. Nevertheless, the values for $(A.h.)$ (which form part of model 3.1.1 (Location model), 3.1.2 (Productive Potential) and 3.1.3 (Input Ratio Intensity model)) fit, in many cases, with the general thesis outlined on p. 470 to p. 478. The main anomalies seem to be Canada Sector (f) and U.S. Northern Sector (f) where, on models 3.1.1 and 3.1.3 one should expect extensive grazing but in fact there is also considerable intensive tillage, alternating and tree fruits; the marked summer moisture deficit and the effect of mountains on L_o may provide partial explanations.

U.S. Southern Sector (d) and Israel Sector (b) have high values of $(A.h.)$ but are hilly, i.e. L_o is large, and are perhaps socio-economically handicapped so input ratios are low. $(A.h.)$ for Israel Sector (c) seems too low—for this covers the rather productive valleys of Galilee, whilst $(A.h.)$ for France Sector (iv) seems too high, partly because the annual P overlooks the hot, dry summer and the 'wasted' winter rains; also L_o is high in the mountainous parts of this sector.

In Japan the high population density (p. 476) and the possible inapplicability of model 3.1.1 to wet rice may account for the dominance of intensive tillage where $(A.h.)$ and L_o would suggest grassland (intensive or extensive).

Finally, the absence of intensive systems in South American sectors (a) and (b) with their high values of $(A.h.)$ is probably attributable to socio-economic factors (p. 332) including L_m (market access).

APPENDIX TO CHAPTER 3.1

Table 3.1.3

Some Values for Location Variables

1 Transect and sector	2 Median T in the T–G–S (in)	3 Median P annual (in)	4 C.V. of P (%)	5 Median A in the T–G–S (in)	6 $\dfrac{T}{P}$	7 $\dfrac{P}{T}$	8 h	9 $A \cdot h$	10 L_m	11 Farming systems
Canada *(a)*	21	35	10–15	21	0·6		0·6	12·6	2	Int. Grazing and Int. Alternating
(b)	23	32	10–15	21	0·7		0·7	14·7	1	Int. Alternating and Int. Tillage
(c)		25	15–20						4	Ext. Tillage and Ext. Grazing
(d)	19	15	15–25	13		0·8	0·8	10·4	3	*Ext. Grazing and some Int. Tree Crops
(e)		25	15–20						3	*Int. Alternating, Int. Tillage and Int. Tree Crops in valleys
(f)	22	50	10–15	19	0·4		0·4	7·6	2	
U.S.A. North *(a)*	28	45	10–15	27	0·6		0·6	16·2	1	Int. Tillage, Int. 'landless' livestock
(b)	28	40	15–20	26	0·7		0·7	18·2	1	Int. Grass and Int. Alternating
(c)	26	30	15–25	25	0·9		0·9	22·5	1	Int. Tillage and Int. Alternating
(d)	25	15	20–25	20		0·6	0·6	12·0	3	*Ext. Grazing and Ext. Tillage
(e)	25	13	15–25	14		0·5	0·5	7·0	3	*Ext. Grazing and Int. Tillage in Valleys
(f)	26	60	10–15	21	0·4		0·4	8·4	2	*Int. Grass, Int. Alternating, Int. Tillage and Tree Crops
U.S.A. South *(a)*	37	45	15–20	36	0·8		0·8	28·8	2	Tillage and some Alternating
(b)	32	60	15–20	31	0·5		0·5	15·5	2	Int. Tillage and some Int. Alternating
(c)	36	50	20–25	33	0·7		0·7	23·1	2	Int. Tillage
(d)	35	38	25–30	32	0·9		0·9	28·8	2–3	Varies
(e)	33	23	30–35	25		0·7	0·7	17·5	3	*Ext. Tillage and Ext. Grazing
(f)	22	15	35–40	16		0·7	0·7	11·2	3	*Ext. Grazing and some Int. Tillage
(g)	50	5	30–40	5		0·1	0·1	0·5	3	*Some Int. Tillage
(h)	35	20	30–50	14		0·6	0·6	8·4	2	*Int. and Ext. Tillage, Tree Fruits, Ext. Grazing
Australia *(a)*	33	33	15–20	28	1·0		1·0	28·0	3	Int. Grazing, Int. Alternating, Ext. Grazing
(b)	36	20	20–25	20		0·6	0·6	12·0	3	Ext. Grazing, Ext. Tillage or Alternating
(c)	36	15	15–25	17		0·4	0·4	6·8		*Int. Tree Fruits, Int. and Ext. Alternating, Ext. Grazing
(d)	36	15	15–25	18		0·4	0·4	7·2	4	Int. Tree Fruits, Ext. Tillage or Alternating, Ext. Grazing
(e)	38	8	15–30	7		0·2	0·2	1·4	5	None
(f)	36	25	15–20	18		0·7	0·7	12·6	4	*Int. Tree Fruits, Ext. Alternating, Ext. Grazing
New Zealand *(a)*	27	60	12–14	27	0·5		0·5	13·5	4	Int. Grassland
(b)	21	70	14	21	0·3		0·3	6·3	5	Ext. Grassland
(c)	27	35	20–24	26	0·8		0·8	20·8	4	Int. and Ext. Grassland and Alternating
(d)	26	110	10–15	26	0·2		0·2	5·2	4	Int. Grassland
(e)	22	200	15–20	21	0·1		0·1	2·1	5	Ext. Grassland
(f)	25	28	15–20	24	0·9		0·9	21·6	4	*Int. Alternating

488

Transect and sector	Median T in the T-G-S (in)	Median P annual (in)	C.V. of P (%)	Median A in the T-G-S (in)	$\frac{T}{P}$	$\frac{P}{T}$	h	$A.h$	L_m	Farming systems
Sweden (a)	21	23	<10	21	0·9		0·9	18·9	1	Int. Tillage and Alternating
(b)	20	21	10–15	19	1·0		1·0	19·0	2	Int. Alternating
(c)	19	21	10–15	17	0·9		0·9	15·3	1	Int. Alternating
(d)	18	21	10–15	16	0·9		0·9	14·4	4	Ext. Alternating
(e)	15	19	15	15	0·8		0·8	12·0	4	Very Ext. Grazing
Netherlands (a)	25	30	<10	24	0·8		0·8	19·2	1	Int. Tillage and Int. Grassland
(b)	25	30	<10	23	0·8		0·8	18·4	1	Int. Tillage and Int. Grassland
(c)	24	30	<10	22	0·8		0·8	17·6	1	Int. Alternating
(d)	23	30	<10	22	0·8		0·8	17·6	1	Int. Tillage and Int. Grassland
(e)	23	25	<10	21	0·9		0·9	18·9	1	Int. Tillage and Int. Grassland
France (i)	25	30	10–15	23	0·8		0·8	18·4	1	Int. Tillage and Int. Alternating
(ii)	23	30	10–15	21	0·8		0·8	16·8	1	Int. Tillage and Int. Alternating
(iii)	26	30	10–15	24	0·9		0·9	21·6	2	Semi-Int. Grazing and Int. Alternating
(iv)	29	30	20	24	1·0		1·0	24·0	1	*Int. Tree Fruits, Int. Tillage, Ext. Grazing
(v)	22	30	10–15	22	0·7		0·7	15·4	2	Ext. Grazing and some Semi-Int. Tillage or Alternating
(vi)	28	50	10–15	25	0·6		0·6	15·0	2	Int. or Semi-Int. Tillage or Alternating
Israel (a)	45	24	25	21		0·5	0·5	10·5	3	*Very Int. Tree Crops, Int. Tillage
(b)	30	30	25	20	1·0		1·0	20·0	3	*Semi-Int. Tillage, Semi-Int. Tree Crops and Ext. Grazing.
(c)	30	17	27	16		0·6	0·6	9·6	3	Int. Tree Crops, Int. Tillage, Int. Alternating and Ext. Grazing
(d)	35	16	34	15		0·5	0·5	7·5	3	*Int. Tree Crops, Int. Tillage and Int. Alternating
(e)	35	14	28	16		0·4	0·4	6·4	3	Semi-Int. Tillage and Ext. Grassland
(f)	40	8	30	9		0·2	0·2	1·8	3	Ext. Tillage and Ext. Grazing
(g)	50	3	35	3		0·06	0·06	0·2	4	*Int. Tillage and Int. Horticulture
South America (a)	35	43	20–25	34	0·8		0·8	27·2	3	Semi-Int. Grazing, Some Alternating or Tillage
(b)	32	30	20–25	27		0·9	0·9	24·3	3	Semi-Int. Grazing and Alternating
(c)	32	18	25–35	15		0·6	0·6	9·0	5	Ext. Grazing and Some Tillage or Alternating
(d)	—	—			—		—		5	Ext. Grazing (Andean)
(e)	29	18	30	18		0·6	0·6	10·8	4	*Semi-Int. Alternating, Tree Crops.
Japan (a)	21	43	10–15	21	0·5		0·5	10·5	2	Int. Tillage and Semi-Int. Grassland
(b)	25	58	10–15	25	0·4		0·4	10·0	2	Int. Tillage and Semi-Int. Grassland
(c)	30	55	10–15	30	0·5		0·5	15·0	2	Int. Tillage
(d)	32	65	10–15	32	0·5		0·5	16·0	2	Int. Tillage
(e)	34	75	10–15	34	0·5		0·5	17·0	2	Int. Tillage

Sources: Transect tables in Part 2, Figs. 3.1.2 to 3.1.10 inclusive.
Definition of terms in Cols. 2 to 11 inclusive, pp. 470–472 of text and Chap. 2.1.
* Wholly or partly irrigated.

Food and Agricultural Trends in Temperate Regions

Keevey (1960) in Burton and Kates 1965 suggests there have, in man's history, been three major technical revolutions. The first was the discovery of tools in the Lower Palaeolithic Age about 1m years ago; the second was the invention of agriculture about 8,000 B.C. or earlier; and the third, the advent of science and industry which began three centuries ago. Just before the second revolution started, the total world population was about 5m and the density of human population per square kilometre was, Keevey suggests, about 0·04 persons. Ten or 15 years ago when the total population was 2,400m the density was 16·4; by the end of the century it may be 46, i.e. nearly three times the density in mid-century (Keevey). This then is the challenge facing food producers, scientists, agriculture teachers and advisers, the input and food-processing industries and governments. Can we use our resources to produce enough food to meet these forecast human needs if they materialise? If not, will biological disasters, natural homeostatic mechanisms or the actions or inactions of governments adjust the population to the available food supplies?

In this and the next chapter we look at current trends in food, agriculture and population in temperate and tropical regions respectively, and attempt to assess the constraints that may inhibit increased food production. In Chap. 3.4 the world food and population problem is examined with particular reference to less-developed countries. World trends and situations in total and *per capita* food and commodity supply and in nutrition are given in Tables 3.2.1, 3.2.2 and 3.4.1.

In temperate regions, not all countries and agricultures are advanced. Apart from China and the northern parts of the Indian sub-continent, there are large less-developed areas in the Middle East and south-west Asia, on the southern rim of the Mediterranean, in the temperate parts of South and Central America (e.g. Chile) where food, agricultural and population problems are in many ways similar to those in the developing tropics. Their problems and prospects are, therefore, outlined in Chaps. 3.3 and 3.4. There are also parts of the temperate world such as Spain and Greece, districts in France, and the non-European areas of South Africa where the under-production of food, poverty or underemployment of farm people are prevalent and where the technological revolution has not progressed as rapidly as it has in such advanced regions as north-western Europe, the U.S.A. and Canada and Australasia. In the advanced economies, under-nutrition is found only in small geographical pockets (e.g. parts of the south-east of the U.S.A.) and malnutrition is confined to the poorer sectors (e.g. to some of the old people and to large young families in the United Kingdom), to those who have the money but not the knowledge or the will to feed properly and to some of the overindulged sedentary rich.

But with these exceptions, the world can be divided into (a) the developed areas, with more or less advanced economies, where total food production is rising more rapidly than population, *per capita* food consumption has been increasing (Table 3.2.1 and Simantov and Tracy 1966) and there is (as shown

Table 3.2.1

World Agricultural Production
(*Note:* the figures in brackets are for 1967.)

Country or Region	Production 1964–66 as % of 1957–59	
	Total	Per capita
U.S.A.	113	102
Canada	131	114
Latin America	121	99
Western Europe	118	110
Eastern Europe	116	110
U.S.S.R.	121	109
Japan	117	109
Communist Asia	96	84
Other Far East	121	102
Western Asia	124	102
Republic of South Africa	126	107
Other Africa	119	101
Australia–New Zealand	124	108
World (including Communist Asia)	115	101
World (excluding Communist Asia)	119 (127)	103 (107)
Developed*	118 (126)	108 (113)
Less Developed†	121 (130)	102 (104)

Source: U.S.D.A. 1967 and 1968.
* U.S.A., Canada, Europe, U.S.S.R., Japan, South Africa and Australia–New Zealand.
† Latin America, Other Far East, Western Asia and Other Africa.

below) a tendency for food production to exceed the growth of demand, and (b) the less-developed countries where total food production has been increasing at about the same or a lower rate as the developed countries, but where population has been growing at the same or a higher rate, as food supply and, as a result, *per capita* food consumption has been static or deteriorating (Table 3.2.1 and Chaps. 3.3 and 3.4). In these countries the *per capita* food consumption in cash or dietary terms is substantially less than in developed countries and there either is a chronic tendency to, or at least an ever-present risk of under-nutrition or malnutrition. The food resource consumption in terms of plant equivalent (K_e) is half that of developed countries (Fig. 1.1.8).

Table 3.2.2

Estimated World Production of Selected Commodities

Commodity	Unit (1 ton=2,000 lb)		Average 1955–59 (A)	Average 1964–66 (B)	(B) as % of (A)
Wheat	million tons		216	259	120
Rye	,,	,,	37	32	86
Rice, rough*	,,	,,	214	247	115
Maize	,,	,,	167	207	124
Barley	,,	,,	71	95	134
Oats	,,	,,	59	43	73
Sorghum and millet	,,	,,	28	36	129
Grain (total)	,,	,,	(792)	(919)	(112)
Sugar, centrifugal	,,	,,	45·0	64·8	144
Sugar, non-cent.	,,	,,	6·7	8·4	125
Fruits, citrus	,,	,,	14·5	20·5	141
Apples and pears	,,	,,	12·8	17·4	136
Potatoes	,,	,,	232·0	236·0	102
Dry beans	,,	,,	4·0	4·9	123
Dry peas	,,	,,	0·6	0·5	88
Hops	,,	,,	0·07	0·09	131
Soya-beans	,,	,,	24·0	32·0	133
Peanuts	,,	,,	13·4	15·9	119
Cotton-seed	,,	,,	19·3	22·4	116
Flax-seed (Linseed)	,,	,,	3·4	3·3	98
Sesame seed	,,	,,	1·5	1·5	102
Castor beans	,,	,,	0·5	0·7	148
Sunflower seed	,,	,,	5·1	7·8	153
Rape-seed	,,	,,	3·5	4·2	121
Olive oil	,,	,,	1·0	1·2	117
Palm oil	,,	,,	1·3	1·3	101
Palm kernel oil	,,	,,	0·4	0·4	89
Coconut oil	,,	,,	2·1	2·2	106
Oil-seeds and oils (total)	,,	,,	(75·5)	(92·9)	(117)
Butter†	,,	,,	4·6	5·2	114
Milk	,,	,,	259·0	292·0	113
Meats	,,	,,	44·5	52·1	117
Eggs	,,	,,	10·8	12·8	119
Lard	,,	,,	3·1	3·1	99·9
Tallow and greases	,,	,,	2·9	4·0	135
Animal products (total)	,,	,,	(324·9)	(369·2)	(116)
Tobacco	,,	,,	3·9	4·5	117
Coffee	,,	,,‡	3·8	4·3	112
Tea	,,	,,	0·7	0·9	128
Cocoa beans	,,	,,	0·9	1·4	155
Cotton	,,	,,§	10·5	12·3	116
Wool	,,	,,	2·4	2·6	108
Jute	,,	,,	2·1	2·3	113
Sisal	,,	,,	0·5	0·7	125
Henequen	,,	,,	0·1	0·2	121
Abaca	,,	,,	0·1	0·1	91
Fibres (total)	,,	,,	(15·7)	(18·2)	(112)

Source: U.S.D.A. 1967 and 1968.
* Includes Mainland China; excludes Nepal, North Korea and North Vietnam.
† Product weight; includes Ghee.
‡ In bags of 132 lb (60 kg).
§ In bales of 480 lb (218 kg).

The present chapter, therefore, considers, in relation to developed countries:

(*a*) the nature of the current technological revolution;
(*b*) the geographical effects of the trends generated by this revolution;
(*c*) the trends in the interactions between population growth, demand and food-production resources;
(*d*) implications for the future.

Analysis of the Technological Revolution

Advanced or developed countries are characterised by superior infra-structures; relative political stability; a high degree of industrialisation or close economic links (for example, as in New Zealand and Eire) with highly industrialised countries; and a relatively high level of science-based technology on farms, in industry and in the public services. Except in periods of deflation, this combination generates a high and increasing gross national product (G.N.P.), growing purchasing power per head, a rising dietary level and better amenities. These trends are accompanied by high and rising inputs of medical, sanitary and other population-conserving services, and by falling death-rates, longer life expectancies and greater *potentialities* for population increase. In practice, however, in many developed countries family limitation, and possibly some other factors, restrict population growth-rates to 1% per annum compared to 3% or more in some less-developed countries.

But the combination of population growth and higher purchasing power per head does not generate an equivalent increase in the total physiological *or* economic demand for food. This is, firstly, because the energy needs of man are satisfied by a mean intake of 3,000/3,250 kcal of K_f (p. 12) per head per day; a level which has been reached in advanced countries. Secondly, although as living standards rise, consumers tend to shift from cereals and starchy foods to animal foods which have lower energy (Mcal) outputs per acre (p. 10) and are expensive to produce, and to fruit and vegetables which, like animal foods, are costly in farm resources per Mcal produced. Plant equivalent (K_e) consumption per head rises but not apparently as fast as farmers' capacity to produce more. Thirdly, as they grow richer, consumers spend, as noted on p. 475, a smaller proportion of their income on food. In fact, in very rich countries, such as the U.S.A., expenditure on food per head may have almost reached its ceiling.*

Against this trend for the rate of increase of food demand to fall away is the tendency for the rate of total food production to rise in developed countries. In most developed countries the acres farmed are static or slowly falling, and manpower on farms is decreasing at 4 to 5% per annum. The technological

* The demand elasticities for food used by Simantov and Tracy 1966 were: North America and Oceania, 0·1; western Europe, 0·3; Japan, 0·45, with a tendency to decrease with time. The comparable figures for less-developed regions are: Latin America, 0·4; Africa, 0·6; Near East, 0·6; Far East, 0·8.

revolution is increasing yields at a rate which more than offsets losses of farmed land or due to pollution, etc. (p. 504). Yields of tillage crops have been rising at about 3% per annum (see Fig. 2.3.3 for the U.S.A. and Figs. 3.2.1 and 3.2.2 for the U.K.). Yields of animal products *per acre* have also been increasing by reason of (*a*) the growing number of animals per feed acre (i.e. per acre of grass and of cereals, etc. fed to livestock) (Duckham 1966),

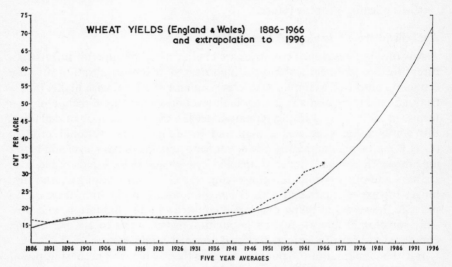

Fig. 3.2.1 Wheat Yields (England and Wales) 1886–1966 and by extrapolation to 1966.

- - - - - Actual five-year average yield. *Source:* Napolitan, 1966
——— Yield from Dent, J. B., where
$$Y = a + 1 \cdot 704x - 0 \cdot 2934x^2 + 0 \cdot 0152x^3$$
  * Three-year average

(*b*) rising yields per animal (Fig. 3.2.3), and (*c*) the improving efficiency of feed conversion, especially with pigs (see Fig. 3.2.4) and poultry. In 12 years the outputs of eggs and of milk *per feed acre*† on the Reading University farms have doubled. But, overall, yields in the United Kingdom are only rising at about 2% per annum (Wibberley 1958). Although the above examples are drawn from only a few countries, they are representative of the trends in advanced economies.

These rising yields are the result of the interactions between:

 (i) technological advances,
 (ii) the institutional and economic consequences of such advances, and
 (iii) governmental actions.

† Feed acre includes acreage equivalent of inter-farm inputs, for example, purchased feeding stuffs, etc.

(i) *Technological advances.* Orwin and Whetham 1964, describe traditional farming in Britain in 1895–1914—a picture which then applied to most developed countries. The tractor; oil fuels; rural electrification; the industrial fixation of atmospheric nitrogen; the discovery of genetics; the sciences of animal health, crop hygiene, plant and animal physiology and nutrition (including the discovery of vitamins and of trace element needs); and the use of organic pesticides and of herbicides; together with many other scientific advances and engineering inventions have, in a lifetime, translated farming in developed countries into an industry increasingly dependent no

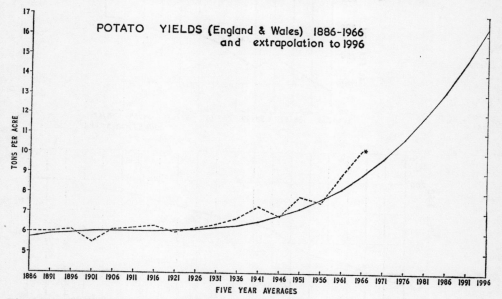

Fig. 3.2.2 Potato Yields (England and Wales) 1886–1966 and by extrapolation to 1996

- - - - - Actual five-year average yield. *Source:* Napolitan, 1966.
———— Yield from Dent, J. B., where
$$Y = a + 0.1741x - 0.0321x^2 + 0.00209x^3$$
* Three-year average

industrial and scientific inputs (Fig. 2.3.3 and Duckham 1966, 1968). Input ratios (R_1, R_2, R_3) have a growing industrial content which has complemented or supplemented, in part replaced, the 'natural' ecological inputs (Chaps. 1.2 and 1.3) of traditional farming. They have created the conditions for higher potential output, partly by improving photosynthetic capacity (G in the model on p. 5) and partly by reducing potential wastages (W_p and W_a) whilst higher input ratios (R_1, R_2, R_3) exploit these potentials (see also models 1.1.1, 3.1.2, 3.1.3).

(ii) *Institutional and economic consequences.* This enhanced output potential and its partial realisation, has, together with the drift of manpower from the

R

FARMING SYSTEMS OF THE WORLD

land (which is simultaneously the result of and cause of many technical and economic changes in agriculture), generated new or accelerated established economic, social and technical forces. These forces have led to or can only be fully exploited by (*a*) *specialisation* in enterprises involving specialised equipment to secure division of labour and greater control of operations and of ecological variables and to reduce the strain of management; (*b*) increasing the *size of the 'firm'* for example, by enlarging the acreage of the average farm, by increasing its investment in consumable inputs such as fertilisers, or by

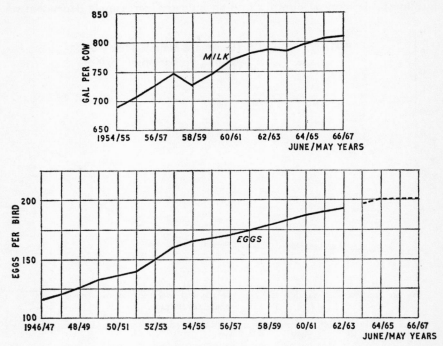

Fig. 3.2.3 Annual Yields 1946/47 to 1966/67 in the U.K. *Source:* H.M.S.O. 1967

co-operation, vertical or horizontal integration, contract milking, etc—each of which bring to farming some of the advantages of scale enjoyed by large urban firms; (*c*) *reducing weather sensitivity*, for example, by irrigation or drainage, by housing livestock, by drying grains and by the use of larger and faster field machinery to give greater operational timeliness (Duckham 1963 p. 419, 1966 and in Taylor 1967); (*d*) the *heavy injection of capital* in 'fixed assets' or durable resources which generate 'overhead' costs.

However, the technical and economic advantages which flow from (*a*), (*b*) and (*c*) are to some extent offset by the cost and by particularly the lack of flexibility imposed by (*d*). For much of this capital is, with the manpower in farming, either functionally or geographically immobile. They are 'trapped' in agriculture (Hathaway 1963). Specialised buildings (e.g. for poultry)

496

and specialised machines (e.g. sugar-beet harvesters) have a low functional mobility (Duckham 1963, p. 413), whilst the former are also locationally fixed. These resources, as well as many general purpose resources, cannot easily be converted to other farming or non-farming uses. The same applies to farm manpower, which may be tied to the land by sentiment, by difficulties of re-training or by housing problems. This lack of mobility is less marked in land use, especially on 'good land' in the hydro-neutral belt (p. 461) which has a wide range of potential uses; whilst technological advance, including irrigation, drainage, and other 'weather-proofing' devices (Duckham in

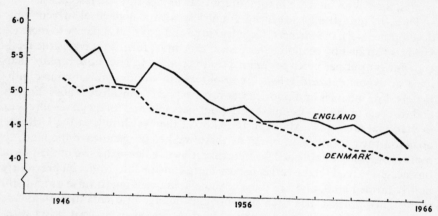

Fig. 3.2.4 Food Conversion Rate for bacon pigs in Denmark and the U.K. (lb meal per lb liveweight gain) *Source:* Sturrock *et al.* 1966

Taylor 1967) has increased the functional mobility of much 'poor' land in temperate areas.

The combination of technological advance and of its institutional and eco-nomic consequences results not only, as noted, in substantially higher yields of crop and animal products per acre, but also in greater yield stability be-tween years. This increased yield stability comes from greater overall ecolo-gical control, especially of pests and diseases and of energy-moisture balances; from reduced weather sensitivity, for example, through operational timeliness; and from the 'buffering' effects of livestock housing and of fertilisers, especi-ally N. Greater yield stability, in combination with governmental and other institutional measures fosters greater income stability (Fig. 2.3.2 and Table 2.2.4). The building, consciously or otherwise, of such 'buffers' (i.e. cushions which reduce yield variability or price fluctuations) into food-supply systems, helps to give the farmer added confidence. This tends to stimulate still further the investment in consumable inputs (i.e. the use of higher input ratios), and in 'trapped' resources, which in their turn tend to generate higher and more reliable yields.

(iii) *Governmental actions and the problems of excess production.* We have a situation, therefore, where yields of crop and animal products per acre are increasing (and will continue to rise) and where total food output tends to rise more rapidly than consumer demand because neither consumable inputs are being decreased nor are durable 'fixed' resources, which are often 'trapped', being withdrawn from farming. Often the only way to meet or to reduce the overhead costs of these trapped resources is to raise input ratios, usually by using more industrial inputs (or inter-farm inputs which comes to much the same thing). As the farmers' reactions to lower prices are often to increase output to try and maintain the value of their gross sales and net incomes, there is an increase in total output per unit of trapped resources.

Thus, in industrially advanced countries, with a high level of economic development (B), the farmers' technical urge and physical capacity to increase yields is fortified by economic necessity; this, in its turn, aggravates the tendency for output per acre, per animal and per man, to rise more rapidly than local or export demand grows, or resources are withdrawn. Of course, ultimately, low incomes and poor career prospects on farms, force those farmers and workers (and their families) with poor personal and financial resources or in marginal areas (e.g. Maine in the U.S.A., the Welsh hills in the U.K.) to abandon their farms. But, often, as these weaker farmers leave the land, so more progressive established farmers take it over, enlarge the size of their farm businesses, spread their overheads, use more consumable inputs on previously poorly farmed land, and are thus, in some areas, tending to raise output still more. This, at least, has been the apparent linkage of events in parts of the U.S.A. and of the United Kingdom. In sum, resources are not being withdrawn fast enough to offset the impact of technology and investment, and, there is a chronic tendency towards food surpluses, lower farm-gate prices and agrarian discontent in the more advanced countries.

Faced with this situation, governments, or their instruments, have, in the last 30 or 40 years, increasingly intervened in ways which are often mutually antagonistic. First, legal (production or acreage) limitations or financial incentives have been used to reduce apparent capacity by withdrawing land from farming or from particular crops. But such restrictions have often been offset by farmers achieving higher yields from fewer acres; thus the cotton acreage in the U.S.A. has been halved since the thirties, but total output remains much the same. Such schemes are, as in the U.S.A., sometimes associated with soil and water conservation, afforestation, etc. Secondly, governments have purchased (or lent money on the security of) price-depressing surpluses, stored them and either subsidised their export to less-developed countries (e.g. U.S. grains) or their conversion into non-food products (e.g. sugar-beet and wine grapes made into industrial alcohol in France). However, such actions tend to stabilise farm-gate prices and regenerate farmers' confidence. This, as noted, may encourage investment and greater output of food and fibre (cotton, flax, wool, etc.)—the fibres facing the growing competition of man-made fibres. Thirdly, in attempts to reduce rural poverty, governments have fostered the amalgamation of farms that are too small to be viable, the consolidation of fragmented holdings and the transfer and training of excess farm or rural population to or in other occupations. The Netherlands, Italy, France, Sweden, the United Kingdom and other countries have such schemes. But, as noted, the potential fall in food production is often more than offset by the greater and more skilful use of inputs by the 'higher' farming of enlarged farms with less manpower.

In brief, the actions of governments, whether inspired by political or humanitarian motives or by economic strategy, appear to have, at most, only a limited effect on checking the trend for farm output to rise. This situation seems likely to continue, in advanced countries, for another fifteen years or so (Duckham 1966). What happens after that seems more likely to depend on population, demand and resource use trends, discussed below, than on the apparently unlimited productive potential of farming in advanced countries (Duckham 1968).

Geographical Trends

The complex situation outlined in (i), (ii) and (iii) above is having long-term geographical effects.

As the availability of scientific and industrial inputs increases so the 'natural' ecological and operational constraints outlined in Chapters 1.2, 1.3, 1.4 and 1.6 are becoming less important as factors influencing the location and intensity of farming systems. Many examples of this trend have been seen in Part 2. *Climatic and weather constraints* are being reduced by better-adapted livestock housing (for example, in the south-west of the U.S.A.), by crop genotypes with greater frost resistance, by frost prevention, by sprinkler or flood irrigation, etc. Fertilisers and minor or trace elements are reducing *soil chemical constraints*. Disease, pest and weed control are reducing *biological constraints*. Bigger, faster and better-adapted machines, including aircraft, are lessening *operational constraints*. Thus one can now take a quick crop between flood risk dates on some flood-liable lowlands, or use aeroplanes (pp. 161, 199) for seeding and fertilising of mountains, or take crops of potatoes and sugar-beet on heavy land that would have been ill suited to these crops 20 or 30 years ago. Technological advances are giving the farmer, especially the farmer in the hydro-neutral zone (Table 3.1.1), a greater choice of enterprises and of input intensity; his land has a greater functional mobility. His choice is now determined more by prices and by his access to industrial inputs and to markets, and rather less by ecological constraints. In non-land-using farm systems, such as intensive poultry or feed-lot beef, it is indeed the same non-ecological factors that effect the location of industrial manufacturers (viz. access to inputs and markets and good infra-structure), that increasingly determine the siting of large, intensive, livestock enterprises.

It is difficult, as yet, to see what effect this growing liberation from the classical ecological constraints is having on the geography of farming systems, on a world scale. The technological revolution is too young and the inertia of agrarian change is too great for generalisations about geographical effects. Locally, however, there are many examples: the increase in potato production when irrigation was introduced to poor sandy soils in Minnesota; the sudden development of really intensive tillage and alternating systems on the chalklands of England; the intensification of cattle and sheep on previously cobalt or copper-deficient land in New Zealand. On the one hand technology is opening up new areas to extensive systems, for example, by correcting mineral deficiencies in areas of low $(A.h)$ (p. 190) to sheep in Australia, or by tractors and herbicides, etc. opening dry cereal farming districts in Israel; on the other hand, technology is favouring, if only by reason of ease of access to inputs, land in the hydro-neutral zone with a good $(A . h)$ which, traditionally, was well farmed, for example, the Corn Belt or the East Anglian Fens have much higher industrial inputs (higher input ratios) and much higher yields than 40 years ago.

Population Growth, Demand and Resource Use

Although in developed countries, production is rising more rapidly than population (Table 3.2.1.), and apparently than effective demand, these rising standards of living, the urban sprawl and the biological, including medical, interactions between urbanisation, industrialisation and technically advanced farming are, in many such countries, actually or potentially creating problems of resource use, which fall into four groups. These are: (a) land use, (b) water resources, (c) cross-pollution, and (d) overall conservation policy and the 'quality of life'. They are most obvious, or at least have been most recognised, in highly developed countries (high B) with large industrialised conurbations and growing populations with increasing net incomes. The Netherlands, the United Kingdom, Belgium, the Ruhr Valley of Western Germany, the north-eastern seaboard states and (to some extent) the western seaboard states of the U.S.A., and perhaps Japan, are examples.

LAND USE. Rising standards of living, based on growing industrial outputs and improving infra-structures (pp. 11, 77, 78), express themselves, in part, in public demand for more land for roadways or facilities for public recreation, and, in most countries, in private demand for lower-density housing. Thus, though the rich (and the poor) may live in multistorey (high rising) flats (or apartments) near the city centre, the general tendency, at least amongst middle-class families in English-speaking countries, is for new housing to be at low densities in suburbs or satellite towns (Fig. 3.2.5). In such districts, the housing density is, in England for instance, often only 100 acres per 1,000 population (i.e. 0·1 acres per person) as against 50 or even 25 acres per 1,000 in older highly urbanised settlements. In the last sixty years, the farmed area (excluding upland ranches) in England and Wales has fallen about $2\frac{1}{2}$m acres which equals nearly 10% of the present total (based on G. P. Wibberley in Yapp and Watson 1958). With current yields and distribution of farming

systems, this irreversibly 'lost' land would be enough to feed 4 to 5m people
on current diets at current yields. Of course yields are rising, but as population
numbers and as real incomes rise, diet tends to contain more animal products,
even though the elasticity of food demand is low in developed countries (p.
493). In other words, plant equivalent (K_e) and food resource use per head
(model 3.1.5) tend to increase.

The situation is summarised in models 3.1.8 and 3.1.9 (p. 477). In order
to maintain or increase dietary kcal (K_f) per head or plant equivalent (K_e)
per head, then, when farmed area per head (S_d) is falling, i.e. population
per farmed acre (D_s) is rising, yields must continuously be increased. Are
they likely to rise fast enough? The answer seems to be 'Yes'. In many

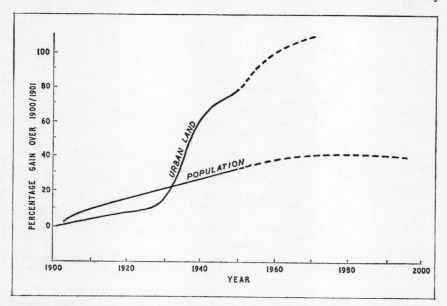

Fig. 3.2.5 Proportionate increase in population and urban land in England and
Wales 1900–2000. *Source:* Wibberley in Yapp *et al.* 1958

advanced countries, there seem to be no technical reasons why, over the next
35 to 40 years, by improved input ratios and better genotypes, photosynthesis
$(R_1.A.G)$ in the models on pp. 5 and 473, should not rise by 50% and
wastages $(R_2.W_p + R_3.W_a)$ should not be reduced by a third. In many commodities this combination would double yields of edible human food per acre
(Table 3.2.3. See also the extrapolations in Figs. 3.1.1 and 3.1.2). Doubling
yields in 35 to 40 years implies a *linear* increase of about $2\frac{1}{2}$% per annum in
overall yields, which is more than twice the 1·28% increase assumed for 1965–
75 in a 1962 Oxford study of the United Kingdom outlook quoted by
Wibberly (1964, 1965).

Another approach is to consider the effects of population increase and the

Table 3.2.3

Energetic and Yield Efficiency in the United Kingdom —
Current and Predicted

		Today	Early 21st century
Percentage of solar radiation receipts retained by photosynthesis	Max. theoretical Max. recorded in field experiments Farm average	8[1] 4[1] 1[1]	$1\frac{1}{2}$
Percentage of Mcal in photo-synthates produced that are utilised in human diets, 5	Potatoes Cereals Grass beef	30[2,3] 33[4,5] 3·6[2,3]	40[2] 44[4] 4·8[2]
Product yields in lb per acre per annum	Potatoes (marketable 'ware') Wheat (grain) Milk (liquid) per feed acre (S.E. England)	20,000[6] 3,700[6] 3,000[6]	37,000[7] 8,000[7] 6,500
Livestock efficiency	Eggs per hen per year Pigs (breeding and fattening): lb meal consumed per lb live- weight sold	200 4·9	350–400 3
Mcal human food produced per 100,000 Mcal solar radiation	Intensive tillage food crops (cereals, potatoes, sugar-beet) (Chain A) Intensive housed livestock (pigs, poultry, 'barley' beef, etc.) (Chain B) Intensive grassland — meat (Chain C) — milk Good 'mixed' farms (Chains A, B, C, D)	200–250[8] 15–30[8] 5–25[8] 30–80[8] 30–50[8]	400–500 30–60 10–50 60–150 60–100

Sources and Notes:
1. Monteith 1966.
2. Excluding Pest and Disease Wastage.
3. From Figs. 1.1.2 and 1.1.3.
4. Excluding plant respiration loss. The pre-1939 percentage was 22.
5. Based on Holliday 1966.
6. Official Statistics and Provincial Agricultural Economists data.
7. Extrapolations of J. B. Dent's equations (Figs. 3.2.1 and 3.2.2).
8. Table 1.1.1 and Duckham 1968.

resultant land loss on the increase in yields required from the remaining land. In the United Kingdom, if population increased linearly from 50m by 1 % per annum to 75m in 50 years and if 0·1 acres of farmed land (S_a) were needed per extra person, the farmed area would fall by $2\frac{1}{2}$m acres or 8 %. This loss could, at present dietary levels, be made good by an increase in yields on the remaining land of about $1\frac{2}{3}$ % linear increase per annum, *without* raising food imports above present levels (Duckham *et al.* 1966).

This projection does not seem impracticable, but it disregards the limited effect on food demand of rising living standards; it also assumes a self-sufficient industrial region. If imports of food and feeding stuffs were increasingly denied to north-west Europe, and particularly to the United Kingdom, by the growth of population and demand elsewhere, or by politico-military events in exporting countries, there would be, however, no cause for complacency. However, in such conditions, the pressures to raise productive potential (for example, by irrigation with desalinised water), to increase farmed area (for example, by reclamation schemes), to use unconventional processes and foods (e.g. leaf protein, microbial proteins from farm wastes, synthetic amino-acids) and to step up input ratios, would be great and probably successful in outcome. The critical problems would probably be the raising of managerial efficiency on farms, the supply and cost of irrigation water, drainage schemes and other 'land-saving' inputs, e.g. fertilisers, and hence the capital for investment in agricultural public works and the input industries (Duckham 1966). It might be necessary to reduce the percentage of animal calories in the diet thus reducing food resource consumption (K_e) and to change dietary habits in other ways (Duckham 1968). But given continued access to basic energy resources (fossil fuels and nuclear energy) and industrial ingenuity, then dietary austerity would probably not have to be excessive.

WATER RESOURCES. Water is the heaviest input in every farming system. 1 in of rain equals roughly 110 short tons per acre. So a 200-acre farm with a mean annual precipitation of 30 in receives over 6,500 tons of water which on intensive alternating farming systems, may be up to 60 times the weight of all other inputs combined. In the U.S.A. where the mean annual precipitation is about 30 in, the mean use of water is roughly (U.S.D.A. 1962): (1) evaporation and crop transpiration, 70%; (2) 'available water supply' for domestic, industrial and irrigation use, 30%; of which some 22% is impounded and used for irrigation, industrial and domestic purposes. But though in the U.S.A. only one-fifth of the 'available water supply' is used, there are serious seasonal or chronic local deficiencies and quality problems. In this the U.S.A. is not exceptional, for nearly all O.E.C.D. countries (which include most of the advanced agricultural countries except the U.S.S.R.) have problems of locally excessive demand and of quality control (O.E.C.D. 1967). As population, industrialisation and living standards rise so does the water demand. It is forecast, for instance, that in the south-east of England the 1964 demand for water will have more than doubled by the end of the century, mainly for household and industrial use. It takes 150 tons of water to manufacture a ton of steel and 800 tons to make one ton of artificial textiles (O.E.C.D. 1967) but much of this water can be reused, whereas the 1,200 or more tons received per 2,000 lb of wheat grain produced in England is largely 'lost' in evaporation and transpiration, and so cannot be reused. In dry areas, e.g. California or Southern Australia, water is used for irrigation; in the Netherlands and Israel, for flushing out saline water. In both cases such 'available water supply' can only be used once.

Further, surface reservoirs or dams across valleys often divert land from farming. Again, unlike flat farm land which can receive and hold storm water until it drains to streams, built-up areas have rapid run-off after sharp rains; this adds to flood risks on low-lying farms, or to river-level control problems, and may waste potentially available water supply. Thus, in temperate advanced countries water conservation and use are likely to become increasing national, and even international, problems. Fortunately, however, research and development in the recycling of industrial water, in the storage of water in coastal barrages or underground aquifers, in desalination by various promising techniques and in water economy generally, is progressing well.

POLLUTION AND CROSS POLLUTION

The impact, on public health and on wild life of residues of the newer persistent, agricultural chemicals (such as DDT) has been much publicised and has attracted much scientific research (Brady 1967), some alarm amongst consumers and some control by governments. Thus in some countries the use of dieldrin (a chlorinated hydrocarbon), which is used as an insecticidal sheep dip or as a seed dressing on cereals, etc., is prohibited or controlled, whilst attempts to limit pollution of water courses by faeces and urine from intensive livestock units are increasingly evident. Such public health problems are more marked in intensively farmed parts of developed countries than elsewhere because sophisticated chemicals are used. The converse problem, viz. the impact of industrial, automotive and household pollutants on the atmosphere and soil, and hence on farming and food supply, has received less publicity, though it has stimulated considerable research (p. 57; A.R.C. 1967, Brady 1967), whilst pollution by domestic sewage and toxic industrial (including radioactive) effluents is engaging both national and international efforts (for example, in the Rhine, Meuse Valley complex (O.E.C.D. 1967)). (Though mostly a hazard to public health, polluted water may sometimes be unsafe for irrigation use.)

RESOURCE USE AND THE 'QUALITY OF HUMAN LIFE'

In developed countries the interactions between industry, urbanisation, land use, farming, food supply, population density, recreation needs and other aspects of human living (see Bracey 1963) are obviously intricate and difficult to quantify; they are also liable to engender great emotions. Food and population problems are not confined to the less-developed countries. In developed areas, however, they appear to be more qualitative or at least difficult to quantify in generally acceptable parameters; this is because they are concerned with some of the less tangible or less measurable needs of human beings. Thus is it worth being well and cheaply fed if it involves exposure to possible toxic risks from farm chemicals? Are recreation parks in areas of great natural beauty a better land use than say forestry or extensive sheep grazing? Is a well-watered coastal area of high sunshine hours, and good A–E–T and

medium quality land, better used for very intensive horticultural production
or for a new conurbation? The fact that well-to-do nations debate such issues
shows that their food and other material living standards are already high by,
say, Asian or African standards. Nevertheless, these quantitative and qualitative problems of resource use, including pollution, are becoming very real
and our grandchildren may regret irreversible decisions taken, for example,
on releasing land for urbanisation, without adequate cost-benefit analysis.*

* Castle, E. N. in Brady 1967 offers at p. 262 a useful initial approach.

Acknowledgements. For comments and criticisms to T. R. Morris. Thanks to
J. B. Dent for the regression equations which we have extrapolated in Figs.
3.2.1 and 3.2.2, and for permission to reproduce Figs. 3.2.3, 3.2.4 and 3.2.5 to
H.M.S.O., School of Agriculture, Cambridge and the Institute of Biology.

References and Further Reading

A.R.C. 1967. *The Effects of Air Pollution on Plants and Soil*. Agricultural Research
Council. London.

*Bracey, H. E. 1963. *Industry and the Countryside*. London.

*Brady, N. C. (Ed.) 1967. *Agriculture and the Quality of our Environment*. American
Ass. Adv. Sci. Washington, D.C. and London.

*Burton, I. and Kates, R. W. (Eds.) 1965. *Readings in Resource Management and
Conservation*. Chicago and London.

*Duckham, A. N. 1963. *Agricultural Synthesis: The Farming Year*. London.

*Duckham, A. N. 1966. *J. Royal Agric. Soc.* **127**. 7–16.

Duckham, A. N. 1968. *Chem. and Ind.* 903–906.

Duckham, A. N., Jennings, R. C. and Morris, T. R. 1966. *Future Trends in and
Interactions between Acreage of Farm Land, Yields, Urban Growth, Population and
Human Nutrition*. Dept. of Agric., Univ. Reading (Unpublished.)

Hathaway, D. E. 1963. *Government and Agriculture: Public Policy in a Democratic
Society*. New York.

*Higbee, E. 1963. *Farms and Farmers in an Urban Age*. Twentieth Century Fund.
New York.

Holliday, R. 1966. *Agric. Progress*. **41**. 24–34.

Monteith, J. L. 1966. *Agric. Progress*. **41**. 9–23.

Napolitan, L. 1966. *J. Royal Agric. Soc.* **127**. 95.

O.E.C.D. 1967. *O.E.C.D. Observer*. **26**. 38–41. O.E.C.D. Paris.

Orwin, C. S. and Whetham, E. H. 1964. *History of British Agriculture 1864–1914*.
341 ff. London.

*P.S.A.C. 1967. Rept. (U.S.) President's Science Advisory Committee. *The World
Food Problem*. 3 vols. Washington, D.C.

Ridgeon, R. F. and Sturrock, F. G. 1969. *Economics of Pig Production*. Univ.
Cambridge. Dept. Land Economy. Agric. Econ. Dept. No. 65.

*Schultz, T. W. 1960. *Transforming Traditional Agriculture*. Studies in Comparative
Economics No. 3. Yale University. New Haven and London.

Conclusions *Sietz, F. (Ed.) 1967. *Applied Science and Technological Progress.* Report to the House Committee on Science and Aeronautics, U.S. Congress, Govt. Printer. Washington, D.C.

Simantov, A. and Tracy, M. 1966. *O.E.C.D. Observer.* **22.** 27–33. O.E.C.D. Paris.

Taylor, J. A. (Ed.) 1967. *Weather and Agriculture.* London.

U.S.D.A. 1962. *Major Uses of Land and Water in the United States. Summary for 1959.* Agric. Econ. Rept. No. 13. U.S. Dept. of Agric., Washington, D.C.

U.S.D.A. 1967, 1968. *The World Agricultural Situation.* Foreign Agric. Econ. Dept. Reports Nos. 33 and 38. U.S. Dept. of Agric., Washington, D.C.

Wibberley, G. P. 1964. *The Changing Rural Economy of Britian.* Univ. London, Wye College, Reprint No. 265. New series.

Wibberley, G. P. 1965. *Pressures in Britain's Land Resources.* Univ. Nottingham. Dept. of Agric. Econ. Tenth Heath Memorial Lecture.

Yapp, W. B. and Watson, D. J. (Eds.) 1958. *The Biological Productivity of Britain.* Institute of Biology. London.

* Asterisked items are for further reading.

CHAPTER 3.3

Trends in Tropical Food and Agriculture

Food

In this chapter, we are only concerned with trends in tropical food production in the recent past; the problems of future food supplies will be examined in the next chapter. There are two aspects to be considered: the quantity, and the quality, of food which tropical agriculture has provided for its consumers. The quantity of world food production has since the Second World War been better documented than ever before by F.A.O. records (F.A.O. 1965 and 1966). This quantity showed, with slight interruptions, a rise from 1945 to 1966 which can largely be attributed to agricultural science and extension work. However, since human populations were also rising fast throughout the period the important figure is food production per caput, and this presents a much more dismal picture. From 1945 to 1959 this figure, again with some interruptions, did rise for the world as a whole; but in the important regions of the Far East and Latin America, food production per caput was still in 1959 lower than before the Second World War. From 1959 to 1966, food production per caput stagnated and in the later years of the period slowly fell; and these overall world figures conceal a position which was better than the average for the highly-developed countries, but worse than average in the tropics where the margin of food supplies was already smallest. However this does not necessarily mean that the inhabitants of all tropical countries ate less; some of these countries became increased importers of food, at an unwelcome cost in foreign exchange.

As against this depressing picture, it can be said that deaths from famine during this period were not allowed to occur on the catastrophic scale which history had sometimes previously recorded. Not since the Bengal famine of 1943–44 in which at least 1½ million people died has there been any comparable loss of life from starvation in one area. It is difficult to be more categorical, because many of the small famines which still occur are in countries whose governments have no adequate recording facilities and sometimes do not wish to advertise the existence of famine. Occasionally figures come to light which administer a shock, as when an Indonesian government statement was issued in October 1966 that 10,000 people had died of starvation in the first six months of that year on the island of Lombok. But there is no doubt that the world was better organised during this period than any previous one to relieve known famines. This was mainly due to the massive amounts of American grain which were available to be poured into India and to a lesser extent other countries at times of food shortage, but partly also to other re-lieving agencies such as the World Food Programme of the United Nations which began operations in 1963. The protection of humanity against cata-strophic famine was however only achieved at the expense of running down the world's surplus stocks of grain, mainly held in North America, which by 1966 no longer provided the massive buffer against disaster that they had represented at their peak around the year 1960.

As regards the quality of food consumed, there is little evidence of any substantial change either for better or worse in tropical diets in recent years. Both the production and consumption of rice increased considerably in South America and Africa, where rice must have been partially or wholly substituted for maize, millets or roots in the diet of many consumers; this probably re-flects a rising standard of living which enabled consumers to afford a more palatable food, which nutritionally is superior to the roots but inferior (be-cause of its lower protein content) to the cereals which it replaced. Wheat products were increasingly consumed in several parts of the tropics. In Africa this was again due to rising standards of living and a preference for bread over other foods, leading to increasingly heavy imports of wheat or wheat flour by many countries in which for climatic reasons the crop cannot be grown. In India, on the other hand, food shortages were only allayed by massive im-ports of American wheat, with the consequence that in successive periods of food rationing large rice-eating populations in southern India were reluctantly forced to accept part of their cereal ration as wheat or to go without. Another effect of increasing population in India was that, owing to pressure on milk supplies, ghee, the cooking fat most prized by the Indian housewife, came in-creasingly to be regarded as a rich man's luxury whilst most people cooked in vegetable oils. A final generalisation which can be made is that throughout the tropics middle-class urban families were increasingly taking to more sophisticated food preparations, especially canned and packaged foods and bottled drinks (both alcoholic and 'soft') which sometimes had to be imported and represented increased expenditure for the sake of convenience rather than any nutritional gain.

Agriculture

Although the acreage of most crops in the tropics increased during this period in rough proportion to the increase in the human race, there were a few for which the increase was conspicuously greater. The most important of these were rice and sugar-cane. Responding to the increased demands for consumption, rice production more than doubled in South America and nearly doubled in Africa during the twenty years to 1965. Neither of these continents is on the world scale a major rice producer, but taken with the con-current increases in rice production in the United States and Australia, these figures do represent a slow swing away from the overwhelming concentration of world rice production in Asia, a continent which nevertheless in 1965 was still estimated to produce 90% of the world's rice crop (Grist 1965). In the same twenty years, world production of sugar-cane more than doubled, the increase again being most conspicuous in South America and Africa, and also high in India and Pakistan. Consumption of sugar in the tropics may not in this case have increased by quite the same amount, for in some of the countries concerned local production represented import substitution, and in others the main object was an increase in exports.

Livestock populations are not by their nature capable of such rapid change as crop acreages, but in general it may be said that output of animal products in the tropics has roughly kept pace with the human population increase. Amongst individual animals, the exception is the camel whose principal functions have been gradually superseded by mechanical transport and whose global numbers have stagnated in recent years.

Turning from individual crops and livestock to a consideration of farming methods, mechanisation in the tropics has not advanced as rapidly as was expected in many quarters when development could be resumed after the Second World War. Tractors have, as might be expected, found their readiest application on the larger tropical farms, and the rate of their adoption has been almost proportional to the common size of farm in the country concerned, with the South American and Australian tropical regions leading, and Africa and Asia lagging behind. Small rotary cultivators have found a particularly useful niche in the small rice fields of Asia and may prove to be of considerably wider value. Smaller items of mechanical equipment, exemplified by such things as knapsack sprayers, rice threshers and maize mills have been steadily increasing in much wider areas than those in which tractors have been adopted, and are helping to familiarise the tropical farmer with mechanical principles. Ox cultivation has been given a fillip by the development of suitable toolbars which can carry a variety of implements (Minto 1966).

The use of fertilisers, like that of tractors, has not spread so widely as had been hoped at the beginning of this period. Nevertheless, great encouragement has been given by the fact that in India, as in Egypt a generation earlier, a peasant population which had hardly heard of fertilisers has been turned into one which insistently demands them, so that the demand in India had by the middle of the 1960s outrun the supply. Many tropical (like many temperate) countries have found it necessary to subsidise fertilisers in order to provide sufficient economic attraction to establish their use. The minds of tropical farmers were also being opened to the use of other agricultural chemicals, particularly pesticides, but also including some rather specialised use of plant hormones on particular crops such as rubber and pineapples. Research on herbicides, which appear to have a particularly great potential in tropical agriculture, has been (apart from plantation use) very inadequately promoted and it was not until the 1960s that literature began to appear (Hocombe and Yates 1963, Kasasian 1964) on which advisory officers could confidently base recommendations to tropical farmers. This lag in development is unfortunate because in many tropical areas the use of herbicides could offer a more immediate reduction than mechanical cultivation in the drudgery of farm labour and thus make it easier to attract intelligent school-leavers into the farming profession. The reluctance of this class to take up farming emerged during the period as one of the greatest weaknesses of tropical agriculture.

The increase of irrigation in these years undoubtedly made a significant contribution to the output of tropical farms, but published information is unfortunately inadequate to assess it statistically. Apart from China, for which particularly little information is available, the chief extensions seem to have been in India and Pakistan which already had very large irrigated areas. In addition to traditional tropical methods of surface irrigation, an important development was the installation on a still small but significant scale of overhead irrigation, chiefly on estate lands.

In the sphere of plant breeding, hybrid maize made its first appearance on the tropical scene in the 1950s, and within a decade had made significant contributions to increased maize yields in a few countries such as Pakistan, India and Rhodesia. Elsewhere, and especially in West Africa, 'synthetic' maize varieties provided a useful advance on older types. Serious attention was given to sorghum breeding, where crossing of types again proved a valuable means of increasing grain yields. Amongst plantation crops, the working out of feasible methods of rooting cocoa cuttings was followed by similar success with tea, and coupled with rigid selection provided a means of improving yields in a crop where they had long remained stagnant.

In crop protection, the period marked a real breakthrough with its demonstration that tropical smallholders, as well as plantation owners, were capable of learning the value and use of pesticides. This movement was spear-headed in West Africa where African farmers utilised sprays delivered from knapsack sprayers to control black pod of cocoa (caused by the fungus *Phytophthora palmivora*) in Nigeria and insect pests of cocoa, mainly Mirid bugs, in Ghana. Elsewhere, this development was perhaps most conspicuous with cotton, a crop which is subject to rather numerous pests and diseases, but which it is now quite common to see being sprayed by small tropical growers in many of the countries where it is cultivated.

In respect of livestock improvement, this period saw the crytallisation of opinion as to which were in fact the outstanding types of indigenous tropical cattle. The Sahiwal and Red Sindhi breeds from India and Pakistan were widely distributed in tropical countries for the improvement of milk production, whilst in eastern Africa time confirmed the outstanding qualities of the Boran as a beef animal. But these developments were rather overshadowed by the increasing introduction into the tropics of European cattle, whether to be bred pure or used for upgrading local stocks, which had largely been made possible by better veterinary control of animal diseases. A rather short list of European breeds, especially the Channel Island breeds, Friesian (Holstein), and Brown Swiss, increasingly emerged as the most popular for this type of operation. Unfortunately these developments have not been matched by a comparable improvement in the feeding of livestock, which remains one of the greatest weaknesses in tropical agriculture. Whilst there has been a slow increase in the feeding of oil-cakes and other concentrates, the vast majority of tropical ruminant livestock are still fed only on poor grazing and on crop residues such as rice straw, sugar-cane tops and maize stover which are of low nutrient value. The successful introduction of Pangola grass (*Digitaria decumbens*) into the Caribbean region to improve pastures is one of the few exceptions to these generalisations.

Progress in the control of tropical animal diseases, which has already been mentioned above as making easier the introduction of exotic animals, has on the whole been more encouraging than the battle against plant diseases, where new threats seem so continually to arise. Outbreaks of rinderpest during this period were more often brought under early control, and the elimination of the disease from Tanzania marked a distinct gain. The areas affected by bovine pleuropneumonia were reduced and some pockets of the disease eliminated. Drugs were for the first time discovered which had significant curative and prophylactic uses against animal trypanosomiasis, and in some cases enabled a milk supply to be maintained in tsetse-infested areas of Africa. The use of spraying against ticks and the diseases carried by them came as an aid and partial replacement to the older practice of dipping, and was cheaper to install. This increasing control over some major diseases was inevitably reflected, as in the more highly developed countries, by the springing into greater prominence of diseases hitherto regarded as minor, particularly infections of the reproductive tract and skin troubles. These are not however killers on the scale which used to be all too familiar with tropical animal diseases.

In the face of these various advances in scientific technique, it became painfully obvious that equivalent advances were not always being made in simpler matters of crop and animal husbandry, which generally demand only understanding and care rather than capital outlay or mechanical equipment. The yield of tropical crops could be considerably increased in almost all regions by more intelligent attention to such matters as planting date, transplanting, spacing, thinning and pruning where appropriate. In animal husbandry and production, higher standards could be attained by greater cleanliness and better observation of the behaviour and performance of individual animals.

Increasing experience in the improvement of tropical agriculture also emphasised a fact which has sometimes been called the 'principle of complementarity' by agricultural economists, or referred to by agronomists accustomed to the statistical analysis of field experiments simply as 'interactions'. These terms mean that the effect of one improvement (e.g. irrigation, fertilisers, pest control) often enhances that of another, so that for example if a crop is irrigated it is profitable to apply more fertiliser to it, and perhaps still more if pests are also controlled; whilst there is also an interaction between such inputs and husbandry, as for example where the full benefit is not obtained from fertilisers or irrigation unless the crop is spaced more closely. Multiple improvements have therefore more than a simple additive value over single improvements. Unfortunately there is in simple societies a limit to the number of changes which peasants can be induced to make at one time, and the scientists must here be guided by the extension worker in the art of the possible. Nevertheless the principle of multiple improvement began to gain official recognition, notably in the Intensive Agricultural District Programme of the Indian community development plan, which came to be popularly known as the 'package programme'.

Food and Agriculture

Turning finally to the widest aspects of the problems of tropical agriculture during this period, the salient fact is the development of the 'population explosion' and the gradual and alarming realisation of its implications for agriculture. This can well be illustrated from the writer's experience of a district in Uganda. In 1936, a slight increase in the population of this district, the first ever recorded since statistics were kept, was hailed as cause for delight and self-congratulation on the part of the authorities concerned with health and agriculture. Twenty years later, the continuing rise was a matter for gloomy head-shaking; and ten years later again, it was regarded as almost a disaster. It was in this same year, 1966, that the Director-General of F.A.O. could write (F.A.O. 1966):

> Any remaining complacency about the food and agriculture situation must surely have been dispelled by the events of the past year. As a result of widespread drought, world food production, according to F.A.O.'s preliminary estimates, was no larger in 1965–66 than the year before, when there were about 70 million less people to feed. But for good harvests in North America, world production would almost certainly have declined. In fact, in each of the developing regions except the Near East, food production is estimated to have fallen by 2 per cent in total and by 4 to 5 per cent on a per capita basis.

One result of this slowly mounting crisis was a belated switch of effort in tropical agricultural science from cash crops to food crops. From the 16th century, when Europeans first entered the field of tropical agriculture, they had been chiefly interested in promoting the production of export products, originally for their own advantage, but in later times also with the altruistic object of increasing the national incomes and standards of living of tropical territories; it was tacitly assumed that tropical populations could continue to feed themselves, as they had always done, without any external assistance. This outlook overlapped the coming of modern agricultural science to the tropics, with such results as that, in the field of genetics, crops like sugar-cane, rubber and cotton were intensively bred whilst selection had hardly started on the local food crops. Such developments in export crops moreover paid speedy cash returns to hard-pressed governments and plantation companies promoting them, whilst the improvement of food crops gave no such dividends. By the end of the Second World War, this imbalance had come to be realised and began to be remedied; the process was made easier by the increasing availability of funds for tropical development (derived, indeed, in part from the success of export production) which meant that research was not so imperatively required to produce immediate cash returns. The change was reinforced also by the increased understanding of nutritional science, which showed how widespread were such conditions of malnutrition as 'kwashiorkor' (first properly understood in the 1940s) and how much they affect the health, happiness and efficiency of tropical populations. The newer emphasis on food production in tropical agriculture is a real revolution, not always understood outside the tropics, from an older outlook to which in some degree our present troubles are due. A consoling feature of the situation is that so little scientific work has yet been done on the tropical food and fodder crops that the scope for improvement is enormous. The yield of cane sugar per acre obtainable on plantations increased by about 1,000% in the century 1840–1940, and gives cause for hope when science is applied on a similar scale to tropical food crops.

A feature of the population explosion which, by contrast, seems not to have attracted its due share of attention is its effect on farm size. There is evidence, which unfortunately there is all too little statistical coverage to evaluate, that in many parts of the tropics average farm sizes are shrinking as population increases. In one fairly densely populated area in Uganda which was surveyed in 1937 and 1953, the mean area of cropped land per family fell between those years from 9·3 to 7·8 acres, an overall decrease of about 1% per year in an area where population was probably increasing by about 2% per year (Wilson and Watson 1956). The rise in population could be contained without fragmentation of farms if it could all be absorbed into industry, but unfortunately this is not happening in most developing countries and the problem of shrinking farm size has to be faced. There are probably still too many extension services which are preaching the replacement of manual cultivation by animal or tractor-drawn implements in a situation where such replacement becomes continuously less feasible or relevant as farms decline in size. Increased productivity per acre rather than per man is the essential need of the tropics, where much farm labour is already underemployed, partly due to the smallness of farms. Developed countries are already worried about the non-viability of small farms, where the 'smallness' is much less in degree. One example where this problem has already engaged official attention in the tropics comes from Kenya. In the first stage of take-over of European farms for settlement by Africans (the 'first million acres'), these farms were parcelled out into very small holdings. But experience of the results of this scheme led, in the second phase of the operation, to the retention of the large farms as units to preserve economies of scale, and their working by African settlers under various types of unitary management.

The problems we have been considering in this section are broad ones, whose threat to world food supplies as a whole is real and urgent. But their incidence falls very unevenly between different countries. There are countries such as India, Pakistan, Indonesia and Egypt where the future maintenance of food supplies is so manifestly precarious that governments and people are at least aware of the situation, and this awareness has in some cases proved a spur to action. But there are also, in most of South America, much of Africa, and parts of south-east Asia, vast areas of uncultivated land still in reserve. In such regions it is difficult to convince governments of the urgency of the world food situation as a whole, and the consequent lack of interest in the problem which tends to clog international action in this field is itself one of the most difficult aspects of current problems in tropical agriculture.

References

F.A.O. 1965. *The State of Food and Agriculture 1965: Review of the Second Postwar Decade*. Rome.

F.A.O. 1966. *The State of Food and Agriculture 1966*. Rome.

Grist, D. H. 1965. *Rice*. London. 4th edn.

Hocombe, S. D. and Yates, R. J. 1963. *A Guide to Chemical Weed Control in East African Crops*. East African Literature Bureau, Dar-es-Salaam.

Kasasian, L. 1964. *Tropical Weed Control Notes*. Univ. of the West Indies, Trinidad.

Minto, S. D. 1966. 'Progress in Agricultural Engineering and Mechanisation in East Africa', *E. Afr. Agr. and For. J.* **32**. 72.

Wilson, P. N. and Watson, J. M. 1956. 'Two Surveys of Kasilang Erony, Teso, 1937 and 1953,' *Uganda J.* **20**. 182.

General Reading

Allan, W. 1965. *The African Husbandman*. Edinburgh.

Drogat, N. 1962. *The Challenge of Hunger*. London.

Hutchinson, J. (Ed.) 1969. *Population and Food Supply*. Cambridge.

Masefield, G. B. 1963. *Famine: Its Prevention and Relief*. London.

Webster, C. C. and Wilson, P. N. 1966. *Agriculture in the Tropics*. London.

Williamson, G. and Payne, W. J. A. 1965. *An Introduction to Animal Husbandry in the Tropics*. London. 2nd edn.

CHAPTER 3.4

World Food and Population Problems

It is unfortunate that any discussion of what may well prove to be mankind's major problem for the rest of this century must be based on inadequate statistics and uncertain facts. On the population side, statistics of the existing human population are notoriously inaccurate for some parts of the world, and projections of future population trends are even more liable to error; many cautious authorities make maximum and minimum projections, and base their calculations on a figure somewhere between these. Figures of current food production throughout the world, although diligently collected and published by F.A.O., are extremely uncertain material on which to work; the statistics of food consumption are no more reliable. The exact levels of human food requirements, and hence the definitions of 'under-nutrition' and 'malnutrition', are still scientifically uncertain. Such factors as the calorie and protein requirements of human populations depend, amongst other things, upon climate, average body weights, the amount of work to be performed and the proportion of different age-groups in the population. These variables have been well discussed in certain publications by United Nations agencies (F.A.O. 1958, W.H.O. 1965) which give us a little more confidence in using suitably processed figures but do not fail to point out that these are still only approximations. It is partly because of the wide range of figures which can be used with some show of plausibility that some authors have been able to take what appears to be an excessively alarmist view of food and population problems whilst others are open to the charge of over-complacency. Most authoritative opinion has hovered somewhere between the two extremes but in recent years has tended more and more to strike the note of alarm.

Rational discussion of this problem may well start with the assessment of the situation as seen in F.A.O.'s 'Third World Food Survey' (F.A.O. 1963). In Table 3.4.1 are presented some figures from this survey which give a synoptic view of the world's population and food position. Taking the first three columns together, it will be seen that the Far East region, containing over half the world's population, has not even sufficient food to supply the computed calorie needs of that population on average. In three other regions (Near East, Africa and Latin America), and for the world as a whole, calorie supplies are so little above requirements that it is obvious that many people must go hungry in years of poor local harvests, and probably some people at all times amongst the poorest classes. In parts of these regions moreover the supply available is partly made up by imports of food which are only precariously paid for by exports or are dependent on the generosity of the governments of exporting countries who supply them under concessionary conditions. In protein nutrition, only the three highly-developed regions (North America, Europe, Oceania) have enough to supply what are considered the proper long-term targets for this class of nutrient; and, of course, in the other regions, many consumers are getting so little protein of animal origin that malnutrition is likely to be caused by a deficiency of some of the essential amino-acids. Finally, the table shows that the world's population in the year 2,000 is expected to be more than double that in 1960, a rate of increase for which history hitherto provides no parallel.

Looking at these things in another way, the survey estimates that from 10 to 15% of the world's population are simply not getting enough food. This may amount to the frightening figure of some 400 million people. However it should be remembered that this total is not made up of the same individuals

Table 3.4.1

World Population and Food Supplies

Region	Population, 1960, in millions	Daily supply of kcal per caput:		Daily protein supply, grammes per caput:		Estimated population year 2000 in millions
		Present	Required	Present	Long-term target	
Far East	1,602	2,060	2,300	56	74	3,753
Near East	132	2,470	2,400	76	79	326
Africa	215	2,360	2,350	61	75	458
Latin America	211	2,510	2,400	67	71	595
Europe	639	3,040	2,600	88	—	954
North America	199	3,110	2,600	93	—	325
Oceania	16	3,250	2,600	94	—	30
WORLD	3,014	2,420	2,400	68	79	6,441

Source: Various tables in 'Third World Food Survey', F.F.H.C., *Basic Study*, No. 11. F.A.O., Rome 1963.

all the time. If it were, they would manifestly not survive for very long; but in fact this class mostly consists of people who suffer alternating periods of a low plane of nutrition and a somewhat better one when a good harvest has been newly gathered. A further proportion of up to 35% of the world's population is estimated to suffer from malnutrition, mostly a deficiency in protein or vitamins. For the less-developed countries, these proportions are estimated to rise to 20% suffering from under-nutrition and a further 40% from malnutrition.

Turning to the future, the survey estimated that merely to maintain the existing inadequate levels of nutrition in face of rising populations would require a world increase in food production of 36% by 1975 and 123% by 2000. If a reasonable minimum level of nutrition were to be aimed at, the required increase by 1975 would be over 50% for the world as a whole, and 80% in the less-developed countries. This would call for an annual increase of some 2% in per caput supplies in the latter group of countries; but in the event, as we have seen in the last chapter, production per caput in these countries has not risen but actually fallen in the subsequent years. This does not provide a very hopeful outlook for the final target suggested by the survey, which is that to provide reasonably adequate nutrition for the population expected in 2000, the world's total food supply would have to be trebled, whilst in the less-developed countries it would need to be quadrupled and the output of animal food increased six times.

Such global figures as have just been quoted inevitably conceal differences between regions and groups of countries which largely condition the way in which their inhabitants view these problems. At one extreme we have the well publicised example of India which has given the impression in recent years of moving into a condition of chronic food shortage even by its own modest standards. There are other countries in a similar position whose problems are much less widely known. In Somalia, for instance, it was estimated in the 1960s that 300,000 people were almost annually at risk of famine with 18,000 in actual danger of death. A second group of countries may find themselves at times severely short of food but have so far possessed the wealth or the influence with friendly nations to increase their food imports sufficiently to contain the situation; both Russia and China came into this class in some years of the 1960s when they greatly increased wheat imports. At the upper end of the scale are the highly-developed Western nations to whose populations the world food situation has only been brought to physical notice as a general shortage of beef, making it more expensive to buy and leading to some substitution for it of pig and poultry meats with no nutritional deprivation. The most politically unfortunate aspect of the situation is that the difference between nutritional intakes in the best and the least well-supplied countries has tended in spite of all efforts to widen rather than to narrow and has thus made the situation even less tolerable.

In general it is evident that the world food situation contains two elements which make unequal demands on agriculture. The first is sheer lack of food per head, a deficiency which can only be made good by increased agricultural production somewhere in the world. The second is malnutrition, a problem whose solution in many cases depends more on education than agriculture. Often the required foods are available, or could be made so without any changes in agricultural technique if people only knew that they needed to eat them. A great deal of protein deficiency could, for example, be relieved by an informed decision on the part of individual farmers to grow one acre more of groundnuts and one less of starchy roots. Vitamin and mineral deficiencies could be much reduced by the consumption of fruits or leafy vegetables which are very often available but ignored, or could be grown in sufficient quantities without difficulty.

The relief of under-nutrition is therefore the prime agricultural problem. This problem centres upon cereals, both because these are the basic suppliers of calories for most human populations, and because the quantity in which they are consumed, combined with their very appreciable protein content, makes them also in quantity the chief suppliers of proteins in poor countries.

517

Amongst the cereals, the problem again narrows itself down mainly to supplies of wheat and rice, one or other of which is the staple food of most of mankind; maize has comparable importance only for some much smaller populations in parts of Africa and the Americas, and millets in some rather sparsely populated arid areas.

Supplies of wheat on the world scale are not at the moment, and do not look like being in the near future, crucial. Most of the wheat-eating countries either grow their own supplies or have long-established channels of import

Table 3.4.2

Exports and Imports of Wheat and Rice, 1965

(A) Wheat
(Figures in short tons; all exports or imports over 1 million tons included)

Country	Exports	Country	Imports
U.S.A.	19,511,550	India	7,201,500
Canada	13,097,540	U.S.S.R.	7,027,340
Argentina	7,342,050	China	5,786,900
Australia	6,299,370	U.K.	4,860,640
France	4,459,950	Japan	4,018,030
U.S.S.R.	1,832,700	Brazil	2,068,260
		Pakistan	1,851,330
		W. Germany	1,826,640
		Poland	1,519,200
		U.A.R.	1,356,280
		E. Germany	1,350,330
		Yugoslavia	1,314,610

(B) Rice
(Figures in short tons: all exports or imports over 400,000 tons included)

Country	Exports	Country	Imports
Thailand	2,146,310	Japan	1,066,260
U.S.A.	1,707,920	India	863,290
Burma	1,485,360	Malaysia	621,590
China	813,950	Philippines	616,850
Cambodia	521,280	Ceylon	584,220

Source: F.A.O. Trade Yearbook, Rome 1966.

from countries with a surplus, of which the United States and Canada are by far the most important with Argentina and Australia next in rank (Table 3.4.2). It is true that there are some wheat-eating populations, as in northern India and some North African countries, who have with difficulty been supplied in recent years, but the difficulty has usually been in payment rather than procurement of wheat. World wheat exports could be increased if they could be paid for, and there is a well-proved potential for expansion in the United States where the acreage of this crop has been artificially limited by government policy following earlier overproduction for the market. There would be no real problem about world wheat supplies if wheat did not have to be used for the emergency feeding of people who normally eat rice.

Rice supplies are at present the Achilles' heel of the world food situation, and rice production probably needs the attention of agriculturalists more urgently than that of any other commodity. This position arises from several different factors. The number of important rice-exporting countries is less than for wheat, and there are, as shown in Table 3.4.2 only three major suppliers: Thailand, the United States and Burma. Any downward fluctuation of the crop in any one of these countries is thus likely to cause immediate distress elsewhere. The chief importing countries are mostly Asian; in recent years the largest importers have usually been Malaysia, Indonesia, India, Ceylon, Hong Kong and the Philippines. In these countries the normal diet of large populations is so barely adequate that even a slight deficiency in the usual rice supplies is likely to throw them quickly into famine conditions. In the words of an F.A.O. study, 'Experience shows that serious food shortages are most likely to occur among rice-eating populations' (F.A.O. 1956); and this is largely due to the extreme variability of rainfall in the monsoon regions of Asia which grow most of the world's rice. Unlike wheat or maize, world rice supplies have never since the Second World War been much more than enough to meet demands, and it has not at any time been possible to accumulate surplus stocks to form a reserve against a bad year as has been accomplished with the other major cereals. In this precarious situation it is heartening that research and extension work on the rice crop has at last though belatedly begun to receive the attention it deserves, and that this work has already had an effect in increased production in some areas, although this has largely been nullified by the increase in population and much more effort will be needed before the situation is less than decidedly alarming.

Apart from the regional deficiencies in calories which we have been considering, the world's other most important need is for more protein. Pulse crops, which have a higher protein content than cereals, are a group whose production must and can be intensified in most of the less-developed countries. Plant breeding can sometimes help, as in the recent development of maize types with a high content of the amino-acid lysine. But animal protein is even more deficient and more needed than vegetable protein, and here a great dilemma presents itself. Put shortly, is it right in a country where millions are undernourished because of shortage of grain, to use some of that grain to feed pigs or poultry which will provide fewer calories but more animal protein for those millions, largely the same ones, who are suffering from protein deficiency? And is it right to use land for grazing animals to produce milk and meat when that land could produce more calories for human consumption if it grew crops? In such a country as India, it is possibly easier to show that the farmer does not lose financially by turning over from tillage to mixed farming—a problem which has been discussed by Finney and Pannikar (1953) and Whyte (1962)—than that there is an overall benefit in food supplies.

*   519

Perhaps all that can be said generally is that any decision between these two systems must be based on the most exhaustive analysis possible of their respective benefits, not overlooking even such small factors in the equation as the value of the animals' excreta for increased crop production or the export of hides and skins which will buy some food imports. And it is clear that, in the kind of nutritional situation we are discussing, livestock production can only be justified for those products which are most economical of land, for example, milk rather than beef, and with the most intensive possible management of grass and fodder crops.

Having tried to define the targets which need to be aimed at, the next question which naturally arises is what are the possibilities of increasing the production of the critical foodstuffs by the required amounts? Here a preliminary point to be clarified is that of storage losses. It is sometimes argued that world losses of grain in store, chiefly by insects and in hot countries, are so large that their elimination alone would alleviate many food shortages and do it more quickly than any practicable increase in field production. Certainly some losses are alarming. A United Nations (1950) report records losses of cereals and pulses in store in various Central American countries of from 25 to 50%. On the other hand, Buyckx and Decelle (1957) quote maize losses in the Congo of only 5–6% and Giles (1964) losses of sorghum and millet in farmers' storage in northern Nigeria of about 4%. It would seem that a general figure of not more than 10% would be nearer the truth than anything more alarming, and that in the most famine-prone countries storage losses are less important than this for the following reasons: (1) in such countries, as we have seen, the critical foodstuff is rice which is one of the least susceptible of the cereals to storage loss; (2) the countries most liable to drought famines generally have climates which are extremely dry for much of the year, and this is favourable for storage; (3) turnover of stocks is very rapid in countries which are short of food, and prolonged storage is therefore rare.

Increased food storage capacity remains nevertheless one of the world's prime needs to even out fluctuations in crop yields and alleviate distress. The achievements of different countries in this respect are very uneven. At one extreme we have Norway, a country of relatively small population, which has succeeded in storing enough grain to feed the population for a year at rationed quantities. At the other is India, where it was announced in 1965 that the Government hoped over a number of years to build up stocks of 6 million tons of cereals, only a tiny proportion of the annual consumption of some 90 million tons.

Turning from storage to production, we have to face the question of what are the possibilities of increasing world food output sufficiently to meet the targets outlined above as desirable for the rest of this century. There are two possible sources of increased agricultural production. One is the use of new lands at present unutilised, and the other is greater yields from land already in use. As earlier chapters have shown, there are quite a number of countries with both developed and less-developed economies which still have large reserves of unused land. However the development of these lands raises several problems. It is only natural that most of them are less favourable to agriculture, usually by reason of poorer climate, soils, terrain or access, than the lands which men have already chosen to develop. An extension of crops into such areas often tends to reduce rather than increase national average yields of crops (Masefield 1963); because of low yields, their production may not be economically competitive with that of older developed areas, and may require special subsidisation, as in the case of some kinds of hill farming in Great Britain. The bringing into production of, for example, desert areas is often very slow, whereas the need for food is urgent; furthermore, it is usually expensive, and the capital available will often produce more extra food if applied to the further intensification of already developed land than if used in the reclamation of hitherto neglected areas. Finally, those countries which are really short of food are usually precisely the ones which have already brought all their land into some kind of use.

This brings us therefore to the problem of increasing yield per acre in areas already farmed. It is here probably that the most immediately practical solution to world food problems is to be found. It is very generally accepted that, by the application of existing knowledge without any further research, the overall yields of crops in the less-developed countries could be at least doubled (more than this for some crops, but less for others) (Richardson 1960). This sort of estimate is based on yields obtained in field experiments which also suggest that about half this increase would be due to improvement in soil conditions (principally by fertilisers and irrigation) and about half to other measures (principally genetic improvement and control of pests, diseases and weeds). It is evident therefore that the problem of eliminating food scarcities in the underdeveloped world is not biologically insuperable; what prevents its being solved is the slow spread of knowledge of improved techniques, and above all the initial and recurrent capital needed to apply them. In the highly-developed countries the situation is rather different, for farmers' techniques have followed more closely behind the advancing frontier of knowledge, and further increase of yield is therefore more dependent upon problematical future advances in science (but see p. 526). There is however another factor here, which is the deliberate discouragement of the expansion of some staple food crops by the governments of certain countries, such as the United States, which have seen the depressing effects of overproduction in the past (p. 498).

This factor therefore raises another problem. There is no doubt that production of certain crops, such as wheat, maize and barley could be greatly and quickly increased in a few of the highly-developed countries. But of how much use would such surpluses be to starving people in distant parts of the globe? The answer to this question involves two main problems, transport and payment. In 1962 surplus wheat could be bought in the United States for about £25 a ton, but the cost of shipping, handling and distributing it for consumption in India amounted to another £10 (Williams 1962). The sea transport of much more grain than is at present moved to countries needing it could not be undertaken without a large increase in the world's shipping. Even more immediately, experience has shown that the harbour and transport facilities of many of the countries in greatest need are already overtaxed bottlenecks which are incapable of handling much greater quantities of food. On the payments side, most of the countries lack foreign exchange for the purchase of further grain, and there are obvious economic limits to the generosity which has already been so liberally displayed by donor countries in supplying grain under various schemes of assistance.

The real solution to world food shortages therefore still lies in increasing production within the regions where food is short. This solution is hardly likely to be reached without considerable changes, not only in farming techniques, but in the agricultural commodities which are produced. It is probable that, as human population densities increase, we shall have to place more and more reliance on those crops which produce a very high yield of food calories per acre. Outstanding in this class are sugar-cane and oil-palms, the latter of which may fill the gap in world fat supplies as pressure of population leads to more replacement of livestock by crop production and hence a drop in the production of animal fats. It is also arguable that bananas yield more calories than tropical root or cereal crops; but the calorie yields of annual crops are not susceptible of strict comparison owing to their variable periods of growth and the fact that in many environments more than one crop can be taken in a year. The enormous world deficiency in meat supplies is likely to call especially for increased production of pigs and poultry, not only because of the rapid multiplication rate of these animals, but because they can be fed in arid regions on grain grown during a single rainy part of the year and do not, like ruminants, demand green herbage at all seasons which can only be provided by well-distributed rainfall or expensive irrigation. A strong theoretical case can also be made for the domestic rabbit, as an extremely cheap meat producer from almost costless waste materials (crop residues and weeds) and very easily housed.

Such developments are however in the future, and in the meantime we are still faced with the situation that, as we have already seen, although agricultural improvement is taking place it is often outpaced by the increase in population. Fortunately it is not necessary to postulate from this fact an inevitably lost battle and a steady decline in food supplies per head; on the contrary, there are factors which, if agricultural science and extension work can contain the position for another generation or two, may come to the rescue of mankind in its predicament.

The first of these factors is the use of non-agricultural foods to supplement the supply of conventionally produced foodstuffs. These include the use as food of plankton, cultured algae, yeasts and leaf protein. All of these have already been experimentally produced and consumed. The obstacles to their

522

further use are in some cases their cost, and the fact that they are strange and are often considered less palatable than conventional foods. All these objections may be expected to diminish as other foods get shorter. More important in the longer term could be the hope that feasible means can be discovered for the mass synthesis of both proteins and carbohydrates. Synthetic amino-acids are already commercially available in small quantities and can be used to 'fortify' natural foods, and with increasing knowledge the synthesis of carbohydrates seems a possibility no longer so remote. Even if such processes prove to involve difficulty in supplying the necessary energy or other factors, the production of even marginal quantities would provide some relief to the demands on agriculture.

The limitation of population through family planning must, as a contribution to stabilising food demands, also be placed amongst the longer-term prospects. It can, even where widely adopted, have little effect on the population explosion for a generation, for the numbers already born are so large that, even if many of them have limited families, population increase will take a long time to slow down. The increasing acceptance of the necessity to limit population is nevertheless a hopeful factor in the situation.

If to these long-term prospects of alleviation of food scarcity we add the still unforeseeable discoveries in science which may help either agricultural output, non-agricultural food production, or family limitation, then the future prospect does not appear quite so dark as the immediate problems suggest. It may be that the span of one more human generation will prove to be the most crucial phase in man's struggle for sufficient nutrients.

References

Buyckx, E. J. and Decelle, J. 1957. 'Résultats d'une Enquête sur la Conservation de Denrées au Congo Belge', *Stored Food Products*. Comm. Tech. Co-op. in Africa South of Sahara. Salisbury.

F.A.O. 1956. *Functions of a World Food Reserve—Scope and Limitations*. Commodity Policy Studies No. 10. Rome.

F.A.O. 1958. *Calorie Requirements*. F.A.O. Nutritional Studies No. 15. Rome.

F.A.O. 1963. *Third World Food Survey*. F.F.H.C. Basic Studies No. 11. Rome.

Finney, D. J. and Panikkar, M. R. 1953. 'Experimental Tests of Mixed Farming in India', *Indian J. agr. Sci.* **23**. 269.

Giles, P. H. 1964. 'The Storage of Cereals by Farmers in Northern Nigeria', *Trop. Agr. Trinidad*. **41**. 197.

Masefield, G. B. 1963. 'Population Increase: A Possible Effect on Crop Yields', *World Crops*. **15**. 135.

Richardson, H. L. 1960. 'Increasing World Food Supplies Through Greater Production'. *Outlook on Agriculture*. **3**. 9.

United Nations. 1950. *Agricultural Requisites in Latin America*. Dept. of Economic Affairs. New York.

FARMING SYSTEMS OF THE WORLD

W.H.O. 1965. *Protein Requirements. Report of a Joint F.A.O./W.H.O. Expert Group. W.H.O. Technical Report Series No. 30.* (Also published as *F.A.O. Nutrition Meetings Report Series No. 37*).

Whyte, R. O. 1962. 'International Development of Grazing and Fodder Resources. XI—India (Sequel)', *J. Brit. Grassl. Soc.* **17.** 287–293.

Williams, P. 1962. 'The Use of World Food Surpluses', *The World Today.* **18.** 304.

General Reading

Burton, I. and Kates, R. W. (Eds.). 1960. *Readings in Resource Management and Conservation.* Univ. of Chicago Press, Chicago and London.

F.A.O. 1963. *Possibilities of Increasing World Food Production.* F.F.H.C. Basic Studies No. 10. Rome.

Sukhatme, P. V. 1961. 'The World's Hunger, and Future Needs in Food Supplies' *J. Roy. Stat. Soc.* (A). **124.** 463.

U.S. Dept. of Agriculture. 1964. *Farmer's World: The Year Book of Agriculture 1964.* Washington.

CHAPTER 3.5

Summary and Conclusions

Each farming system consists of one or more of the four food chains, viz. (*a*) Tillage Crops/Man; (*b*) Tillage Crops/Livestock/Man; (*c*) Grassland/Ruminant Livestock/Man; or (*d*) Tillage Crops and Grassland/Ruminant Livestock /Man (Chap. 1.1). Although there are innumerable variations on the main themes, there are, basically, only four farming systems, viz. *Permanent Plantation* or Tree Crops: *Tillage* with or without livestock; *Grassland/ Grazing* with ruminant livestock; and *Alternating* between tillage and grassland. Each system can be subdidivded, on the basis of input intensity, into extensive, semi-intensive and intensive (Chap. 2.1). Two or more systems, often at different levels of intensity, may be found on the same farm or local area (Chaps. 1.1 and 3.1) especially in the temperate hydro-neutral zone (p. 461).

The influents of the location and intensity of farming systems are ecological, operational and socio-economic (Chaps. 1.1, 1.4, and 1.7). Of these the first and third are the most important, whilst in advanced economies, the ecological group has less impact than in developing ones. The ecological group comprises climate, soils, 'natural' vegetation and 'natural' fauna, with climate as the dominant factor. Agriculturally, there are six major climates (p. 42) viz. cool humid, cool dry, warm humid, warm dry (including Mediterranean), dry tropical and wet tropical. When socio-economic factors are held constant, ecological factors and market access (Chaps. 1.7 and 3.1) are dominant influents. But their importance is receding before the advance of science and technology; and socio-economic influents, including scientific and industrial inputs, increasingly dominate the use of the environment (Chaps. 3.2, 3.3 and 3.4).

From man's viewpoint, a biologically efficient food and farming system must exploit the productive potential (p. 473) of a given land area by the optimal use, for the given site, of ecological, social and economic resources; it must also be reliable between years and persistent over decades and centuries (Chaps. 1.1 and 3.1). But, even in advanced economies, such optimal use, i.e. the use of optimal input ratios (p. 473) is rare. This is not usually because the current actual farming systems are ill suited to the climatic, soil, operational and topographical (relief) factors which determine (Chap. 1.4) the feasible enterprises and systems for a given site. In fact, on a world basis the 'goodness of fit' between actual farming systems and local ecological/operational constraints is not unsound, either ecologically or economically, as Part 2 and Chap. 3.1 demonstrate. Nor is the failure to achieve better use of productive potential due to inadequate scientific knowledge; if man *now* applied what he *now* knows food supply could probably be doubled (Bawden 1967, Chaps. 3.2 and 3.4) *if* the needed industrial inputs were made available. Moreover, though the 'yield' from future research must be speculative, the increase of present maximal levels of photosynthesis (*A.G.*) and the minimisation of present crop (W_p) and animal (W_a) wastages (p. 5) might reasonably be expected to double, within 30 or 40 years, the present theoretical ceiling of food supply from conventional agriculture provided this was aided by the development of 'synthetics' based on fossil fuels and even by full synthesis from atmospheric nitrogen and carbon dioxide, etc. (Chap. 3.2).

The main reasons for failure to optimise the use of current and potential resources are now, and will probably go on being, social (including political) and economic (Chaps. 1.5, 1.7, 3.2, 3.3 and 3.4). In countries with advanced economies, this failure has few adverse effects on human nutrition and health. Indeed, in such countries, food production per head and food resource use per head, when measured as plant (or wheat) equivalent output or consumption per head is, in some cases, two or three times higher than is nutritionally strictly necessary (pp. 11, 12). Many advanced countries show, in fact, a chronic tendency to national overproduction (Chaps. 3.2 and 3.4); they could, with national advantage, encourage the transfer of manpower, capital and other resources on farms to other sectors of the community. Several countries have, in fact, such resource transfer policies, mainly for manpower (Chaps. 2.5, 2.6).

But, in less-developed countries, whether tropical or temperate, the margin of nutritional safety is much narrower or even non-existent (Chaps. 3.3 and 3.4). Plant equivalent consumption per head is nearly the same as the consumption of total dietary food kcal per head, and there may not be enough energy or total protein or both in the diet; in particular, dietary animal protein (or more specifically the amounts of lysine, methionine, and possibly other specific amino-acids) may be insufficient (Chaps. 3.3 and 3.4). Moreover, in the developing countries, the rate of increase in food production is only too often handicapped by unreliable climates and by the difficulty of controlling pests, disease and other wastages in tropical eco-systems. In many such cases, increased food supply is not, or is only just, keeping pace with the rate of increase in human population (Chaps. 3.3 and 3.4).

There are strong, or at least marked, positive, and apparently causally related, correlations between the level of socio-economic development; the level of farm technology and input intensity; cereal yields per acre; and the level and quality of human diets per head (Figs. 1.1.7, 1.1.8, 1.1.9 and Table 1.1.2). Well-fed advanced industrial countries mostly have advanced productive food and agricultural systems, largely dependent on industrial and scientific inputs. But it is unlikely that, in the next 30 or 40 years, developing countries will, by rapidly industrialising and by building the elaborate and often biologically wasteful (*qua* man) complex of a socially and economically highly developed state, such as the Netherlands or the U.S.A., be able to overcome the twin problems of marginal nutrition and of rapidly growing populations. *Prima facie*, however, adequate food and agricultural systems capable of supplying adequate diets to rapidly expanding populations, ought to be attainable without most of the non-biological trappings (e.g. television, high motor-car populations) of a socio-economically advanced society.

Research, therefore, is needed (*a*) to test the validity of the assumed causal relationships between economic development, advanced farming, high yields and good nutrition (Chaps. 1.1 and 3.1), and (*b*) in particular, to determine what are the minimal levels of socio-economic development, of farming technology and of ancillary food and agricultural industries that are needed to ensure the consistent supply of adequate diets for growing populations in less-developed areas. This research must depend, in part, on creating more trustworthy statistical data (obtained by census or sampling) and better technical and social information on many aspects of food and agricultural systems (Chaps. 3.3, 3.4). On the quantitative side, there is urgent need, in particular, for better probability statistics on the incidence and possible causal factors of famine and other instabilities (Masefield 1963). There is also a need for quantitative research, on the lines now being applied by ecologists in other fields (Watt 1967), into the *total* food and agricultural systems of developing countries. Such research should firmly identify (*a*) the location and size of the vulnerable links in the food chains, and (*b*) those links most likely to respond, economically, to scientific and technical research, to advisory and administrative treatment or to capital injection. What seems to be needed is, in fact, a combination of systems analysis and cost/benefit analysis applied with vision to real situations and problems.

Such analyses should help, *inter alia*, to focus attention on, and to evaluate the critical technical problems and to indicate the comparative advantages of alternative solutions. Thus, in a given rural or urban situation, can protein deficiency most effectively be remedied by changes in cropping practice, for example, more legumes and fewer carbohydrate crops, by genetic improvement of the amino-acid content of traditional crops (e.g. rice), by the fortification of the diet by the addition of synthetic amino-acids, by unconventional processes (e.g. the use of leaf protein, microbial protein from petroleum waxes), *or* by conventional means, i.e. by increased livestock production, possibly using non-protein nitrogen (e.g. urea) to augment the low protein content of local livestock feeds? (Incidentally, if non-protein nitrogen compounds such as diammonium phosphate can be used as partial protein substitutes in poultry feeding, have they a possible role in human nutrition?)

In the meantime, on the social side, programmes of general education, of technical advice and of government action intended to raise efficiency and, where necessary, to control population (Chaps. 3.3 and 3.4) should be firmly applied. Such application might be helped by more international interchange on methods of inducing both agricultural change (Bunting 1969) and shifts in human dietary habits and by further research into extension and 'promotion'

Conclusions methods suitable for farmers with little formal education, limited land and little or no capital, and for domestic consumers.

To outline the specific items of scientific and other food and agricultural research which the text of this book suggests to us are needed, would be to repeat many fields in which we have noted research is already progressing. But it does seem to us appropriate to add, to the research suggested above, the need to check critically and to quantify the models and the arguments in Chaps. 1.1 and 3.1. Pending such critical examination, we suggest that the location and intensity of the farming systems of the world, despite their infinite variety, have a logical pattern (Chaps. 1.4 and 1.7, the transects in Part 2, Chap. 3.1) which can be tentatively expressed, for advanced temperate areas, as algebraic models that are partly, and potentially fully, quantifiable. If the statistical data mentioned above as needed for developing areas become available, comparable models could probably be applied to the tropics.

In the natural sciences and the physical technologies logical patterns and simple models, even if they are not mathematically expressable, facilitate international communication, help to overcome language barriers and thus help to promote international co-operation between professionals. Agriculture, unlike mathematics, has today no such international language; we hope that our efforts at least hold out the prospect of such interchange. Secondly, experience in other technologies suggests that, though such patterns or models may, in their first forms, have many defects, they nevertheless help to clarify thinking, and thus should ultimately, in our case, ease (*a*) the analysis of food and farming problems, (*b*) the identification and quantification of alternative solutions and (*c*) ultimately, the choice of effective and economic courses of action.

To conclude, we may perhaps repeat our thanks, expressed in the dedication of this book, to all those who have helped us in many lands.

References and Further Reading

Bawden, F. C. 1967. *Intl. J. Agrarian Affairs.* 5. 2. 116–29.

Masefield, G. B. 1963. *Famine: Its Prevention and Relief* at p. 94. London.

Bunting, A. H. (Ed.) 1969. *Proc. Intl. Symposium on Change in Agriculture.* (Univ., Reading, 1968). Duckworth, London. (In the press.)

Watt, K. E. F. (Ed.) 1967. *Systems Analysis in Ecology.* New York and London.

Jurion, F. and Henry, J. 1969. *Can Primitive Farming be Modernised?* Transl. from the French. L'Institut National pour L'Etude Agronomique du Congo. Brussels.

Glossary and Subject Index

NOTES

1. Pages on which a term is defined or on which there is an important reference are shown in bold type.

2. Brackets in the Index below indicate a definition or description.